PRACTICAL MATHEMATICS

SKILLS AND CONCEPTS

Holt, Rinehart and Winston, Inc.

Austin New York San Diego Chicago Toronto Montreal

Authors
Marguerite M. Fredrick
Computer Consultant, K–12
Pleasantville Union Free School District
Pleasantville, New York
Formerly Math Chairperson, 7–12
The Ursuline School
New Rochelle, New York

Steven J. Leinwand
Mathematics Consultant, K–12
Connecticut State Dept. of Education
Hartford, Connecticut

Robert D. Postman
Professor and Chairperson
Mathematics Education
Mercy College
Westchester, New York

Laurence R. Wantuck
Mathematics Supervisor, K–12
Broward County Schools
Fort Lauderdale, Florida

For photo and art credits, see page 486.

Requests for permission to make copies of any part of the work should be mailed to:
Permissions, Holt, Rinehart and Winston, Inc., 1627 Woodland Avenue, Austin, Texas 78741

Printed in the United States of America

ISBN: 0-03-012757-2

90 032 9876543

CONTENTS

CHAPTER 1 NUMERATION

CHAPTER 2 WHOLE NUMBER COMPUTATION

CHAPTER 3 DECIMAL COMPUTATION

CHAPTER 4 STATISTICS

CHAPTER 5 FRACTIONS

CHAPTER 6 FRACTION COMPUTATION

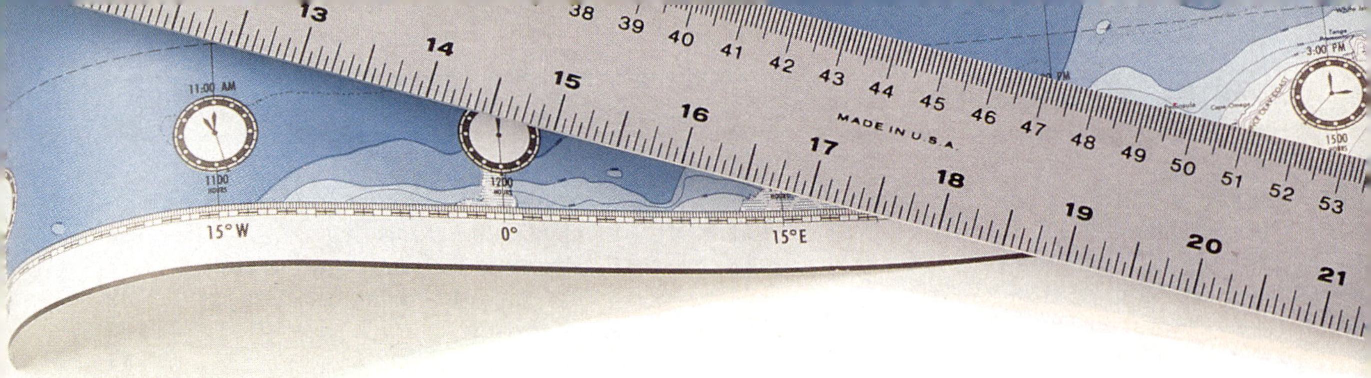

CHAPTER 7 MEASUREMENT

CHAPTER 8 RATIO AND PROPORTION

CHAPTER 9 PERCENT

CHAPTER 10 PROBABILITY

CHAPTER 11 GEOMETRY

CHAPTER 12 PERIMETER, AREA, AND VOLUME

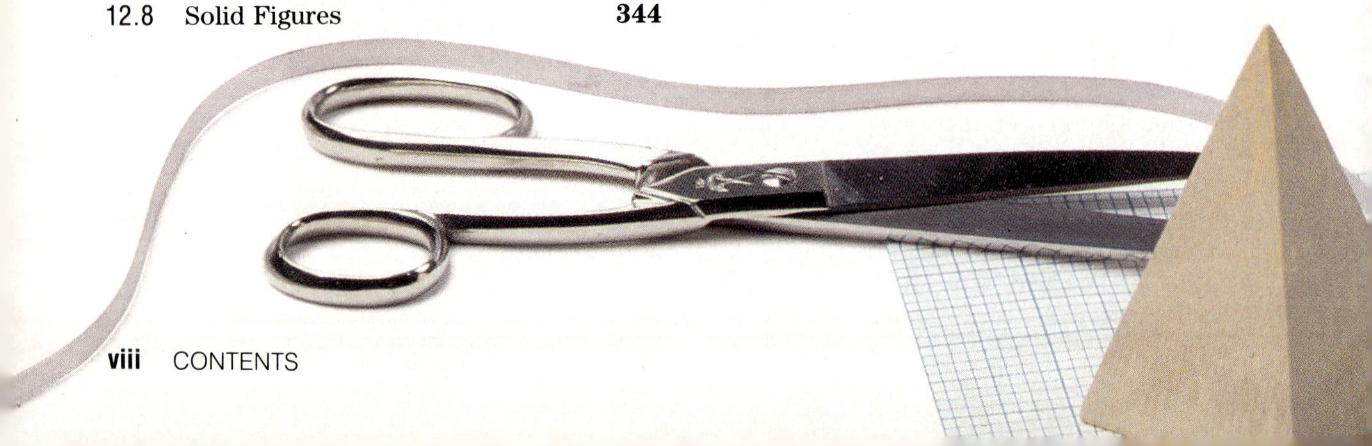

CHAPTER 13 RATIONAL NUMBERS

CHAPTER 14 EQUATIONS

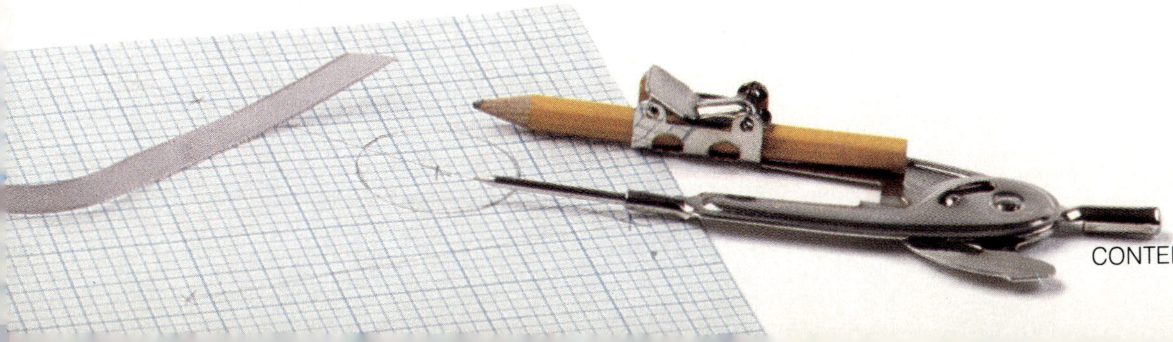

PreSkills TEST

Use the number line to identify the point using a letter.

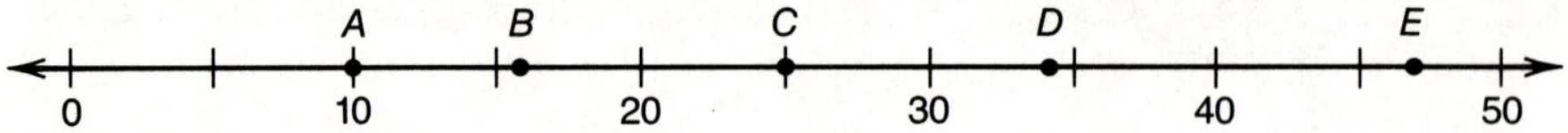

1. 10 **2.** 25 **3.** 34 **4.** 16 **5.** 47

Complete the sequence.

6. 25, 30, 35, 40, ■, ■, ■

7. 46, 56, 66, ■, ■, ■

8. 100, 200, 300, ■, ■, ■

9. 2,000, 4,000, 6,000, ■, ■

10. 0.1, 1.1, 2.1, ■, ■, ■

11. 5.4, 5.3, ■, ■, 5.0

Write the number.

12. eleven **13.** twenty-one **14.** thirty-three

Complete.

15. 15 = 1 ten ■ ones **16.** 27 = ■ tens 7 ones **17.** 47 = 4 tens ■ ones

Round to the nearest 10. Use the number line.

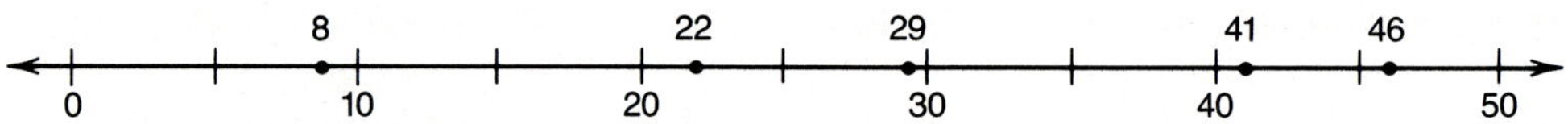

18. 8 **19.** 22 **20.** 29 **21.** 41 **22.** 46

Find the answer.

23. $7 + 2$	**24.** $9 + 5$	**25.** $6 + 7$	**26.** $8 + 4$	**27.** $7 + 7$	**28.** $9 + 6$	**29.** $5 + 4$
30. $12 - 6$	**31.** $14 - 9$	**32.** $15 - 7$	**33.** $13 - 4$	**34.** $10 - 8$	**35.** $16 - 7$	**36.** $17 - 8$
37. 4×3	**38.** 5×5	**39.** 8×3	**40.** 7×6	**41.** 5×9	**42.** 8×7	**43.** 6×4
44. $6\overline{)54}$	**45.** $8\overline{)72}$	**46.** $7\overline{)56}$	**47.** $9\overline{)81}$	**48.** $6\overline{)42}$	**49.** $5\overline{)40}$	**50.** $3\overline{)24}$

Chapter 1
NUMERATION

We use numbers in many different ways. What would happen if all numbers disappeared one day?

1.1 NUMBERS EVERYWHERE

We use numbers in many different ways. Does 45 Pearl Street mean there are 45 houses on Pearl Street? Is Highway 95 longer than Route 4?

Numbers answer the questions, "how many," "how much," "where," "when," and "which one."

EXAMPLE 1

Read the following story. Does the number tell "how many," "how much," "where," "when," or "which one"?

The Grade 9 class trip to Washington, D.C. cost $25 a student. On September 18th, 37 students toured 1600 Pennsylvania Avenue.

THINK: 9 tells "which one." $25 tells "how much." 18th tells "when." 37 tells "how many." 1600 tells "where."

A number can be classified as a count, a measure, a location, or a code.

Use	Question	Classification
17 albums	How many?	Count
88 km, $2.15	How much?	Measure
5th Avenue, Oct. 12, 1492	Where?/When?	Location
555-7676, ITX 5871	Which one?	Code

EXAMPLE 2

Classify each number on the check below.

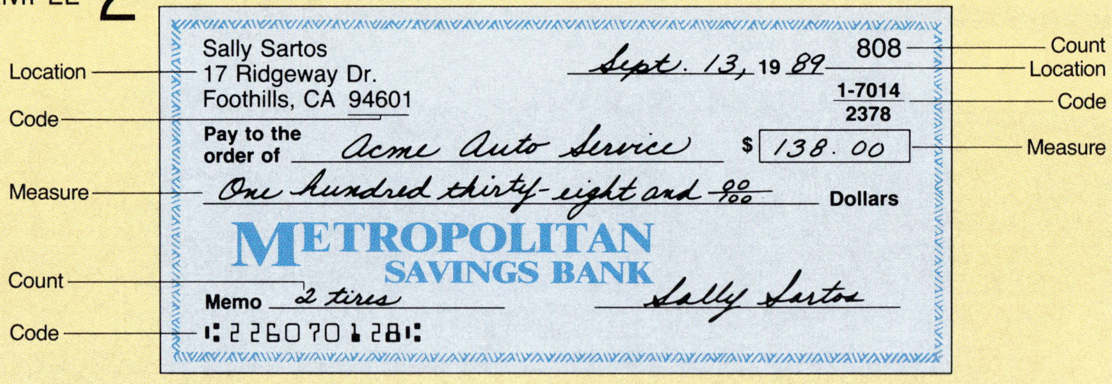

FOR DISCUSSION

Explain how 10 could be used as a count, a measure, a location, or a code.

PRACTICE EXERCISES

Complete the table. Does the number tell "how many," "how much," "where," "when," or "which one"? Then, classify the number.

Use	Question	Classification
9 kg	1. ■	2. ■
409-1467	3. ■	4. ■
15 books	5. ■	6. ■
June 30, 1989	7. ■	8. ■
$38.99	9. ■	10. ■

Use	Question	Classification
12th Street	11. ■	12. ■
Model number	13. ■	14. ■
Birth date	15. ■	16. ■
Height	17. ■	18. ■
Latitude	19. ■	20. ■

Use the bill to classify the number.

21. 284-4916-75
22. 35
23. 02874394
24. 5.67
25. 9-27-89
26. 3
27. 12459-3801
28. 12.68
29. 67912358
30. 1

Quality Stores

Acct. No. 284-4916-75
Martha Rumple
35 Ranchero Drive
Los Pedros, NM 12459-3801

Date	Reference	Quantity	Amount
9-13-89	02874394	3	16.23
9-27-89	67912358	1	5.67
9-29-89	03491726	4	12.68

Solve.

31. Class 10a took a trip to the museum on October 3rd. The bus cost $5 and admission to the museum was $2. Which number is a code?

32. The historical site at 125 Main St. was built in 1738. It is open from 10 to 3 every day. Admission is $0.75. Which numbers are locations?

TIME OUT

Numbers represented by dots as a geometric figure are called **figurate numbers.** Here are the first four triangular numbers.

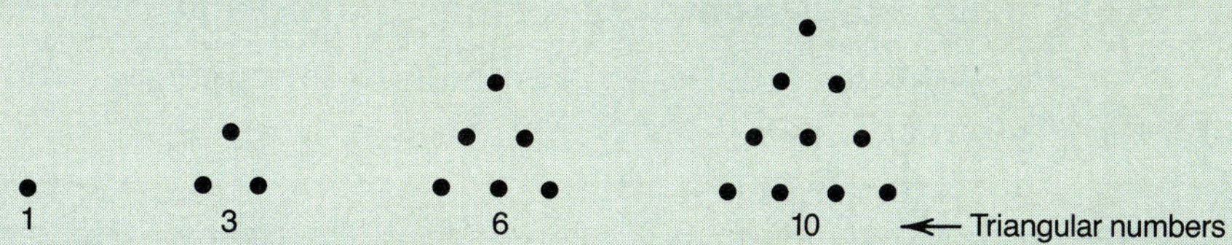

1. Arrange dots to form the first four square numbers.

2. Arrange dots to form the first four rectangular numbers.

1.2 PLACE VALUE: WHOLE NUMBERS

Our number system is made up of 10 digits: 0, 1, 2, 3, 4, 5, 6, 7, 8, and 9.

To write numbers greater than 9, a place-value system has been developed. Study the following place-value chart. Notice that each digit has a value 10 times greater than the value of the place to its right.

Periods	Billions			Millions			Thousands			Ones		
Places	Hundred Billions 100,000,000,000	Ten Billions 10,000,000,000	Billions 1,000,000,000	Hundred Millions 100,000,000	Ten Millions 10,000,000	Millions 1,000,000	Hundred Thousands 100,000	Ten Thousands 10,000	Thousands 1,000	Hundreds 100	Tens 10	Ones 1
Digits		4	2	0	1	9	3	8	5	6	2	7

You can find the value of a digit in a number by multiplying the digit by its place value.

EXAMPLE 1 Find the place and the value of 4 in the number 42,019,385,627.

THINK: In the chart above, the 4 is in the ten-billions place.

DIGIT × PLACE VALUE = VALUE

$4 \times 10{,}000{,}000{,}000 = 40{,}000{,}000{,}000$

You can read and write word names for numbers using the period names.

EXAMPLE 2 Read 42,019,385,627. Then write the word name.

READ: 42 billion, 19 million, 385 thousand, 627

WORD NAME: forty-two billion, nineteen million, three hundred eighty-five thousand, six hundred twenty-seven

To write a number from a word name, start with the digit in the greatest place. Then write the digit for each place to the right. Write zeros as placeholders if necessary. Use commas to separate the periods.

EXAMPLE 3 Write the number for two million, six thousand, three hundred twenty-nine.

	Millions	Hundred thousands	Ten thousands	Thousands	Hundreds	Tens	Ones
THINK:	Millions	Hundred thousands	Ten thousands	Thousands	Hundreds	Tens	Ones
WRITE:	2	0	0	6	3	2	9

NUMBER: 2,006,329

PRACTICE EXERCISES

Find the place and the value of the underlined digit.

1. $3\underline{4}1$ **2.** $\underline{6}25$ **3.** $2{,}\underline{8}47$ **4.** $\underline{9}{,}564$

5. $\underline{3}2{,}419$ **6.** $7\underline{8}{,}627$ **7.** $80{,}\underline{4}39$ **8.** $\underline{9}3{,}205$

9. $3\underline{2}1{,}604$ **10.** $459{,}\underline{3}28$ **11.** $\underline{8}45{,}169$ **12.** $89\underline{6}{,}437$

13. $3{,}002{,}15\underline{8}$ **14.** $7{,}\underline{3}25{,}893$ **15.** $\underline{7}0{,}138{,}625$ **16.** $11\underline{3}{,}864{,}018$

17. $2{,}\underline{5}89{,}411{,}007$ **18.** $3\underline{9}{,}425{,}687{,}990$ **19.** $7\underline{3}9{,}435{,}689{,}101$ **20.** $\underline{9}63{,}004{,}211{,}357$

Write the word name.

21. 175 **22.** 358 **23.** 4,568 **24.** 8,604

25. 23,961 **26.** 48,543 **27.** 67,801 **28.** 73,496

29. 325,406 **30.** 462,001 **31.** 580,102 **32.** 706,005

33. 1,890,000 **34.** 2,340,010 **35.** 3,400,009 **36.** 8,060,123

Write the number.

37. 336 thousand, 154

38. 893 thousand, 201

39. 4 million, 4 thousand, 4

40. 7 million, 16 thousand, 89

41. 5 billion, 28 thousand, 50

42. 85 billion, 347

43. forty-seven thousand, two hundred six

44. two hundred twenty-five thousand, sixty

45. sixty-five million, forty-six thousand, one hundred fifty

46. thirty-four billion, nine hundred five million, sixty-two

47. Copy this blank check. Use the information provided to fill it out.

Amount: $109.58

Date: 8-15-89

Check number: 562

Memo: Tune-up

Pay to: Gregg's Garage

48. Design your own check. Then fill out one of your checks for a down payment on a car.

1.3 PLACE VALUE: DECIMALS

You can extend the place-value chart to write decimals less than one.

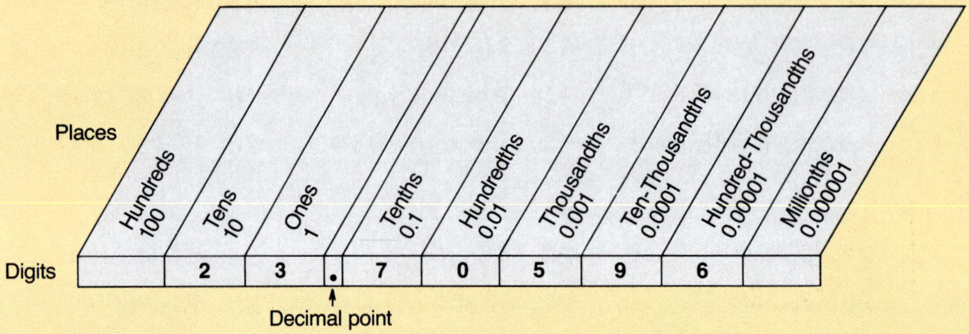

You can find the value of a digit in a decimal by multiplying the digit by its place value.

EXAMPLE 1 Find the place and the value of 5 in the number 23.70596.

THINK: In the chart above, the 5 is in the thousandths place.

DIGIT	×	PLACE VALUE	=	VALUE
5	×	0.001	=	0.005

The decimal point that appears between the ones and tenths place is read as *"and."*

EXAMPLE 2 Read 23.70596. Then write the word name.

READ: 23 *and* 70,596 hundred-thousandths

WORD NAME: twenty-three and seventy thousand, five hundred ninety-six hundred-thousandths

Sometimes zeros are needed as placeholders when writing decimals.

EXAMPLE 3 Write the number for seventy thousand and thirty-seven hundred-thousandths.

THINK: Hundred-thousandths means 5 decimal places.

WRITE: 70,000.00037

FOR DISCUSSION

You have two 5-digit decimals. The middle digit in each number is the same. Do the middle digits have the same value? Use examples to explain your answer.

PRACTICE EXERCISES

Find the place and the value of the underlined digit.

1. $23.\underline{1}$
2. $45.3\underline{6}$
3. $61.\underline{4}7$
4. $93.8\underline{5}$
5. $1.0\underline{4}6$
6. $8.19\underline{3}$
7. $14.\underline{9}87$
8. $63.79\underline{5}$
9. $5.00\underline{3}6$
10. $17.\underline{8}934$
11. $59.701\underline{2}$
12. $125.49\underline{8}7$
13. $6.3987\underline{2}$
14. $10.\underline{1}0106$
15. $18.76\underline{3}04$
16. $78.0\underline{4}921$
17. $0.529\underline{7}16$
18. $2.940\underline{3}72$
19. $11.6937\underline{2}4$
20. $41.10084\underline{6}$

Write the word name.

21. 0.5
22. 1.8
23. 16.9
24. 59.6
25. 0.04
26. 0.29
27. 1.51
28. 2.89
29. 0.007
30. 0.039
31. 0.241
32. 1.372
33. 0.0002
34. 0.0017
35. 0.0356
36. 0.1245

Write the number.

37. 4 and 27 hundredths
38. 13 and 63 hundredths
39. 195 thousandths
40. 66 thousandths
41. 12 and 249 thousandths
42. 49 and 7 thousandths
43. 1,259 ten-thousandths
44. 48 ten-thousandths
45. 83 and 3,459 ten-thousandths
46. 56 and 23 ten-thousandths
47. 129 hundred-thousandths
48. 315 millionths
49. 1,238 billionths
50. 342 and 25 billionths
51. four and five thousand three ten-thousandths
52. six thousand and sixty-six hundred-thousandths

Solve.

53. A car's odometer reads 2,105.75. Write the word name for this number.

54. The manual for your new car says, "Bring the car in for a check at five thousand miles." What would your odometer show for this number?

55. Classify the following numbers:

Power Steering Fluid

Part No.:	1050017
Size:	32 oz
Usage:	Power steering

56. The manual states that your tires should be rotated every 15,000 miles. Your odometer reads:

1 4 5 8 7 . 6

Are you overdue in rotating your tires?

1.4 COMPARING AND ORDERING NUMBERS

The air distance between New York and Honolulu is 4,964 mi. Is this distance less than (<), equal to (=), or greater than (>) the air distance between New York and Istanbul?

Air Distance (in miles)

Cities	Moscow	New York	Paris
Hong Kong	4,439	8,054	5,985
Honolulu	7,037	4,964	7,438
Istanbul	1,091	4,975	1,400

To solve, you need to compare the air distances. To compare numbers, start at the left and move right until the digits in the same place are different.

EXAMPLE 1 Compare: 4,964 ● 4,975. Write =, <, or >.

THINK: The digits are the same in the thousands and hundreds places. The digits differ in the tens place.

4,964
4,975

Since $6 < 7$; $4{,}964 < 4{,}975$. (< **is less than**)

The air distance between New York and Honolulu is less than the air distance between New York and Istanbul.

When comparing decimals, you sometimes have to write additional zeros.

EXAMPLE 2 Compare:

a. 0.293 ● 0.2

THINK: 0.2 = 0.200

0.293
0.200

Since $9 > 0$, $0.293 > 0.2$. (> **is greater than**)

b. 0.293 ● 0.2936

THINK: 0.293 = 0.2930

0.2930
0.2936

Since $0 < 6$, $0.293 < 0.2936$. (< **is less than**)

Three or more numbers can be arranged in order by repeated comparisons.

EXAMPLE 3 Arrange 748.25, 749.12, 748.04 in order from greatest to least.

THINK: The digits are the same in the hundreds and tens places. The digits differ in the ones place.

748.25
749.12
748.04

Since 9 is the greatest, 749.12 is the greatest.

Now compare: 748.25
748.04

Since $2 > 0$, $748.25 > 748.04$.

748.04 is the least.

The numbers in order from greatest to least are 749.12, 748.25, 748.04.

PRACTICE EXERCISES

Compare. Write =, <, or >.

1. 342 ● 915 **2.** 1,000 ● 999 **3.** 5,638 ● 5,836 **4.** 27,825 ● 27,826

5. 493,926 ● 943,962 **6.** 540,500 ● 540,050

7. 67,989,192 ● 67,990,192 **8.** 78,340,982 ● 78,340,892

9. 0.02 ● 0.2 **10.** 0.9 ● 0.90 **11.** 649.81 ● 650 **12.** 70.5 ● 7.05

13. 96.31 ● 936.1 **14.** 1.902 ● 1.934 **15.** 0.71 ● 0.671 **16.** 12.012 ● 12,012

17. 10,056.95 ● 10,060.05 **18.** 580.632 ● 581.002

19. 25,000.465 ● 25,000 **20.** 1.3842 ● 1.38421

21. 35.7004 ● 35.70040 **22.** 823,918.67205 ● 823,918.6725

23. 9.084153 ● 9.085143 **24.** 29,621,000.93 ● 29,621,001.03

25. 439,708.0023 ● 439,780.0203 **26.** 37,294.5400 ● 37,294.54

27. 852,001.3205 ● 852,001.32005 **28.** 300,001,094.56 ● 300,001,094.553

Choose the number that is least.

29. **a.** 4,521 **b.** 541 **c.** 1,425

30. **a.** 6,739 **b.** 6.739 **c.** 67.39

31. **a.** 5.0095 **b.** 5.0059 **c.** 5.00095

32. **a.** 10.6803 **b.** 10.8603 **c.** 10.8630

Arrange in order from greatest to least.

33. 4,279; 2,972; 9,742; 7,492

34. 0.0863; 0.863; 0.6308; 0.3086

35. 42,915; 0.937; 4.2915; 9.730

36. 78.625; 82.67; 208; 60.28

Arrange in order from least to greatest.

37. 36,081; 36,180; 306,810; 38,016

38. 0.527; 0.57; 20.75; 0.7

39. 3,428,865; 3,428.8605; 3,428.8065; 3,428,856

Use the table at the top of page 8 to solve.

40. Which is greater, the distance between Moscow and Istanbul or the distance between Paris and Istanbul?

41. Which is less, the distance between Paris and Hong Kong or the distance between New York and Istanbul?

42. Would you classify the numbers in the table as counts, measures, locations, or codes?

43. The greatest distance is between which two cities?

PROBLEM Solving APPLICATION

1.5 CONSUMER: READING METERS

Electric energy is measured in **kilowatt-hours (kWh)**. One kilowatt is 1,000 **watts (W)**. When an appliance uses 1,000 W in an hour, 1 kWh of energy is used.

The more electrical energy that is used, the greater the cost. Meters record the amount of electricity used.

Read the dials from left to right. If the pointer is between two numbers, read the lower number. When the pointer is between 9 and 0, read 9.

Find the number of kilowatt-hours represented on this electric meter.

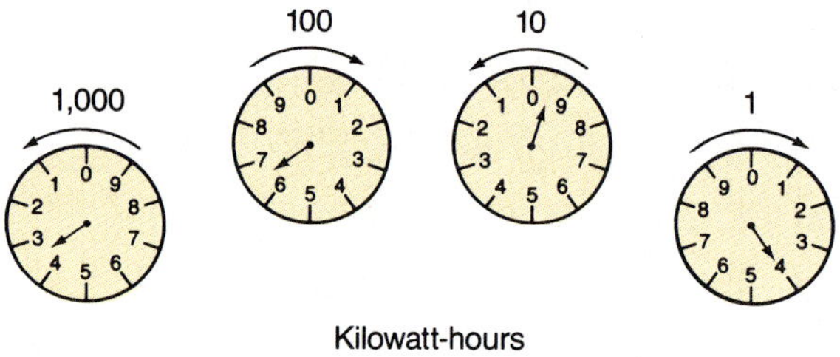

Thousands and tens dials move counterclockwise. Hundreds and ones dials move clockwise.

1. Read the thousands dial.
   THINK: The pointer is between 3 and 4. Read 3.
2. Read the hundreds dial.
   THINK: The pointer is between 6 and 7. Read 6.
3. Read the tens dial.
   THINK: The pointer is between 9 and 0. Read 9.
4. Read the ones dial.
   THINK: The pointer is at 4. Read 4.

So, the reading on the electric meter is 3,694 kWh.

FOR DISCUSSION

If the pointer on a dial is between 9 and 0, why would you read 9 and not 0?

Write the number of kilowatt-hours represented on the electric meter.

1.

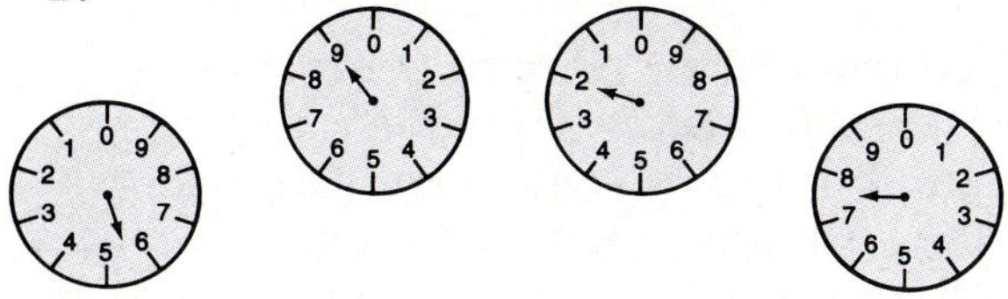

2.

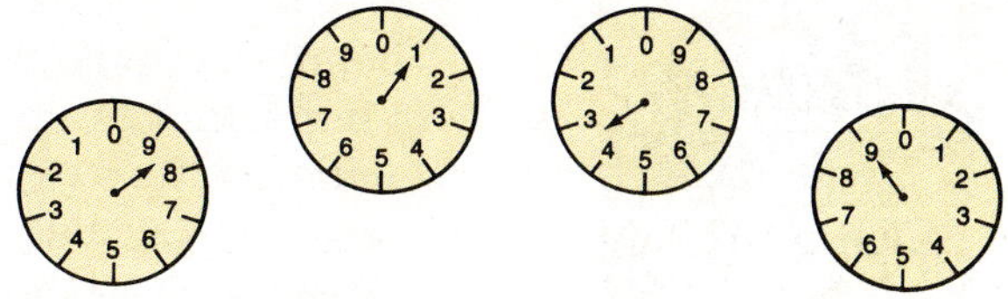

3.

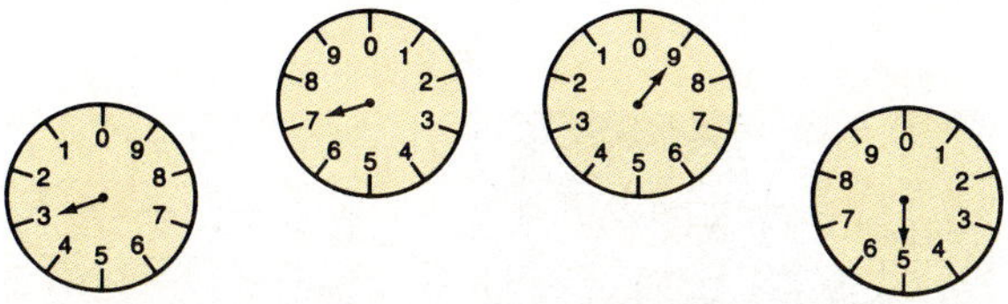

4.

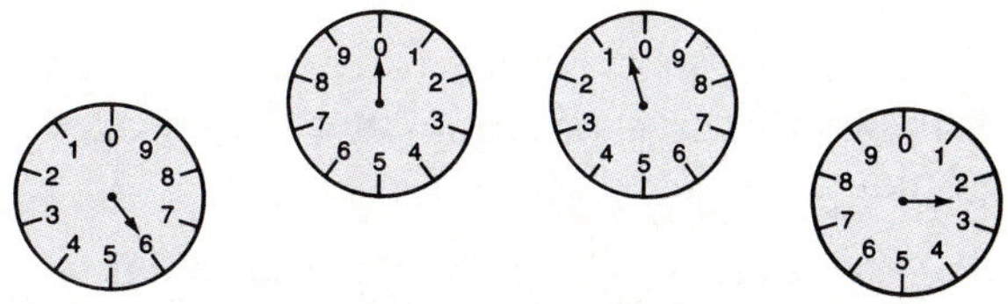

Use the electric meter at the right for Exercises 5–8.

January 31 Reading

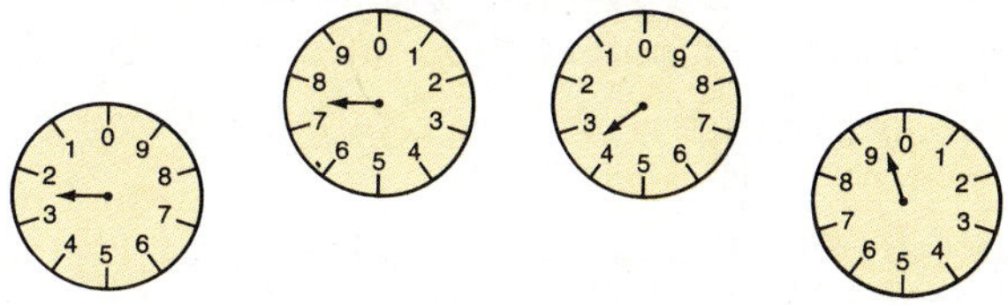

5. Write the number that is 10 more than the January 31 reading.

6. Write the number that is 100 more than the January 31 reading.

7. Write the number that is 1,000 more than the January 31 reading.

8. Write the number that is 1 more than the January 31 reading.

This electric meter has five dials. The dial at the left is for tens of thousands of kilowatts. Use this meter for Exercises 9–15.

September 15 Reading

9. Write the number of kilowatt-hours represented in the September 15 reading.

10. Write the number that is 20 more than the September 15 reading.

11. Write the number that is 500 more than the September 15 reading.

12. Write the number that is 20,000 more than the September 15 reading.

13. Write the number that is 6,000 more than the September 15 reading.

14. What is the greatest number that can be represented on this meter?

15. Draw a meter with five dials. Then draw pointers to represent 70,806.

PROBLEM Solving APPLICATION

1.6 CAREER: POSTAL CLERK

Nicholas Tutela is a postal clerk. The ability to read tables is important in his job. The tables below show the rates for mailing letters or packages first class. Notice that if a package weighs more than 2 lb, the cost depends on the number of zones it will travel as well as on its weight.

What is the cost of shipping a 2 lb 8 oz package to Zone 6?

THINK: 2 lb 8 oz is over 2 lb, but does not exceed 3 lb.

Find the entry across row 3 and under Zone 6 of the second table.

The cost is $3.74.

FIRST CLASS

LETTER RATES:
1st ounce. . . . . 22¢
Each additional ounce . . . . 17¢

For Pieces Not Exceeding (oz)	*The Rate is*	For Pieces Not Exceeding (oz)	*The Rate is*
1	$0.22	7	$1.24
2	0.39	8	1.41
3	0.56	9	1.58
4	0.73	10	1.75
5	0.90	11	1.92
6	1.07	12	2.09

FOR PIECES OVER 12 OUNCES SEE FIRST-CLASS ZONE RATED (PRIORITY) MAIL RATES

FIRST-CLASS ZONE RATED (PRIORITY) MAIL

Weight over 12 ounces and not exceeding-pound(s)	Rate					
	Local zones 1, 2, and 3	Zone 4	Zone 5	Zone 6	Zone 7	Zone 8
1	$2.40	$2.40	$2.40	$2.40	$ 2.40	$ 2.40
2	2.40	2.40	2.40	2.40	2.40	2.40
3	2.74	3.16	3.45	3.74	3.96	4.32
4	3.18	3.75	4.13	4.53	4.92	5.33
5	3.61	4.32	4.86	5.27	5.81	6.37
6	4.15	5.08	5.71	6.31	6.91	7.66
7	4.58	5.66	6.39	7.09	7.80	8.67
8	5.00	6.23	7.07	7.87	8.68	9.68
9	5.43	6.81	7.76	8.66	9.57	10.69
10	5.85	7.39	8.44	9.44	10.45	11.70

Find the cost of mailing, first class, a package with the given weight.

1. 3 oz
2. 8 oz
3. 10 oz
4. 15 oz
5. 1 lb 3 oz
6. 1 lb 8 oz

Find the cost of mailing the package.

7. 2 lb 6 oz, Zone 3
8. 2 lb 6 oz, Zone 5
9. 3 lb 4 oz, Zone 4
10. 3 lb 4 oz, Zone 8
11. 4.5 lb, Zone 2
12. 4.5 lb, Zone 7

Solve.

13. A package weighs 6.25 lb. What is the least amount and the greatest amount that the first-class postage might be?

14. For which package is the first-class postage greater?

Package A: 7.5 lb, Zone 6
Package B: 8.2 lb, Zone 5

MIDCHAPTER REVIEW

Use the store receipt to classify each number. [2]

1. 1

2. 9/8/89

3. 123

4. 4

5. (303) 555-3535

6. 9.00

7. 80739

8. 2

9. 19.90

10. CR26

ABC Stationers, Inc. Date 9/8/89
123 Main Street • Laid, CO 80739 • Phone (303) 555-3535

1	ream paper	$ 5.95
2	greeting cards @ 1.50	3.00
4	bx CR26 @ 2.25	9.00
	wrapping paper	1.95
	Total	$ 19.90

Find the place and the value of the underlined digit. [4, 6]

11. 4,$\underline{9}$25

12. 6$\underline{7}$,819

13. $\underline{3}$42,896

14. $\underline{7}$,105,264

15. 16.$\underline{9}$

16. 38.7$\underline{2}$

17. 5.0$\underline{5}$75

18. 89.13706$\underline{6}$

Write the word name for the number. [4, 6]

19. 426

20. 279,300

21. 10.562

22. 0.0129

Write the number. [4, 6]

23. 205 thousand, 506

24. four hundred twenty-nine billion, eighty-four thousand, seventy-five

25. ten and one hundred seventy-seven thousandths

26. forty-five and thirty-two hundred-thousandths

Compare. Write =, <, or >. [8]

27. 627 ● 617

28. 6,892 ● 6,829

29. 3,000 ● 30,000

30. 152 ● 15.2

31. 26.9 ● 26.900

32. 8.05 ● 8.005

33. 9.5624 ● 9.5723

34. 3.70956 ● 4.71

35. 325.81922 ● 325.81923

Arrange in order from least to greatest. [8]

36. 9,528; 5,928; 9,285; 9,852

37. 562; 9.532; 0.565; 0.562

Solve. [10]

38. Write the number of kilowatt-hours represented on the electric meter.

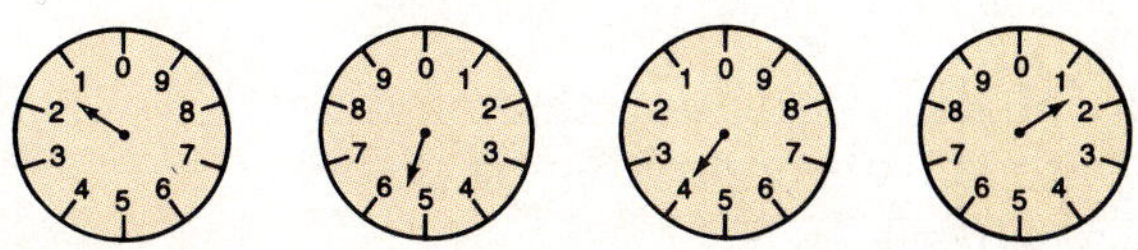

PROBLEM *Solving* STRATEGY

1.7 IDENTIFYING NEEDED INFORMATION

Situation:

Niagara Falls consists of two waterfalls, the Horseshoe Falls and the American Falls. The Horseshoe Falls is about 173 ft high and 2,600 ft wide at its widest point. The American Falls is about 182 ft high.

Strategy:

Identify only the information you need to solve a problem. Sometimes you will not have enough information.

Applying the Strategy:

A. Which falls is higher?

THINK: The Horseshoe Falls is 173 ft high.

The American Falls is 182 ft high.

Compare the heights. $173 < 182$

The American Falls is higher.

B. Which falls is wider?

THINK: The Horseshoe Falls is 2,600 ft wide.

The width of the American Falls is not given.

This problem cannot be solved.

PRACTICE EXERCISES

Which sentence describes the problem?

1. The ledge of the Horseshoe Falls wears away at the rate of about 3 in. a year. The ledge of the American Falls erodes more slowly. How much does the ledge of the American Falls erode each year?

a. Not enough information is given to solve the problem.

b. More information than you need to solve the problem is given.

2. The Horseshoe Falls is about 2,600 ft wide at its widest point. The American Falls is about 182 ft high and 1,000 ft wide at its widest point. Which falls is the widest?

a. Not enough information is given to solve the problem.

b. More information than you need to solve the problem is given.

Solve if possible. If not, identify the needed information.

3. The Niagara River plunges into a gorge. The gorge consists of 120 ft of layers of different stone and 80 ft of limestone. Which part of the gorge is thicker, the layers of different stone or the limestone?

4. Niagara Falls was originally at Lewiston, New York, but it gradually moved about 7 mi back upstream toward Lake Erie. How long ago was Niagara Falls at Lewiston, New York?

5. There are four observation towers so that tourists can view the falls. One tower is 282 ft high. Another tower is 500 ft high. Which of the four towers is the highest?

6. Niagara Falls Park was established in 1885. The park covers 430 acres. Queen Victoria Park was established in 1886. It covers 196 acres. Which park was larger when it was established?

7. Rock slides have changed the appearance of Niagara Falls. In 1931, about 80,000 tons of rock fell from the American Falls. Then years later, about 185,000 tons of rock fell. In which year did a greater amount of rock fall?

8. Three sightseeing boats took tourists to the base of the American Falls twice in 1 day. Boat A took 186 tourists. Boat B took 168 tourists. Boat C took 196 tourists. List the boats in order from the greatest number of tourists on board to the least number of tourists on board.

1.8 ROUNDING NUMBERS

Video Views rented 3,972 movies during the summer season. About how many movies did they rent?

You can find *about how many* by rounding.

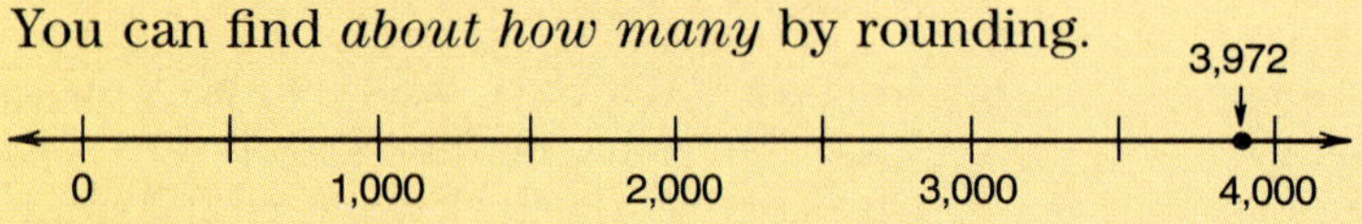

3,972 is closer to 4,000 than to 3,000. So, about 4,000 movies were rented.

> When you round to a given place, look at the digit to the right of that place. If the digit is:
>
> Less than 5 → round down   5 or greater → round up

EXAMPLE 1

Round 67,438 to the nearest hundred.

1. Find the hundreds place. — 67,438 (↑ under the 4)

   THINK: 3 is to the right. $3 < 5$, so round down.

2. To round down, keep the digit in the hundreds place. Change the digits to the right to zeros. — 67,400

67,438 rounded to the nearest hundred is 67,400.

When rounding decimals, you may have to drop some digits.

EXAMPLE 2

Round 5.291 to the nearest tenth.

1. Find the tenths place. — 5.291 (↑ under the 2)

   THINK: 9 is to the right. $9 > 5$, so round up.

2. To round up, increase the digit in the tenths place by 1. Drop the digits to the right. — 5.3

5.291 rounded to the nearest tenth is 5.3.

When working with money, sometimes you have to round to the nearest cent.

EXAMPLE 3

The tax on an item is $2.195. Round the tax to the nearest cent.

1. Find the cents. — $2.195 (↑ under the 9)

   THINK: 5 is to the right. $5 = 5$, so round up.

2. To round up, increase the cents digit by 1. Drop the digit to the right. — $2.20

The tax is $2.20.

PRACTICE EXERCISES

Round to the nearest hundred and to the nearest thousand.

1. 2,782 **2.** 3,984 **3.** 5,283 **4.** 6,096
5. 8,599 **6.** 17,851 **7.** 63,800 **8.** 74,210
9. 89,954 **10.** 93,150 **11.** 354,167 **12.** 489,649
13. 666,111 **14.** 899,521 **15.** 942,235 **16.** 999,999

Round to the nearest tenth and to the nearest hundredth.

17. 0.0193 **18.** 0.077 **19.** 1.549 **20.** 4.051
21. 16.951 **22.** 39.945 **23.** 85.563 **24.** 0.045
25. 0.4811 **26.** 2.1673 **27.** 11.1598 **28.** 55.1231
29. 0.35432 **30.** 5.12345 **31.** 23.66666 **32.** 48.98326

Round to the nearest dollar.

33. $5.15 **34.** $39.50 **35.** $65.39 **36.** $483.98
37. $8.358 **38.** $17.163 **39.** $32.843 **40.** $133.518

Round to the nearest cent.

41. $0.858 **42.** $1.899 **43.** $16.455 **44.** $24.296
45. $0.1462 **46.** $8.3586 **47.** $50.8192 **48.** $138.1955

Round to the underlined place.

49. $\underline{3},586$ **50.** $49.\underline{5}27$ **51.** $87,\underline{8}41$ **52.** $24.\underline{0}9$
53. $64.9\underline{3}8$ **54.** $\$36.8\underline{6}5$ **55.** $601,\underline{4}98$ **56.** $\$1\underline{6}.324$

Solve.

57. Tax on a video tape came to $0.3594. What is the tax to the nearest cent?

58. Video View made $1,484.54 last week. This week it made $1,470.18. During which week did it make more money?

59. Brian spends about $8.97 a month to rent movies. Last year he spent $107.64. How much did he spend last year to the nearest dollar?

60. A number was rounded to 150,000. Give three possible numbers that could have been rounded to 150,000.

1.9 EXPONENT FORM

A short way to show repeated multiplication is **exponent** form. For example, 3^2 is a number written in exponent form. The exponent, 2, tells how many times 3 is used as a factor; 3 is called the **base.**

Base $\rightarrow 3^2 = 3 \times 3 = 9$

$\uparrow$ **Exponent** (points to the 2 in 3^2); $\uparrow$ **Product** (points to 9)

3^2 is read as three to the second power, or three squared.

EXAMPLE 1

Write $2 \times 2 \times 2$ in exponent form.

$2 \times 2 \times 2$ **THINK:** The base is 2.
2 is used as a factor 3 times.
The exponent is 3.

$2 \times 2 \times 2 = 2^3$

2^3 is read as two to the third power, or two cubed.

To find the product of a number in exponent form, first write it as a repeated multiplication.

EXAMPLE 2

Write the product of 2^4.

2^4 **THINK:** 2 is used as a factor 4 times.

$2^4 = 2 \times 2 \times 2 \times 2 = 16$

Study the table below. Do you see a pattern?

Repeated Factors	Product	Exponent Form
10×10	100	10^2
$10 \times 10 \times 10$	1,000	10^3
$10 \times 10 \times 10 \times 10$	10,000	10^4
$10 \times 10 \times 10 \times 10 \times 10$	100,000	10^5
$10 \times 10 \times 10 \times 10 \times 10 \times 10$	1,000,000	10^6

Notice that when 10 is raised to any power, the exponent tells the number of zeros in the product.

EXAMPLE 3

a. Write 1,000,000 in exponent form.

1,000,000 **THINK:** 6 zeros
$1{,}000{,}000 = 10^6$

b. Write the product of 10^8.

10^8 **THINK:** 8 zeros
$10^8 = 100{,}000{,}000$

FOR DISCUSSION

Look at the patterns. What do you notice?

Pattern 1: $3^1 = 3$ $5^1 = 5$ $7^1 = 7$ $8^1 = 8$

Pattern 2: $2^0 = 1$ $4^0 = 1$ $6^0 = 1$ $9^0 = 1$

PRACTICE EXERCISES

Write in exponent form.

1. 2 squared **2.** 7 cubed **3.** 9 to the third power

4. 10 to the fifth power **5.** 12 to the sixth power **6.** 18 to the second power

7. $2 \times 2 \times 2 \times 2 \times 2$ **8.** $3 \times 3 \times 3$ **9.** 4×4

10. $8 \times 8 \times 8 \times 8$ **11.** $12 \times 12 \times 12$ **12.** $13 \times 13 \times 13 \times 13$

Write as a product.

13. 2^2 **14.** 7^0 **15.** 3^2 **16.** 12^0 **17.** 3^3 **18.** 6^2 **19.** 1^3

Write in exponent form.

20. 100 **21.** 10,000 **22.** 10,000,000 **23.** 1,000,000,000

Write as a product.

24. 10^0 **25.** 10^3 **26.** 10^5 **27.** 10^2 **28.** 1^8 **29.** 10^4 **30.** 10^1

Solve.

31. X-rays have a frequency of more than 10,000,000,000 vibrations per second. Write this number in exponent form.

32. The diameter of Venus is 7,700 mi. The diameter of Earth is 7,926 mi. Which number is greater?

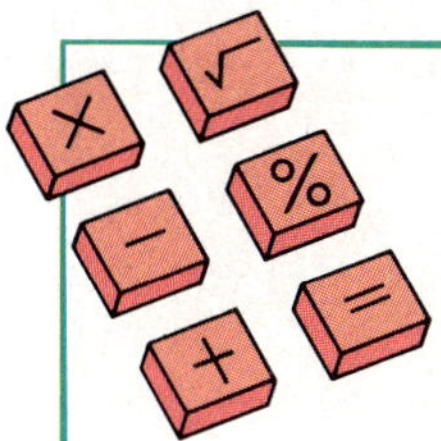

CALCULATOR

Most calculators have an x^2 key and $\sqrt{\ }$ key. The x^2 key is called the **square** key. The $\sqrt{\ }$ is called the **square root** key. Square and square root are called inverse operations. This means one undoes the other. For example, enter 6, then press the square key. What does the display show? Now press the square root key. What does the display show? Now repeat the procedure for 8. What is the definition of square root? Guess the square root of the number. Use the calculator to check your answer.

1. 4 **2.** 9 **3.** 16 **4.** 25 **5.** 49

PROBLEM Solving APPLICATION

1.10 CONSUMER: UNDERSTANDING MENUS

You and a friend decide to eat at the Hearthstone Restaurant. The waiter gives you a menu. There are some words and phrases you must understand. **A la carte** indicates a separate price for each item not bought as a part of a complete dinner. An **entrée** is a main dish.

HEARTHSTONE RESTAURANT

MENU

Dinners include entrée, side order, beverage, and dessert.

ENTRÉE

	A la Carte	Complete Dinner
Filet of sole	$ 9.95	$11.50
Salmon steak	$10.95	$12.50
Spaghetti and meatballs	$ 6.25	$ 7.75
Roast chicken	$ 8.75	$10.25
Chopped sirloin	$ 7.50	$ 8.95

SIDE ORDERS

French fried potatoes	$0.85	*Cole slaw*	$0.75
Vegetable	$0.85	*Salad*	$1.10

DESSERTS

Fruit and cheese	$1.15	*Jello*	$0.85
Yogurt	$1.10	*Rice pudding*	$0.95

BEVERAGES

Milk	$0.75	*Herbal tea*	$0.85
Lemonade	$0.95	*Orange juice*	$0.90

FOR DISCUSSION

Why would the total cost for the filet of sole complete dinner be more than $11.50?

Use the Hearthstone Restaurant menu on page 20. Write the cost.

1. A complete dinner with filet of sole, salad, milk, and jello
2. Roast chicken if a person does not order a complete dinner
3. A complete dinner with chopped sirloin, vegetable, herbal tea, and yogurt

Solve.

4. In the a la carte column for entrées, how many items cost less than \$9.00?
5. In the a la carte column for entrées, how many items cost more than \$10.00?
6. In the complete dinner column, how many items cost less than \$9.00?
7. In the complete dinner column, how many items cost more than \$11.00?
8. Which of the a la carte entrées is the least expensive?
9. Which dessert is the most expensive?

In Exercises 10–15, give the answer to the nearest dollar.

10. Salmon steak ordered a la carte
11. Spaghetti and meatballs ordered a la carte
12. A complete dinner with filet of sole as the entrée
13. Yogurt ordered a la carte
14. If you order a la carte, about how much more does roast chicken cost than chopped sirloin?
15. If you order a complete dinner, about how much more does filet of sole cost than spaghetti and meatballs?

CHAPTER REVIEW

Vocabulary Choose the letter of the word(s) that completes each statement.

1. A number that tells "where" is classified as a ■. [2]
a. count **b.** measure **c.** location **d.** code

2. You can find the value of a digit by multiplying the digit by its ■. [4]
a. word name **b.** period **c.** place value **d.** numeral

3. 5^3 is read as five to the third ■. [18]
a. power **b.** base **c.** exponent **d.** factor

Skills Find the value of the underlined digit. [4, 6]

78,$\underline{2}$47 *200*

4. 9$\underline{2}$4,518 **5.** $\underline{4}$,806,231 **6.** 23,$\underline{6}$27,856

7. 0.5$\underline{6}$7 **8.** 0.034$\underline{2}$ **9.** 3.25$\underline{6}$7 **10.** 14.003$\underline{5}$6

Compare. Write =, <, or >. [8]

765 ● 76 >

11. 5,408 ● 4,379 **12.** 23,456 ● 23,546 **13.** 80,060 ● 80,006

14. 0.56 ● 0.68 **15.** 4.57 ● 4.569 **16.** 0.129 ● 0.1290 **17.** 34.876 ● 34.872

Arrange in order from least to greatest. [8]

567; 470; 322 *322; 470; 567*

18. 2,908; 3,541; 3,154

19. 74,876; 68,999; 86,098

20. 5,098; 5,456; 4,978; 4,899

21. 0.456; 0.57; 0.392

22. 3.9; 3.29; 2.98; 2.098

Round the number to the underlined place. [16]

$\underline{3}$4,876 *30,000*

23. 70$\underline{9}$,543 **24.** 45,$\underline{8}$76 **25.** 97$\underline{6}$,568

26. 0.$\underline{4}$56 **27.** 3.4$\underline{6}$8 **28.** 2.6$\underline{7}$19 **29.** 15.$\underline{6}$666

Write in exponent form. [18]

4 cubed *4^3*

30. 3 to the ninth power **31.** $8 \times 8 \times 8 \times 8$

32. 5 squared **33.** 6 to the fourth power **34.** $7 \times 7 \times 7 \times 7 \times 7$

CHAPTER TEST

Use the ad to classify each number.

1. 2 **2.** 1989 **3.** 12

4. 6810 **5.** 555-1020 **6.** 6 p.m.

For Sale:

2 tires for a 1989 Comet 12-speed bike, Model 6810
Call 555-1020 after 6 p.m.

Write the word name for the number.

7. 12,489 **8.** 240,007 **9.** 5.841 **10.** 0.0487

Write the number.

11. twelve billion, nine thousand, five hundred two

12. nineteen and one hundred forty-one ten-thousandths

Find the value of the underlined digit.

13. $\underline{2}89{,}650$ **14.** $2\underline{7}8{,}598{,}643$ **15.** $1{,}01\underline{5}{,}734$ **16.** $\underline{8}4{,}073{,}219$

17. $4.92\underline{9}$ **18.** $2.094\underline{4}06$ **19.** $577.\underline{7}73$ **20.** $7.0367\underline{8}5$

Compare. Write =, <, or >.

21. 6,739 ● 6,719 **22.** 58,326 ● 85,362 **23.** 4.0056 ● 4.0506

24. 0.0048 ● 0.00480 **25.** 123,995 ● 123,895 **26.** 90.937 ● 90.9371

Arrange in order from least to greatest.

27. 48.321; 48.32; 40.8832 **28.** 586.17; 508.617; 586.107; 561.87

Round 87,915.0649 to the nearest:

29. thousand. **30.** tenth. **31.** hundredth. **32.** hundred.

Write in exponent form.

33. $7 \times 7 \times 7$ **34.** 3 squared **35.** $4 \times 4 \times 4 \times 4$ **36.** 14 cubed

Write as a product.

37. 2^3 **38.** 10^{10} **39.** 1^{46} **40.** 8 squared

Solve if possible. If not, identify the needed information.

41. At Magic Mountain park in California, the Sky Tower is 385 ft high. At Six Flags Over Georgia, the parachute jump is 225 ft high. Which is taller?

42. The October 15 reading of an electric meter was 41,347 kWh. Write the number that is 500 kWh more.

ENRICHMENT ANCIENT NUMBER SYSTEMS

The Egyptian number system is based on the number 10. Each symbol stands for a different power of 10.

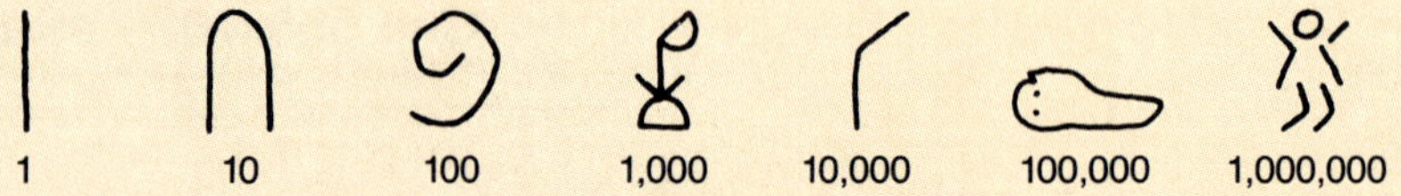

The Egyptian system uses addition. The value of a number is found by adding the values of the symbols.

Example

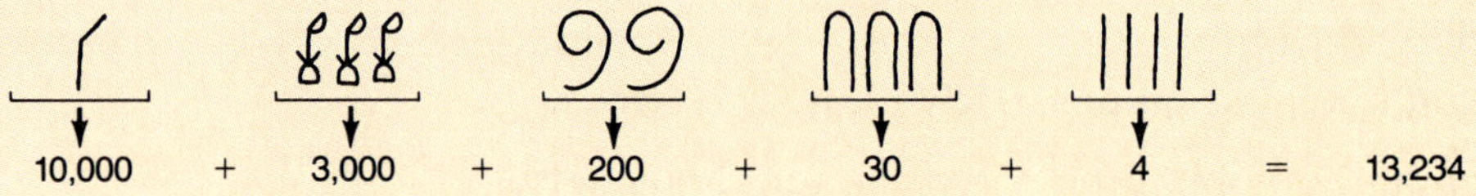

The Roman number system uses the following symbols.

I	V	X	L	C	D	M
1	5	10	50	100	500	1,000

This system uses addition and subtraction. If a symbol has a value greater than or equal to the value of the symbol to the right, add. If a symbol has a value less than the value of the symbol to the right, subtract.

Examples M CM LXX IV: 1,000 + 900 + 70 + 4 = 1,974; MM XL IX: 2,000 + 40 + 9 = 2,049

Write our numbers.

1. 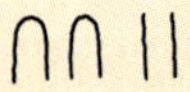 **2.** 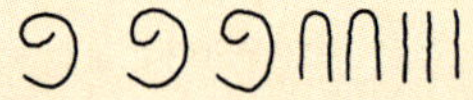 **3.** 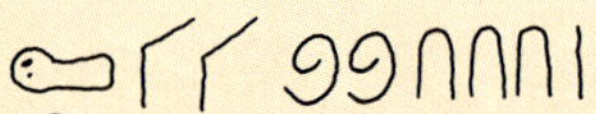

4. 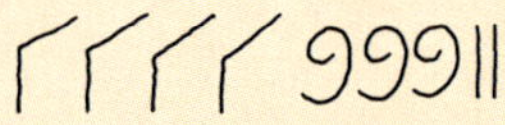 **5.** 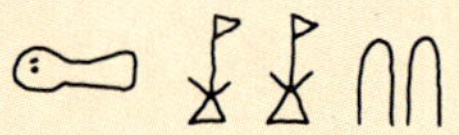 **6.** 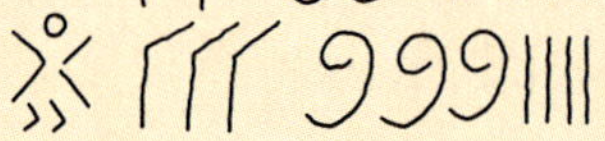

Write Egyptian numbers.

7. 25 **8.** 46 **9.** 131 **10.** 158 **11.** 1,254

Write our numbers.

12. L X X V I **13.** C C L X V **14.** D C C C X X X I I

15. X C V **16.** C D L I V **17.** M C M L I X

Write Roman numbers.

18. 13 **19.** 38 **20.** 156 **21.** 509 **22.** 944

CALCULATOR

BASIC OPERATIONS ON A CALCULATOR

A calculator can save you time and increase the accuracy of your computations. The four operations are performed as shown below.

Operation	Calculator Entry	Display
Add: 6,936 + 7,098	[6] [9] [3] [6] [+] [7] [0] [9] [8] [=]	14,034.
Subtract: 32.715 − 9.846	[3] [2] [.] [7] [1] [5] [−] [9] [.] [8] [4] [6] [=]	22.869
Multiply: 542 × 1,714	[5] [4] [2] [×] [1] [7] [1] [4] [=]	928,988.
Divide: 3.66 ÷ 0.48	[3] [.] [6] [6] [÷] [.] [4] [8] [=]	7.625

Several operations can be performed in sequence. You do not have to use the = key after each operation.

Compute: 729 × 38 ÷ 18

Operation	Calculator Entry	Display
1. Multiply.	[7] [2] [9] [×] [3] [8] [÷] ← Get ready to divide.	27,702.
2. Divide.	[1] [8] [=]	1,539.

If a number is entered incorrectly, the **CE** (**Clear Entry**) key can be used to make the necessary correction.

Correct, then complete the following computation on a calculator.

Add: 13.95 + 4.78 (Error)

		Display
Calculator entry:	[1] [3] [.] [9] [5] [+] [4] [7] [.] [8]	47.8
1. Press the CE key to clear 47.8 from the calculator.		0.
2. Complete the computation. Enter:	[4] [.] [7] [8] [=]	18.73

Use a calculator to compute. Use the Clear Entry key if you enter a number incorrectly.

1. 35,248 + 9,675

2. 8,200 − 7,957

3. 97 × 964

4. 5,280 ÷ 176

5. 4.738 + 19.824

6. 342.15 − 18.76

7. 0.034 × 4,673

8. 193.2 ÷ 34.5

9. 478 + 9,732 + 5,286

10. 378.4 − 59.7 + 18.9

11. 36 × 207 × 188

12. 36.3 × 48 ÷ 7.2

PreSkills TEST

Round to the nearest hundred and to the nearest thousand. [16]

1. 4,839 **2.** 5,999 **3.** 16,573 **4.** 238,421

Complete. [4]

5. 56 = ■ tens 6 ones

6. 87 = 8 tens ■ ones

7. 748 = ■ hundreds 4 tens 8 ones

8. 650 = 6 hundreds ■ tens 0 ones

Add.

9. 5 + 8	**10.** 8 + 6	**11.** 6 + 4	**12.** 3 + 8	**13.** 8 + 7	**14.** 7 + 9
15. 30 + 60	**16.** 40 + 50	**17.** 800 + 800	**18.** 1 + 4 + 3	**19.** 80 + 70 + 20	**20.** 400 + 300 + 500

Subtract.

21. 17 − 9	**22.** 11 − 6	**23.** 15 − 7	**24.** 9 − 3	**25.** 14 − 7	**26.** 15 − 6
27. 80 − 30	**28.** 90 − 40	**29.** 800 − 700	**30.** 140 − 70	**31.** 120 − 50	**32.** 160 − 80

Multiply.

33. 7 × 3	**34.** 9 × 4	**35.** 5 × 8	**36.** 2 × 6	**37.** 4 × 8	**38.** 5 × 3
39. 6 × 7	**40.** 8 × 7	**41.** 9 × 3	**42.** 7 × 9	**43.** 6 × 6	**44.** 9 × 8

Divide.

45. 63 ÷ 9	**46.** 28 ÷ 7	**47.** 54 ÷ 6	**48.** 72 ÷ 8	**49.** 49 ÷ 7
50. 27 ÷ 5	**51.** 37 ÷ 4	**52.** 60 ÷ 8	**53.** 56 ÷ 6	**54.** 53 ÷ 9

Chapter 2

WHOLE NUMBER COMPUTATION

Community centers offer a variety of activities and services for people of all ages. What would the ideal community center provide for your community?

2.1 ESTIMATING SUMS AND DIFFERENCES

For a 3-day bike-a-thon, the Big Wheels Cycle Club raised \$2,492, \$3,096, and \$2,987. About how much money did they raise in all?

To find *about how much*, use estimation.

You can estimate a sum or difference by rounding each number to its greatest place. Then, add or subtract.

EXAMPLE 1 Estimate: \$2,492 + \$3,096 + \$2,987

\$2,492 + \$3,096 + \$2,987
↓ ↓ ↓
\$2,000 + \$3,000 + \$3,000 = \$8,000

So, the Cycle Club raised about \$8,000.

Sometimes you must round the numbers to different places.

EXAMPLE 2 Estimate: 11,157 − 8,301

1. Round each number to the greatest place.
2. Subtract.

11,157	→	10,000
− 8,301	→	8,000
		2,000

The estimated difference is 2,000.

You can also estimate using **front-end estimation.** Add or subtract only the digits in the greatest place.

EXAMPLE 3 Estimate:

a. 5,479 + 6,890 + 239

THINK: The greatest place is thousands. Add only the digits in the thousands place.

5 + 6 = 11

5,479
6,890
+ 239
11,000

b. 78,152 − 6,583

THINK: The greatest place is ten thousands. Subtract only the digit in the ten-thousands place.

7 − 0 = 7

78,152
− 6,583
70,000

FOR DISCUSSION

Why will front-end estimation for addition always be an underestimation?

PRACTICE EXERCISES

Estimate by rounding.

1. 739 + 421	**2.** 1,583 + 6,291	**3.** 2,986 + 1,123	**4.** 43,023 + 29,436	**5.** 183,426 + 158,559
6. 468 + 37	**7.** 2,804 + 657	**8.** 6,239 + 118	**9.** 11,219 + 2,658	**10.** 287,999 + 17,823
11. 862 942 + 871	**12.** 3,682 4,286 + 9,870	**13.** 4,765 5,821 + 8,601	**14.** 11,006 29,132 + 32,446	**15.** 666,231 476,003 + 123,459
16. 361 802 + 54	**17.** 4,362 399 + 213	**18.** 5,673 4,986 + 211	**19.** 29,014 981 + 3,025	**20.** 28,697 186,760 + 34,876
21. 627 − 374	**22.** 9,428 − 5,037	**23.** 6,829 − 2,268	**24.** 57,342 − 19,820	**25.** 379,005 − 124,198
26. 109 − 67	**27.** 3,711 − 994	**28.** 5,468 − 967	**29.** 12,287 − 7,594	**30.** 107,812 − 45,485

Estimate by using front-end estimation.

31. 739 + 421	**32.** 6,291 + 1,583	**33.** 6,823 + 3,104	**34.** 43,123 + 12,999	**35.** 586,036 + 228,451
36. 836 + 22	**37.** 4,223 + 349	**38.** 5,982 + 411	**39.** 34,097 + 1,789	**40.** 607,098 + 12,569
41. 316 250 + 123	**42.** 4,361 2,102 + 1,254	**43.** 5,786 1,102 + 2,006	**44.** 21,009 33,436 + 42,985	**45.** 613,890 506,231 + 438,506
46. 482 41 + 512	**47.** 1,056 2,213 + 872	**48.** 912 3,069 + 6,730	**49.** 31,986 82,411 + 6,121	**50.** 212,765 17,645 + 5,680
51. 356 − 127	**52.** 6,436 − 4,672	**53.** 5,208 − 1,341	**54.** 42,118 − 34,882	**55.** 379,005 − 124,198
56. 984 − 26	**57.** 5,434 − 267	**58.** 9,896 − 398	**59.** 26,402 − 3,857	**60.** 704,012 − 13,564

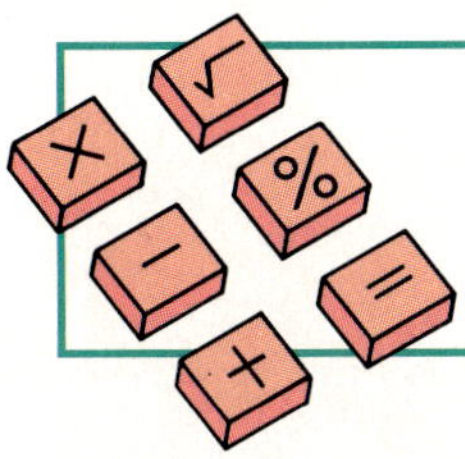

CALCULATOR

Use a calculator to find the exact answers to Exercises 6 through 25. Compare these answers to your estimates. Were they close?

2.2 ADDING

There are 840 miles of coastline in California, 296 miles in Oregon, and 157 miles in Washington. How many miles of coastline are there in all?

To find the total number of miles, add the three **distances**.

EXAMPLE 1 Add: 840 + 296 + 157

1. Add the ones. Regroup.

$$\begin{array}{r} {}^{1} \\ 840 \\ 296 \\ +157 \\ \hline 3 \end{array}$$

2. Add the tens. Regroup.

$$\begin{array}{r} {}^{11} \\ 840 \\ 296 \\ +157 \\ \hline 93 \end{array}$$

3. Add the hundreds. Regroup.

$$\begin{array}{rl} {}^{11} & \\ 840 & \leftarrow \\ 296 & \leftarrow \textbf{Addends} \\ +157 & \leftarrow \\ \hline 1{,}293 & \leftarrow \textbf{Sum} \end{array}$$

So, there are 1,293 miles of coastline.

To determine if an exact answer is reasonable, compare the answer to an estimate. If they are close, the answer is reasonable.

EXAMPLE 2 Add: 87,950 and 71,285. Is your answer reasonable?

1. Estimate.

$$\begin{array}{rcr} 87{,}950 & \rightarrow & 90{,}000 \\ +71{,}285 & \rightarrow & +70{,}000 \\ \hline & & 160{,}000 \end{array}$$

2. Add.

$$\begin{array}{r} 87{,}950 \\ +71{,}285 \\ \hline 159{,}235 \end{array}$$

The estimate, 160,000, and the exact answer, 159,235, are close. The answer is reasonable.

To check your addition, you can add in the reverse order.

EXAMPLE 3 Add: 4,935 + 138,861 + 72,049 + 6,113. Check your answer.

1. Add down.

$$\begin{array}{r} 4{,}935 \\ 138{,}861 \\ 72{,}049 \\ +\quad 6{,}113 \\ \hline 221{,}958 \end{array} \downarrow$$

2. Check up.

$$\begin{array}{r} 4{,}935 \\ 138{,}861 \\ 72{,}049 \\ +\quad 6{,}113 \\ \hline 221{,}958 \end{array} \uparrow$$

FOR DISCUSSION

Explain how you can tell just by looking that the given sum is incorrect: 976 + 854 + 792 = 4,087.

PRACTICE EXERCISES

Estimate by rounding. Which exact answer is reasonable?

1. 673 + 985 + 261	**a.** 119	**b.** 1,919	**c.** 1,119
2. 506 + 742 + 328	**a.** 1,576	**b.** 2,076	**c.** 14,476
3. 5,291 + 8,702 + 4,095	**a.** 29,088	**b.** 26,088	**c.** 18,088
4. 7,936 + 298 + 1,031	**a.** 9,265	**b.** 14,265	**c.** 22,265

Add. Check your answer.

5. 24 + 86
6. 19 + 32
7. 75 + 48
8. 94 + 7
9. 89 + 3

10. 372 + 549
11. 503 + 467
12. 728 + 360
13. 831 + 69
14. 967 + 64

15. 3,942 + 5,607
16. 9,314 + 6,812
17. 6,982 + 5,436
18. 7,873 + 951
19. 8,324 + 807

20. 87,024 + 13,142
21. 75,087 + 49,932
22. 83,012 + 14,624
23. 4,863 + 94,057
24. 90,816 + 7,643

25. 854,017 + 513,946
26. 760,912 + 841,536
27. 982,413 + 116,908
28. 359,806 + 67,238
29. 613,501 + 39,489

30. 86 + 57 + 23
31. 139 + 243 + 478
32. 4,236 + 5,113 + 8,982
33. 15,982 + 26,875 + 33,483
34. 383,051 + 127,987 + 431,856

35. 107 + 936 + 641 + 285
36. 2,076 + 8,341 + 6,593 + 1,845
37. 516 + 4,389 + 63 + 9,482
38. 92,037 + 6,139 + 804,625 + 39
39. 7,256 + 511,833 + 76,032 + 568,912

First estimate by rounding. Then find the exact answer.

40. 684 + 9,705 + 734,281
41. 17,827 + 3,496 + 28,092
42. 28,045 + 396 + 7,198
43. 7,831 + 64,189 + 24,815

Solve. Use the table.

44. Estimate the length of the total Pacific coastline.

45. Find the actual length of the total Pacific coastline.

46. Rank the states by the length of coastline from longest to shortest.

PACIFIC COASTLINE

State	Length (mi)
California	840
Oregon	296
Washington	157
Alaska	5,580
Hawaii	750

2.3 SUBTRACTING

The ROBO Inc. car manufacturing plant made 5,682 cars. The records show 729 cars were not assembled by robots. How many cars were assembled by robots?

To find the difference, subtract 729 from 5,682.

EXAMPLE 1 Subtract: 5,682 − 729

1. Regroup. Subtract the ones.
2. Subtract the tens.
3. Regroup. Subtract the hundreds.
4. Subtract the thousands.

1.	2.	3.	4.
7 12	7 12	4 16 7 12	4 16 7 12
5,682	5,682	5,682	5,682
− 729	− 729	− 729	− 729
3	53	953	4,953

So, 4,953 cars were assembled by robots.

When you regroup across zeros, you have to regroup more than once.

EXAMPLE 2 Subtract: 600 − 187

1. Regroup.
   THINK: 6 hundreds = 5 hundreds, 10 tens
2. Regroup.
   THINK: 10 tens = 9 tens, 10 ones
3. Subtract.

1.	2.	3.
	9	9
5 10	5 10 10	5 10 10
600	600	600
−187	−187	−187
		413

So, 600 − 187 = 413.

Remember, you can determine if an exact answer is reasonable by comparing the answer to an estimate.

EXAMPLE 3 Subtract: 9,693 − 2,874. Is your answer reasonable?

1. Estimate.

9,693 →	10,000
−2,874 →	− 3,000
	7,000

2. Subtract.

9,693
−2,874
6,819

Since 7,000 and 6,819 are close, the exact answer is reasonable.

You can use addition to check subtraction.

EXAMPLE 4 Subtract: 82,603 − 16,571. Check your answer.

1. Subtract.

$$\begin{array}{r} 82{,}603 \\ -16{,}571 \\ \hline 66{,}032 \end{array}$$

2. Check.

$$\begin{array}{r} 66{,}032 \\ +16{,}571 \\ \hline 82{,}603 \end{array}$$

CHECKPOINT

Write the letter of the correct answer.

Subtract.

1. 50,000 − 19,381	**a.** 41,729	**b.** 31,711	**c.** 49,381	**d.** 30,619
2. 16,428 − 7,523	**a.** 8,905	**b.** 9,105	**c.** 9,945	**d.** 8,945

PRACTICE EXERCISES

Estimate by rounding. Which exact answer is reasonable?

1. 912 − 234	**a.** 368	**b.** 678	**c.** 858
2. 5,709 − 2,457	**a.** 2,252	**b.** 3,252	**c.** 5,252
3. 8,178 − 605	**a.** 3,573	**b.** 7,573	**c.** 8,573
4. 19,936 − 6,292	**a.** 8,644	**b.** 9,464	**c.** 13,644

Subtract. Check your answer.

5. 97 − 47	**6.** 43 − 29	**7.** 51 − 33	**8.** 82 − 7	**9.** 70 − 9
10. 837 − 245	**11.** 653 − 491	**12.** 924 − 828	**13.** 501 − 76	**14.** 900 − 45
15. 3,419 − 2,652	**16.** 8,738 − 2,804	**17.** 3,651 − 1,836	**18.** 5,017 − 849	**19.** 3,802 − 956
20. 56,957 − 34,817	**21.** 26,815 − 13,906	**22.** 71,657 − 65,738	**23.** 90,508 − 6,591	**24.** 34,090 − 904
25. 452,311 − 109,533	**26.** 611,092 − 272,041	**27.** 923,187 − 867,059	**28.** 260,900 − 59,654	**29.** 501,087 − 43,852

First estimate by rounding. Then find the exact answer.

30. 611 − 421	**31.** 7,986 − 2,814	**32.** 38,999 − 12,437
33. 383 − 42	**34.** 8,822 − 396	**35.** 9,123 − 436
36. 36,123 − 8,041	**37.** 59,001 − 4,124	**38.** 683,421 − 36,923

MIXED REVIEW

Find the value of the underlined digit.

39. 765,034 **40.** 18,345,097 **41.** 674,091,254

42. 2.00865 **43.** 0.671294 **44.** 19.3850014

Compare. Write =, >, <.

45. 9,457 ● 9,449 **46.** 74,653 ● 74,660 **47.** 201,567 ● 199,567

48. 8.43 ● 8.430 **49.** 2.0056 ● 2.056 **50.** 243.9 ● 243.8876

Solve.

51. Robo's goal for the year is to build 25,000 cars. To date, 9,476 cars have been built. How many more cars are needed to reach the goal?

52. The Bion car plant produces 12,493 cars. All but 2,874 are made by robots. How many more cars are assembled by robots than by people?

53. Gross sales of the Jetstar from September to November were 6 thousand, 104 thousand, and 83 thousand. Write the number for the total gross sales.

54. The Zap model car has 3,194 moving parts. The Zing model has 23,842, and the Zoom model has 9,375. How many moving parts are there in all three models?

TIME OUT

You can make an interesting computing device called a **nomograph.**

1. On a piece of graph paper, draw three horizontal lines as shown. Label them *A*, *B*, and *C*.
2. Mark the scales as shown.

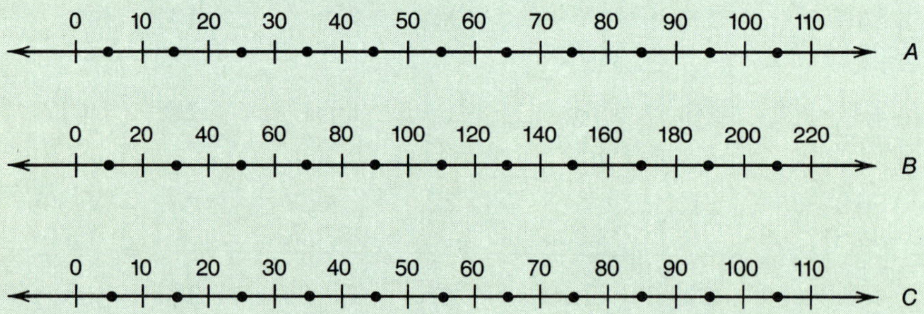

To find 70 − 25, place a straightedge so that it crosses *A* at 25 and *B* at 70. The answer is where it crosses *C* at 45.

Use your nomograph to find the difference.

1. 80 − 15 **2.** 110 − 5 **3.** 180 − 70 **4.** 125 − 40

PROBLEM *Solving* APPLICATION

2.4 CONSUMER: MAKING CHANGE

In daily life people often need to count money or make change. Change is generally given using the least number of coins and bills. Start with the coin of least value. Add on until the original amount of money given is reached.

Suppose you work in a stationery shop. A customer buys an item that costs 63 cents. He hands you a dollar bill. Count out the change and identify the amount of money he receives.

Give: 2 pennies	**THINK:** 63 + 2 = 65	Say: 65 cents
Give: 1 dime	**THINK:** 65 + 10 = 75	Say: 75 cents
Give: 1 quarter	**THINK:** 75 + 25 = 100	Say: 1 dollar (100 cents)

So, the customer receives 2 pennies, 1 dime, and 1 quarter.
The total amount of money he receives is $0.37.

FOR DISCUSSION

Sometimes you may not have all types of coins. Suppose you did not have quarters. How could you have given the change in the situation above?

List the least amount of coins or coins and bills you would use to make change for the purchase. Then identify the total amount of change the customer would receive.

	Cost of Item	Money Paid
1.	79 cents	$1 bill
2.	17 cents	$1 bill
3.	$0.28	$1 bill
4.	$0.66	$1 bill
5.	$1.24	two $1 bills
6.	$1.53	two $1 bills
7.	$3.91	$5 bill
8.	$7.53	$20 bill

Solve.

9. Mrs. Carson buys an item for $0.32. She hands the cashier a $1 bill. The cashier has only pennies and dimes to use as change. List the least amount of coins and the total amount she would receive.

10. In problem 9, suppose the cashier had only pennies, nickels, quarters, and half-dollars to use as change. List Mrs. Carson's change in the fewest coins possible.

2.5 MULTIPLYING AND DIVIDING: POWERS OF 10

It is estimated that Americans receive 5,000 tons of advertising mail each day. How many tons do they receive in 5 days?

To find how many tons, multiply.

EXAMPLE 1 Multiply: 5 × 5,000

THINK: 5 × 5 thousands = 25 thousands
WRITE: 5 × 5,000 = 25,000

So, Americans receive 25,000 tons of advertising mail in 5 days.

Discover a shortcut for multiplying multiples of powers of 10.

9 × 10 = 90	2 × 30 = 60	70 × 40 = 2,800
9 × 100 = 900	2 × 300 = 600	70 × 400 = 28,000
9 × 1,000 = 9,000	2 × 3,000 = 6,000	70 × 4,000 = 280,000

To multiply multiples of powers of 10:

1. Multiply the nonzero digits.
2. Write as many zeros as there are zeros in each factor.

EXAMPLE 2 Multiply:

a. 5,000 × 30

THINK: 5 × 3 = 15

So, 5,000 × 30 = 150,000.

b. 8,000 × 500

THINK: 8 × 5 = 40

So, 8000 × 500 = 4,000,000.

Patterns also show a shortcut for dividing multiples of powers of 10.

80 ÷ 8 = 10	70 ÷ 10 = 7	270 ÷ 30 = 9
800 ÷ 8 = 100	700 ÷ 10 = 70	2,700 ÷ 30 = 90
8,000 ÷ 8 = 1,000	7,000 ÷ 10 = 700	27,000 ÷ 30 = 900

To divide multiples of powers of 10:

1. Cross out the same number of zeros in each of the numbers.
2. Divide the result.

EXAMPLE 3 Divide:

a. 720,000 ÷ 800

THINK: $\frac{720{,}000}{800}$ **← divided by** (two zeros crossed out in each)

72 ÷ 8 = 9
7,200 ÷ 8 = 900

So, 720,000 ÷ 800 = 900.

b. 400,000 ÷ 8,000

THINK: $\frac{400{,}000}{8{,}000}$ (three zeros crossed out in each)

40 ÷ 8 = 5
400 ÷ 8 = 50

So, 400,000 ÷ 8,000 = 50.

CHECKPOINT

Write the letter of the correct answer.

Multiply or divide.

1. 769 × 1,000	**a.** 7,690,000	**b.** 769,000	**c.** 76,900
2. 20 × 50,000	**a.** 1,000,000	**b.** 100,000	**c.** 10,000
3. 3,600 ÷ 40	**a.** 9	**b.** 90	**c.** 900
4. 3,000,000 ÷ 600	**a.** 50	**b.** 500	**c.** 5,000

PRACTICE EXERCISES

Multiply.

1. 20 × 10 **2.** 70 × 60 **3.** 800 × 80 **4.** 600 × 20

5. 900 × 100 **6.** 600 × 800 **7.** 400 × 700 **8.** 800 × 900

9. 6,000 × 700 **10.** 800 × 5,000 **11.** 400 × 600

12. 2,000 × 9,000 **13.** 4,000 × 6,000 **14.** 5,000 × 6,000

15. 5,000 × 70,000 **16.** 4,596 × 10,000 **17.** 3,000 × 500,000

18. 4,387 × 100,000 **19.** 70,000 × 400,000 **20.** 300 × 6,000,000

Divide.

21. 800 ÷ 10 **22.** 600 ÷ 30 **23.** 1,200 ÷ 20 **24.** 2,100 ÷ 70

25. 54,000 ÷ 90 **26.** 72,000 ÷ 80 **27.** 120,000 ÷ 40 **28.** 150,000 ÷ 50

29. 7,000 ÷ 100 **30.** 9,000 ÷ 300 **31.** 15,000 ÷ 500

32. 28,000 ÷ 7,000 **33.** 420,000 ÷ 6,000 **34.** 300,000 ÷ 5,000

35. 270,000 ÷ 90,000 **36.** 930,000 ÷ 10,000 **37.** 200,000 ÷ 40,000

38. 3,690,000 ÷ 10,000 **39.** 4,000,000 ÷ 50,000 **40.** 6,300,000 ÷ 90,000

Solve.

41. Advertising time for a television special cost $140,000 a minute. How much did it cost for 4 min?

42. A poll showed that 2,000 teenagers in each of 5,000 cities chose a product because of its ad. How many teenagers chose it?

TIME OUT

As a promotion, a local record store ran a contest. The contest was to guess how many pennies there are in a stack that is 1 yd high. Jenna guessed 100 pennies, Frank guessed 1,000 pennies, and Ben guessed 10,000 pennies. They knew that 20 pennies in a stack is about 1 in. high. Whose guess was closest to the actual answer?

2.6 ESTIMATING PRODUCTS AND QUOTIENTS

An animal shelter takes care of 374 stray animals a week. About how many animals does it care for in 8 weeks? in a year (52 weeks)?

To find *about how many,* use estimation.

You can estimate a product by rounding each number to its greatest place. Then multiply.

EXAMPLE 1 Estimate:

a. 8×374

$$\begin{array}{rcr} 374 & \rightarrow & 400 \\ \underline{\times \quad 8} & \rightarrow & \underline{\times \quad 8} \\ & & 3{,}200 \end{array}$$

About 3,200 animals are cared for in 8 weeks.

b. 52×374

$$\begin{array}{rcr} 374 & \rightarrow & 400 \\ \underline{\times \ 52} & \rightarrow & \underline{\times \ 50} \\ & & 20{,}000 \end{array}$$

About 20,000 animals are cared for in a year.

You can also estimate a quotient by rounding each number to its greatest place. Then divide.

EXAMPLE 2 Estimate:

a. $773 \div 8$

$$\begin{array}{ccccc} 773 & \div & 8 & & \\ \downarrow & & \downarrow & & \\ 800 & \div & 8 & = & 100 \end{array}$$

The estimated quotient is 100.

b. $56{,}214 \div 32$

$$\begin{array}{ccccc} 56{,}214 & \div & 32 & & \\ \downarrow & & \downarrow & & \\ 60{,}000 & \div & 30 & = & 2{,}000 \end{array}$$

The estimated quotient is 2,000.

Another way you can estimate a quotient is by using **compatible numbers,** numbers that divide evenly.

EXAMPLE 3 Estimate:

a. $196 \div 7$

THINK: $21 \div 7 = 3$

$$\begin{array}{ccccc} 196 & \div & 7 & & \\ \downarrow & & \downarrow & & \\ 210 & \div & 7 & = & 30 \end{array}$$

b. $1{,}102 \div 58$

THINK: $12 \div 6 = 2$

$$\begin{array}{ccccc} 1{,}102 & \div & 58 & & \\ \downarrow & & \downarrow & & \\ 1{,}200 & \div & 60 & = & 20 \end{array}$$

FOR DISCUSSION

Estimate $743 \div 8$ by rounding each number to its greatest place, and then by using compatible numbers. Which method was easier? Explain.

PRACTICE EXERCISES

Estimate by rounding.

1. 49×7	**2.** 56×8	**3.** 792×9	**4.** 539×6	**5.** 816×7
6. $2{,}134 \times 4$	**7.** $6{,}982 \times 6$	**8.** $7{,}316 \times 9$	**9.** $8{,}012 \times 5$	**10.** $9{,}986 \times 7$
11. 94×43	**12.** 76×58	**13.** 84×17	**14.** 67×32	**15.** 73×41
16. 390×69	**17.** 496×23	**18.** 695×48	**19.** $2{,}394 \times 78$	**20.** $6{,}221 \times 11$
21. $5\overline{)235}$	**22.** $7\overline{)749}$	**23.** $8\overline{)408}$	**24.** $6\overline{)594}$	**25.** $5\overline{)215}$
26. $3\overline{)5{,}925}$	**27.** $4\overline{)7{,}604}$	**28.** $5\overline{)2{,}195}$	**29.** $6\overline{)3{,}120}$	**30.** $7\overline{)7{,}385}$
31. $8\overline{)37{,}904}$	**32.** $9\overline{)94{,}986}$	**33.** $4\overline{)39{,}126}$	**34.** $3\overline{)28{,}666}$	**35.** $4\overline{)36{,}436}$
36. $19\overline{)834}$	**37.** $48\overline{)986}$	**38.** $27\overline{)592}$	**39.** $39\overline{)19{,}426}$	**40.** $23\overline{)59{,}611}$

Estimate using compatible numbers.

41. $352 \div 8$	**42.** $527 \div 9$	**43.** $652 \div 7$	**44.** $133 \div 4$
45. $4{,}123 \div 6$	**46.** $8{,}236 \div 9$	**47.** $5{,}802 \div 83$	**48.** $1{,}391 \div 22$
49. $3{,}597 \div 41$	**50.** $4{,}814 \div 87$	**51.** $30{,}752 \div 43$	**52.** $29{,}456 \div 39$

Solve.

53. A kennel uses 387 cases of dog food a month. There are 24 cans in each case. About how many cans does the kennel use?

54. Last year, the number of registered German shepherds dropped from 65,073 to 59,450. How many fewer registrations were there last year?

55. In 1983, 64,389 Labrador retrievers were registered in 936 kennel clubs. About the same number were registered in each club. About how many were registered in each club?

56. In a recent year, there were 1,085,248 registrations in the American Kennel Club. Round the number of registrations to the nearest million.

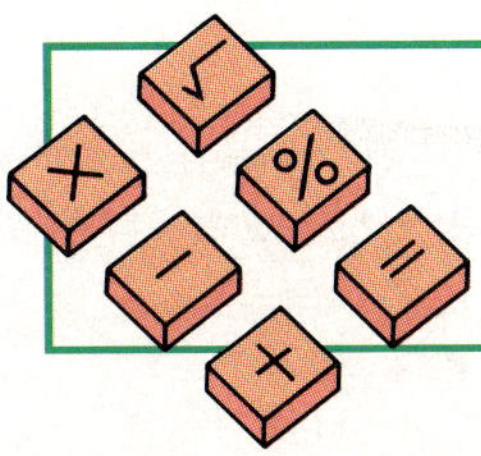

CALCULATOR

Use a calculator to find the exact answers to Exercises 21 through 40. Compare these answers to your estimates. Were they close?

2.7 MULTIPLYING BY 1- AND 2-DIGIT NUMBERS

The school cafeteria sells an average of 2,049 sandwiches each week. How many sandwiches does it sell in a month (4 weeks)?

To find the number of sandwiches, multiply 2,049 by 4.

EXAMPLE 1

Multiply: $4 \times 2{,}049$

1. Multiply the ones. Regroup.

$$\begin{array}{r} \scriptstyle 3 \\ 2{,}049 \\ \times \quad 4 \\ \hline 6 \end{array}$$

2. Multiply the tens. Regroup.

$$\begin{array}{r} \scriptstyle 1\,3 \\ 2{,}049 \\ \times \quad 4 \\ \hline 96 \end{array}$$

3. Multiply the hundreds.

$$\begin{array}{r} \scriptstyle 1\,3 \\ 2{,}049 \\ \times \quad 4 \\ \hline 196 \end{array}$$

4. Multiply the thousands.

$$\begin{array}{r} \scriptstyle 1\,3 \\ 2{,}049 \\ \times \quad 4 \\ \hline 8{,}196 \end{array} \begin{array}{l} \left.\right\} \text{Factors} \\ \leftarrow \text{Product} \end{array}$$

So, it sells an average of 8,196 sandwiches in a month.

You can use the same steps to multiply greater numbers.

EXAMPLE 2

Multiply: 38×97

1. Multiply by the ones.

$$\begin{array}{r} 97 \\ \times 38 \\ \hline 776 \end{array} \leftarrow \mathbf{8 \times 97}$$

2. Multiply by the tens.

$$\begin{array}{r} 97 \\ \times 38 \\ \hline 776 \\ 2\,910 \end{array} \leftarrow \mathbf{30 \times 97}$$

3. Add the partial products.

$$\begin{array}{r} 97 \\ \times 38 \\ \hline 776 \\ 2\,910 \\ \hline 3{,}686 \end{array}$$

So, $38 \times 97 = 3{,}686$.

To determine if an exact answer is reasonable, compare the answer with an estimate.

EXAMPLE 3

Multiply: 29×814. Is your answer reasonable?

1. Estimate.

$$\begin{array}{r} 814 \\ \times\ 29 \\ \hline \end{array} \begin{array}{c} \rightarrow \\ \rightarrow \end{array} \begin{array}{r} 800 \\ \times\ 30 \\ \hline 24{,}000 \end{array}$$

2. Multiply.

$$\begin{array}{r} 814 \\ \times\ 29 \\ \hline 7\,326 \\ 16\,280 \\ \hline 23{,}606 \end{array}$$

Since the estimate, 24,000, and the exact answer, 23,606, are close, the answer is reasonable.

PRACTICE EXERCISES

Estimate by rounding. Which answer is reasonable?

1. 6 × 241	**a.** 146	**b.** 1,446	**c.** 2,446
2. 5 × 3,096	**a.** 150,480	**b.** 25,480	**c.** 15,480
3. 42 × 78	**a.** 1,876	**b.** 3,276	**c.** 6,876
4. 71 × 409	**a.** 29,039	**b.** 19,309	**c.** 2,930

Multiply.

5. 13 × 6	**6.** 72 × 4	**7.** 86 × 5	**8.** 51 × 6	**9.** 97 × 9
10. 142 × 6	**11.** 217 × 7	**12.** 344 × 8	**13.** 563 × 9	**14.** 893 × 9
15. 1,236 × 5	**16.** 3,542 × 3	**17.** 4,615 × 2	**18.** 2,538 × 4	**19.** 5,173 × 6
20. 6,094 × 7	**21.** 5,607 × 9	**22.** 3,940 × 6	**23.** 7,815 × 8	**24.** 6,009 × 7
25. 16,423 × 4	**26.** 29,415 × 5	**27.** 36,017 × 6	**28.** 41,009 × 8	**29.** 67,149 × 9
30. 83 × 27	**31.** 56 × 19	**32.** 92 × 53	**33.** 78 × 16	**34.** 65 × 38
35. 367 × 25	**36.** 290 × 56	**37.** 648 × 32	**38.** 815 × 24	**39.** 509 × 47
40. 1,123 × 28	**41.** 2,983 × 32	**42.** 4,786 × 41	**43.** 5,980 × 56	**44.** 7,004 × 82

45. 3 × 95	**46.** 7 × 908	**47.** 9 × 5,602	**48.** 7 × 8,432
49. 37 × 52	**50.** 60 × 318	**51.** 32 × 806	**52.** 28 × 629
53. 41 × 1,456	**54.** 53 × 2,004	**55.** 64 × 6,559	**56.** 80 × 6,117

First estimate by rounding. Then find the exact answer.

57. 2,308 × 5	**58.** 4,879 × 8	**59.** 38 × 57	**60.** 82 × 32	**61.** 216 × 74

Solve.

62. 279 students each raised $46 for a community fund raiser. How much money did they raise?

63. Class enrollments are 987, 893, 905, and 936. The school gym seats 3,500. Can all the classes fit in the gym at the same time?

2.8 MULTIPLYING BY 3-DIGIT NUMBERS

Dan's Discount Stores have VCRs on sale for $298. In 1 week, 367 VCRs were sold. What was the total amount of sales?

To find the total amount of sales, multiply $298 by 367.

EXAMPLE 1 Multiply: 367 × $298

1. Multiply.

```
   $298
 ×  367
  2 086 ←     7 × 298
 17 880 ←    60 × 298
 89 400 ←   300 × 298
```

2. Add the partial products.

```
    $298
  ×  367
   2 086
  17 880
  89 400
$109,366
```

So, the total amount of sales was $109,366.

Work carefully when the bottom factor has one or more zeros.

EXAMPLE 2 Multiply: 803 × 4,906

```
    4,906
  ×   803
   14 718 ←     3 × 4,906
3 924 800 ←   800 × 4,906
3,939,518
```

It is often helpful to choose the number you will use as the bottom factor. Try to make the multiplication easier.

EXAMPLE 3 Multiply:

a. 839 × 2,004

THINK: 2,004 has two nonzero digits.

```
      839
  × 2,004
    3 356 ← Two partial products
1 678 000 ←
1,681,356
```

b. 419 × 737

THINK: In 737, there are two 7s.

```
    419
  × 737
  2 933 ← Same nonzero
 12 570   digits in the
293 300 ← partial products
308,803
```

FOR DISCUSSION

Look at the following example: 892 × 635 = 56,640.
Is the answer reasonable? Explain.

PRACTICE EXERCISES

Multiply.

1. 729 × 132 **2.** 563 × 245 **3.** 612 × 328 **4.** 435 × 517 **5.** 347 × 193

6. 951 × 376 **7.** 816 × 439 **8.** 712 × 635 **9.** 967 × 249 **10.** 658 × 427

11. 5,416 × 325 **12.** 4,176 × 193 **13.** 3,284 × 267 **14.** 5,318 × 241 **15.** 4,935 × 564

16. 3,618 × 127 **17.** 4,589 × 236 **18.** 5,912 × 348 **19.** 6,142 × 459 **20.** 7,196 × 517

21. 495 × 703 **22.** 360 × 580 **23.** 704 × 600 **24.** 527 × 203 **25.** 803 × 450

26. 1,639 × 200 **27.** 3,258 × 460 **28.** 7,503 × 305 **29.** 6,479 × 506 **30.** 7,857 × 308

31. 218 × 677 **32.** 241 × 303 **33.** 167 × 414 **34.** 479 × 808 **35.** 825 × 626

36. 3,007 × 213 **37.** 4,256 × 848 **38.** 7,700 × 541 **39.** 4,114 × 623 **40.** 6,006 × 589

41. 288 × 673 **42.** 764 × 222 **43.** 345 × 7,000 **44.** 6,060 × 539

45. 707 × 123 **46.** 163 × 528 **47.** 404 × 209 **48.** 127 × 5,004

49. 8,006 × 412 **50.** 919 × 246 **51.** 4,182 × 936 **52.** 636 × 8,923

53. 321 × 865 **54.** 159 × 827 **55.** 803 × 507 **56.** 821 × 936

57. 5,006 × 127 **58.** 962 × 141 **59.** 6,431 × 322 **60.** 239 × 9,002

61. 8,623 × 428 **62.** 5,002 × 128 **63.** 7,514 × 424 **64.** 165 × 8,008

Use the table to solve.

Find the total amount of sales for:

65. televisions.

66. stereos.

67. computers.

68. How many items were sold in all?

69. What was the total amount of sales for all the items?

DAN'S DISCOUNT SALES

Item/Price	Number Sold	Total Sales
VCR/$298	367	$109,366
TV/$507	129	■
Stereo/$249	243	■
Computer/$628	68	■
Total	■	■

MIDCHAPTER REVIEW

Estimate by rounding. [28]

1. 842 +396	**2.** 3,719 +5,478	**3.** 623 2,534 +6,195	**4.** 4,857 −3,149	**5.** 56,493 − 7,346

Estimate using front-end estimation. [28]

6. 893 −346	**7.** 9,863 −1,347	**8.** 6,982 − 783	**9.** 59,007 −45,113	**10.** 623,456 − 54,897

Add or subtract. Check your answer. [30, 32]

11. 536 +342	**12.** 819 +675	**13.** 5,093 +9,847	**14.** 26,534 +37,628	**15.** 629,845 +356,971
16. 729 15 +648	**17.** 935 194 +382	**18.** 5,497 2,386 4,713 +8,565	**19.** 98 8,567 481 +6,005	**20.** 725,347 52,485 600,721 + 8,340
21. 786 −357	**22.** 629 − 35	**23.** 8,596 −5,738	**24.** 5,774 − 865	**25.** 7,000 −3,829
26. 3,000 − 537	**27.** 94,617 −28,369	**28.** 93,648 −74,956	**29.** 529,400 −132,675	**30.** 600,000 −250,785

Multiply. [36]

31. 20 × 10 **32.** 300 × 40 **33.** 500 × 30

34. 600 × 30 **35.** 500 × 600 **36.** 400 × 6,000

Divide. [36]

37. 600 ÷ 20 **38.** 900 ÷ 30 **39.** 8,000 ÷ 40

40. 50,000 ÷ 50 **41.** 6,000 ÷ 300 **42.** 20,000 ÷ 400

Solve. [35]

43. Joe Moser bought an album for $7.98. He gave the cashier a $10 bill. List the least amount of coins and bills he would receive as change.

44. Ellen Kamp bought a record for $1.79. She gave the clerk two $1 bills. List the least amount of coins she would receive as change.

PROBLEM Solving APPLICATION

2.9 CAREER: NURSE

Luveen Smith is a nurse in a hospital. She helps people who are sick. She also helps people avoid sickness through immunization programs. Often in her work she must keep records of the people she has attended to and helped.

This table shows the number of people immunized for various diseases at three hospitals. Use the table to complete Exercises 1–10. Remember to estimate whenever you use your calculator.

Types of immunization	Number of people immunized			Types of immunization	Number of people immunized		
	Hospital A	Hospital B	Hospital C		Hospital A	Hospital B	Hospital C
Diphtheria	805	427	608	Polio	1,674	923	1,527
Flu	3,349	1,103	1,649	Rubella	248	189	256
Measles	405	389	426	Smallpox	637	420	489
Mumps	297	187	288	Tuberculin	479	342	461

What was the total number of people immunized for:

1. diphtheria?

2. polio?

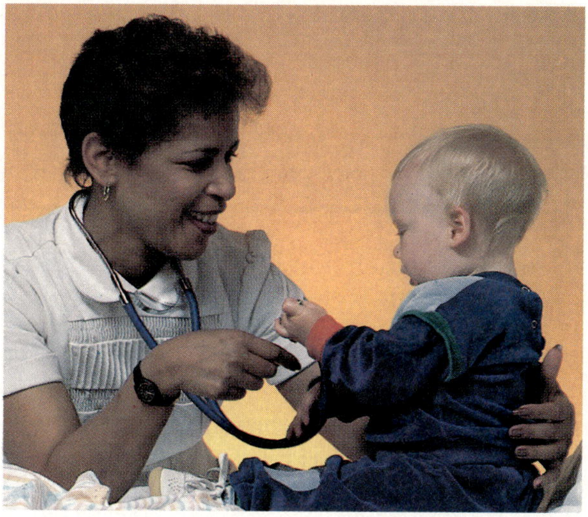

How many more people were immunized for measles than for mumps:

3. at Hospital A?

4. at Hospital C?

What was the total number of immunizations for:

5. rubella and smallpox?

6. mumps and measles?

Solve.

7. Which two hospitals immunized the greatest number of people?

8. Which two hospitals immunized the least number of people?

9. The number of diphtheria immunizations at Hospital A was about how many times as great as the number immunized for diphtheria at Hospital B?

10. The number of flu immunizations at Hospital A was about how many times as great as the number of flu immunizations at Hospital B?

2.10 DIVIDING BY A 1-DIGIT DIVISOR

The Computer Corner stores had a sale on software. The warehouse had 317 software packages to distribute equally to 5 different stores. How many software packages did each store receive? How many packages were left?

To find the number of packages delivered to each store, divide 317 by 5.

EXAMPLE 1

Divide: $5\overline{)317}$

1. Determine the first place that can be divided.

   $5\overline{)317}$ **THINK:** $5 > 3$ Can't divide the hundreds.
   $5 < 31$ Divide the tens.

2. Divide the tens.

```
   6
5)317
  30
   1
```

3. Divide the ones. Write the remainder.

```
  63 R2
5)317
  30
   17
   15
    2
```

Remainder

So, each store received 63 software packages and 2 were left over.

Sometimes when you divide, you need a zero placeholder in the quotient.

EXAMPLE 2

Divide: $3\overline{)2{,}418}$

1. Determine the first place that can be divided.

   $3\overline{)2{,}418}$ **THINK:** $3 > 2$ Can't divide the thousands.
   $3 < 24$ Divide the hundreds.

2. Divide the hundreds.

```
    8
3)2,418
  2 4
    0
```

3. Divide the tens.

```
    80
3)2,418
  2 4
    01
```

4. Divide the ones.

```
    806
3)2,418
  2 4
    018
     18
      0
```

You can use multiplication to check division.

EXAMPLE 3 Divide: 3,412 ÷ 7. Check your answer.

1. Divide.

Quotient → 487 R3 ← **Remainder**
Divisor → $7\overline{)3{,}412}$ ← **Dividend**

$$\begin{array}{r} 487 \text{ R3} \\ 7\overline{)3{,}412} \\ \underline{2\,8} \\ 61 \\ \underline{56} \\ 52 \\ \underline{49} \\ 3 \end{array}$$

2. Check.

$$\begin{array}{rl} 487 & \leftarrow \textbf{Quotient} \\ \times \quad 7 & \leftarrow \textbf{Divisor} \\ \hline 3{,}409 & \\ + \quad 3 & \leftarrow \textbf{Remainder} \\ \hline 3{,}412 & \leftarrow \textbf{Dividend} \end{array}$$

PRACTICE EXERCISES

Divide.

1. $6\overline{)824}$	**2.** $4\overline{)729}$	**3.** $9\overline{)459}$	**4.** $5\overline{)830}$	**5.** $8\overline{)548}$
6. $7\overline{)636}$	**7.** $3\overline{)180}$	**8.** $5\overline{)354}$	**9.** $9\overline{)918}$	**10.** $6\overline{)653}$
11. $8\overline{)842}$	**12.** $5\overline{)249}$	**13.** $7\overline{)161}$	**14.** $6\overline{)590}$	**15.** $9\overline{)837}$
16. $3\overline{)630}$	**17.** $5\overline{)700}$	**18.** $8\overline{)440}$	**19.** $7\overline{)500}$	**20.** $6\overline{)200}$
21. $3\overline{)1{,}296}$	**22.** $7\overline{)3{,}584}$	**23.** $4\overline{)8{,}092}$	**24.** $6\overline{)3{,}726}$	**25.** $8\overline{)7{,}632}$
26. $5\overline{)4{,}520}$	**27.** $9\overline{)6{,}395}$	**28.** $7\overline{)2{,}429}$	**29.** $8\overline{)1{,}523}$	**30.** $3\overline{)8{,}427}$
31. $7\overline{)2{,}358}$	**32.** $4\overline{)7{,}012}$	**33.** $6\overline{)3{,}288}$	**34.** $8\overline{)5{,}028}$	**35.** $9\overline{)6{,}327}$
36. $4\overline{)8{,}400}$	**37.** $5\overline{)6{,}500}$	**38.** $8\overline{)6{,}000}$	**39.** $5\overline{)3{,}600}$	**40.** $7\overline{)4{,}000}$
41. $5\overline{)15{,}825}$	**42.** $7\overline{)21{,}357}$	**43.** $4\overline{)37{,}296}$	**44.** $9\overline{)63{,}828}$	**45.** $6\overline{)45{,}630}$
46. $6\overline{)57{,}486}$	**47.** $8\overline{)65{,}368}$	**48.** $9\overline{)38{,}271}$	**49.** $5\overline{)90{,}355}$	**50.** $7\overline{)58{,}135}$
51. $6\overline{)24{,}750}$	**52.** $7\overline{)78{,}491}$	**53.** $4\overline{)68{,}624}$	**54.** $7\overline{)36{,}157}$	**55.** $8\overline{)49{,}017}$

Divide.

56. 372 ÷ 6	**57.** 896 ÷ 4	**58.** 215 ÷ 7	**59.** 567 ÷ 3
60. 500 ÷ 4	**61.** 200 ÷ 8	**62.** 600 ÷ 7	**63.** 900 ÷ 5
64. 4,028 ÷ 4	**65.** 3,692 ÷ 4	**66.** 5,348 ÷ 7	**67.** 27,229 ÷ 9

Solve.

68. There are 8,750 sheets of computer paper divided evenly into 7 boxes. How many sheets are in each box?

69. The computer work center sells for \$149. If each piece is bought separately, it costs \$174. How much is saved by buying the complete center?

2.11 DIVIDING BY A 2-DIGIT DIVISOR

Sarah O'Brien has 115 record albums in her collection. She can put 21 albums in a case for display. How many cases can she fill? How many albums will be left over?

To find how many cases, divide 115 by 21.

EXAMPLE 1

Divide: 21)115

1. Determine the first place that can be divided.

   21)115 **THINK:**

$21 > 1$	Can't divide the hundreds.
$21 > 11$	Can't divide the tens.
$21 < 115$	Divide the ones.

2. Estimate the first digit. **THINK:** 21)115 or 2)11 Try 5.

3. Divide the ones.
   Write the remainder.

```
     5 R10
21)115        Remainder
   105
    10
```

Sarah can fill 5 display cases, and 10 albums will be left over.

Sometimes your estimate of the first digit of a quotient may be too large.

EXAMPLE 2

Divide: 69)3,105

1. Determine the first place that can be divided.

   69)3,105 **THINK:**

$69 > 3$	Can't divide the thousands.
$69 > 31$	Can't divide the hundreds.
$69 < 310$	Divide the tens.

2. Estimate the first digit. **THINK:** 69)310 or 6)31 Try 5.

3. Divide the tens.

```
       5
69)3,105
   3 45     345 > 310
            5 is too large.
            Try 4.

       4
69)3,105
   2 76
     34
```

4. Divide the ones.

```
      45
69)3,105
   2 76
     345
     345
       0
```

You can estimate to determine if your answer is reasonable.

EXAMPLE 3 Divide: $34\overline{)17,238}$. Is your answer reasonable?

1. Estimate using compatible numbers.

$$34\overline{)17,238} \rightarrow \begin{array}{r} 600 \\ 30\overline{)18,000} \end{array}$$

2. Divide.

$$\begin{array}{r} 507 \\ 34\overline{)17,238} \\ \underline{17\,0} \\ 238 \\ \underline{238} \\ 0 \end{array}$$

Since the estimate, 600, and the exact answer, 507, are close, the answer is reasonable.

PRACTICE EXERCISES

Divide.

1. $32\overline{)268}$	**2.** $51\overline{)397}$	**3.** $71\overline{)651}$	**4.** $32\overline{)109}$	**5.** $76\overline{)542}$
6. $60\overline{)415}$	**7.** $58\overline{)406}$	**8.** $64\overline{)510}$	**9.** $43\overline{)297}$	**10.** $24\overline{)127}$
11. $97\overline{)291}$	**12.** $34\overline{)284}$	**13.** $42\overline{)336}$	**14.** $98\overline{)600}$	**15.** $86\overline{)430}$
16. $23\overline{)4,600}$	**17.** $40\overline{)8,805}$	**18.** $67\overline{)3,618}$	**19.** $18\overline{)9,342}$	**20.** $56\overline{)1,960}$
21. $19\overline{)1,292}$	**22.** $75\overline{)6,000}$	**23.** $96\overline{)2,692}$	**24.** $27\overline{)3,726}$	**25.** $81\overline{)5,675}$
26. $68\overline{)1,872}$	**27.** $35\overline{)2,253}$	**28.** $79\overline{)3,950}$	**29.** $90\overline{)5,130}$	**30.** $26\overline{)9,724}$
31. $30\overline{)67,020}$	**32.** $25\overline{)15,050}$	**33.** $54\overline{)16,524}$	**34.** $16\overline{)77,000}$	**35.** $82\overline{)29,274}$
36. $47\overline{)33,135}$	**37.** $91\overline{)57,057}$	**38.** $60\overline{)93,960}$	**39.** $74\overline{)38,702}$	**40.** $36\overline{)84,996}$
41. $18\overline{)73,836}$	**42.** $26\overline{)29,199}$	**43.** $36\overline{)20,989}$	**44.** $34\overline{)30,334}$	**45.** $54\overline{)35,208}$

First estimate by using compatible numbers. Then find the exact answer.

46. $459 \div 51$	**47.** $837 \div 27$	**48.** $4,760 \div 68$	**49.** $1,890 \div 54$
50. $57,792 \div 48$	**51.** $59,328 \div 72$	**52.** $498 \div 93$	**53.** $6,408 \div 92$

Solve.

54. Delta Jazz sold 43,884 albums of their new hit in 12 weeks. On an average, how many did they sell each week?

55. Buzz Recording Studio made $96,252 from 26 records sold or played last year. On an average, how much did it make from each record?

2.12 DIVIDING BY A 3-DIGIT DIVISOR

The Small Car Parts Company makes 3,500 spark plugs an hour.
A box of spark plugs contains 432 spark plugs.
How many boxes will 3,500 spark plugs fill?
How many spark plugs will be left over?

You can divide to find the number of boxes.

EXAMPLE 1 Divide: $432\overline{)3,500}$. Check your answer.

1. Determine the first place that can be divided.

$432\overline{)3,500}$ **THINK:**

432 > 3	Can't divide the thousands.
432 > 35	Can't divide the hundreds.
432 > 350	Can't divide the tens.
432 < 3,500	Divide the ones.

2. Estimate the first digit. **THINK:** $432\overline{)3,500}$ or $4\overline{)35}$ Try 8.

3. Divide the ones.

$$\begin{array}{r} 8 \text{ R}44 \\ 432\overline{)3,500} \\ \underline{3\,456} \\ 44 \end{array}$$

4. Check.

$$\begin{array}{r} 432 \\ \underline{\times \quad 8} \\ 3,456 \\ \underline{+ \quad 44} \\ 3,500 \end{array}$$

So, 8 boxes will be filled, and 44 spark plugs will be left over.

Sometimes you will need a zero placeholder in the quotient.

EXAMPLE 2 Divide: $318\overline{)95,405}$

1. Determine the first place that can be divided.

$318\overline{)95,405}$ **THINK:**

318 > 9	Can't divide the ten thousands.
318 > 95	Can't divide the thousands.
318 < 954	Divide the hundreds.

2. Estimate the first digit. **THINK:** $318\overline{)954}$ or $3\overline{)9}$ Try 3.

3. Divide the hundreds.

$$\begin{array}{r} 3 \\ 318\overline{)95,405} \\ \underline{95\,4} \\ 0 \end{array}$$

4. Divide the tens.

$$\begin{array}{r} 30 \\ 318\overline{)95,405} \\ \underline{95\,4} \\ 00 \end{array}$$

5. Divide the ones.

$$\begin{array}{r} 300 \text{ R}5 \\ 318\overline{)95,405} \\ \underline{95\,4} \\ 005 \\ \underline{0} \\ 5 \end{array}$$

FOR DISCUSSION

When dividing by 467, what is the greatest possible remainder? Explain.

PRACTICE EXERCISES

Divide.

1. 9,804 ÷ 258
2. 8,480 ÷ 200
3. 9,216 ÷ 156
4. 5,440 ÷ 136
5. 5,042 ÷ 168
6. 33,748 ÷ 418
7. 25,599 ÷ 179
8. 28,341 ÷ 308
9. 25,761 ÷ 515
10. 341,440 ÷ 231
11. 505,891 ÷ 569
12. 119,179 ÷ 587
13. 450)9,008
14. 256)8,960
15. 652)5,868
16. 328)9,840
17. 176)8,096
18. 563)8,000
19. 298)7,158
20. 765)6,200
21. 592)4,144
22. 397)5,955
23. 418)5,435
24. 205)8,200
25. 896)8,095
26. 143)26,169
27. 326)16,300
28. 228)17,784
29. 135)40,500
30. 600)90,000
31. 159)16,695
32. 252)32,870
33. 739)72,422
34. 946)61,500
35. 479)99,632
36. 857)68,600
37. 118)107,262
38. 468)131,508
39. 645)598,560
40. 113)469,628

MIXED REVIEW

Find the answer.

41. 675 + 387
42. 1,209 − 97
43. 20 × 600
44. 9 × 132
45. 400,000 ÷ 50
46. 6,400 ÷ 80
47. 9)2,720
48. 7)1,854

Solve.

49. 480 fan belts are packed in a box. How many boxes will 45,600 fan belts fill?
50. Gloria works a 35-hour week. If she earns $420 a week, how much does she earn an hour?
51. Rosella's car averages 41 miles per gallon (mi/gal). How many gallons of gas will be needed for a trip of 943 mi?
52. Adam's gas tank holds 20 gal of gas. His car averages 42 mi/gal. How far can he travel if his gas tank is full?

TIME OUT

If it takes 8 min to cut a log into 3 pieces, how long will it take to cut a log into 4 pieces? (*Hint:* How many cuts are needed to make 3 pieces?)

PROBLEM *Solving* STRATEGY

2.13 WRITING SUBPROBLEMS

Situation:

Ted Cohen wanted to have a 300-calorie (cal) breakfast. He ate 1 cup of oatmeal, an orange, and a glass of skimmed milk. How many fewer calories did he have than he wanted?

Strategy:

Find subproblems you can solve to help you arrive at a final solution.

Applying the Strategy:

Subproblem 1: How many calories were in Ted's breakfast? $134 + 60 + 90 = 284$ cal

THINK: Oatmeal: 134 cal Orange: 60 cal
Skimmed milk: 90 cal

Subproblem 2: How many fewer calories did he have than he wanted? $300 - 284 = 16$ cal

CALORIC COUNT OF SOME BASIC FOODS

Milk group	Meat group	Bread-cereal group	Vegetable-fruit group
1 cup whole milk 165	1 hard-boiled egg 75	1 slice whole wheat bread 55	1 medium orange 60
1 cup skimmed milk 90	1 serving tuna fish 172	1 cup brown rice 648	1 medium apple 72
1 slice Swiss cheese 105	1 small piece chicken 220	1 cup oatmeal 134	1 serving broccoli 45
1 cup plain yogurt 125	1 cup beans 242	1 small roll 122	1 medium baked potato 95

PRACTICE EXERCISES

Use the calorie table on page 52 to solve.

A casserole serves six people. It has 160 cal from the vegetable-fruit group, 648 cal from the bread-cereal group, and 680 cal from the meat group.

1. What is the total number of calories in this casserole?
2. How many calories are in each of six servings?

Ted made a casserole out of two cups of brown rice, four slices of Swiss cheese, and one serving of tuna fish.

3. How many calories were in two cups of brown rice?
4. How many calories were in four slices of Swiss cheese?
5. What was the total number of calories in the casserole?

Solve the problem by writing subproblems. Use the calorie table.

6. Ted wanted to have a 500-cal lunch. He ate one small piece of chicken, a roll, and a cup of whole milk. How many more calories were in the lunch than he wanted?
7. Ted is preparing dinner for his friends. He bought one bag of potatoes for $1.95, one bunch of broccoli for $0.89, and one package of chicken for $4.80. How much change did Ted receive from a $10 bill?
8. Amy had one cup of baked beans and a slice of whole wheat bread for lunch. Ted ate 1 cup of oatmeal and an apple for lunch. How many fewer calories did Ted's lunch contain?
9. Ted wants to prepare a 380-cal lunch. He chose two slices of whole wheat bread, two slices of Swiss cheese, and one piece of fruit. Should he choose an apple or an orange?

DECISION MAKING

2.14 CHOOSING AN ACTIVITY

Many high school students participate in extracurricular activities. Some of them have a real interest in the sport or club. Others get involved because their friends are.

PROBLEM

Lisa is trying to choose an after-school activity. She cut out the following from the school bulletin and added some notes of her own. She has numbered each activity in order of her interest in it.

BAND
Practice: Wed., Fri., 3:00–5:00PM
Play at all games, local parades, and rallies. Schedule varies.
Must audition for membership and submit a letter of recommendation from student's music teacher.

1. I can read music and play the flute. Maybe my brother's music teacher will write a letter for me. SCOTT is in the band!

DANCE CLUB
Rehearsals: Mon., Thurs., 3:00–5:00PM
Performances: Last Friday of the month, 8:00PM
Must audition for membership.

2. I've studied dance, I need the exercise, I know many of the people in the club.

FUTURE TEACHERS' CLUB
Meetings: Thurs., 3:00PM
For all students planning to enter the field of teaching.

3. That's me.
I wonder what they do at meetings??

SCIENCE CLUB
Meetings: Wed., 3:00PM
For all students interested in Science.
Compete in Science Fairs!
Must have at least a B average in Science.

4. Science is my best subject.
This might be fun, but it could be a lot of work.

DECISION-MAKING FACTORS

Interest Qualifications Schedule

DECISION-MAKING COMPARISONS

Complete the table.

	Band	Dance Club	Future Teachers' Club	Science Club
Interest	Very interested	**1.** ■	**2.** ■	**3.** ■
Qualifications	**4.** ■	**5.** ■	An interest in teaching	**6.** ■
Schedule	**7.** ■	**8.** ■	**9.** ■	Wed., 3 P.M.

MAKING THE DECISIONS

10. If interest were the only criteria, which activity would Lisa choose?

11. Based on her qualifications, which groups would most likely accept Lisa as a member?

12. Lisa forgot to consider her part-time job. If she works Tuesdays and Wednesdays from 3:00–6:00 P.M. and Saturdays from noon–8:00 P.M., which activities can she fit into her schedule?

13. If Lisa's friend Gary is willing to trade his Friday afternoon work hours for Lisa's Wednesday hours, how does this change her options?

14. If Lisa is serious about being a dancer, what should she consider before deciding to work on Fridays?

15. Which activities would be most helpful to Lisa in her future career?

16. If Lisa's part-time job is office work, which activity would be best for her health?

17. Which activities would you recommend to Lisa? Why?

CHAPTER REVIEW

Vocabulary Choose the letter of the word(s) that completes the statement.

1. The answer in addition is called the ■. [30]
   a. addend **b.** sum **c.** estimate **d.** product
2. Compatible numbers are numbers that ■. [38]
   a. end in zero **b.** divide evenly **c.** are rounded **d.** are great
3. The product of two whole numbers is greater than either ■. [40]
   a. addend **b.** sum **c.** factor **d.** quotient

Skills Estimate by rounding. [28, 38]

3,796 + 169 *4,200*

4. 567 + 345 **5.** 1,237 + 4,813 **6.** 176,482 + 563,201

7. 821 − 436 **8.** 3,243 − 1,726 **9.** 36,265 − 27,401 **10.** 471,673 − 323,851

11. 46 × 9 **12.** 681 × 4 **13.** 721 × 15 **14.** 4,396 × 21

15. 329 ÷ 2 **16.** 4,385 ÷ 5 **17.** 905 ÷ 33 **18.** 53,744 ÷ 46

Estimate by using compatible numbers. [38]

71)22,104 *300*

19. 6)410 **20.** 8)376 **21.** 7)692

22. 40)3,331 **23.** 62)4,624 **24.** 36)28,341 **25.** 24)74,116

Add or subtract. [30, 32]

456 + 870 = *1,326*

26. 62 + 43 **27.** 94 − 63 **28.** 642 − 314

29. 843 + 19 **30.** 43,201 + 842 **31.** 7,699 − 4,021 **32.** 90,468 − 37,491

Multiply or divide. [40, 42, 46, 48, 50]

357 × 19 = *6,783*

33. 32 × 9 **34.** 5)524 **35.** 43)788

36. 40 × 900 **37.** 5,000 × 80 **38.** 60)5,400 **39.** 900)270,000

40. 784 × 21 **41.** 140 × 683 **42.** 87)5,682 **43.** 190)36,401

CHAPTER TEST

Estimate by rounding.

1. $\begin{array}{r} 705 \\ +364 \\ \hline \end{array}$

2. $\begin{array}{r} 4,235 \\ 2,563 \\ +\ \ 337 \\ \hline \end{array}$

3. $\begin{array}{r} 8,317 \\ -2,852 \\ \hline \end{array}$

4. $\begin{array}{r} 9,604 \\ -\ \ 715 \\ \hline \end{array}$

5. $\begin{array}{r} 2,364 \\ \times\ \ \ \ 7 \\ \hline \end{array}$

6. $\begin{array}{r} 4,864 \\ \times\ \ 15 \\ \hline \end{array}$

Estimate by using front-end estimation.

7. $\begin{array}{r} 2,426 \\ +\ \ 951 \\ \hline \end{array}$

8. $\begin{array}{r} 38,499 \\ +11,987 \\ \hline \end{array}$

9. $\begin{array}{r} 25,012 \\ -\ \ 1,683 \\ \hline \end{array}$

Estimate by using compatible numbers.

10. $9\overline{)7,352}$

11. $19\overline{)6,514}$

12. $58\overline{)17,589}$

Add.

13. $\begin{array}{r} 9,643 \\ +5,717 \\ \hline \end{array}$

14. $\begin{array}{r} 43,869 \\ +69,454 \\ \hline \end{array}$

15. $\begin{array}{r} 1,264 \\ 38 \\ +\ \ 815 \\ \hline \end{array}$

16. $\begin{array}{r} 56,729 \\ 1,845 \\ +66,083 \\ \hline \end{array}$

Subtract.

17. $\begin{array}{r} 5,044 \\ -2,381 \\ \hline \end{array}$

18. $\begin{array}{r} 7,000 \\ -\ \ 965 \\ \hline \end{array}$

19. $\begin{array}{r} 60,236 \\ -28,675 \\ \hline \end{array}$

20. $\begin{array}{r} 21,812 \\ -\ \ 9,023 \\ \hline \end{array}$

Multiply.

21. 50×300

22. 70×700

23. 800×400

24. $900 \times 1,200$

25. $\begin{array}{r} 6,432 \\ \times\ \ \ \ 8 \\ \hline \end{array}$

26. $\begin{array}{r} 847 \\ \times\ \ 56 \\ \hline \end{array}$

27. $\begin{array}{r} 4,573 \\ \times\ \ \ 99 \\ \hline \end{array}$

28. $\begin{array}{r} 805 \\ \times 614 \\ \hline \end{array}$

Divide.

29. $6,300 \div 70$

30. $9,600 \div 80$

31. $2,400 \div 800$

32. $4,200 \div 600$

33. $8\overline{)3,784}$

34. $9\overline{)5,042}$

35. $29\overline{)8,816}$

36. $152\overline{)4,560}$

Solve.

37. On Monday, Maria sold 12 novels, 11 mysteries, and 4 cookbooks. On Tuesday, she sold 14 novels, 6 cookbooks, and 3 mysteries. On which day did she sell more books?

38. Bob Tompkins bought a paperback mystery for \$3.50. He gave the clerk a \$10 bill. List the least amount of coins and bills he would receive as change.

ENRICHMENT DIVISIBILITY

When you say that 15 **is divisible by** 3, it means that when you divide 15 by 3, the remainder is 0. You can use the following rules to determine if a number is divisible by 2, 3, 5, 9, or 10.

A number is divisible by:

2 if the ones digit is even.
3 if the sum of the digits is divisible by 3.
5 if the ones digit is 0 or 5.
9 if the sum of the digits is divisible by 9.
10 if the ones digit is 0.

Example Is 336 divisible by 2, 3, 5, 9, 10, or none of these?

1. Look at the ones digit.

   THINK: The ones digit is 6, an even number.

   336 is divisible by 2, but not divisible by 5 or 10.

2. Look at the sum of the digits.

   $3 + 3 + 6 = 12$

   THINK: 12 is divisible by 3, but not divisible by 9.

   336 is divisible by 3, but not divisible by 9.

In some cases you can combine the rules to make up other divisibility rules. For example, a number is divisible by 6 if it is divisible by 2 and 3.

Example Is 252 divisible by 6?

1. Look at the ones digit.

   THINK: The ones digit is 2, an even number.

   252 is divisible by 2.

2. Look at the sum of the digits.

   $2 + 5 + 2 = 9$

   THINK: 9 is divisible by 3.

   252 is divisible by 3.

So, 252 is divisible by 6.

Is the number divisible by 2, 3, 5, 9, 10, or none of these?

1. 42 **2.** 60 **3.** 91 **4.** 105 **5.** 162
6. 301 **7.** 350 **8.** 450 **9.** 1,611 **10.** 3,627

Is the number divisible by 6, 15, or neither of these?

11. 60 **12.** 72 **13.** 109 **14.** 135 **15.** 1,422
16. 2,463 **17.** 4,500 **18.** 6,750 **19.** 13,500 **20.** 2,088

ESTIMATION SKILLS

ADJUSTING FRONT-END ESTIMATES

You can usually adjust an estimated sum in order to bring it closer to the actual sum. To adjust a front-end estimate, consider the digits in the next-to-greatest place.

Estimate by using front-end estimation: 8,216,450 + 3,735,420

1. Estimate.

THINK: The greatest place is millions. Add the digits in the millions place.

```
  8,216,450
+ 3,735,420
 11,000,000 ← First estimate
```

2. Adjust the estimate.

THINK: Add the digits in the hundred-thousands place.

```
  8,|216,450 }
+ 3,|735,420 } about 1 million
 12,000,000 ← Adjusted estimate
```

Estimate by using front-end estimation: 9,135 + 616 + 5,315 + 1,478

1. Estimate.

THINK: The greatest place is thousands. Add the digits in the thousands place.

```
  9,135
    616
  5,315
+ 1,478
 15,000 ← First estimate
```

2. Adjust the estimate.

THINK: Add the digits in the hundreds place.

```
  9,|135 }
    |616 } about 1 thousand
  5,|315 }
+ 1,|478 } about ½ thousand
 16,500 ← Adjusted estimate
```

Estimate by using front-end estimation. Then adjust your estimate.

1. 865 + 632	**2.** 1,256 + 5,317	**3.** 35,408 + 25,320	**4.** 424,375 + 91,429	**5.** 9,433,698 + 8,052,728
6. 942 + 628 + 136	**7.** 3,848 + 1,204 + 462	**8.** 12,426 + 2,342 + 1,068	**9.** 738,420 + 903,725 + 367,416	**10.** 5,802,624 + 1,629,283 + 8,541,739
11. 6,281 + 846 + 1,592 + 3,468	**12.** 38,604 + 20,318 + 52,125 + 65,277	**13.** 11,764 + 21,068 + 4,283 + 3,265	**14.** 880,425 + 691,076 + 316,351 + 822,075	**15.** 236,914 + 874,003 + 556,947 + 463,081
16. 92,143 + 205,526 + 9,187 + 312,416	**17.** 2,438,007 + 1,204,318 + 938,409 + 1,072,361	**18.** 1,946,146 + 2,038,308 + 457,162 + 615,519	**19.** 7,249,196 + 85,156 + 714,039 + 2,445,136	**20.** 46,193 + 12,468,148 + 5,156,749 + 34,476,374

PreSkills TEST

Round to the given place. [16]

Number	Tenths	Thousandths	Ones	Hundreds
27.0938	**1.** ■	**2.** ■	**3.** ■	**4.** ■
0.85032	**5.** ■	**6.** ■	**7.** ■	**8.** ■

Estimate by rounding. [28, 38]

9. 428 + 369

10. 23,109 + 42,063

11. 3,491 − 807

12. 73,804 − 9,531

13. 32 × 76

14. 692 × 841

15. 42)375

16. 645)29,871

Estimate using compatible numbers. [38]

17. 3)27,014

18. 86)47,301

19. 57)326,085

Add or subtract. [30, 32]

20. 459 + 456

21. 896 + 37

22. 787 − 89

23. 923 + 787

24. 5,296 + 743

25. 74,802 − 5,064

26. 88,843 + 10,529

27. 19,042 − 12,894

28. 23,003 + 34,991

29. 50,339 − 5,892

30. 17,859 + 2,789

31. 45,006 − 4,869

32. 19,004 − 8,994

33. 99,018 − 39,896

34. 165,565 + 13,006

35. 644,887 − 78,998

Multiply or divide. [36, 40, 42, 46, 48, 50]

36. 62 × 100

37. 3,401 × 1,000

38. 732 × 10,000

39. 2,300 ÷ 100

40. 100,000 ÷ 100

41. 8,400 ÷ 10,000

42. 27 × 83

43. 192 × 68

44. 731 × 596

45. 4,085 × 637

46. 8)4,232

47. 49)34,398

48. 301)807,425

Chapter 3

DECIMAL COMPUTATION

People often hold fund-raising events for special purchases. How could you and your classmates raise enough money to buy a personal computer for your class?

3.1 ADDING DECIMALS

In New York City, the average rainfall for October is 2.85 in.
In November, it is 4.14 in. On the average, how much rain falls in New York during these 2 months?

Add 2.85 and 4.14 to find the total rainfall for these months.

EXAMPLE 1 Add: 2.85 + 4.14

1. Line up the decimal points.

$$\begin{array}{r} 2.85 \\ +4.14 \\ \hline . \end{array}$$

2. Add as with whole numbers.

$$\begin{array}{r} 2.85 \\ +4.14 \\ \hline 6.99 \end{array}$$

So, the total rainfall for October and November is 6.99 in.

Sometimes it is helpful to use equivalent decimals so that each addend has the same number of places.

EXAMPLE 2 Add: 4.06 + 17.9 + 35.082

1. Line up the decimal points.

$$\begin{array}{r} 4.06 \\ 17.9 \\ +35.082 \\ \hline . \end{array}$$

2. Add.

THINK: 4.06 = 4.060
17.9 = 17.900

$$\begin{array}{r} 4.060 \\ 17.900 \\ +35.082 \\ \hline 57.042 \end{array}$$

As with whole numbers, you can estimate to determine if a sum is reasonable.

EXAMPLE 3 Add: 0.84 + 0.7 + 29.082. Is your answer reasonable?

1. Estimate.

$$\begin{array}{rcr} 0.84 & \rightarrow & 0.8 \\ 0.7 & \rightarrow & 0.7 \\ +29.082 & \rightarrow & 30.0 \\ \hline & & 31.5 \end{array}$$

2. Add.

$$\begin{array}{r} 0.840 \\ 0.700 \\ +29.082 \\ \hline 30.622 \end{array}$$

Since 31.5 and 30.622 are close, the answer is reasonable.

CHECKPOINT

Write the letter of the correct answer.

Add.

1. 0.6 + 3 + 45.08 **a.** 48.68 **b.** 54.08 **c.** 45.98 **d.** 45.17

2. \$64.29 + \$203.75 **a.** \$268.04 **b.** \$26.80 **c.** \$2680.04 **d.** \$2.68

PRACTICE EXERCISES

Add.

1. 49.6 + 27.2

2. 83.1 + 51.9

3. 70.59 + 18.64

4. 0.832 + 0.513

5. 0.0052 + 0.9614

6. 72.083 + 19.5

7. 47.6 + 82.054

8. 6.0738 + 5.1

9. 7.602 + 1.8573

10. 38.0006 + 97

11. 0.82 + 6.1

12. 3.4 + 17.092

13. 54 + 3.019

14. 680.07 + 4.529

15. $493.00 + 86.87

16. $6.83 4.51 + 0.39

17. 3.71 0.48 + 9.06

18. $42.63 78 + 51.85

19. 63.07 92.48 + 14.65

20. 13.784 4.063 + 0.905

21. 0.0019 + 0.0263

22. 3.024 + 17.8 + 5

23. 65.3 + 0.94 + 2.071

24. 48.079 + 0.846

25. $4.63 + $0.93

26. 0.901 + 0.099

First estimate by rounding, and then find the exact answer.

27. 806.09 + 27.614

28. 798.14 + 12.80

29. 8.005 + 5.7319

30. 0.9216 + 0.4738

31. 38.06 + 8.0807

32. 25.1 + 13.84

33. 108.65 + 329.57

34. $64.87 + 19.63

35. 0.83 + 0.417

36. 0.059 + 0.4026

Solve for n.

37. $76.4 + 3.05 + 28 = n$

38. $0.05 + 0.839 + 7 = n$

39. $0.505 + 0.04 + 0.055 = n$

40. $6.29 + 18 + 0.075 = n$

Use the table for Exercises 41–44.
Solve.

YEARLY TOTAL PRECIPITATION (in.)

	Bismarck	Boston	Chicago	Denver	Reno
Rain	14.77	50.24	34	16.49	4.28
Snow	72.5	44.6	38.7	65.73	16.5

41. During the year, did more rain or snow fall altogether in the cities? Snow

42. Which city had more rain than snow?

43. Estimate the total precipitation for each city by rounding.
80 in.; 90 in.; 70 in.; 90 in.; 20 in.

44. Round the snowfall for each city to the nearest inch. About how many feet of snow fell in each city?

3.2 SUBTRACTING DECIMALS

During the 1972 Olympics, Monika Zehrt from East Germany ran the 400-m dash in 51.08 seconds. In 1984, Valerie Brisco-Hooks of the United States finished it in 48.83 seconds. How much faster was the 1984 record?

Subtract 48.83 from 51.08 to find the difference between the winning times.

EXAMPLE 1 Subtract: 51.08 − 48.83

1. Line up the decimal points.

$$\begin{array}{r} 51.08 \\ -48.83 \\ \hline . \end{array}$$

2. Subtract as with whole numbers.

$$\begin{array}{r} \scriptstyle 10 \\ \scriptstyle 4\ 11\ 10 \\ \not{5}\not{1}.\not{0}8 \\ -48.83 \\ \hline 2.25 \end{array}$$

So, the 1984 record was 2.25 seconds faster.

Sometimes it is helpful to use equivalent decimals to subtract.

EXAMPLE 2 Subtract: 23 − 7.94

1. Line up the decimal points.

$$\begin{array}{r} 23. \\ -\ 7.94 \\ \hline . \end{array}$$

2. Subtract.

THINK: 23 = 23.00

$$\begin{array}{r} \scriptstyle 12\ \ 9 \\ \scriptstyle 1\ 13\ 10\ 10 \\ \not{2}\not{3}.\not{0}\not{0} \\ -\ 7.94 \\ \hline 15.06 \end{array}$$

As with whole numbers. you can estimate the difference to determine if your answer is reasonable.

EXAMPLE 3 Subtract: 0.7916 − 0.027. Is your answer reasonable?

1. Estimate.

$$\begin{array}{rcr} 0.7916 & \rightarrow & 0.80 \\ -0.027 & \rightarrow & 0.03 \\ \hline & & 0.77 \end{array}$$

2. Subtract.

$$\begin{array}{r} 0.7916 \\ -0.0270 \\ \hline 0.7646 \end{array}$$

Since 0.77 and 0.7646 are close, the answer is reasonable.

FOR DISCUSSION

Is 5.25 a reasonable answer for 18.67 − 0.1342? Explain.

PRACTICE EXERCISES

Subtract.

1. $\begin{array}{r} 3.9 \\ -1.6 \\ \hline \end{array}$ **2.** $\begin{array}{r} 73.8 \\ -26.2 \\ \hline \end{array}$ **3.** $\begin{array}{r} 63.18 \\ -47.09 \\ \hline \end{array}$ **4.** $\begin{array}{r} 17.21 \\ -12.59 \\ \hline \end{array}$ **5.** $\begin{array}{r} 5.989 \\ -0.849 \\ \hline \end{array}$

6. $\begin{array}{r} 28.67 \\ -17.041 \\ \hline \end{array}$ **7.** $\begin{array}{r} 71.8 \\ -19.638 \\ \hline \end{array}$ **8.** $\begin{array}{r} 92 \\ -56.082 \\ \hline \end{array}$ **9.** $\begin{array}{r} 6.38 \\ -4.7094 \\ \hline \end{array}$ **10.** $\begin{array}{r} 0.6834 \\ -0.21 \\ \hline \end{array}$

11. $\begin{array}{r} 973.061 \\ -\ \ 85.042 \\ \hline \end{array}$ **12.** $\begin{array}{r} 5.2897 \\ -0.8245 \\ \hline \end{array}$ **13.** $\begin{array}{r} 5.0004 \\ -2.917 \\ \hline \end{array}$ **14.** $\begin{array}{r} 2 \\ -0.87 \\ \hline \end{array}$ **15.** $\begin{array}{r} 37 \\ -\ \ 0.235 \\ \hline \end{array}$

16. 0.9007 − 0.5819 **17.** 7.003 − 4.08 **18.** 13 − 8.49 4.51 **19.** 0.63 − 0.327

20. \$24.10 − \$6.00 **21.** \$12.24 − \$7.45 **22.** 84.3 − 7.62 **23.** 16 − 4.7026

First estimate by rounding, then find the exact answer.

24. $\begin{array}{r} 7 \\ -3.43 \\ \hline \end{array}$ **25.** $\begin{array}{r} 0.8061 \\ -0.782 \\ \hline \end{array}$ **26.** $\begin{array}{r} 64.2 \\ -\ \ 7.928 \\ \hline \end{array}$ **27.** $\begin{array}{r} \$143.00 \\ -\ \ \ \ 5.16 \\ \hline \end{array}$ **28.** $\begin{array}{r} 492.06 \\ -\ \ 8.527 \\ \hline \end{array}$

Solve for n.

29. $12.864 - 5.017 = n$ **30.** $6.731 - 4.98 = n$ **31.** $4.805 - 0.3906 = n$

32. $n - 5.83 = 16$ **33.** $0.1064 - n = 0.0089$ **34.** $n - 2.048 = 7$

Use the table to the right for Exercises 35–37.

35. What is the difference between men's and women's 200-m times?

36. What is the difference between men's and women's 400-m times?

37. Write the men's and women's times for the 800-m event in minutes and seconds.

WORLD TRACK RECORDS (in seconds)

	100 m	200 m	400 m	800 m
Men's time	9.93	19.72	43.86	101.87
Women's time	10.76	21.71	47.99	113.26

TIME OUT

During a field day competition, one of the events was the Ladder Lag. The goal was to get to the twelfth step of a ladder in the shortest time. The "lag" was that for every four steps you climbed, you had to go back three steps.

Alex took 2 seconds to climb up four steps and lag back three steps. How long did it take Alex to get to the twelfth step?

PROBLEM Solving APPLICATION

3.3 CONSUMER: NET PAY

Helga Schmidt works for the Omega Tax Company. Each week she is paid by check. The stub of her paycheck gives information about her earnings and deductions.

Earnings			
Period Ending	Regular Pay	Overtime	Gross Earnings
5_15_88	280.00	42.00	322.00

Deductions				
F.I.C.A.	Federal Withholding Tax	State Withholding Tax	Medical Insurance	Total Deductions
24.18	40.00	6.44	2.56	73.18
				Net Pay
				248.82

The **gross earnings,** or **gross pay,** is the total amount of money earned from regular pay and overtime pay.

The **deductions** are all the various amounts subtracted for taxes and other purposes. F.I.C.A. indicates Social Security taxes.

The **net pay** is the amount that Helga receives after the deductions are subtracted.

As the stub of the paycheck shows, Helga's gross earnings are $322.00. (Dollar signs are omitted on the stub.) There are four types of deductions, and the total deductions are $73.18. Helga's net pay is $248.82.

$$\$322.00 - \$73.18 = \$248.82$$

Len Borden works part-time in a factory. His gross earnings are $225.00 per week. The deductions from his gross earnings are shown below. What is his net pay?

F.I.C.A.	Federal Withholding Tax	State Withholding Tax	Medical Insurance
16.80	25.00	4.50	1.15

1. Add to find the total of the deductions. $\$16.80 + \$25.00 + \$4.50 + \$1.15 = \$47.45$
2. Subtract $47.45 from the gross earnings to find the net pay. $\$225.00 - \$47.45 = \$177.55$

So, Len Borden's net pay is $177.55 per week.

Use the stub from Helga's paycheck (page 66) to answer Exercises 1–4.

1. What is Helga's regular pay?

2. How much greater are Helga's gross earnings than her regular pay?

3. Which of the four types of deductions is the greatest amount?

4. How much greater is the deduction for F.I.C.A. than the deduction for state withholding tax?

Solve. Remember to estimate whenever you use your calculator.

5. Tina Adams works part-time. Her gross earnings are $185.50 per week. Total deductions are $40.81. What is her net pay?

6. Raul Cortez has a full-time job. His gross earnings are $395.00 per week. Total deductions are $94.86. What is his net pay?

7. One week Carl Bennett earns $385.00 in regular pay and $52.00 in overtime pay. Total deductions that week are $113.62. What is his net pay?

8. One week Jill Donnelly earns $412.00 in regular pay and $83.00 in overtime pay. Total deductions that week are $130.48. What is her net pay?

9. Paul Herson works part-time. His gross earnings are $255.00. The deductions are shown below. Find the total deductions and his net pay.

F.I.C.A.	Federal Withholding Tax	State Withholding Tax	Medical Insurance
19.15	30.00	5.10	1.27

10. Anna Zagat works full-time. Her gross earnings are $473.00. The deductions are shown below. What is her net pay?

F.I.C.A.	Federal Withholding Tax	State Withholding Tax	Medical Insurance	Retirement
35.52	76.00	9.46	4.12	24.80

Estimate.

11. Suppose you work full-time and your gross earnings are $519.00 per week. Total deductions from your gross earnings are $124.78. About how much is your net pay?

12. Your friend says that his net pay is $387.62 per week. He also says that the total deductions are $131.12. About how much are his gross earnings?

3.4 MULTIPLYING AND DIVIDING BY POWERS OF 10

The cafeteria at the Widgit Corporation sells yogurt in servings of 3.25 oz. How many ounces would be needed for 10 servings? 100 servings? 1,000 servings?

To multiply a decimal by a power of 10, move the decimal point as many places to the right as there are zeros in the power of 10.

EXAMPLE 1

Multiply: 3.25 by 10, by 100, and by 1,000.

10 × 3.25 = 3.2 5 → 32.5

THINK: One zero, so move the decimal point one place to the right.

100 × 3.25 = 3.2 5 → 325

THINK: Two zeros, so move the decimal point two places to the right.

1,000 × 3.25 = 3.2 5 0 → 3,250

THINK: Three zeros, so move the decimal point three places to the right. Write a zero in the product to place the decimal point correctly.

The cafeteria needs 32.5 oz of yogurt for 10 people, 325 oz for 100 people, and 3,250 oz for 1,000 people.

To divide a decimal by a power of 10, move the decimal point as many places to the left as there are zeros in the power of 10.

EXAMPLE 2

Divide: 61.7 by 10, by 100, and by 1,000.

61.7 ÷ 10 → 6 1. 7 → 6.17

THINK: One zero, so move the decimal point one place to the left.

61.7 ÷ 100 → 0 6 1. 7 → 0.617

THINK: Two zeros, so move the decimal point two places to the left.

61.7 ÷ 1,000 → 0 6 1. 7 → 0.0617

THINK: Three zeros, so move the decimal point three places to the left. Write a zero in the quotient to place the decimal point correctly.

CHECKPOINT

Write the letter of the correct answer.

Multiply or divide.

1. 5.68 × 100 **a.** 0.0568 **b.** 568 **c.** 5,680 **d.** 56,800

2. 46.3 × 10,000 **a.** 0.00463 **b.** 4,630 **c.** 463,000 **d.** 4,630,000

3. 795 ÷ 100 **a.** 0.00795 **b.** 7.95 **c.** 79.50 **d.** 79,500

4. 1.04 ÷ 1,000 **a.** 0.000104 **b.** 0.00104 **c.** 1,040 **d.** 104,000

PRACTICE EXERCISES

Multiply.

1. 50.4×10
2. 7.2×10
3. 79.7×10
4. 0.652×100
5. $4.280 \times 1{,}000$
6. $3.67 \times 1{,}000$
7. $58.9 \times 10{,}000$
8. $0.0084 \times 1{,}000$
9. 6.273×100
10. $6.42 \times 10{,}000$
11. 7.3×100
12. $4.009 \times 1{,}000$
13. $89 \times 10{,}000$
14. $3.46 \times 1{,}000$
15. 5.8×100
16. 0.61×10
17. $49.01 \times 1{,}000$
18. 814.29×100
19. $0.56 \times 10{,}000$
20. $57.2 \times 1{,}000$

Divide.

21. $5.5 \div 10$
22. $9 \div 10$
23. $43.17 \div 100$
24. $734 \div 10$
25. $6.52 \div 100$
26. $0.029 \div 100$
27. $86.4 \div 1{,}000$
28. $43.07 \div 10$
29. $586.29 \div 100$
30. $2.7 \div 100$
31. $79.2 \div 10$
32. $30.65 \div 100$
33. $69.32 \div 10$
34. $851 \div 100$
35. $203.1 \div 100$
36. $14 \div 1{,}000$
37. $4.83 \div 1{,}000$
38. $10.5 \div 10{,}000$
39. $0.39 \div 10{,}000$
40. $0.065 \div 100$

Complete the table.

Number	×100	÷1,000	×10,000	÷100,000
6.002	**41.** ■	**42.** ■	**43.** ■	**44.** ■
2.58	**45.** ■	**46.** ■	**47.** ■	**48.** ■
0.037	**49.** ■	**50.** ■	**51.** ■	**52.** ■

MIXED REVIEW

Estimate by rounding.

53. $9{,}468 + 7{,}512$
54. $4{,}614 - 1{,}638$
55. $7{,}142 + 8{,}813 + 849$
56. 4×623
57. 29×802
58. $2{,}846 \div 59$

Solve.

59. Ms. Sloan, the cafeteria manager, paid $0.23 each for 100 grapefruit. What was the total cost?

60. Ms. Sloan paid $22.50 for 15 lb of prepared tuna salad. What was the cost per pound?

61. Lucy Davis spent $3.25 for a sandwich, $2.85 for salad, and $0.95 for a carton of juice. How much did she spend altogether?

62. Ms. Sloan needs 80 apples. Each apple costs $0.35. If she buys 100 apples, the cost drops to $0.25 each. Should she buy 100 apples? Explain.

PROBLEM *Solving* STRATEGY

3.5 GUESSING AND CHECKING

Situation:

The Eloni family belongs to a record and tape club. Mr. Eloni collected \$4.20 from Bryan, who took quarters and dimes from his bank to pay for the tapes he ordered. Bryan gave his father 21 coins. How many of each coin did Bryan give his father?

Strategy:

Guessing and checking can often help you to solve a problem.

Applying the Strategy:

Guess 1	Check	
10 quarters	$10 \times \$0.25 = \$2.50$	THINK: $\$3.60 < \$4.20$
11 dimes	$11 \times \$0.10 = +1.10$	Make another guess. Try more
	\$3.60	quarters to raise the total.

Guess 2	Check	
15 quarters	$15 \times \$0.25 = \$3.75$	THINK: $\$4.35 > \$4.20$
6 dimes	$6 \times \$0.10 = +0.60$	Make another guess. Try fewer
	\$4.35	quarters to lower the total.

Guess 3	Check	
14 quarters	$14 \times \$0.25 = \$3.50$	
7 dimes	$7 \times \$0.10 = +0.70$	Bryan gave his father 14
	\$4.20	quarters and 7 dimes.

PRACTICE EXERCISES

Check each guess. Then choose the correct guess.

1. The Eloni family ordered some records and tapes. They ordered four more tapes than records. If they ordered 14 items this month, how many were records?
   a. Guess: 3 records
   b. Guess: 5 records
   c. Guess: 6 records

2. Bryan paid his father for some tapes. Mr. Eloni gave his son $1.90 in quarters and nickels for change. He gave him 18 coins. How many coins of each kind were there?
   a. Guess: 3 quarters, 15 nickels
   b. Guess: 4 quarters, 14 nickels
   c. Guess: 5 quarters, 13 nickels

Solve the problem by using the guess-and-check method.

3. Dee Eloni gave her mother $4.90 in quarters and dimes to pay for a record she ordered. Dee gave her 22 coins. How many of each coin did Dee give her mother?

4. In September, the Eloni family ordered records and tapes. They ordered five more tapes than records. If they ordered 17 items, how many were records?

5. During the past 3 years, the Eloni family bought 108 new tapes. If they bought 15 tapes more each year than the year before, how many tapes did the Elonis buy last year?

6. During a special sale, tapes sell for $6 each and records for $4 each. The Eloni family spent $60 in all for tapes and records. They bought the same number of tapes as records. How many of each item did they buy?

7. In October, the Elonis ordered 23 records and tapes. There were seven fewer tapes than records. How many of each did the Elonis order?

8. Ben's Record Store ordered records and tapes. They ordered 584 more tapes than records. If they ordered 2,200 items, how many were records?

MIDCHAPTER REVIEW

Add. [62]

1. $\begin{array}{r} 36.04 \\ +17.95 \\ \hline \end{array}$ **2.** $\begin{array}{r} 5.693 \\ +8.042 \\ \hline \end{array}$ **3.** $\begin{array}{r} 0.617 \\ +0.028 \\ \hline \end{array}$ **4.** $\begin{array}{r} 0.0053 \\ +0.0271 \\ \hline \end{array}$ **5.** $\begin{array}{r} 0.054 \\ +0.738 \\ \hline \end{array}$

6. $\begin{array}{r} 6.39 \\ +18.6 \\ \hline \end{array}$ **7.** $\begin{array}{r} 11.2 \\ +\ 4.835 \\ \hline \end{array}$ **8.** $\begin{array}{r} 0.0935 \\ +0.008 \\ \hline \end{array}$ **9.** $\begin{array}{r} 0.7605 \\ +0.09 \\ \hline \end{array}$ **10.** $\begin{array}{r} 52.4 \\ +\ 0.736 \\ \hline \end{array}$

11. $\begin{array}{r} 4.62 \\ 0.095 \\ +57.8 \\ \hline \end{array}$ **12.** $\begin{array}{r} 63 \\ 8.401 \\ +\ 0.5 \\ \hline \end{array}$ **13.** $\begin{array}{r} 6.903 \\ 0.72 \\ +0.006 \\ \hline \end{array}$ **14.** $\begin{array}{r} 136.5 \\ 0.837 \\ +\ 25.06 \\ \hline \end{array}$ **15.** $\begin{array}{r} 0.037 \\ 72.9 \\ +\ 4.85 \\ \hline \end{array}$

16. 38.16 + 5.29 **17.** 5.06 + 0.984 **18.** 149.6 + 0.0037

19. 58.64 + 9.87 + 126.34 **20.** 4 + 0.867 + 15.24 **21.** 0.0407 + 3 + 9.5

Subtract. [64]

22. $\begin{array}{r} 738.1 \\ -596.8 \\ \hline \end{array}$ **23.** $\begin{array}{r} 42.09 \\ -23.86 \\ \hline \end{array}$ **24.** $\begin{array}{r} 6.431 \\ -2.948 \\ \hline \end{array}$ **25.** $\begin{array}{r} 0.807 \\ -0.594 \\ \hline \end{array}$ **26.** $\begin{array}{r} 9.801 \\ -2.572 \\ \hline \end{array}$

27. $\begin{array}{r} 300 \\ -\ 62.37 \\ \hline \end{array}$ **28.** $\begin{array}{r} 0.72 \\ -0.493 \\ \hline \end{array}$ **29.** $\begin{array}{r} 0.0829 \\ -0.043 \\ \hline \end{array}$ **30.** $\begin{array}{r} 0.71 \\ -0.6208 \\ \hline \end{array}$ **31.** $\begin{array}{r} 4.8 \\ -1.756 \\ \hline \end{array}$

32. 86.54 − 27.36 **33.** 395.06 − 184.97 **34.** 703.01 − 96.28

35. 26.07 − 8.351 **36.** 19 − 0.8065 **37.** 0.6 − 0.5792

Multiply or divide. [68]

38. 8.365 × 10,000 **39.** 0.693 × 1,000 **40.** 0.00562 × 100

41. 231.09 × 1,000 **42.** 9.72 × 10,000 **43.** 0.000093 × 100,000

44. 75.69 ÷ 1,000 **45.** 0.463 ÷ 100 **46.** 607.094 ÷ 10,000

47. 531 ÷ 10,000 **48.** 8.67 ÷ 1,000 **49.** 0.00234 ÷ 100,000

Solve for n. [68]

50. $6.37 \times 100 = n$ **51.** $0.80547 \times 10{,}000 = n$ **52.** $75.69 \div 100 = n$

Solve. [66]

53. Marilyn Holmes works in a bakery. Her gross earnings are $325.50 per week. Total deductions are $97.75. What is her net pay?

54. John Jacobson earns $430.25 in regular pay. He earns $37.00 in overtime pay. Total deductions that week are $175.50. What is his net pay?

PROBLEM
Solving
APPLICATION

3.6 CONSUMER: CHECKING ACCOUNTS

If you have a checking account, it is important to keep an accurate record of your balance. The **balance** is the amount of money that is in the account.

When you write a check, subtract that amount to find the new balance. When you make a deposit, add that amount to find the new balance.

JANE CALDWELL CHECKBOOK RECORD

Check Number	Date	Description of Transaction	Amount of Check	Amount of Deposit	Balance	
					250	00
1	11/6	Ted's Market	34.75		34	75
					215	25
	11/8	Deposit		95.87	95	87
					311	12

The checkbook record shows a beginning balance of $250.00.

On 11/6, Jane wrote a check for $34.75 to Ted's Market. The new balance was $215.25.

On 11/8, Jane deposited $95.87. The new balance was $311.12.

Find each new balance after the given transaction.

Check Number	Date	Description of Transaction	Amount of Check	Amount of Deposit	Balance	
					375	00
	1/29	Deposit		50.85		
					1. ■	
1	2/3	John Forbes	55.25			
					2. ■	
2	2/12	Book Mart	19.75			
					3. ■	
	2/15	Deposit		50.85		
					4. ■	
3	2/21	Pat Simms, Inc.	165.32			
					5. ■	

Solve.

6. On 3/1, Sue opened a checking account with $425.00. On 3/8, she deposited $65.38 and wrote checks for $27.60 and $120.95. On 3/22, she deposited $75.32. What was Sue's new balance?

3.7 MULTIPLYING DECIMALS

Tom Rosicki's car goes 24.4 miles on a gallon of gasoline.
When his fuel tank is full, it holds 18 gallons.
How many miles can his car travel on a full tank of gas?

Multiply 24.4 by 18 to find the total number of miles.

EXAMPLE 1 Multiply: 18 × 24.4

1. Multiply as with whole numbers.

```
   244
 ×  18
  1952
  244
  4392
```

2. Count the decimal places in the factors.

THINK: The product has the same number of decimal places as the factors.

```
  24.4  ← 1 decimal place
×   18  ← 0 decimal places
 439.2  ← 1 decimal place
```

So, Tom's car can travel 439.2 miles on a full tank of gasoline.

Sometimes extra zeros are needed as placeholders.

EXAMPLE 2 Multiply: 0.0092 × 0.53

1. Multiply as with whole numbers.

```
   00092
 ×   053
     276
    4600
    4876
```

2. Count the decimal places in the factors. Use zeros as place holders.

```
  0.0092  ← 4 decimal places
×   0.53  ← 2 decimal places
0.004876  ← 6 decimal places
```

So, 0.092 × 0.53 = 0.004876.

Estimating your answer first can help you determine the reasonableness of your answer.

EXAMPLE 3 Multiply: $7.32 × 0.08. Check the reasonableness of your answer.

1. Estimate.

```
$ 7.32 → $    7  ← 0 decimal places
×0.08  →   0.08  ← 2 decimal places
         $ 0.56  ← 2 decimal places
```

2. Multiply.

```
$  7.32  ← 2 decimal places
 ×0.08   ← 2 decimal places
$0.5856  ← 4 decimal places
```

Since $0.56 is close to $0.5856, the answer is reasonable.
When the product represents money, round to the nearest cent.

So, $7.32 × 0.08 = $0.5856 —**rounds to**→ $0.59.

PRACTICE EXERCISES

Multiply.

1. $\begin{array}{r} 5.6 \\ \times 7.9 \\ \hline \end{array}$	**2.** $\begin{array}{r} 1.04 \\ \times \ 32 \\ \hline \end{array}$	**3.** $\begin{array}{r} 42 \\ \times 3.7 \\ \hline \end{array}$	**4.** $\begin{array}{r} 28.6 \\ \times \ 5.9 \\ \hline \end{array}$	**5.** $\begin{array}{r} 3.69 \\ \times 0.14 \\ \hline \end{array}$
6. $\begin{array}{r} 43.28 \\ \times 0.016 \\ \hline \end{array}$	**7.** $\begin{array}{r} 2.73 \\ \times 0.004 \\ \hline \end{array}$	**8.** $\begin{array}{r} 0.0084 \\ \times \ 0.7 \\ \hline \end{array}$	**9.** $\begin{array}{r} 0.0034 \\ \times \ 0.071 \\ \hline \end{array}$	**10.** $\begin{array}{r} 0.237 \\ \times 0.008 \\ \hline \end{array}$
11. $\begin{array}{r} \$52.68 \\ \times \ 8.37 \\ \hline \end{array}$	**12.** $\begin{array}{r} \$27.02 \\ \times \ 0.04 \\ \hline \end{array}$	**13.** $\begin{array}{r} \$0.24 \\ \times 0.83 \\ \hline \end{array}$	**14.** $\begin{array}{r} \$35.70 \\ \times \ 0.09 \\ \hline \end{array}$	**15.** $\begin{array}{r} \$80.92 \\ \times \ 3.06 \\ \hline \end{array}$
16. $\begin{array}{r} 0.62 \\ \times 0.38 \\ \hline \end{array}$	**17.** $\begin{array}{r} 94.2 \\ \times 0.083 \\ \hline \end{array}$	**18.** $\begin{array}{r} 76.1 \\ \times \ 15 \\ \hline \end{array}$	**19.** $\begin{array}{r} 306 \\ \times 8.4 \\ \hline \end{array}$	**20.** $\begin{array}{r} \$93.56 \\ \times \ 37 \\ \hline \end{array}$
21. $\begin{array}{r} 1.005 \\ \times \ 7.3 \\ \hline \end{array}$	**22.** $\begin{array}{r} 62.19 \\ \times 0.008 \\ \hline \end{array}$	**23.** $\begin{array}{r} 8.007 \\ \times 0.119 \\ \hline \end{array}$	**24.** $\begin{array}{r} 3.921 \\ \times 0.018 \\ \hline \end{array}$	**25.** $\begin{array}{r} 85 \\ \times 0.8 \\ \hline \end{array}$
26. $\begin{array}{r} 63 \\ \times 0.4 \\ \hline \end{array}$	**27.** $\begin{array}{r} 418 \\ \times 0.5 \\ \hline \end{array}$	**28.** $\begin{array}{r} 12.3 \\ \times \ 8.6 \\ \hline \end{array}$	**29.** $\begin{array}{r} 1.085 \\ \times 0.036 \\ \hline \end{array}$	**30.** $\begin{array}{r} 140 \\ \times 5.8 \\ \hline \end{array}$

First estimate by rounding, and then find the exact answer.

31. $\begin{array}{r} 792 \\ \times 0.6 \\ \hline \end{array}$	**32.** $\begin{array}{r} 6.95 \\ \times 0.18 \\ \hline \end{array}$	**33.** $\begin{array}{r} 52 \\ \times 1.3 \\ \hline \end{array}$	**34.** $\begin{array}{r} 4.09 \\ \times \ 3.7 \\ \hline \end{array}$	**35.** $\begin{array}{r} \$60.08 \\ \times \ 0.51 \\ \hline \end{array}$
36. $\begin{array}{r} 418 \\ \times 0.5 \\ \hline \end{array}$	**37.** $\begin{array}{r} 603 \\ \times 2.9 \\ \hline \end{array}$	**38.** $\begin{array}{r} 87 \\ \times 3.4 \\ \hline \end{array}$	**39.** $\begin{array}{r} 83.2 \\ \times 0.74 \\ \hline \end{array}$	**40.** $\begin{array}{r} 0.7083 \\ \times \ 0.045 \\ \hline \end{array}$

41. 7.1×0.008 **42.** 0.92×0.005 **43.** 5.9×0.26

Solve for n.

•**44.** $3.08 \times 0.65 = n$ **45.** $0.52 \times \$81.30 = n$ **46.** $(0.407 - 0.111) \times 0.28 = n$

Solve.

•**47.** In 1940, a reort showed that consumers spent \$16.6 billion on food. In 1980, a report showed that they spent 18.25 times that amount. How much money was spent on food in 1980?

48. In 1920, a Census report showed that the population per square mile in Illinois was 115.7 people. In 1980, it was reported to be 205.3 people. What was the increase in population?

49. In 1950, a government report showed that \$21.7 billion was spent on housing. Write that number in standard form.

50. A report showed that a car could travel 25.4 mi on a gallon of gasoline. How far could a car go on 20 gal of gas?

PROBLEM Solving APPLICATION

3.8 CONSUMER: COST OF ELECTRICITY

The amount of electrical power used by appliances varies. The table below shows the approximate power used in an hour by electrical appliances. If the cost of electricity per kilowatt-hour is known, as well as the number of hours an appliance is used, the electrical costs can be computed.

Appliance	Power used (kW)
Air conditioner	1.250
Clothes dryer	3.750
Radio	0.125
Refrigerator	0.250
Television set	0.225
Toaster	0.775
Vacuum cleaner	0.550

The television set is used an average of 4 hours per day. If the cost of electricity is $0.12 per kilowatt-hour, what is the cost of electricity for the television set for a year (365 days)?

1. Multiply to find the number of hours the set is used per year.

HOURS USED PER DAY	×	NUMBER OF DAYS	=	HOURS USED PER YEAR
4	×	365	=	1,460

2. Multiply to find the cost of using the set for 1 hour.

POWER IN KILOWATTS	×	COST PER KILOWATT-HOUR	=	COST OF USE PER HOUR
0.225	×	$0.12	=	$0.027

3. Multiply the number of hours by the cost per hour.

1,460 × $0.027 = $39.42

So, the cost of electricity for the television set is $39.42 per year.

Use the table to find the cost. Assume the cost of electricity is $0.12 per kilowatt-hour. One year is 365 days, or 52 weeks. Remember to estimate whenever you use your calculator.

1. The air conditioner is used 8 hours per day for 60 days.

2. The vacuum cleaner is used an average of 2 hours per week for 1 year.

3. The refrigerator is used 24 hours per day for 1 year.

4. The radio is used 6 hours per day for 1 year.

Use $0.14 as the cost of electricity per kilowatt-hour in Exercises 5–8.

5. The clothes dryer is used an average of 4 hours per week for 1 year.

6. The air conditioner is used 9 hours per day for 60 days. What is the cost?

7. The toaster is used 10 minutes a day for 60 days. What is the cost?

8. Which appliance costs about 53 cents an hour to use?

PROBLEM *Solving* APPLICATION

3.9 CAREER: LAB TECHNICIAN

Carol Hormel is a lab technician. She performs experiments, makes measurements, and records the results of the experiments.

The table below shows the amount of three compounds produced in a series of five experiments.

AMOUNT OF COMPOUND PRODUCED (LITERS)

Experiment number	Compound A	Compound B	Compound C
101	0.342	0.175	0.675
102	0.287	0.148	0.551
103	0.625	0.332	1.209
104	0.536	1.078	0.261
105	0.858	1.736	0.403

Solve. Remember to estimate whenever you use your calculator.

How many liters of the three compounds were produced:

1. in Experiment 101?

2. in Experiment 103?

In Experiment 102, how much more was produced of:

3. Compound A than Compound B?

4. Compound C than Compound A?

List the amounts in order from greatest to least.

5. The compounds produced in Experiment 103

6. The compounds produced in Experiment 105

7. Did Experiment 101 produce more than five times as much Compound C as Compound B?

8. Did Experiment 104 produce more than four times as much Compound B as Compound C?

9. Which experiments produced about twice as much Compound A as Compound B?

10. Which experiments produced about twice as much Compound B as Compound A?

3.10 DECIMAL QUOTIENTS

Video Vibes magazine had an introductory offer of $33.75 for nine issues. What was the cost of a single issue?

To find the cost per issue, divide $33.75 by 9.

EXAMPLE 1

Divide: $33.75 ÷ 9

1. Write a decimal point in the quotient directly above the decimal point in the dividend.

```
      .
9)$33.75
```

2. Divide as with whole numbers.

```
  $ 3.75
9)$33.75
   27
    6 7
    6 3
      45
      45
       0
```

So, each issue costs $3.75.

Estimating a quotient first is a good way to check the reasonableness of your answer. Sometimes you will need to insert zeros in the quotient as placeholders.

EXAMPLE 2

Divide: 0.3924 ÷ 18. Check the reasonableness of your answer.

1. Estimate by using compatible numbers.

2. Divide.

Zero placeholder

```
                   .0200               .0218
18)0.3924 → 20)0.4000          18)0.3924
                                   36
                                    32
                                    18
                                    144
                                    144
                                      0
```

Since 0.0218 is close to 0.0200, the answer is reasonable.

Sometimes you need to write one or more zeros in the dividend to complete the division.

EXAMPLE 3 Divide:

a. $13.4 \div 4$

$$\begin{array}{r} 3.35 \\ 4\overline{)13.40} \\ \underline{12} \\ 1\,4 \\ \underline{1\,2} \\ 20 \\ \underline{20} \\ 0 \end{array}$$

b. $3 \div 8$

$$\begin{array}{r} 0.375 \\ 8\overline{)3.000} \\ \underline{2\,4} \\ 60 \\ \underline{56} \\ 40 \\ \underline{40} \\ 0 \end{array}$$

PRACTICE EXERCISES

Divide.

1. $4\overline{)27.32}$	**2.** $6\overline{)\$45.06}$	**3.** $35\overline{)634.55}$	**4.** $14\overline{)44.24}$
5. $5\overline{)8}$	**6.** $4\overline{)37}$	**7.** $8\overline{)7}$	**8.** $12\overline{)9}$
9. $5\overline{)0.0095}$	**10.** $6\overline{)0.0042}$	**11.** $2\overline{)0.48}$	**12.** $12\overline{)0.0132}$
13. $7\overline{)0.63}$	**14.** $11\overline{)0.264}$	**15.** $94\overline{)3.478}$	**16.** $7\overline{)\$21.42}$
17. $5\overline{)\$43.05}$	**18.** $8\overline{)0.0592}$	**19.** $3\overline{)23.46}$	**20.** $67\overline{)348.4}$
21. $49\overline{)4.067}$	**22.** $38\overline{)36.48}$	**23.** $3\overline{)\$12.39}$	**24.** $7\overline{)142.8}$
25. $4\overline{)5.308}$	**26.** $7\overline{)1.057}$	**27.** $9\overline{)48.6027}$	**28.** $8\overline{)0.2448}$
29. $4\overline{)54}$	**30.** $5\overline{)23}$	**31.** $8\overline{)123}$	**32.** $12\overline{)117}$
33. $17\overline{)1.02}$	**34.** $36\overline{)619.2}$	**35.** $32\overline{)\$26.88}$	**36.** $23\overline{)4.37}$

First estimate using compatible numbers. Then find the exact quotient.

37. $173.6 \div 28$	**38.** $69.16 \div 76$	**39.** $3.54 \div 59$	**40.** $168.3 \div 45$
41. $0.2368 \div 64$	**42.** $52.44 \div 92$	**43.** $3.504 \div 48$	**44.** $0.1159 \div 19$
45. $0.2257 \div 37$	**46.** $130.208 \div 26$	**47.** $1.6274 \div 79$	**48.** $48.6567 \div 81$

Solve.

49. Daisy Comics prints 2.44 million comic books every month. The same number of comics are sent to each of eight distribution centers. How many comics are sent to each center?

50. In 1985, Daisy Comics had 3.6 million readers. By 1990, the number of readers will increase by 1.26 million. What will the readership be in 1990?

3.11 DIVIDING BY DECIMALS

In a laboratory experiment, the temperature of a solution being heated rose 1° every 0.5 min. How many degrees did the temperature rise after 22.5 min?

To find the number of degrees, divide 22.5 by 0.5.

When the divisor is a decimal, multiply both divisor and dividend by the same power of 10 to make the divisor a whole number.

EXAMPLE 1 Divide: 22.5 ÷ 0.5

1. Make the divisor a whole number.

   THINK: Multiply the divisor and dividend by 10.

   0.5)22.5

2. Divide.

```
      45.
05.)225.
    20
     25
     25
      0
```

So, the temperature rose 45°.

Sometimes you need to round the quotient.

EXAMPLE 2 Find 49.308 ÷ 7.05 to the nearest hundredth.

1. Make the divisor a whole number.

   THINK: Multiply the divisor and dividend by 100.

   7.05)49.308

2. Divide to thousandths. Round to the nearest hundredth.

```
             6.994 → 6.99
705.)4,930.800
     4,230
       700 8
       634 5
        66 30
        63 45
         2 850
         2 820
            30
```

So, 49.308 ÷ 7.05 to the nearest hundredth is 6.99.

To estimate with a decimal divisor, first multiply to make the divisor a whole number.

EXAMPLE 3 Estimate: $0.73\overline{)0.2628}$

1. Make the divisor a whole number.

$$0.73\overline{)0.2628} \rightarrow 73.\overline{)26.28}$$

2. Estimate by using compatible numbers.

$$\begin{array}{r} 0.4 \\ 70\overline{)28.00} \end{array}$$

CHECKPOINT

Write the letter of the correct answer.

Divide.

1. $0.9\overline{)0.225}$	**a.** 0.025	**b.** 0.25	**c.** 2.5	**d.** 25
2. $3.6\overline{)164.52}$	**a.** 0.457	**b.** 4.57	**c.** 45.7	**d.** 457
3. $6.47\overline{)\$17.83}$	**a.** \$0.28	**b.** \$2.76	**c.** \$27.60	**d.** \$276
4. $0.0089\overline{)0.658}$	**a.** 0.739	**b.** 7.39	**c.** 73.9	**d.** 739

PRACTICE EXERCISES

Name the power of 10 needed to make the divisor a whole number. Write 10; 100; 1,000; or 10,000.

1. $83.6\overline{)4.92}$	**2.** $5.39\overline{)87.4}$	**3.** $0.0072\overline{)5.68}$	**4.** $9.063\overline{)2.7}$
5. $0.376\overline{)98.506}$	**6.** $0.843\overline{)52}$	**7.** $4.0815\overline{)28.6}$	**8.** $2.861\overline{)7.69}$

Divide.

9. $0.9\overline{)1.8}$	**10.** $0.4\overline{)4.8032}$	**11.** $0.3\overline{)16.38}$	**12.** $0.5\overline{)5.95}$
13. $0.06\overline{)4.6728}$	**14.** $0.08\overline{)0.06952}$	**15.** $0.07\overline{)63.504}$	**16.** $0.02\overline{)38.6}$
17. $0.67\overline{)0.00201}$	**18.** $0.21\overline{)0.147}$	**19.** $2.3\overline{)14.03}$	**20.** $5.7\overline{)23.94}$
21. $6.8\overline{)0.0238}$	**22.** $1.9\overline{)6.232}$	**23.** $0.037\overline{)0.2405}$	**24.** $7.2\overline{)22.176}$
25. $0.205\overline{)1.23}$	**26.** $0.714\overline{)22.848}$	**27.** $0.832\overline{)79.872}$	**28.** $0.603\overline{)4.2813}$
29. $4.83\overline{)3.5259}$	**30.** $0.510\overline{)1621.8}$	**31.** $18.5\overline{)5.6425}$	**32.** $3.59\overline{)3.1233}$

Divide. Round to the nearest hundredth.

33. $0.94\overline{)3.50432}$	**34.** $0.041\overline{)0.37187}$	**35.** $0.58\overline{)0.05568}$
36. $0.025\overline{)0.014575}$	**37.** $0.307\overline{)12.28921}$	**38.** $51.9\overline{)1074.33}$

First estimate each quotient by using compatible numbers, and then find the exact answer.

39. $0.007\overline{)51.94}$	**40.** $0.04\overline{)1.638}$	**41.** $0.8\overline{)3.07}$	**42.** $0.005\overline{)0.04998}$
43. $4.2\overline{)97.44}$	**44.** $0.15\overline{)3.75}$	**45.** $6.1\overline{)14.64}$	**46.** $0.038\overline{)1.7005}$
47. $53.8\overline{)28.245}$	**48.** $0.704\overline{)2.64}$	**49.** $3.05\overline{)65.88}$	**50.** $0.619\overline{)0.3095}$
51. $0.02\overline{)8.36}$	**52.** $0.06\overline{)0.0582}$	**53.** $0.008\overline{)0.00424}$	**54.** $0.04\overline{)0.29}$
55. $8.4\overline{)5.04}$	**56.** $0.76\overline{)0.2204}$	**57.** $6.11\overline{)2.444}$	**58.** $0.473\overline{)0.2838}$

MIXED REVIEW

Find the answer.

59. $\begin{array}{r} 4{,}899 \\ +6{,}559 \\ \hline \end{array}$ **60.** $\begin{array}{r} 5{,}238 \\ +9{,}784 \\ \hline \end{array}$ **61.** $\begin{array}{r} 56{,}896 \\ 36{,}998 \\ +\ 7{,}596 \\ \hline \end{array}$ **62.** $\begin{array}{r} 101{,}237 \\ 987{,}786 \\ +\ 89{,}867 \\ \hline \end{array}$

63. $\begin{array}{r} 9{,}697 \\ -\ 899 \\ \hline \end{array}$ **64.** $\begin{array}{r} 78{,}213 \\ -\ 8{,}989 \\ \hline \end{array}$ **65.** $\begin{array}{r} 87{,}657 \\ -78{,}986 \\ \hline \end{array}$ **66.** $\begin{array}{r} 785{,}433 \\ -699{,}850 \\ \hline \end{array}$

67. $\begin{array}{r} 679 \\ \times\ 76 \\ \hline \end{array}$ **68.** $\begin{array}{r} 188 \\ \times\ 86 \\ \hline \end{array}$ **69.** $\begin{array}{r} 997 \\ \times 877 \\ \hline \end{array}$ **70.** $\begin{array}{r} 5{,}785 \\ \times\ 749 \\ \hline \end{array}$

71. $0.096 - 0.078$ **72.** $45.998 + 4.31$ **73.** 894.03×100 **74.** $73.906 \div 1{,}000$

Solve.

75. Jan spends 4.9 hours a week working in the lab. She has logged 189.63 hours. How many weeks has she worked?

76. A box of test tubes weighs 12.8 lb. A shipment of test tubes arrives that weighs 307.2 lb. How many boxes of test tubes are inside?

77. Bart typed reports at 57.4 words per minute. By using a word processing program, his speed increased to 61.2 words per minute. What was the increase in words per minute?

78. A printer in the lab prints 210 characters per second. It is in use from 7:30 A.M. until 4:30 P.M. How many characters does it print in a day?

TIME OUT

Complete the following puzzle to find the "starting number."

Starting number **Finishing number**

■ + 2.4 = ■ ÷ 3 = 7.2

PROBLEM Solving APPLICATION

3.12 CONSUMER: FINDING MILES PER GALLON

It is useful for a car owner to know how many miles a car can travel per gallon of gasoline. A car will usually travel more **miles per gallon (mpg)** in highway driving than in city driving.

In highway driving, a car travels 567 mi on 18 gal of gasoline. How many miles does it travel per gallon?

THINK: **NUMBER OF MILES** ÷ **NUMBER OF GALLONS USED** = **MPG**

567 ÷ 18 = 31.5

So, the car travels 31.5 mpg in highway driving.

Sometimes, the answer is rounded to the nearest whole number. In the situation above, the car travels about 32 mpg.

Complete. Remember to estimate whenever you use your calculator.

Number of miles (highway)	Number of gallons used	mpg	Number of miles (city)	Number of gallons used	mpg
1. 520	20	■	**2.** 280	16	■
3. 513	18	■	**4.** 270	20	■
5. 272	10	■	**6.** 129	8.6	■
7. 288	9.6	■	**8.** 183	12.2	■

Solve.

9. In city driving, Ricardo's car travels 189 mi on 14 gal of gasoline. How many miles does his car travel per gallon?

10. Ricardo drove 346 mi on the highway and used 16.4 gal of gasoline. What is the mpg?

11. A Sunburst KX travels 32 mpg in city driving. How far will it travel on 11.5 gal of gasoline?

12. The Sunburst travels 42 mpg on the highway and 28 mpg in the city. How much farther will it go on 12.5 gal on the highway than in the city?

3.13 WHERE TO SHOP

DECISION MAKING

Most people have several stores to choose from when they go shopping. Choosing the right store can save time and money, but there are other benefits to be gained from shopping wisely.

PROBLEM

Carla and Tony have just moved from Ohio to Texas. They each need a new suit for the warmer climate that they can wear on job interviews. They also want to replace some of the items they sold before they moved. They made a list and checked department store ads for prices. They also made some notes about the stores and their locations.

	Downtown Department Store	Exclusive Department Store	Discount Department Store
Tony:			
Suit	\$159 – \$239	\$265 and up	\$75
Tie	\$12 – \$29	\$19 – \$79	2 for \$15
desklamp	\$39	\$39	not available
19" Television	\$297	\$499	\$249
Carla			
Suit	\$100 – \$250	\$265 – \$500	\$50
briefcase	\$150	\$275	not available
clock radio	\$79	\$79	\$13
Calculator	\$19	\$49	\$5
Location	2 mi away	30 mi away at Marvelous Mall	10 mi away at Mini-Mall
Hours	Mon.-Wed. 10 am - 6 pm Thurs. – Sat. 10 am - 9 pm	7 days 10 am – 9 pm	Mon.- Sat. 9 am - 9 pm
Parking	none	3,000 Cars	100 Cars
Public Transportation	yes	none	none
Number of stores in area	30	150	10
Selection	average	extensive	limited
Quality	Well-Known moderate-priced brands, some better labels	Well-known brands, designer fashions, latest trends	Some brand names, mostly store labels
Other	Free delivery, free alternations on all clothing	Free delivery, free alterations on men's suits	Free delivery for furniture and major appliances, no alterations

DECISION-MAKING FACTORS

Cost Selection Convenience Quality

DECISION-MAKING COMPARISONS

Compare the three stores by completing the table.

	Downtown	Exclusive	Discount
Cost	1. ■	2. ■	Inexpensive
Selection	Average	3. ■	4. ■
Convenience			
Location	5. ■	Inconvenient	6. ■
Parking	7. ■	Extensive	8. ■
Shopping hours	9. ■	10. ■	Convenient
Quality	11. ■	Excellent	12. ■

MAKING THE DECISIONS

Where should Tony and Carla shop:

13. if cost were the only factor?

14. if selection were the only factor?

15. if convenience were the only factor?

16. if quality were the only factor?

17. How much could Carla save on a suit by shopping at Discount rather than Exclusive?

18. If Carla is looking for a suit that is very fashionable, where should she shop?

19. Tony wants a suit that will wear well and look good. If his total budget is $500, where should he shop for the suit?

20. If Carla and Tony didn't have a car, where would they have to shop?

21. If the television set is the same brand and model at all three stores, where should Tony buy it? Why?

22. Would it make sense to buy the desk lamp and clock radio at Downtown or at Exclusive? Why?

23. The briefcases vary in price and quality. What other factors should Carla consider?

24. Why do people, even if they can't afford to shop there, visit large malls with expensive stores?

25. If you were Tony or Carla, where would you buy each of the items on his or her list?

CHAPTER REVIEW

Vocabulary Choose the letter of the word that completes the statement.

1. When multiplying decimals, the number of decimal places in the product is equal to the number of decimal places in the ▒. [74]
   a. addends **b.** factors **c.** quotient **d.** sum

2. The first step in dividing by a decimal is to move the decimal point in the ▒. [80]
   a. divisor **b.** quotient **c.** dividend **d.** remainder

Skills Add. [62]

$$\begin{array}{r} 2.87 \\ +6.453 \\ \hline \textit{9.323} \end{array}$$

3. $\begin{array}{r} 48.66 \\ +20.05 \\ \hline \end{array}$ **4.** $\begin{array}{r} 0.86 \\ +3.99 \\ \hline \end{array}$ **5.** $\begin{array}{r} \$17.84 \\ 3.08 \\ +\ 9.02 \\ \hline \end{array}$ **6.** $\begin{array}{r} 6.76 \\ 0.0025 \\ +100.3 \\ \hline \end{array}$

7. 0.632 + 1.987 **8.** 8.362 + 0.0119 **9.** 0.3437 + 3.3347

Subtract. [64]

$$\begin{array}{r} 73.090 \\ -38.145 \\ \hline \textit{34.945} \end{array}$$

10. $\begin{array}{r} 39.32 \\ -32.46 \\ \hline \end{array}$ **11.** $\begin{array}{r} 48.06 \\ -\ 7.333 \\ \hline \end{array}$ **12.** $\begin{array}{r} 86.4 \\ -\ 0.003 \\ \hline \end{array}$ **13.** $\begin{array}{r} 12.345 \\ -11.9834 \\ \hline \end{array}$

14. 1.0934 − 0.679 **15.** 20.004 − 11.441 **16.** 1.0348 − 0.9697

Multiply or divide by powers of 10. [68]

82.5 × 100	*8,250*
77.82 ÷ 10	*7.782*

17. 9.347 × 100 **18.** 74.2 × 10,000 **19.** 14.3681 × 1,000

20. 1,674.3 ÷ 1,000 **21.** 45,098 ÷ 10,000 **22.** 899,763 ÷ 100,000

Multiply. [74]

$$\begin{array}{r} 8.76 \\ \times 0.08 \\ \hline \textit{0.7008} \end{array}$$

23. $\begin{array}{r} 9.89 \\ \times 0.32 \\ \hline \end{array}$ **24.** $\begin{array}{r} 14.64 \\ \times\ \ 1.1 \\ \hline \end{array}$ **25.** $\begin{array}{r} 128.3 \\ \times\ 0.89 \\ \hline \end{array}$ **26.** $\begin{array}{r} 100.06 \\ \times\ 1.066 \\ \hline \end{array}$

27. 1,660.2 × 0.75 **28.** 14,000.3 × 3.03 **29.** 0.6897 × 10.3

Divide. [78, 80]

$$\begin{array}{r} \textit{4.5} \\ 8.2\overline{)36.9} \end{array}$$

30. 118.4 ÷ 4 **31.** 108.8 ÷ 8 **32.** 91.5 ÷ 5

33. $0.476\overline{)0.119}$ **34.** $7.32\overline{)4.8312}$ **35.** $1.03\overline{)\$15.45}$ **36.** $2.002\overline{)10.8108}$

CHAPTER TEST

Estimate by rounding to the greatest place.

1. $16.83 + 5.07$ **2.** $304.87 + 69.14$ **3.** $48.05 - 26.93$ **4.** $300.62 - 46.38$

5. 42.6×2.1 **6.** 0.409×2.006 **7.** $4.68\overline{)19.83}$ **8.** $0.003\overline{)0.91}$

Compute.

9. $27.65 + 18.09$ **10.** $2.09 + 86.5$ **11.** $7.395 - 2.086$ **12.** $93.7 - 6.082$

13. $75.06 + 3.94$ **14.** $8.052 + 13.09$ **15.** $4.0086 + 0.029$

16. $403.8 - 56.7$ **17.** \$400 − \$67.29 **18.** $238 - 0.923$

19. $63.74 \times 1{,}000$ **20.** $0.809 \times 10{,}000$ **21.** $5.2064 \times 10{,}000{,}000$

22. $18.951 \div 100$ **23.** $0.031 \div 1{,}000$ **24.** $620 \div 100{,}000$

25. 8.3×7 **26.** 95×0.6 **27.** 4.7×1.5 **28.** 3.0617×0.005

29. 64.7×1.8 **30.** 0.0049×6.3 **31.** 0.0671×0.48 **32.** 5.009×0.082

33. $0.08\overline{)0.376}$ **34.** $6.3\overline{)15.12}$ **35.** $0.47\overline{)0.02397}$ **36.** $0.051\overline{)0.0001836}$

Divide. Round to the nearest hundredth.

37. $5.109 \div 0.004$ **38.** $72.08 \div 0.19$ **39.** $0.01362 \div 0.54$

Solve.

40. Juan collects baseball and football cards. He has 44 more baseball cards than football cards. If he has 112 cards, how many are football cards?

41. Geno scored 38 more points at basketball than Bill. Together they scored 96 points. How many points did Bill score?

42. Sarah drove 378 mi on the highway. She used 14 gal of gasoline. What was the mpg?

43. Victoria's motorbike gets 48 mpg. How many miles can she travel on 3.8 gal of gasoline?

CUMULATIVE REVIEW

Choose the letter of the correct answer.

1. What is the standard numeral for 50,000 + 400 + 9?

a. 549 **b.** 5,409 **c.** 50,409 **d.** none of these

2. What is the standard numeral for three hundred nine ten-thousandths?

a. 309.0001 **b.** 0.0390 **c.** 0.3090 **d.** none of these

3. What is 4.5068 rounded to the nearest hundredth?

a. 4.507 **b.** 4.506 **c.** 4.51 **d.** none of these

4. Which of the following decimals is greater than 0.0784?

a. 0.00999 **b.** 0.0613 **c.** 0.08 **d.** none of these

5. What is the total amount of money earned on regular and overtime pay?

a. net pay **b.** gross pay **c.** deduction **d.** none of these

Select the best estimated answer.

6. $43.6 - 2.79$	**a.** 15	**b.** 31	**c.** 37	**d.** 45
7. 9.6×50.17	**a.** 540	**b.** 500	**c.** 400	**d.** 60
8. $22.5 + 9.45 + 10.6$	**a.** 30	**b.** 40	**c.** 110	**d.** 120
9. $3.53 \div 0.09$	**a.** 4	**b.** 5	**c.** 40	**d.** 50

Compute.

10. $45 + 219 + 804$

a. 1,068 **b.** 1,078 **c.** 2,068 **d.** none of these

11. $45{,}098 - 3{,}789$

a. 42,209 **b.** 42,309 **c.** 41,309 **d.** none of these

12. 43.6×5.7

a. 24.852 **b.** 237.42 **c.** 228.42 **d.** none of these

13. $6.7\overline{)13.668}$

a. 0.024 **b.** 2.04 **c.** 2.4 **d.** none of these

14. $80{,}000 \div 100$

a. 80 **b.** 800 **c.** 8,000 **d.** none of these

Solve.

15. A train goes from Miami to Jacksonville, a distance of 652 km; from Jacksonville to Savannah, 225 km; and from Savannah to Richmond, 805 km. What is the total distance by train from Miami to Richmond?

a. 2,452 km **b.** 1,682 km **c.** 982 km **d.** none of these

16. A stockholder owns 244.3 shares of stock in a mutual fund. If each share is worth $2.15, what is the value of her stock to the nearest cent?

a. $525.25 **b.** $425.25 **c.** $200.25 **d.** none of these

17. Pearl's gross earnings for the month are $1,575.00. A total of $439.63 for taxes is deducted. An additional $6.32 is deducted for insurance. What is Pearl's net pay?

a. $1,135.37 **b.** $1,129.05 **c.** $2,020.95 **d.** none of these

18. Just off the coast of California is an undersea mountain called San Juan Seamount. How far below the surface of the water is the top of the mountain if the ocean floor depth is 3,658 m and the mountain rises 3,105 m?

a. 7,863 m **b.** 3,553 m **c.** 653 m **d.** none of these

19. At $1.19 a dozen, how many dozen oranges can you buy for $6.25?

a. 10 dozen **b.** 8 dozen **c.** 6 dozen **d.** none of these

20. A new apartment complex has 542 units. Each unit is furnished with a refrigerator that costs $329. What is the total cost of the refrigerators?

a. $178,318 **b.** $202,680 **c.** $168,308 **d.** none of these

THINKING ABOUT MATH

1. Why will front-end estimation for addition always result in an underestimate?

2. If you wanted to estimate the quotient of 7.45 ÷ 2.9, what compatible numbers would you use?

3. Look at the following number sentence: 976 + 876 + 776 = 4,087. Explain how you can tell by looking that the sum is incorrect.

4. The pointer on an electric meter is between 9 and 0. Which number would you read? Why?

PreSkills Test

Add or subtract. [30, 32, 62, 64]

1. $\begin{array}{r} 7 \\ 6 \\ 8 \\ 5 \\ 3 \\ +6 \\ \hline \end{array}$ **2.** $\begin{array}{r} 3.2 \\ 4.9 \\ 2.7 \\ 3.5 \\ 3.1 \\ +4.2 \\ \hline \end{array}$ **3.** $\begin{array}{r} 382 \\ 387 \\ 376 \\ 372 \\ 383 \\ +385 \\ \hline \end{array}$ **4.** $\begin{array}{r} 72.1 \\ 71.9 \\ 72.6 \\ 72.2 \\ 71.9 \\ +72.3 \\ \hline \end{array}$

5. $\begin{array}{r} 48 \\ -\ 6 \\ \hline \end{array}$ **6.** $\begin{array}{r} 861 \\ -\ 29 \\ \hline \end{array}$ **7.** $\begin{array}{r} 540 \\ -176 \\ \hline \end{array}$ **8.** $\begin{array}{r} 9{,}863 \\ -2{,}794 \\ \hline \end{array}$

9. $\begin{array}{r} 8{,}645 \\ -1{,}295 \\ \hline \end{array}$ **10.** $\begin{array}{r} 80{,}000 \\ -69{,}257 \\ \hline \end{array}$ **11.** $\begin{array}{r} 60.1 \\ -27.8 \\ \hline \end{array}$ **12.** $\begin{array}{r} 39.6 \\ -15.9 \\ \hline \end{array}$

Multiply or divide. [36, 40, 46, 68, 78]

13. $\begin{array}{r} 37 \\ \times 10 \\ \hline \end{array}$ **14.** $\begin{array}{r} 2.7 \\ \times\ 10 \\ \hline \end{array}$ **15.** $\begin{array}{r} 8.75 \\ \times\ \ 10 \\ \hline \end{array}$ **16.** $\begin{array}{r} 14.87 \\ \times\ \ 10 \\ \hline \end{array}$

17. $\begin{array}{r} 54 \\ \times\ 3 \\ \hline \end{array}$ **18.** $\begin{array}{r} 74 \\ \times\ 7 \\ \hline \end{array}$ **19.** $\begin{array}{r} 17 \\ \times 15 \\ \hline \end{array}$ **20.** $\begin{array}{r} 36 \\ \times 21 \\ \hline \end{array}$

21. $10\overline{)6{,}900}$ **22.** $10\overline{)581}$ **23.** $10\overline{)5.1}$ **24.** $10\overline{)4.83}$

25. $6\overline{)486}$ **26.** $8\overline{)2{,}056}$ **27.** $7\overline{)48.3}$ **28.** $4\overline{)35.16}$

Order from least to greatest. [8]

29. 421, 514, 96, 105, 557 **30.** 15.47, 59.65, 58.7, 454.5

Round to the nearest hundred. [16]

31. 705 **32.** 428 **33.** 547 **34.** 8,492

Round to the nearest thousand. [16]

35. 6,027 **36.** 2,490 **37.** 8,664 **38.** 4,583

Round to the nearest hundred thousand. [16]

39. 457,658 **40.** 963,872 **41.** 5,672,541 **42.** 28,678,994

Chapter 4 STATISTICS

One of your most important decisions is what you will do after you graduate. How can you find out if more students in your school are planning to find jobs or continue their education after graduation?

4.1 COLLECTING DATA

Clark Watson has a part-time job at the West Side Recreation Center. His boss wants to know the ages of the teenagers in the center's judo class. Clark records the ages of everyone in the class. Here is the information, or **data.**

AGES OF STUDENTS IN THE JUDO CLASS

14	15	17	15	16
13	15	14	17	18
14	16	15	16	13
17	18	14	15	16

Clark organizes the data into a **frequency table** to make it easier to understand.

EXAMPLE 1 Make a frequency table.

1. Arrange the data in descending order.
2. Make a tally for each age.
3. Count the tallies. Write the frequencies.
4. Total all the frequencies.

AGES: JUDO CLASS

Age	Tally	Frequency
18	II	2
17	III	3
16	IIII	4
15	~~IIII~~	5
14	IIII	4
13	II	2
Total		20

When the data contains many different numbers, group the data into equal intervals.

EXAMPLE 2 Make a grouped-frequency table for the data.

1. Group the data in intervals.
2. Total the frequencies for each interval.
3. Total all the frequencies.

Age	Frequency
17–18	5
15–16	9
13–14	6
Total	20

You can show the data from a grouped-frequency table in a **histogram.**

EXAMPLE 3

NUMBER OF DAYS BETWEEN VISITS TO THE CENTER

Number of days	0–4	5–9	10–14	15–19
Frequency	3	8	6	4

Make a histogram for the data.

1. Draw a horizontal axis and a vertical axis. Label the horizontal axis "Number of Days." Label the vertical axis "Frequency."
2. Mark the lowest value of each interval on the horizontal axis.
3. Make a bar as wide as the interval and as high as the frequency.
4. Write a title for the histogram.

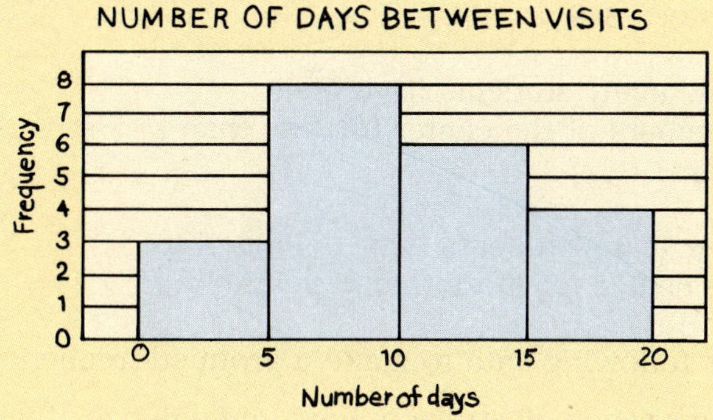

PRACTICE EXERCISES

Use the frequency table at the right to answer Exercises 1–4.

DISTANCE FROM THE RECREATION CENTER

Number of blocks	Tally	Frequency
5	HHT HHT HHT III	18
4	HHT HHT III	13
3	HHT HHT HHT IIII	19
2	HHT HHT HHT HHT II	22
1	HHT HHT HHT I	16

1. How many members live two blocks from the recreation center?

2. How many members live four or more blocks from the center?

3. How many blocks from the center does the greatest number of members live?

4. How many members were included in the survey?

Use the following data to make a frequency table.

5. Clark took a survey of 15 teenagers to see how many times a month they visited the recreation center.
Here is the data:

15, 13, 12, 14, 11, 16, 15, 12,
10, 14, 17, 15, 16, 11, 13

Use the grouped-frequency table at the right to answer Exercises 6–9.

MEMBERSHIP: RECREATION CENTER

Number of years	Frequency
0–1	37
2–3	58
4–5	52
6–7	21
8–9	3

6. For how many years has the greatest number of teenagers belonged to the center?

7. For how many years has the fewest number of teenagers belonged to the center? 8–9

8. How many students have been members of the center for less than 2 years?

9. How many students have belonged to the center for more than 3 years?

Use the following data to make a grouped-frequency table.

10. Clark asked 30 students how many dances they attended in one year at the West Side Recreation Center.
Here are the results:

15, 20, 18, 12, 10, 17, 24, 21, 18, 8, 12, 24, 18, 16, 9,
15, 21, 24, 13, 17, 22, 7, 19, 23, 16, 14, 10, 8, 9, 22

11. Use the following data to make a histogram.

AGE OF ATHLETES: WEST SIDE RECREATION CENTER

Age	5–7	8–10	11–13	14–16	17–19
Frequency	12	18	27	36	24

12. The West Side Recreation Club held a raffle to raise money. Fifteen club members sold raffle tickets. Here is the number of tickets each member sold. Use the data to make a histogram.

4, 8, 14, 26, 7, 9, 16, 30, 2, 24, 15, 5, 28, 11, 44

MIXED REVIEW

Multiply or divide.

13. 28×1.3 **14.** 4.7×2.6 **15.** 3.01×0.02 **16.** 0.003×0.09 **17.** 83.04×0.75

18. $5\overline{)23.4}$ **19.** $0.48\overline{)0.8736}$ **20.** $8\overline{)6}$ **21.** $6.7\overline{)2.3316}$ **22.** $0.26\overline{)185.12}$

23. 1.7×0.55 **24.** $0.9\overline{)3.87}$ **25.** $0.53\overline{)0.02756}$ **26.** 843×0.4 **27.** $4.8\overline{)1.728}$

PROBLEM Solving APPLICATION

4.2 CONSUMER: FINDING THE PRICE OF A NEW CAR

The prices of new cars can vary. There is a base price which is the dealer's price for the car without any optional equipment. If a car includes extra equipment, or if you want certain types of optional equipment, there are extra charges for the equipment.

OPTIONAL EQUIPMENT

Air conditioning (manual)	$795	Electronic instrument cluster	$357
Air conditioning (deluxe)	$956	Power door locks	$195
Rear window defroster	$149	Illuminated entry system	$ 87
Automatic transmission	$482	Extended range fuel tank	$ 54

The table above lists the prices of some optional equipment. You are considering buying a car that has a base price of $9,245. Automatic transmission and power steering/brakes are standard equipment in the car. You want this optional equipment: air-conditioning (manual), rear window defroster, and power door locks. What is the price of the car?

THINK: Add the base price and the prices for the optional equipment.

$\$9{,}245 + \$795 + \$149 + \$195 = \$10{,}384$

So, the price of your car with the optional equipment is $10,384.

Find the price of the car with the optional equipment.
Remember to estimate whenever you use your calculator.

	Car	Base Price	Optional Equipment
1.	Panther	$10,850	Air-conditioning (deluxe), rear window defroster, extended range fuel tank
2.	Eagle	$11,930	Electronic instrument cluster, extended range fuel tank
3.	Dynamo	$ 9,770	Air-conditioning (deluxe), electronic instrument cluster, power door locks, illuminated entry system
4.	TMX	$ 5,860	Automatic transmission, air-conditioning (manual), rear window defroster
5.	Meteor	$12,580	Electronic instrument cluster, rear window defroster, illuminated entry system

4.3 READING AND INTERPRETING DATA

You can see comparisons on a **bar graph.**

EXAMPLE 1 The horizontal bar graph at the right shows the costs of some breakfast foods. Which food is the most expensive?

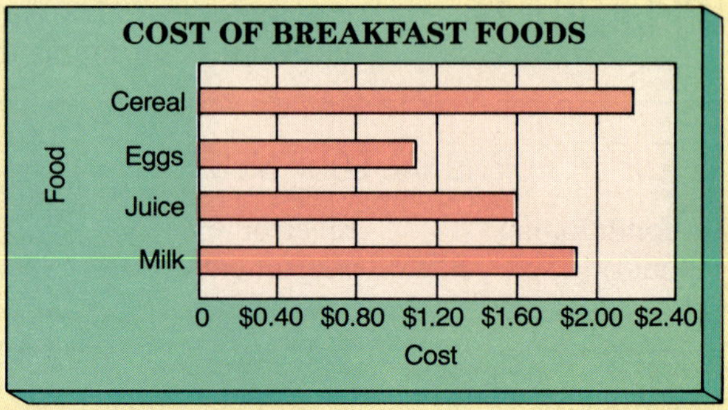

To find the most expensive food, look for the longest bar.

Cereal costs the most.

You can compare two sets of data on a **double-bar graph.**

EXAMPLE 2 The vertical double-bar graph at the right shows the favorite brands of cereal for two different age groups. Which cereal is equally popular with both groups?

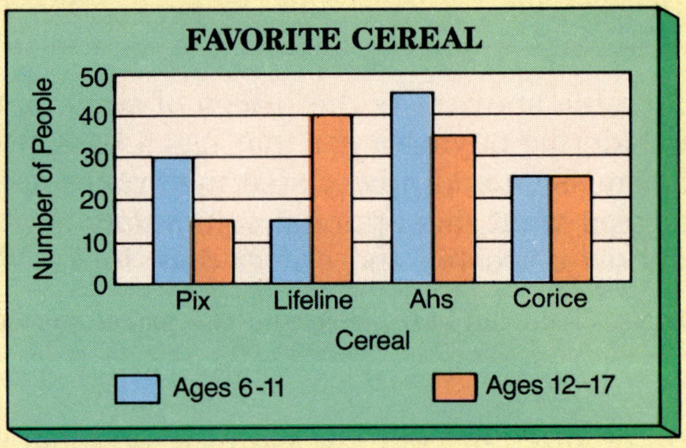

To find the cereal that is equally liked by both groups, look for bars that have the same height.

Corice is equally liked by both age groups.

You can see changes over time on a **broken-line graph.**

EXAMPLE 3 The broken-line graph at the right shows egg production in Maryland from 1980 to 1984. Did egg production increase or decrease between 1980 and 1981?

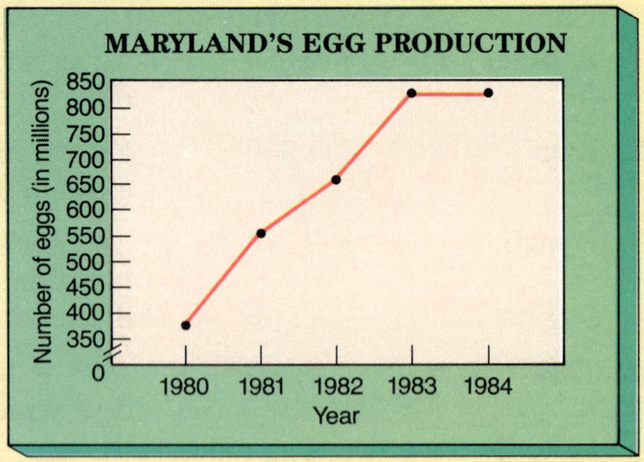

Look at the line segment between 1980 and 1981. The line segment rises.

Egg production increased between 1980 and 1981.

FOR DISCUSSION

What are the advantages of showing data on a graph instead of a table? Think of a situation in which showing data on a graph would be more effective than using a table.

PRACTICE EXERCISES

Use the graph in Example 1 to answer Exercises 1–8.

1. Which food costs the least?

2. Which is more expensive, juice or milk?

3. About how much does a box of cereal cost?

4. About how much more does milk cost than eggs?

5. If you buy cereal and eggs, about how much would you spend?

6. Which item costs about twice the price of eggs?

7. How much would all the items cost?

8. If you pay for all the items with a $10 bill, how much change would you receive?

Use the graph in Example 2 to answer Exercises 9–18.

9. How many people in the 6–11 age group like Pix?

10. How many people in the 12–17 age group like Ahs?

11. Which age group prefers Lifeline?

12. About how many more 6- to 11-year-olds prefer Ahs than 12- to 17-year-olds?

13. How many people chose Corice as their favorite cereal?

14. About how many more 12- to 17-year-olds prefer Lifeline to Pix?

15. How many more people liked Lifeline than Corice?

16. How many people chose Ahs as their favorite cereal?

17. How many more 6- to 11-year-olds prefer Ahs to Pix?

18. How many people are included in the survey?

Use the graph in Example 3 to answer Exercises 19–20.

19. Did egg production increase or decrease between 1982 and 1983?

20. What happened to egg production between 1983 and 1984?

21. Use an almanac to find out if egg production in Maryland increased, decreased, or stayed the same after 1984.

4.4 MEAN

Randy Johnson manages a fast-food restaurant called Smiley's. The head office wants to know the average length of employment at Randy's restaurant. Randy compares the starting date and ending date for every employee who left during the past 6 months. Here is the number of months that each of those employees worked at Smiley's.

6, 14, 8, 3, 12, 9, 15, 18, 14

To find the average length of employment, Randy finds the mean of this data. The **mean** of a set of data is the same as the average.

EXAMPLE 1

Find the mean of the number of months worked by the former employees.

1. Add. $6 + 14 + 8 + 3 + 12 + 9 + 15 + 18 + 14 = 99$
2. Divide by the number of addends. $\frac{99}{9} = 11$

So, the average length of employment at Randy's restaurant was 11 months.

When computing with decimals, you may need to round the mean to the nearest tenth.

EXAMPLE 2

Find the mean to the nearest tenth: 8.8, 17.6, 11.4, 18.9, 7.3, 10.7

1. Add. $8.8 + 17.6 + 11.4 + 18.9 + 7.3 + 10.7 = 74.7$
2. Divide by the number of addends. $\frac{74.7}{6} = 12.45$
3. Round the answer to the nearest tenth. 12.45 rounds to 12.5.

The mean is 12.5 to the nearest tenth.

FOR DISCUSSION

Find the average number of times that the milk shake machine has broken down during the past six months.

Jan.—4 breakdowns; Feb.—1 breakdown; Mar.—4 breakdowns; Apr.—6 breakdowns; May—9 breakdowns; June—13 breakdowns

If you were the manager of Smiley's, would you buy a new milkshake machine? Explain.

PRACTICE EXERCISES

Find the mean. Round to the nearest tenth.

1. 4, 6, 8, 10

2. 19, 15, 26, 38, 32

3. 63, 72, 58, 67, 41, 77

4. 476, 938, 217, 557

5. 702, 149, 678, 350, 836

6. 4, 795; 2,618; 997; 1,008; 6,242

7. 3.5, 4.1, 6.7, 7.9

8. 24.6, 57.3, 44.1, 39.8, 64.5

9. 1.8, 6.2, 9.3, 4.5, 10.2, 7.7, 3.6

10. 5.46, 12.7, 17.36, 8.4, 9.03

Find the missing data.

11. Data: 10, 6, 9, 14, ■
Mean: 10

12. Data: 27.3, 14.6, 32.7, ■
Mean: 24.7

MIXED REVIEW

Add or subtract.

13. $\begin{array}{r} 4.56 \\ 0.02 \\ +114.3 \\ \hline \end{array}$

14. $\begin{array}{r} 0.006 \\ 17.1 \\ +\ 5.27 \\ \hline \end{array}$

15. $\begin{array}{r} 92.5 \\ 0.007 \\ +\ 4.93 \\ \hline \end{array}$

16. $\begin{array}{r} 467 \\ 3.21 \\ 18.6 \\ +\ 0.005 \\ \hline \end{array}$

17. $\begin{array}{r} 9.26 \\ 15.3 \\ 0.004 \\ +891.5 \\ \hline \end{array}$

18. $\begin{array}{r} 72.6 \\ -\ 4.8 \\ \hline \end{array}$

19. $\begin{array}{r} 9.03 \\ -0.7 \\ \hline \end{array}$

20. $\begin{array}{r} 45.1 \\ -28.63 \\ \hline \end{array}$

21. $\begin{array}{r} 884.76 \\ -235.08 \\ \hline \end{array}$

22. $\begin{array}{r} 9.02 \\ -0.005 \\ \hline \end{array}$

23. $\begin{array}{r} 4.26 \\ 19.073 \\ +\ 0.8 \\ \hline \end{array}$

24. $\begin{array}{r} 71.3 \\ -68.51 \\ \hline \end{array}$

25. $\begin{array}{r} 19.48 \\ -\ 0.807 \\ \hline \end{array}$

26. $\begin{array}{r} 5.3 \\ +0.857 \\ \hline \end{array}$

27. $\begin{array}{r} 80 \\ -\ 5.27 \\ \hline \end{array}$

Solve.

Randy asks 20 customers how often they visit the restaurant each month. Here are the results.

2, 5, 1, 3, 0, 4, 8, 6, 5, 2,
1, 9, 2, 4, 8, 7, 3, 0, 5, 5

28. Make a grouped-frequency table for this data.

29. Find the mean number of times these customers visit the restaurant in 1 month.

TIME OUT

Take a survey of 25 people to find out how many times they go to a fast-food restaurant in a month. Find the mean. Compare it with those of your classmates. Are the results about the same?

4.5 RANGE, MEDIAN, AND MODE

Animals have been clocked for maximum speeds. This table shows the speeds of some animals. The speeds vary from 12 mph (miles per hour) to 70 mph.

Animal	Speed (mph)
Cheetah	70
Lion	50
Cat	30
Bear	30
Squirrel	12

The **range** of a set of data shows how much the data varies. It is the difference between the greatest number and the least number.

$70 - 12 = 58$ The range for this set of data is 58 mph.

EXAMPLE 1 Find the range: 45, 79, 21, 86, 55, 94

$$\begin{array}{r} 94 \\ -21 \\ \hline 73 \end{array} \begin{array}{l} \leftarrow \textbf{Greatest number} \\ \leftarrow \textbf{Least number} \\ \leftarrow \textbf{Range} \end{array}$$

The range for this set of data is 73.

The **median** gives the middle value when the numbers are listed in order. The median of the data given in the table is 30 mph.

Sometimes a set of data has two middle numbers. To find the median, find the average of the middle numbers.

EXAMPLE 2 Find the median: 26, 15, 48, 53, 74, 39

1. List the numbers in order from greatest to least. 74, 53, 48, 39, 26, 15
2. Find the average of the middle numbers. $\frac{48 + 39}{2} = \frac{87}{2} = 43.5$

The median of this set of data is 43.5.

The **mode** is the value that occurs most often. The mode of the data given in the table is 30 mph.

A set of data may have no mode, or it may have more than one mode.

EXAMPLE 3 Find the mode:

a. 46, 3, 57, 89, 2, 17, 101, 33

THINK: Every value appears only once.

This set of data has no mode.

b. 19.4, 89.6, 75.34, 19.4, 67.85, 0.01, 89.6, 35.8

THINK: The values 19.4 and 89.6 each appear twice.

This set of data has two modes: 19.4 and 89.6.

FOR DISCUSSION

Here are the speeds at which a racehorse ran 1 mile on 5 different days.

Day	1	2	3	4	5
mph	35	31	28	35	35

Which gives a better idea of the speed at which the horse usually runs, the mean or the mode? Explain.

PRACTICE EXERCISES

Find the statistical measures for the set of data. Round the answer to the nearest tenth.

Data	Range	Median	Mode	Mean
4, 6, 10, 11, 6, 2, 6, 3	**1.** ■	**2.** ■	**3.** ■	**4.** ■
22, 18, 4, 27, 35	**5.** ■	**6.** ■	**7.** ■	**8.** ■
108, 73, 196, 110, 73	**9.** ■	**10.** ■	**11.** ■	**12.** ■
12, 9, 3, 11, 9, 13, 2, 10, 9, 5	**13.** ■	**14.** ■	**15.** ■	**16.** ■
5, 3.7, 8.6, 7.5, 3.7, 7.5, 6	**17.** ■	**18.** ■	**19.** ■	**20.** ■
17, 54.6, 28.2, 47, 32.9	**21.** ■	**22.** ■	**23.** ■	**24.** ■
589.7, 400, 475, 65.4, 475, 523.9	**25.** ■	**26.** ■	**27.** ■	**28.** ■

Solve.

29. Scientists measured the speed at which six sharks swam in the ocean.

Shark	1	2	3	4	5	6
mph	37	32	36	34	39	38

Find the mean.

30. At the World Championship Tortoise Races, the six fastest turtles moved at the following speeds.

0.2 mph, 0.3 mph, 0.4 mph,
0.5 mph, 0.46 mph, 0.34 mph

Find the median.

31. Greyhounds are racing dogs. Here are the speeds, in miles per hour, at which 12 greyhounds finished a 1-mi race:

38, 35, 38, 36, 37, 38, 34, 37, 33, 36, 39, 32

Make a frequency table to show this data.

PROBLEM *Solving* STRATEGY

4.6 MAKING A TABLE

Situation:

Bill, Tom, Mary, and Sue each belong to one of the following after-school clubs: stamp, music, tennis, and hiking. Mary and Tom tried out for the tennis club, but did not make it. Sue does not like to participate in sports activities. Sue and Mary went to hear the music club perform. Which club does each person belong to?

Strategy:

Making a table can help you solve some problems.

Applying the Strategy:

Clue 1: Mary and Tom tried out for the tennis club but did not make it. Write *No* to show that Mary and Tom are not in the tennis club.

	Clubs			
	Stamp	**Music**	**Tennis**	**Hiking**
Bill				
Tom			No	
Mary			No	
Sue				

Clue 2: Sue does not like to participate in sports. Write *No* to show that Sue is not in the tennis or hiking clubs. Write *Yes* to show that Bill belongs to the tennis club.

	Clubs			
	Stamp	**Music**	**Tennis**	**Hiking**
Bill	No	No	Yes	No
Tom			No	
Mary			No	
Sue			No	No

Clue 3: Sue and Mary went to hear the music club perform. Write *No* to show that Sue and Mary are not in the music club. Why do you know that:

- Sue is in the stamp club?
- Mary is in the hiking club?
- Tom is in the music club?

	Clubs			
	Stamp	**Music**	**Tennis**	**Hiking**
Bill	No	No	Yes	No
Tom	No	Yes	No	No
Mary	No	No	No	Yes
Sue	Yes	No	No	No

PRACTICE EXERCISES

Read the problem. Then decide which statement is true.

1. Jamie collects stamps either from foreign countries or from the United States. He does not collect stamps from foreign countries.
   a. Jamie collects U.S. stamps.
   b. Jamie collects stamps from foreign countries.

2. Some hikers like to camp. Amelia is a camper.
   a. Amelia is a hiker.
   b. Amelia may or may not be a hiker.

Solve by making a table.

3. Eli, Bert, and June exercise regularly. One jogs, one swims, and one plays golf. The jogger and Bert work together. Eli and the jogger went to see the golfer play in a tournament. Which kind of exercise does each person do?

4. Willy, Julia, Anne, and Sean all have carpentry as their hobby. They won the first four prizes in a carpentry contest. Sean won third prize. Julia did not win second prize. If Ann won first prize, which prize did Willy win?

5. Jason, Eve, Ben, and Kerry belong to a cooking club. They each baked different muffins: blueberry, banana, apple, and bran. Kerry needed to wash some berries for her muffins. Eve needed to remove the seeds from her fruit. Ben's muffins did not have any fruit. Which kind of muffins did each person bake?

6. Mr. Moore, Mrs. Donato, and Mr. Alen are the teachers for these clubs: photography, science, and acting. Mr. Moore and the photography club teacher are cousins. The science club teacher and Mr. Alen have lunch together. Mr. Alen is not the photography club teacher. What does each person teach?

MIDCHAPTER REVIEW

Use the following data to make a frequency distribution table. [92]

1. Number of children in some families:

2, 3, 2, 1, 5, 2, 6, 3, 1, 1, 2, 4,

3, 5, 2, 1, 8, 2, 4, 4, 3, 1, 1, 6, 5

Use the data below to complete the grouped-frequency table. [92]

Data	**Interval**	**Frequency**
74, 133, 4, 127, 20, 30,	0–30	**2.** ■
5, 147, 100, 27, 142, 121,	31–60	**3.** ■
141, 125, 67, 150, 130, 68,	61–90	**4.** ■
140, 132, 76, 86, 42, 55,	91–120	**5.** ■
59, 67, 102, 9, 124, 95	121–150	**6.** ■

Use the graph at the right to answer the questions. [96]

7. Which station is the most popular?

8. Which stations are preferred by more than 40,000 people?

9. How many more people chose WJBN than WLQZ?

10. How many chose WGPX and WCLH?

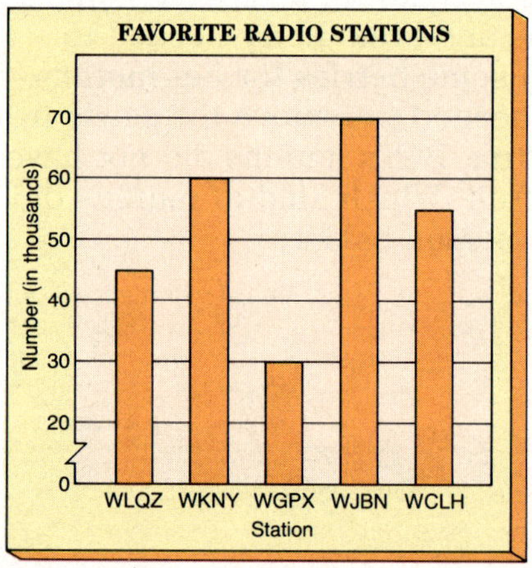

Find the statistical measures for each set of data. [98, 100]

	Range	**Mean**	**Median**	**Mode**
20, 15, 18, 22, 10, 12, 16, 17, 19, 21	**11.** ■	**12.** ■	**13.** ■	**14.** ■
9, 11, 10, 8, 13, 12, 11, 6, 7, 9	**15.** ■	**16.** ■	**17.** ■	**18.** ■

Solve. [95]

19. A 1987 Towne Coupe costs $12,735. Cruise control is optional for $934. What is the total price of the car?

PROBLEM Solving APPLICATION

4.7 CAREER: ECOLOGIST

Anton Kluzewski is an ecologist. He is interested in the environment and the ways in which we can reduce the dangers of pollution. At present he is working for an environmental protection agency.

The map below gives information about the distribution of hazardous waste sites in a recent year.

Hazardous Waste Sites (Final and Proposed) — Number of Sites on National Priority List

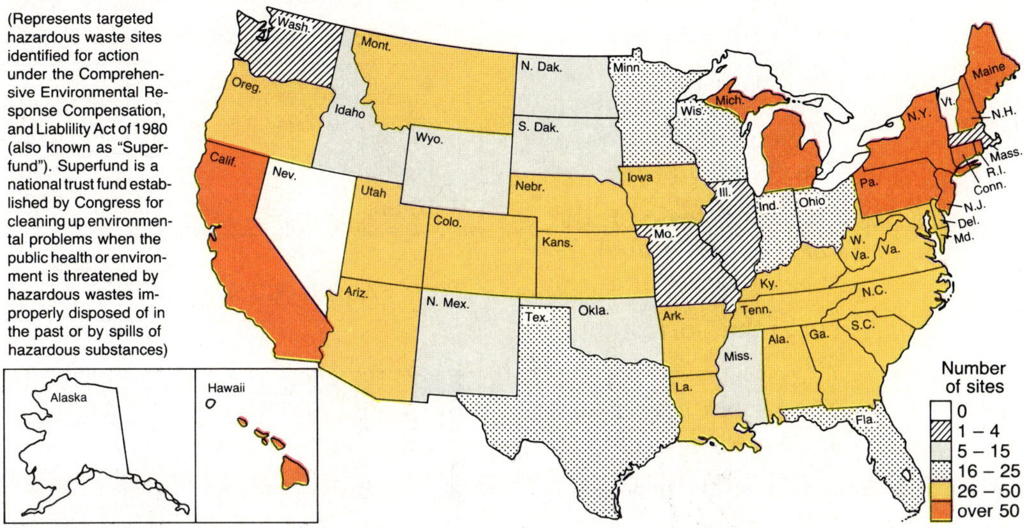

How many sites are in:

1. Texas?

2. California?

3. Nevada?

4. Illinois?

Which states have:

5. over 50 of the sites?

6. 26–50 of the sites?

Which state has more of the sites?

7. Nevada or Utah?

8. Georgia or Florida?

9. South Dakota or Montana?

10. Minnesota or Arkansas?

Solve.

11. Are there more states that have sites than have 16–25 sites?

12. Make a tally that shows the number of states that have these numbers of sites: 0, 1–4, 5–15, 16–25, 26–50, over 50.

4.8 BAR GRAPHS

A **bar graph** helps you compare data quickly and easily. Each bar has a title, a horizontal axis, and a vertical axis. A scale on one of the axes helps you find the number represented by the bar.

EXAMPLE 1 The table shows the sales of computer games in 1 year at the ZAP Computer Stores. Show this data on a vertical bar graph.

1. Round the data to a convenient place to determine a scale.
2. Show the scale on the vertical axis. Label the scale.
3. Label the horizontal axis.
4. Draw a bar to show the sales of each game.
5. Give the graph a title.

TOP-SELLING COMPUTER GAMES

Game	Copies sold (in thousands)	
	Actual	Rounded
York	78	80
Zeon	57	60
Hotshot	52	50
Radar	38	40

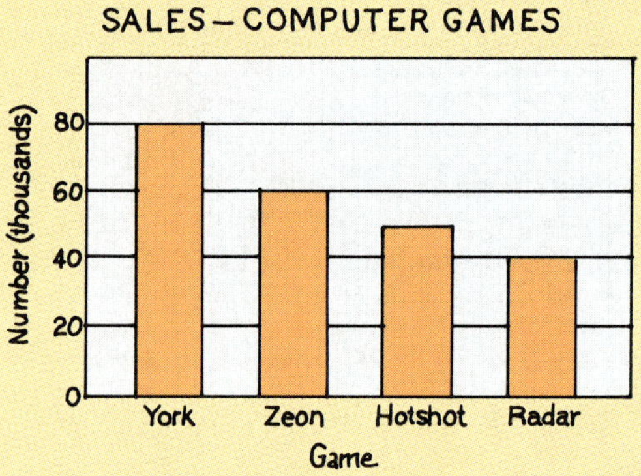

A survey was taken to find out what accessories customers are buying for their computers.

EXAMPLE 2 Make a horizontal bar graph to show this data.

1. Round the data to the nearest million. Choose a scale.

COMPUTER ACCESSORIES

Item	Units sold (in millions)	
	Actual	Rounded
Floppy disk	8.118	8
Hard disk	1.836	2
Impact printer	4.429	4
Modem	2.052	2

2. Show and label the scale on the horizontal axis.
3. Label the vertical axis.
4. Draw the bars.
5. Write a title for the graph.

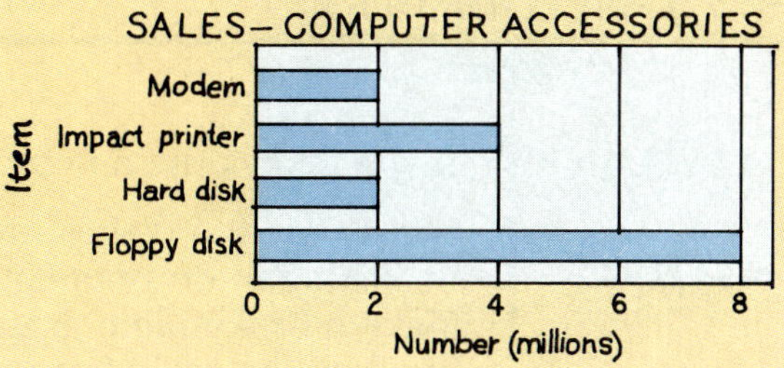

PRACTICE EXERCISES

1. Make a vertical bar graph to show the numbers of magazine subscriptions sold.

SUBSCRIPTIONS SOLD—COMPUTER MAGAZINES

Magazine	Subscriptions sold
Munch	10,980
PC Life	9,870
PRINT!	6,400
T+	12,460

2. The ZAP Computer Stores asked their customers how they used their computers most often. Make a horizontal bar graph to show the primary uses of the computer.

PRIMARY USES—PERSONAL COMPUTERS

Function	Number
Database	4,902
Games	2,487
Spreadsheet	10,053
Word processing	12,685

3. The advertising manager of SOLDOUT Computer Company wants to include a graph of software sales in his monthly report. Use the scales below to make two vertical bar graphs. Which graph gives a better representation of the data?

Scale 1: 0 to 40,000; intervals of 10,000
Scale 2: 30,000 to 40,000; intervals of 2,000

SOFTWARE SALES—AUGUST

Software	Number sold
WRITE RIGHT	30,956
FIGURE-OUT	39,493
INFO FAST	31,611
JAZZ IV	35,082

4. A store sells four different brands of computer diskettes. They want to sell only two brands. Make two horizontal bar graphs using the scales below to show the sales for each brand. Which graph gives a better representation of the data?

Scale 1: 0 to 5,000 with intervals of 1,000
Scale 2: 500 to 4,500 with intervals of 500

BRANDS OF DISKETTES

Brand	Sales
4KB	3,826
COPY-IT	634
GUARDIAN	2,529
DISK III	996

4.9 LINE GRAPHS

A **broken-line graph** shows the amount of change over time.

EXAMPLE 1 Home satellite dishes are becoming more popular. Make a broken-line graph to show the sales from 1981 to 1984.

1. Round the data to a convenient place to determine a scale.
2. Draw and label the horizontal and vertical axes.
3. Locate the point for each piece of data.
4. Draw line segments to connect the points in order.
5. Write a title for the graph.

SALES—HOME SATELLITE DISHES

Year	Number sold (in thousands)	
	Actual	Rounded
1981	32	30
1982	44	40
1983	53	50
1984	114	110

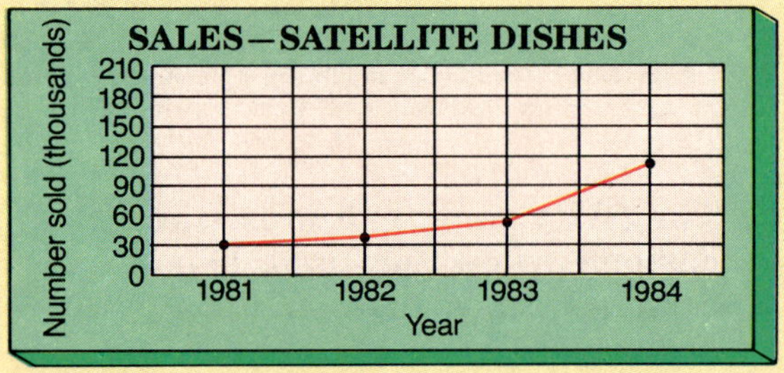

Sometimes you need to break the scale to fit the data on the graph.

EXAMPLE 2 Make a broken-line graph to show the sales of video-cassette recorders for the period shown.

1. Round the data to a convenient place to determine the scale.
2. Draw and label the scale on the vertical axis. Use a jagged line to show the break in the scale from 0 to 2,000.
3. Draw and label the horizontal axis.
4. Locate and connect the points.
5. Write a title for the graph.

SALES—VIDEO-CASSETTE RECORDERS

Month	Number sold	
	Actual	Rounded
October	2,058	2,100
November	2,404	2,400
December	2,348	2,300

Scale: 2,000 to 2,500; intervals of 100

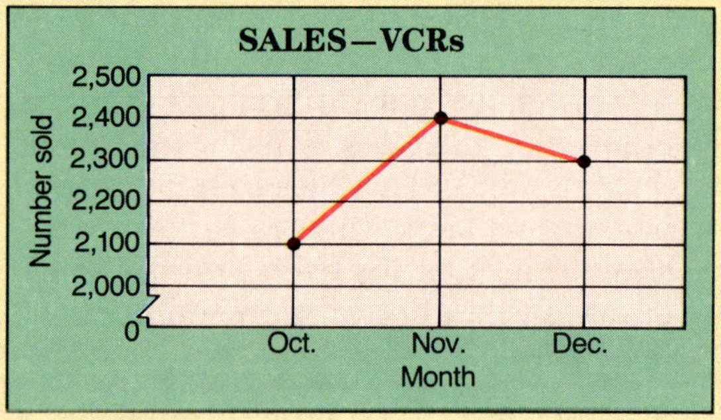

FOR DISCUSSION

Would a bar graph or a broken-line graph be better to show each of the following sets of data? Explain.

Favorite colors of different age groups
Hourly temperatures during a 24-h period

PRACTICE EXERCISES

Use each set of data to make a broken-line graph.

1. Jarene kept a record of the number of calls she received on her answering machine in 1 week.

DAILY CALLS: ANSWERING MACHINE

Day	Mon.	Tues.	Wed.	Thurs.	Fri.	Sat.	Sun.
Number of calls	3	2	6	8	5	10	2

2. The manager of the rock band "STAR" tracked the sales of the group's latest music video.

MONTHLY SALES—"FIREBALL" MUSIC VIDEO

Month	July	Aug.	Sept.	Oct.	Nov.	Dec.
Number sold	7,834	12,642	25,869	46,008	43,683	37,304

3. An airline company has gathered information about the number of mobile telephones installed in airplanes.

MOBILE TELEPHONE INSTALLATIONS

Year	1981	1982	1983	1984	1985	1986
Number installed	24	15	65	89	100	895

4. For the past 3 years, Green Ridge High School has made a videotape of the highlights of the school year in addition to the yearbook. Make a double broken-line graph to compare the sales of the video and the yearbook over the past 3 years. Use - - - - - to show the sales of the video. Use ——— to show the sales of the yearbook.

YEARBOOK/VIDEO SALES

Year	Yearbook	Video
1987	363	52
1988	287	101
1989	167	198

PROBLEM *Solving* APPLICATION

4.10 CONSUMER: ORDERING FROM A CATALOG

Mail-order shopping is a large business in the United States. Often, catalogs provide a wide selection of different items. When ordering by mail, the consumer must give the correct information on the order form.

In this order, compute the cost of the shirts, the cost of the socks, and the total cost of the two types of items.

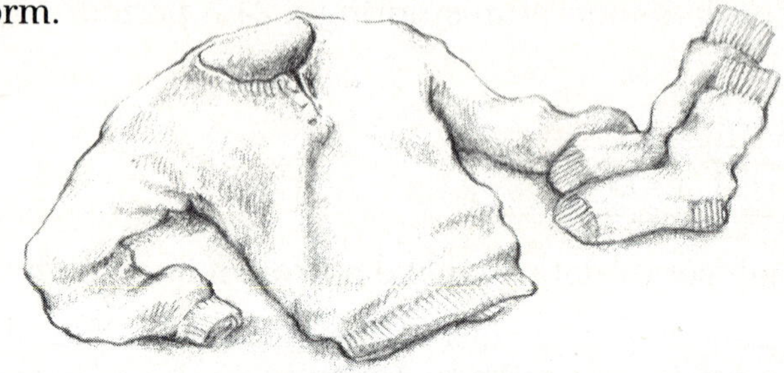

STOCK NO.	ITEM	COLOR NO.	SIZE	HOW MANY	UNIT PRICE	TOTAL AMOUNT
3899RR	Shirts	94	M	3	$14.50	■
1076RR	Socks	42	M	5 pair	$ 3.95	■
					TOTAL AMOUNT FOR ALL ITEMS	■
				Regular shipping and handling FREE within U.S.	Paid by M.M. Green, Inc.	

	NUMBER BOUGHT	×	UNIT PRICE	=	COST OF THE ITEMS
1. Multiply to find the cost of the shirts.	3	×	$14.50	=	$43.50
2. Multiply to find the cost of the socks.	5	×	$ 3.95	=	$19.75
3. Add to find the total cost.	$43.50	+	$19.75	=	$63.25

So, the total cost of the two types of items is $63.25.

FOR DISCUSSION

What services may customers want that M. M. Green, Inc. would charge extra for?

Use the information in this order form to answer Exercises 1–10. Remember to estimate whenever you use your calculator.

STOCK NO.	ITEM	COLOR NO.	SIZE	HOW MANY	UNIT PRICE	TOTAL AMOUNT
2759RR	Solid shirt	45	M	4	$18.50	■
2760RR	Striped shirt	43	M	2	$20.50	■
3064NN	Shorts	42	32	1 pair	$31.25	■
1076RR	Socks	44	M	3 pair	$ 3.95	■
2853RR	Sweater	43	M	2	$11.95	■

What is the cost of the:

1. solid shirts?
2. striped shirts?
3. shorts?
4. socks?
5. sweaters?
6. What is the total cost of all the items?

Find the total cost of the order.

7. 3 shirts, stock no. 2760RR; 4 sweaters, stock no. 2853RR
8. 6 pairs of socks, stock no. 1076RR; 2 shirts, stock no. 2759RR; 2 pairs of shorts, stock no. 3064NN

Estimate the answer.

9. About how much is the cost of 5 striped shirts, stock no. 2760RR?
10. Jan has $24 to spend on socks, stock no. 1076RR. About how many pairs of socks can she buy?

Solve.

11. Five types of slacks cost these amounts per pair: $27.95; $23.95; $25.50; $23.95; $32.50. What is the range of the amounts?
12. Which is the greatest—the mean, the median, or the mode of the costs of the five types of slacks? (See Exercise 11.)

4.11 PICTOGRAPHS

A **pictograph** uses a picture or a symbol to represent the data. A **key** shows how many each symbol represents.

EXAMPLE 1 How many robots are used in the computer industry?

To find how many robots are used, count the symbols.

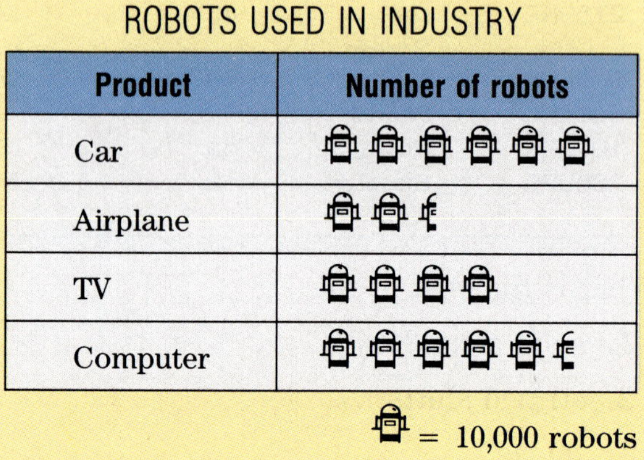

 = 50,000

 = 5,000

55,000

So, 55,000 robots are used in the computer industry.

When you make a pictograph, choose an appropriate symbol for the data.

EXAMPLE 2 Make a pictograph to show the number of employees in each U.S. industry.

EMPLOYEES IN U.S. INDUSTRIES—1980

Industry	Number of employees
Mining	1,207,000
Construction	4,346,000
Finance	5,160,000
Government	2,866,000

1. Round the numbers to an appropriate place.
2. Choose a symbol. Decide how many it represents.

    = 1,000,000

    = 500,000
3. Determine the number of symbols needed.

   1,000,000 = 

   4,500,000 = 
4. Label the vertical axis.
5. Draw the symbols for each number.
6. Write a title for the graph.

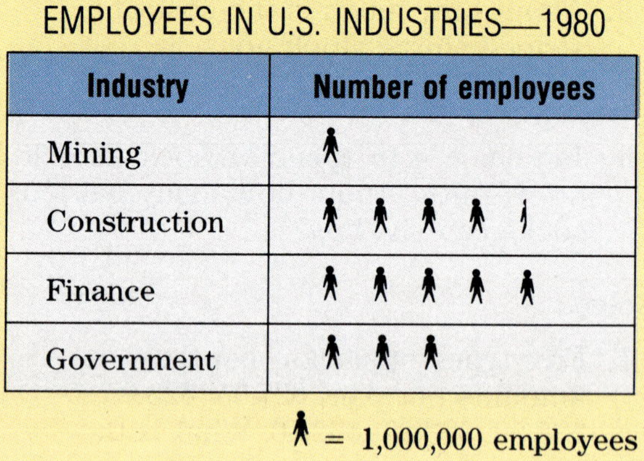

PRACTICE EXERCISES

Use the pictograph to answer Exercises 1–6.

1. How many pairs of sunglasses does each symbol represent?
2. Which lens had sales of 35,000?
3. How many pairs of mirrored lenses were sold?
4. Which lenses had sales of more than 40,000 pairs?
5. How many more pairs of gray lenses were sold than green lenses?
6. Which type of lens had sales three times as high as another?

SALES: CLUBLAND SUNGLASSES

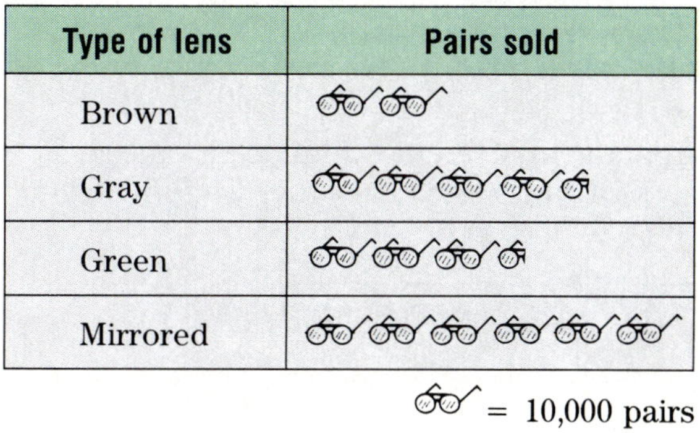

Use the given data to make a pictograph.

7. WORLD SHARK RECORDS

Type	Weight (lb)
Blue	437
Hammerhead	991
White	2,664
Mako	1,080
Tiger	1,780

8. U.S. PUBLIC LIBRARIES

City	Volumes
Erie, PA	420,355
Tucson, AZ	760,000
Austin, TX	817,338
Tampa, FL	1,324,262
Baltimore, MD	1,892,972

9. WORLD CORN PRODUCTION

Country	Tons
Italy	6,820,000
Brazil	21,098,000
Mexico	12,215,000
France	9,100,000

10. LEADING MERCHANT FLEETS

Country	Number of ships
Liberia	2,019
Greece	2,454
Japan	1,712
Panama	3,290

TIME OUT

Find the numbers in this set of data:

Clues:
My mean is 18.4.
My mode is 18.
My median is 18.
My range is 8.
My greatest number is 22.

PROBLEM Solving APPLICATION

4.12 CONSUMER: RENTING APARTMENTS

Statistics show that most young adults live in rental apartments.

When renting an apartment, you need to understand all the costs that are involved. The monthly rent is not your only cost. Usually, you also need to pay a security deposit. The security deposit may be equal to 1 or 2 months' rent.

In some apartment buildings, a charge for gas and electricity is included as part of the rent. In other buildings, you must pay the companies for the gas and electricity used.

WASHINGTON APARTMENTS	
Studio	\$245/month
Two rooms	\$325/month
Three rooms	\$410/month
Four rooms (two bedrooms)	\$508/month
Gas and electricity not included.	
Security deposit: 1 month's rent.	

JEFFERSON MANOR	
Studio	\$286/month
Two rooms	\$374/month
Three rooms	\$468/month
Four rooms (two bedrooms)	\$572/month
Gas and electricity included.	
Security deposit: 1 month's rent.	

You decide to rent a studio apartment in Washington Apartments. How much money will you pay the owner in the first year?

THINK: In the first year, you will need to pay the security deposit plus 12 months' rent.

1. Multiply to find the cost of 12 months' rent. $12 \times \$245 = \$2{,}940$
2. Add to find the total amount. $\$245 + \$2{,}940 = \$3{,}185$
   THINK: The security deposit is equal to 1 month's rent, or \$245.

So, you will pay the owner \$3,185 in the first year.

Use the signs on page 114 to answer the following exercises. Remember to estimate whenever you use your calculator.

1. What is the monthly rent for a two-room apartment in Washington Apartments?

2. What is the monthly rent for a two-room apartment in Jefferson Manor?

3. How much more is the rent for a two-room apartment in Jefferson Manor than in Washington Apartments?

4. Does the rent for a two-room apartment in Jefferson Manor include anything that is not included in the rent for a two-room apartment in Washington Apartments? If so, what?

5. Ann Ramirez has a two-room apartment in Washington Apartments. Her gas and electricity bills average about $53/month. Are her monthly expenses for rent, gas, and electricity less than they would be if she had a two-room apartment in Jefferson Manor?

6. How much money does Ann Ramirez pay for rent, gas, and electricity in 6 months?

Find the total amount that is paid in the first year for the apartment.

7. A studio apartment in Jefferson Manor

8. A two-room apartment in Washington Apartments

9. A three-room apartment in Jefferson Manor

10. A four-room apartment in Washington Apartments

Solve.

11. Jack O'Hara rents a two-room apartment in Jefferson Manor. He stays for 8 months. When he moves, the security deposit is returned to him. How much rent does he pay in the 8 months?

12. Della Truro rents a three-room apartment in Washington Apartments. Her gas and electricity bills average about $62.50 per month. How much money does she pay for rent, gas, and electricity in 10 months?

13. In 6 months, Glen Forbes pays a total of $2,262 for rent, gas, and electricity. Of this amount, $312 is paid for gas and electricity. What is the monthly rental for his apartment? In which building does he live?

CHAPTER REVIEW

Vocabulary Write the letter of the correct answer.

1. Which of the following is not a measure of central tendency? [100]
a. mean **b.** histogram **c.** median **d.** mode

2. What is the difference between the greatest number and the least number in a set of data? [100]
a. mode **b.** mean **c.** range **d.** frequency table

Skills Find the statistical measures for the set of data. [98, 100]

> Data: 55, 76, 67, 47, 55
> Range: *29* Median: *55* Mode: *55* Mean: *60*

	Range	**Median**	**Mode**	**Mean**
Data: 21.7, 35.5, 49.5, 16.3, 35.5	**3.** ■	**4.** ■	**5.** ■	**6.** ■
Data: 108, 93, 426, 766, 518, 210	**7.** ■	**8.** ■	**9.** ■	**10.** ■
Data: 1,506; 8,028; 4,441; 1,507	**11.** ■	**12.** ■	**13.** ■	**14.** ■

Use the frequency table to the right to answer Exercises 15–17. [92]

> How many walkers participated? *98*

MILES COMPLETED IN WALK-A-THON

Miles	Walkers
0–5	16
6–10	25
11–15	39
16–20	18

15. How many people completed more than 10 miles?

16. How many people walked less than 11 miles?

17. How many miles did the least number of people complete?

18. The table at the right shows the number of listeners at WJJK during a 5-month period. Use the data to make a broken-line graph. [108]

LISTENING AUDIENCE—WJJK

Month	Number of listeners
January	53,000
February	54,000
March	52,500
April	55,000
May	53,500

CHAPTER TEST

Use the following data to make a grouped-frequency table.

1. Number of library books checked out per person per year:

48, 10, 89, 63, 21, 15, 41, 10, 32, 18, 48, 3, 19, 25, 55

Find the statistical measures for the set of data in Exercise 1. Round to the nearest hundredth.

2. Range ▒ **3.** Median ▒ **4.** Mode ▒ **5.** Mean ▒

Use the bar graph at the right to answer Exercises 6 and 7.

6. Which route has the most buses?

7. About how many more buses run on the Seaview route than on the 39 St. route?

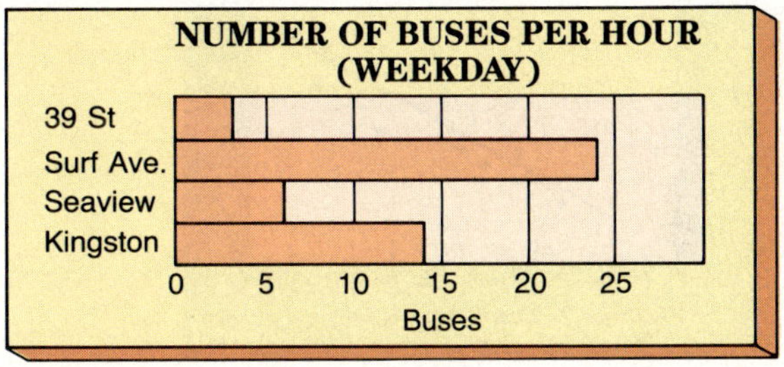

Make a broken-line graph by using the following data.

8. The Tidal Waves have released a new album. Here are the sales for the first 4 weeks.

Week 1: 189,764 Week 3: 500,794
Week 2: 385,618 Week 4: 412,308

Solve.

9. Clara, Leo, Jim, and Heather each belong to one of the following after-school clubs: photography, music, arts and crafts, and drama. Which club does each person belong to? Use the following clues to find out.

Clue 1: Leo and Jim do not like to perform.
Clue 2: Jim and Heather do not own cameras.
Clue 3: Heather watched a performance of the drama club.

Solve.

10. The Kunderas rent an apartment for $815 per month. The security deposit is 1 month's rent. Gas and electricity are furnished. What will the Kunderas pay for their apartment in the first year?

11. A Sidewider sells for $8,349. Power steering/power brakes are standard equipment. Air-conditioning costs an extra $855. What is the total cost of the Sidewider?

ENRICHMENT SCATTERGRAMS

A **scattergram** is a graph that shows whether a correlation exists between two sets of data.

Simone graphed the two scattergrams below. The graph on the left shows the number of hours she studied for each test and her grade for each test. The graph on the right shows the total time each student spent working on a test and the total number of each student's errors.

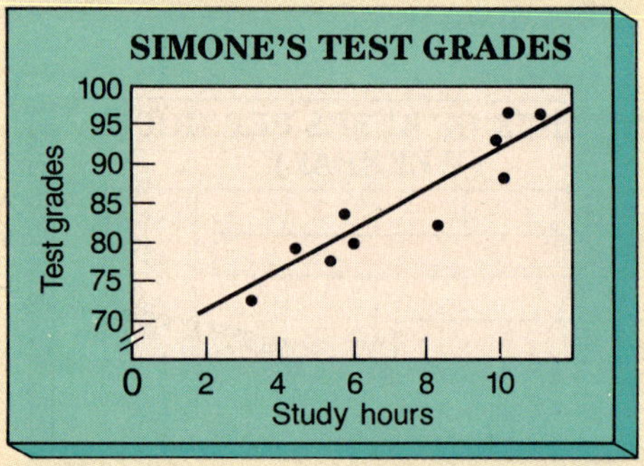

Notice that the line slants upward. This graph shows a **positive correlation** between test grades and study hours. This means Simone's test grades increased as her study hours increased.

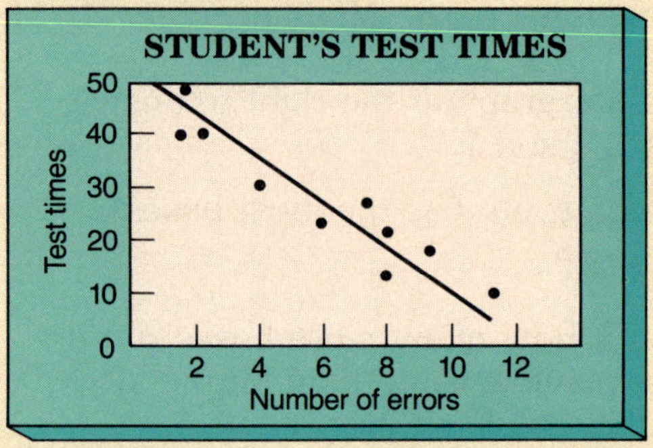

Notice that the line slants downward. This graph shows a **negative correlation** between test time and number of errors. This means their number of errors decreased as test time increased.

If a straight line cannot be drawn close to the points, there is **no correlation** between the data.

Examples Tell which kind of correlation you would probably see in each scattergram.

A scattergram that shows speed versus distance.

THINK: As speed increases, distance increases.

So, this scattergram would show a positive correlation.

A scattergram that shows speed versus fuel economy.

THINK: As speed increases, fuel economy decreases.

So, this scattergram would show a negative correlation.

Tell which kind of correlation you would probably see in the scattergram.

1. A scattergram that shows speed versus stopping distance

2. A scattergram that shows a person's weight versus salary

3. A scattergram that shows practice hours versus number of base hits

4. A scattergram that shows a baby's age in months versus number of sleeping hours

MENTAL MATH CLUSTERING

Sometimes several addends are "clustered" around each other. That is, they are all close to the same value. In such cases, you can multiply to find an estimate. Using the estimate, you add and subtract mentally to find the actual sum.

Add mentally: 95 + 87 + 96

1. Estimate by multiplying.

   THINK: Each of the 3 numbers is about 90.

   $3 \times 90 = 270$ ⟵ **Estimated sum**

2. Adjust the estimate.

   THINK:

95 is 5 more than 90.	5 more than 270 is 275.
87 is 3 less than 90.	3 less than 275 is 272.
96 is 6 more than 90.	6 more than 272 is 278.

The sum is 278.

Add mentally: 62 + 58 + 59 + 56 + 61

1. Estimate by multiplying.

   THINK: Each of the 5 numbers is about 60.

   $5 \times 60 = 300$ ⟵ **Estimated sum**

2. Adjust the estimate.

   THINK:

62 is 2 more than 60.	2 more than 300 is 302.
58 is 2 less than 60.	2 less than 302 is 300.
59 is 1 less than 60.	1 less than 300 is 299.
56 is 4 less than 60.	4 less than 299 is 295.
61 is 1 more than 60.	1 more than 295 is 296.

The sum is 296.

Find the sum mentally. Use the clustering method.

1. 22 + 25 + 16

2. 75 + 80 + 83

3. 21 + 19 + 16 + 24

4. 35 + 44 + 42 + 38

5. 27 + 25 + 33 + 30

6. 85 + 85 + 79 + 81

7. 51 + 60 + 64 + 54 + 69

8. 69 + 71 + 78 + 65 + 62

9. 56 + 41 + 45 + 51 + 55

10. 104 + 98 + 102 + 95 + 96

11. 90 + 98 + 88 + 92 + 85

12. 13 + 24 + 17 + 23 + 27 + 22

PreSkills TEST

Use the number line to identify each point. [8]

1. *A* **2.** *B* **3.** *C* **4.** *D*

Arrange in order from greatest to least. [8]

5. 45, 76, 28, 66, 70 **6.** 54, 60, 98, 65, 56

7. 508, 58, 85, 580, 805 **8.** 917, 179, 197, 719, 791

Complete the sequence.

9. 4, 8, 12, ■, ■, ■, ■ **10.** 7, 14, 21, ■, ■, ■, ■

11. 12, 24, 36, ■, ■, ■, ■ **12.** 15, 30, 45, ■, ■, ■, ■

Complete.

13. 1 × ■ = 12
2 × ■ = 12
3 × ■ = 12

14. 1 × ■ = 15
3 × ■ = 15

15. 1 × ■ = 18
2 × ■ = 18
3 × ■ = 18

16. 1 × ■ = 27
3 × ■ = 27

17. 1 × ■ = 30
2 × ■ = 30
3 × ■ = 30
5 × ■ = 30

18. 1 × ■ = 42
2 × ■ = 42
3 × ■ = 42
6 × ■ = 42

Compute.

19. $(6 \times 7) + 2$ **20.** $(4 \times 5) + 3$ **21.** $(3 \times 8) + 6$

22. $(4 \times 10) + 7$ **23.** $(5 \times 11) + 1$ **24.** $(6 \times 12) + 8$

Divide. Name the quotient and remainder. [46, 48]

25. $9 \div 4$ **26.** $15 \div 6$ **27.** $25 \div 8$ **28.** $40 \div 7$

29. $20 \div 10$ **30.** $37 \div 15$ **31.** $47 \div 12$ **32.** $50 \div 16$

Divide. Round the answer to the nearest hundredth. [78]

33. $9\overline{)5}$ **34.** $8\overline{)7}$ **35.** $6\overline{)8}$ **36.** $7\overline{)23}$

37. $24\overline{)13}$ **38.** $12\overline{)7}$ **39.** $15\overline{)34}$ **40.** $24\overline{)55}$

Chapter 5 FRACTIONS

Teamwork is necessary in baseball and many other sports. In what situations can working as a team be useful?

5.1 INTRODUCTION TO FRACTIONS

You can write a fraction to name part of a whole.

The top number, the **numerator,** identifies the number of equal parts being considered.

The bottom number, the **denominator,** identifies the total number of equal parts.

EXAMPLE 1 A pizza was sliced into 8 equal pieces. Jim ate 3 slices. What fraction of the pie did Jim eat?

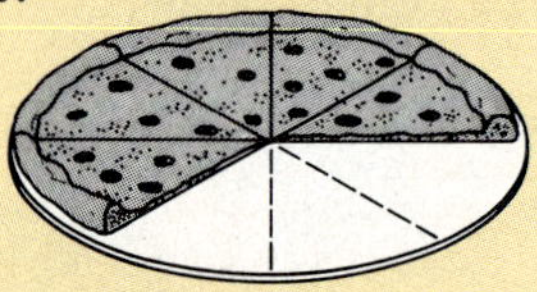

THINK: **Numerator** → 3 ← **Slices eaten**
Denominator → 8 ← **Total number of slices**

Jim ate $\frac{3}{8}$ of the pie.

You can write a fraction to name part of a group of objects.

EXAMPLE 2 What fraction of the circles are red? What fraction of the circles are blue?

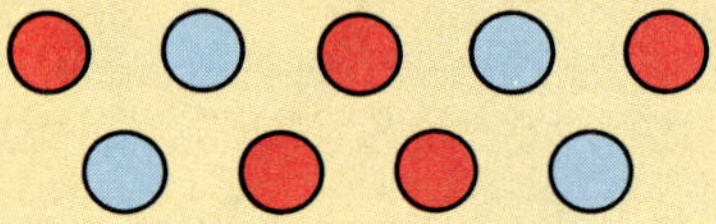

5 ← **Number of red circles**
9 ← **Total number of circles**

$\frac{5}{9}$ of the circles are red.

4 ← **Number of blue circles**
9 ← **Total number of circles**

$\frac{4}{9}$ of the circles are blue.

You can also write a fraction to identify a point on a number line.

EXAMPLE 3 What is point A on the number line?

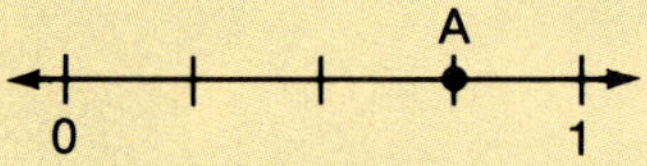

THINK: The distance from 0 to 1 is divided into 4 equal parts. The distance from 0 to A is 3 of these parts.

Point A is $\frac{3}{4}$ on the number line.

FOR DISCUSSION

Can the denominator of a fraction ever be zero? Explain. (*Hint:* What is the least number of parts that you could divide an object into?)

PRACTICE EXERCISES

Write a fraction for the shaded part.

1. 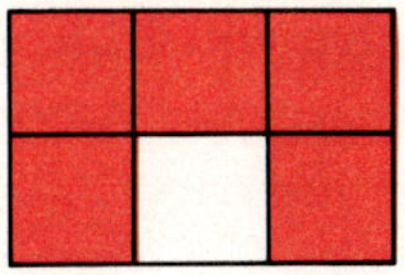

2. 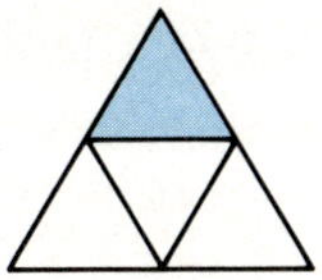

3. 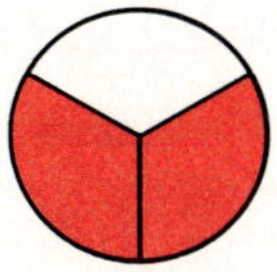

4. 

5. 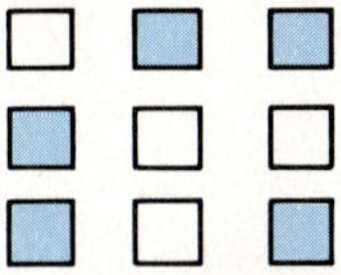

6. 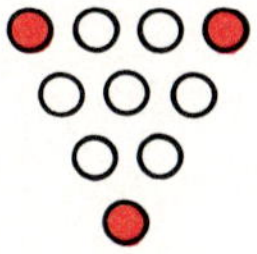

7. 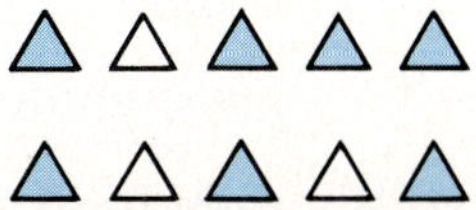

8. 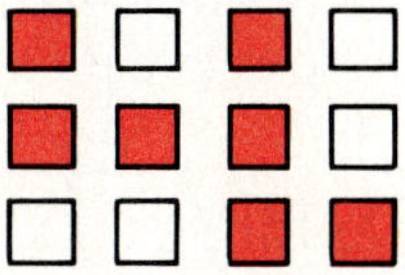

9. 

10. 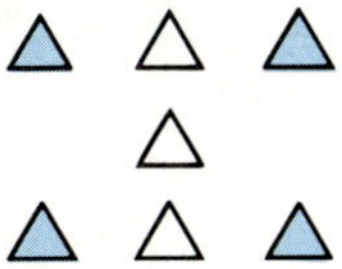

11. 

12. 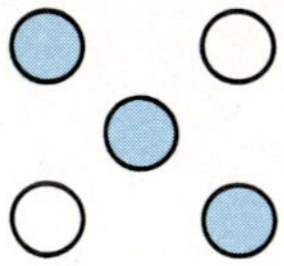

Write a fraction for the point.

13. 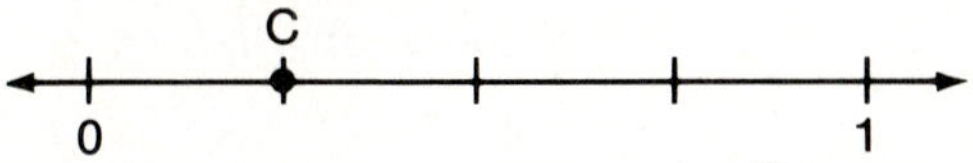

14. 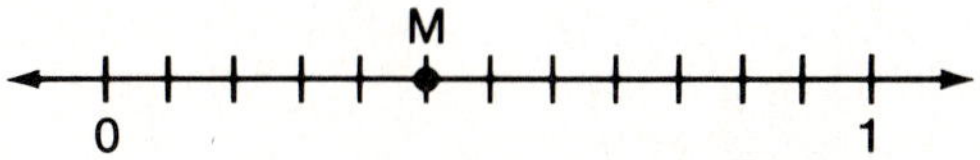

15. 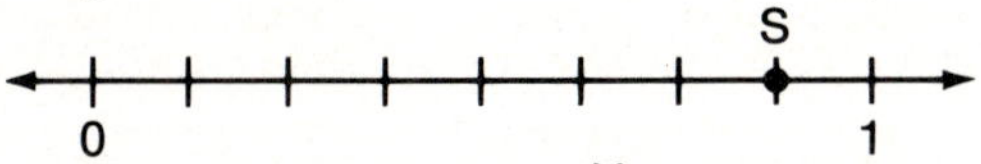

16. 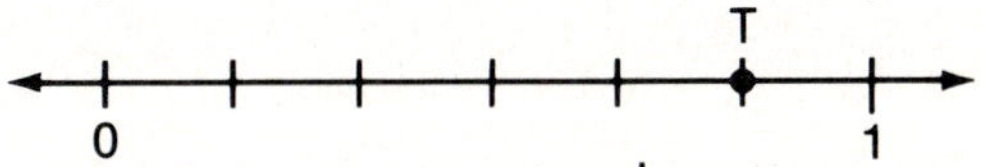

17. 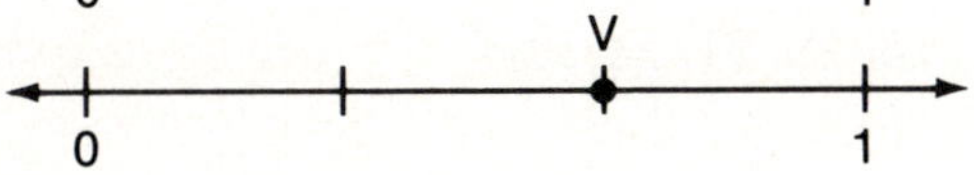

18. 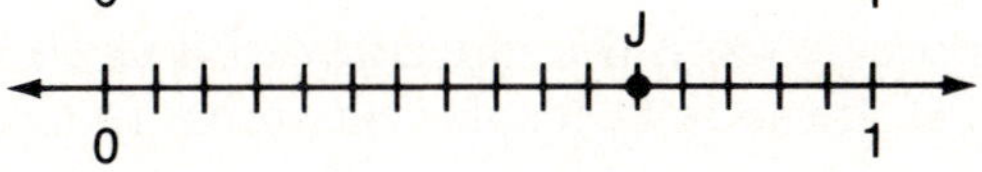

Solve.

19. There are six compartments in a freezer. If five of the compartments are filled, what fraction of the freezer is filled?

20. A pizza costs $5.99. Each extra topping costs $0.88. How much will it cost to buy a pizza with meatballs, peppers, and onions?

21. James ordered a pizza that was divided into 10 equal slices. Three of the slices were topped with mushrooms. What fraction of the pizza did not have mushrooms?

22. Ellen has nine friends at a party. She plans to serve each friend three slices of pizza. If each pizza has eight slices, how many pizzas does she need?

TIME OUT

The world's largest pizza was baked in 1978 at the Oma Pizza Restaurant in Glen Falls, New York. It weighed 18,664 pounds and was cut into 60,318 slices. About how many years would it take one very hungry person to eat the entire pie if that person eats three slices every day?

5.2 GCF AND LCM

To find the factors of 16, list all the numbers which when multiplied give you 16.

$16 = 1 \times 16 \qquad 16 = 2 \times 8 \qquad 16 = 4 \times 4$

So, the factors of 16 are: 1, 2, 4, 8, 16.

Some numbers have common factors. The greatest of these common factors is called the **greatest common factor, GCF.**

EXAMPLE 1 Find the greatest common factor (GCF) of 16 and 24.

1. List all the factors of each number.
   16: 1, 2, 4, 8, 16
   24: 1, 2, 3, 4, 6, 8, 12, 24
2. Identify the common factors. 1, 2, 4, 8
3. Choose the greatest common factor. 8

So, 8 is the greatest common factor (GCF) of 16 and 24.

The product of any number and 3 is a **multiple** of 3. Here are some multiples of 3.

$0 \times 3 = 0$	$3 \times 3 = 9$	$6 \times 3 = 18$
$1 \times 3 = 3$	$4 \times 3 = 12$	$7 \times 3 = 21$
$2 \times 3 = 6$	$5 \times 3 = 15$	$8 \times 3 = 24$

You can write the multiples of 3 as: 0, 3, 6, 9, 12, 15, 18, 21, 24, ...
(The three dots mean *and so on.*)

Some numbers have common multiples. The least of these common multiples, other than 0, is called the **least common multiple, LCM.**

EXAMPLE 2 Find the least common multiple (LCM) of 3 and 4.

1. List some multiples of each number.
   3: 0, 3, 6, 9, 12, 15, 18, 21, 24, . . .
   4: 0, 4, 8, 12, 16, 20, 24, . . .
2. Identify the common multiples. 0, 12, 24, . . .
3. Choose the least common multiple, other than 0. 12

So, 12 is the least common multiple (LCM) of 3 and 4.

CHECKPOINT

Write the letter of the correct answer.

1. Find the greatest common factor (GCF) of 24 and 32.
a. 4 **b.** 8 **c.** 12 **d.** 96

2. Find the least common multiple (LCM) of 24 and 32.
a. 8 **b.** 48 **c.** 64 **d.** 96

PRACTICE EXERCISES

Use the table to determine the greatest common factor (GCF).

Number	Factors	Number	Factors
6	1, 2, 3, 6	16	1, 2, 4, 8, 16
8	1, 2, 4, 8	18	1, 2, 3, 6, 9, 18
9	1, 3, 9	24	1, 2, 3, 4, 6, 8, 12, 24
12	1, 2, 3, 4, 6, 12	36	1, 2, 3, 4, 6, 9, 12, 18, 36

1. 6, 8 **2.** 8, 12 **3.** 9, 18 **4.** 6, 16 **5.** 8, 18
6. 9, 24 **7.** 12, 6 **8.** 18, 24 **9.** 12, 18 **10.** 24, 36

Find the greatest common factor (GCF).

11. 4, 6 **12.** 9, 12 **13.** 5, 7 **14.** 4, 12 **15.** 10, 15
16. 12, 14 **17.** 18, 30 **18.** 12, 20 **19.** 15, 21 **20.** 14, 35
21. 11, 21 **22.** 24, 32 **23.** 18, 27 **24.** 24, 42 **25.** 27, 45
26. 56, 72 **27.** 36, 54 **28.** 42, 60 **29.** 15, 75 **30.** 56, 64

Use the table to determine the least common multiple (LCM).

Number	Multiples	Number	Multiples
2	0, 2, 4, 6, 8, 10, 12, 14, . . .	8	0, 8, 16, 24, 32, 40, 48, 56, . . .
3	0, 3, 6, 9, 12, 15, 18, 21, . . .	9	0, 9, 18, 27, 36, 45, 54, 63, . . .
4	0, 4, 8, 12, 16, 20, 24, 28, . . .	12	0, 12, 24, 36, 48, 60, 72, 84, . . .
5	0, 5, 10, 15, 20, 25, 30, 35, . . .	16	0, 16, 32, 48, 64, 80, 96, 112, . . .

31. 2, 3 **32.** 2, 4 **33.** 3, 5 **34.** 3, 9 **35.** 2, 5
36. 4, 5 **37.** 9, 12 **38.** 8, 12 **39.** 8, 16 **40.** 12, 16

Find the least common multiple (LCM).

41. 4, 6 **42.** 3, 15 **43.** 4, 3 **44.** 6, 9 **45.** 9, 15
46. 14, 21 **47.** 12, 15 **48.** 9, 12 **49.** 24, 8 **50.** 5, 6
51. 12, 18 **52.** 27, 18 **53.** 15, 20 **54.** 12, 16 **55.** 10, 12
56. 24, 36 **57.** 20, 25 **58.** 28, 21 **59.** 15, 18 **60.** 64, 48

5.3 EQUIVALENT FRACTIONS

Equivalent fractions name the same amount using different terms.
$\frac{2}{3}$, $\frac{4}{6}$, and $\frac{8}{12}$ are equivalent fractions.

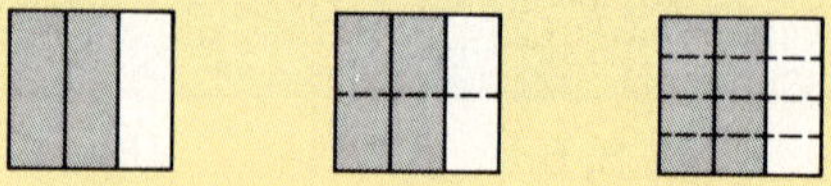

$$\frac{2}{3} = \frac{4}{6} = \frac{8}{12}$$

You can find an equivalent fraction by multiplying or dividing the numerator and denominator of a fraction by the same nonzero number.

To find fractions equivalent to $\frac{4}{6}$, multiply or divide the numerator and denominator by 2.

Multiplying by 2	**Dividing by 2**
$\frac{4}{6} = \frac{4 \times 2}{6 \times 2} = \frac{8}{12}$	$\frac{4}{6} = \frac{4 \div 2}{6 \div 2} = \frac{2}{3}$

$\frac{2}{3}$ is a fraction in lowest terms. The greatest common factor (GCF) of the numerator and denominator is 1. You can write a fraction in lowest terms by dividing the numerator and denominator by the GCF.

EXAMPLE 1 Write $\frac{10}{25}$ in lowest terms.

THINK: GCF of 10 and 25 is 5. Divide by 5. $\frac{10 \div 5}{25 \div 5} = \frac{2}{5}$

So, $\frac{10}{25}$ in lowest terms is $\frac{2}{5}$.

Sometimes you need to find the missing term of two equivalent fractions.

EXAMPLE 2 Complete: $\frac{3}{4} = \frac{\blacksquare}{12}$

THINK: $4 \times \blacksquare = 12$
$4 \times 3 = 12$

Multiply the numerator by 3.

$$\frac{3}{4} \overset{\times 3}{=} \frac{9}{12} \quad (\times 3)$$

So, 9 is the missing term.

You can compare cross products to determine if two fractions are equivalent.

EXAMPLE 3 Using cross products, determine if the fractions are equivalent.

a. $\frac{3}{5}$ and $\frac{9}{12}$

$\frac{3}{5} \times \frac{9}{12}$ **THINK:** $3 \times 12 = 36$
$9 \times 5 = 45$
$36 \neq 45$

So, $\frac{3}{5}$ and $\frac{9}{12}$ are not equivalent.

b. $\frac{4}{8}$ and $\frac{7}{14}$

$\frac{4}{8} \times \frac{7}{14}$ **THINK:** $4 \times 14 = 56$
$7 \times 8 = 56$
$56 = 56$

So, $\frac{4}{8}$ and $\frac{7}{14}$ are equivalent.

FOR DISCUSSION

Is it possible for a fraction to be in lowest terms if the numerator and denominator are both even numbers? Explain.

PRACTICE EXERCISES

Write the fraction in lowest terms.

1. $\frac{3}{9}$ **2.** $\frac{9}{18}$ **3.** $\frac{8}{24}$ **4.** $\frac{7}{21}$ **5.** $\frac{8}{12}$ **6.** $\frac{4}{14}$ **7.** $\frac{12}{15}$ **8.** $\frac{6}{8}$

9. $\frac{12}{18}$ **10.** $\frac{10}{15}$ **11.** $\frac{15}{18}$ **12.** $\frac{18}{24}$ **13.** $\frac{12}{16}$ **14.** $\frac{9}{24}$ **15.** $\frac{14}{21}$ **16.** $\frac{9}{15}$

17. $\frac{12}{21}$ **18.** $\frac{20}{35}$ **19.** $\frac{18}{21}$ **20.** $\frac{15}{20}$ **21.** $\frac{16}{24}$ **22.** $\frac{15}{21}$ **23.** $\frac{21}{28}$ **24.** $\frac{20}{24}$

25. $\frac{12}{36}$ **26.** $\frac{24}{36}$ **27.** $\frac{24}{27}$ **28.** $\frac{15}{24}$ **29.** $\frac{16}{20}$ **30.** $\frac{12}{28}$ **31.** $\frac{36}{45}$ **32.** $\frac{50}{75}$

Complete.

33. $\frac{1}{3} = \frac{\blacksquare}{9}$ **34.** $\frac{3}{5} = \frac{\blacksquare}{10}$ **35.** $\frac{2}{3} = \frac{\blacksquare}{6}$ **36.** $\frac{4}{7} = \frac{\blacksquare}{21}$ **37.** $\frac{5}{6} = \frac{\blacksquare}{18}$

38. $\frac{3}{4} = \frac{\blacksquare}{16}$ **39.** $\frac{7}{8} = \frac{\blacksquare}{24}$ **40.** $\frac{8}{9} = \frac{\blacksquare}{27}$ **41.** $\frac{6}{9} = \frac{12}{\blacksquare}$ **42.** $\frac{5}{7} = \frac{35}{\blacksquare}$

43. $\frac{12}{15} = \frac{\blacksquare}{5}$ **44.** $\frac{9}{18} = \frac{\blacksquare}{6}$ **45.** $\frac{16}{48} = \frac{\blacksquare}{12}$ **46.** $\frac{24}{36} = \frac{\blacksquare}{9}$ **47.** $\frac{18}{24} = \frac{\blacksquare}{8}$

48. $\frac{20}{32} = \frac{\blacksquare}{16}$ **49.** $\frac{16}{40} = \frac{\blacksquare}{10}$ **50.** $\frac{30}{54} = \frac{\blacksquare}{18}$ **51.** $\frac{12}{42} = \frac{4}{\blacksquare}$ **52.** $\frac{24}{48} = \frac{3}{\blacksquare}$

Using cross products, determine if the fractions are equivalent. Write $=$ or $\neq$.

53. $\frac{2}{3}$ ● $\frac{6}{9}$ **54.** $\frac{3}{4}$ ● $\frac{9}{16}$ **55.** $\frac{1}{5}$ ● $\frac{4}{20}$ **56.** $\frac{5}{6}$ ● $\frac{10}{12}$ **57.** $\frac{8}{15}$ ● $\frac{4}{5}$

58. $\frac{9}{12}$ ● $\frac{15}{20}$ **59.** $\frac{12}{20}$ ● $\frac{15}{25}$ **60.** $\frac{8}{12}$ ● $\frac{12}{18}$ **61.** $\frac{10}{16}$ ● $\frac{18}{24}$ **62.** $\frac{12}{21}$ ● $\frac{20}{35}$

Solve. Use the picture at the right.

63. Alice Tompkins bought several jars of baby food. What fraction of the jars are filled with pears? Give two equivalent fractions.

64. What fraction of the jars are filled with vegetables? Give three equivalent fractions.

5.4 COMPARING FRACTIONS

Bill and Carol were asked to paint school murals. Their murals were to be the same size. After a week, Bill had completed $\frac{9}{16}$ of his mural, and Carol had completed $\frac{5}{16}$ of her mural. Who had completed a greater portion of their mural?

$\frac{9}{16}$ $\frac{5}{16}$

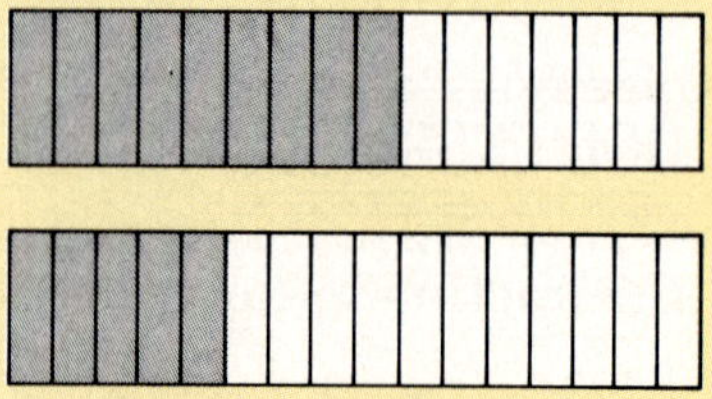

You can compare two fractions that have the same denominator by comparing their numerators.

EXAMPLE 1 Compare: $\frac{9}{16}$ and $\frac{5}{16}$

THINK: Since $9 > 5$, $\frac{9}{16} > \frac{5}{16}$.

Bill painted a greater portion of his mural.

You can compare two fractions that have different denominators.

EXAMPLE 2 Compare: $\frac{3}{8}$ and $\frac{2}{3}$

1. Write equivalent fractions using a common denominator. $\frac{3}{8} = \frac{3 \times 3}{8 \times 3} = \frac{9}{24}$ $\frac{2}{3} = \frac{2 \times 8}{3 \times 8} = \frac{16}{24}$

THINK: The LCM of 8 and 3 is 24.
Use 24 as the denominator.

2. Compare the numerators. Since $9 < 16$, $\frac{9}{24} < \frac{16}{24}$.

So, $\frac{3}{8} < \frac{2}{3}$.

Another way you can compare fractions is by comparing cross products.

EXAMPLE 3 Compare: $\frac{7}{9}$ and $\frac{5}{8}$

1. Find the first cross product. $\frac{7}{9} \searrow \frac{5}{8}$ $7 \times 8 = 56$

2. Find the second cross product. $\frac{7}{9} \nwarrow\!\!\!\!\swarrow \frac{5}{8}$ $5 \times 9 = 45$

3. Compare cross products. Since $56 > 45$, $\frac{7}{9} > \frac{5}{8}$.

So, $\frac{7}{9} > \frac{5}{8}$.

You can order three or more fractions by comparing fractions two at a time.

EXAMPLE 4 Order $\frac{2}{5}$, $\frac{1}{4}$, and $\frac{5}{8}$ from greatest to least.

1. Write equivalent fractions using a common denominator. $\frac{2}{5} = \frac{16}{40}$ $\frac{1}{4} = \frac{10}{40}$ $\frac{5}{8} = \frac{25}{40}$

2. Compare. $\frac{16}{40} > \frac{10}{40}$ $\frac{25}{40} > \frac{16}{40}$

3. List the equivalent fractions in order from greatest to least. $\frac{25}{40}, \frac{16}{40}, \frac{10}{40}$

4. List the original fractions in order from greatest to least. $\frac{5}{8}, \frac{2}{5}, \frac{1}{4}$

FOR DISCUSSION

Look at the following set of fractions: $\frac{1}{4}, \frac{1}{8}, \frac{1}{16}$.
Can you identify the greatest fraction by just looking? Explain.

PRACTICE EXERCISES

Compare using equivalent fractions. Write =, <, or >.

1. $\frac{3}{5}$ ● $\frac{4}{5}$ **2.** $\frac{7}{9}$ ● $\frac{15}{9}$ **3.** $\frac{9}{10}$ ● $\frac{3}{10}$ **4.** $\frac{16}{25}$ ● $\frac{24}{25}$ **5.** $\frac{11}{12}$ ● $\frac{3}{12}$

6. $\frac{4}{6}$ ● $\frac{2}{3}$ **7.** $\frac{3}{8}$ ● $\frac{1}{4}$ **8.** $\frac{7}{10}$ ● $\frac{3}{5}$ **9.** $\frac{5}{6}$ ● $\frac{11}{12}$ **10.** $\frac{3}{5}$ ● $\frac{9}{15}$

11. $\frac{1}{8}$ ● $\frac{1}{3}$ **12.** $\frac{1}{4}$ ● $\frac{1}{5}$ **13.** $\frac{2}{3}$ ● $\frac{1}{5}$ **14.** $\frac{1}{2}$ ● $\frac{3}{7}$ **15.** $\frac{4}{5}$ ● $\frac{7}{8}$

16. $\frac{4}{9}$ ● $\frac{5}{6}$ **17.** $\frac{2}{9}$ ● $\frac{4}{5}$ **18.** $\frac{5}{7}$ ● $\frac{2}{3}$ **19.** $\frac{2}{9}$ ● $\frac{3}{8}$ **20.** $\frac{9}{10}$ ● $\frac{3}{4}$

Compare using cross products. Write =, <, or >.

21. $\frac{4}{7}$ ● $\frac{5}{8}$ **22.** $\frac{7}{8}$ ● $\frac{8}{9}$ **23.** $\frac{3}{5}$ ● $\frac{12}{20}$ **24.** $\frac{8}{16}$ ● $\frac{5}{10}$ **25.** $\frac{4}{9}$ ● $\frac{3}{8}$

26. $\frac{5}{7}$ ● $\frac{8}{11}$ **27.** $\frac{9}{12}$ ● $\frac{3}{4}$ **28.** $\frac{2}{3}$ ● $\frac{17}{25}$ **29.** $\frac{7}{9}$ ● $\frac{10}{12}$ **30.** $\frac{11}{15}$ ● $\frac{7}{10}$

31. $\frac{9}{10}$ ● $\frac{17}{20}$ **32.** $\frac{7}{10}$ ● $\frac{30}{100}$ **33.** $\frac{2}{11}$ ● $\frac{3}{10}$ **34.** $\frac{11}{20}$ ● $\frac{17}{30}$ **35.** $\frac{15}{16}$ ● $\frac{4}{5}$

Arrange in order from greatest to least.

36. $\frac{3}{8}, \frac{7}{8}, \frac{5}{8}$

37. $\frac{7}{10}, \frac{4}{10}, \frac{8}{10}$

38. $\frac{1}{3}, \frac{5}{6}, \frac{7}{12}$

39. $\frac{2}{5}, \frac{9}{15}, \frac{1}{3}$

40. $\frac{1}{2}, \frac{3}{8}, \frac{5}{6}$

41. $\frac{7}{8}, \frac{3}{5}, \frac{9}{10}$

42. $\frac{3}{6}, \frac{3}{4}, \frac{3}{8}, \frac{3}{12}$

43. $\frac{1}{2}, \frac{7}{9}, \frac{2}{3}, \frac{5}{6}$

Arrange in order from least to greatest.

44. $\frac{1}{6}, \frac{0}{6}, \frac{5}{6}$

45. $\frac{17}{24}, \frac{9}{24}, \frac{11}{24}$

46. $\frac{1}{2}, \frac{3}{5}, \frac{7}{10}$

47. $\frac{2}{3}, \frac{7}{8}, \frac{11}{24}$

48. $\frac{1}{3}, \frac{2}{5}, \frac{1}{6}$

49. $\frac{5}{6}, \frac{10}{11}, \frac{2}{3}$

50. $\frac{5}{6}, \frac{5}{9}, \frac{5}{8}, \frac{5}{12}$

51. $\frac{1}{2}, \frac{3}{4}, \frac{4}{5}, \frac{7}{12}$

MIXED REVIEW

Find the answer.

52. 4,329 + 6,820

53. 617 + 4,832 + 9,590

54. 81.6 + 21.9

55. 36.25 + 18.95

56. 8,639 − 259

57. 16,007 − 9,625

58. 29.8 − 16.9

59. 84.63 − 29.87

60. 500 × 300

61. 436 × 28

62. 16.8 × 24

63. 36.83 × 4.9

64. 72,000 ÷ 90

65. 3,264 ÷ 8

66. 88.2 ÷ 18

67. 22.31 ÷ 2.3

Solve.

68. On Monday, Carol used $\frac{1}{3}$ can of red paint and $\frac{1}{2}$ can of blue paint. Did Carol use more red paint or blue paint?

69. Nails at least $\frac{3}{4}$ inches long are needed for a construction project. Can nails that are $\frac{5}{8}$ inches in length be used?

70. Bill used a can of paint for every 210 square feet of the mural. If the mural covers 472.5 square feet, how many cans of paint did Bill use? Write your answer as a decimal.

71. Of 75 students surveyed, 65 approved of the school murals. What fraction of the students approved of the murals? Write the answer in lowest terms.

72. To frame each mural, Sam used 90.25 feet of wood and 26 nails. How much wood does Sam need for two murals?

73. Bill was asked to paint a mural for the town. He was paid $8.50 an hour. If the mural took him 40.5 hours, how much money did Bill earn?

MIDCHAPTER REVIEW

Write a fraction for the shaded part. [122]

1. 

2. 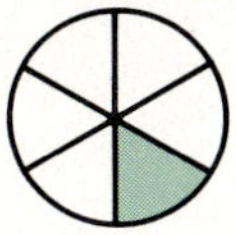

3. 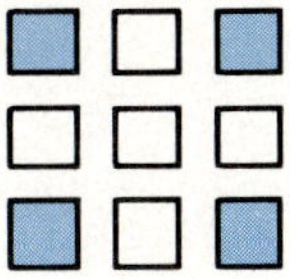

4. 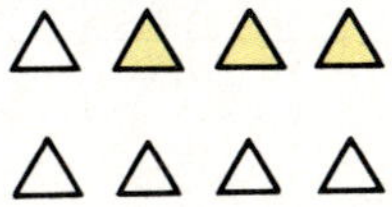

Write a fraction for the point. [122]

5. 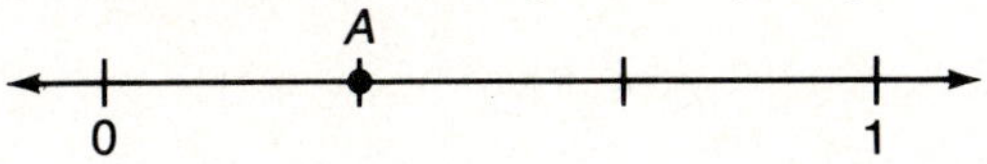

6. 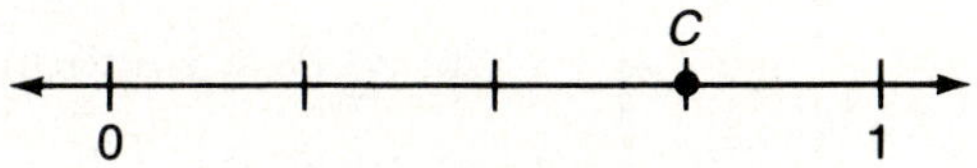

7. 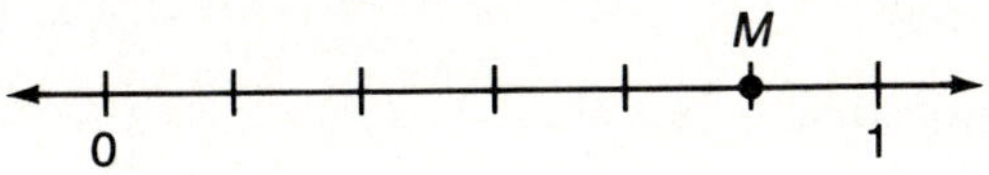

8. 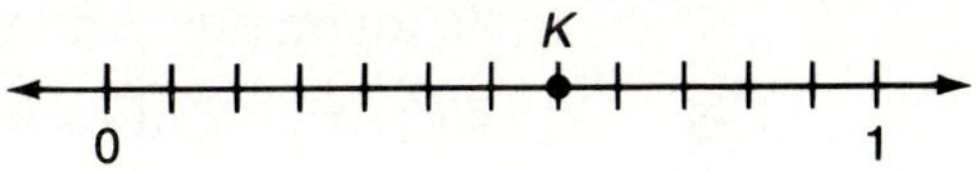

Find the greatest common factor. [124]

9. 7 and 14 **10.** 6 and 8 **11.** 14 and 16 **12.** 16, 20, and 28

Find the least common multiple. [124]

13. 2 and 4 **14.** 6 and 9 **15.** 8 and 18 **16.** 5, 9, and 15

Write the fraction in lowest terms. [126]

17. $\frac{3}{9}$ **18.** $\frac{7}{21}$ **19.** $\frac{8}{36}$ **20.** $\frac{10}{12}$ **21.** $\frac{21}{30}$ **22.** $\frac{54}{81}$

Complete. [126]

23. $\frac{1}{4} = \frac{■}{12}$ **24.** $\frac{1}{3} = \frac{3}{■}$ **25.** $\frac{9}{15} = \frac{■}{5}$ **26.** $\frac{35}{63} = \frac{■}{9}$

Arrange in order from greatest to least. [128]

27. $\frac{3}{10}, \frac{7}{10}, \frac{1}{10}$ **28.** $\frac{5}{12}, \frac{5}{6}, \frac{5}{9}$ **29.** $\frac{3}{4}, \frac{1}{22}, \frac{5}{8}$ **30.** $\frac{1}{4}, \frac{7}{16}, \frac{3}{8}, \frac{1}{2}$

Solve. [122, 128]

31. Susan earned \$48 last weekend. She spent \$8 on an album. What fraction of her earnings did she spend on the album? Write the answer in simplest form.

32. Joseph collects sheet music. $\frac{1}{4}$ of his collection is pop, $\frac{5}{12}$ is jazz, and $\frac{1}{3}$ is rock. Which type of music makes up the largest part of his collection?

5.5 MIXED NUMBERS

A **mixed number** consists of a whole number and a fraction.

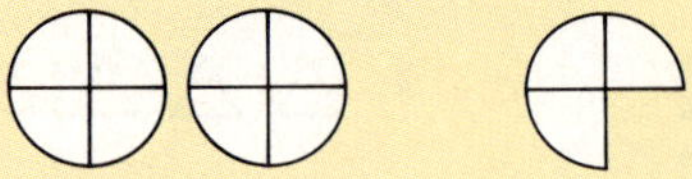

$2 \quad + \quad \frac{3}{4} \quad = \quad 2\frac{3}{4}$

You can write a mixed number as a fraction.

EXAMPLE 1

Write $2\frac{3}{4}$ as a fraction.

1. Multiply the whole number and the denominator. Add the numerator. $2\frac{3}{4}$ **THINK: $4 \times 2 = 8$** **$8 + 3 = 11$**
2. Write the sum over the denominator. $\frac{11}{4}$

So, $2\frac{3}{4} = \frac{11}{4}$.

Fractions with numerators equal to or greater than the denominators are called **improper fractions.** You can write an improper fraction as a mixed number.

EXAMPLE 2

Write $\frac{28}{8}$ as a mixed number in simplest form.

1. Divide the numerator by the denominator. Write the remainder as a fraction.

$$\begin{array}{r} 3 \\ 8\overline{)28} \\ \underline{24} \\ 4 \end{array}$$

$3\frac{4}{8}$ **THINK: 4 is the remainder. 8 is the divisor.**

2. Write the fraction in lowest terms. $3\frac{4}{8} = 3\frac{1}{2}$ ← **In simplest form**

So, $\frac{28}{8} = 3\frac{1}{2}$.

CHECKPOINT

Write the letter of the correct answer.

Complete. Choose the answer written in simplest form.

1. $3\frac{2}{3} =$ ■ **a.** $\frac{18}{3}$ **b.** $\frac{9}{3}$ **c.** $\frac{15}{3}$ **d.** $\frac{11}{3}$

2. $4\frac{1}{8} =$ ■ **a.** $\frac{33}{8}$ **b.** $\frac{8}{8}$ **c.** $\frac{12}{8}$ **d.** $\frac{3}{18}$

3. $\frac{13}{5} =$ ■ **a.** $2\frac{1}{5}$ **b.** $3\frac{1}{5}$ **c.** $2\frac{3}{5}$ **d.** $5\frac{3}{5}$

4. $\frac{16}{6} =$ ■ **a.** $2\frac{4}{6}$ **b.** $3\frac{3}{4}$ **c.** $2\frac{2}{3}$ **d.** $3\frac{1}{2}$

PRACTICE EXERCISES

Write as an improper fraction.

1. $2\frac{1}{3}$ **2.** $1\frac{2}{3}$ **3.** $3\frac{1}{2}$ **4.** $1\frac{3}{4}$ **5.** $2\frac{3}{5}$ **6.** $1\frac{5}{8}$ **7.** $2\frac{4}{7}$ **8.** $5\frac{2}{5}$

9. $7\frac{4}{5}$ **10.** $4\frac{3}{7}$ **11.** $6\frac{5}{6}$ **12.** $5\frac{2}{9}$ **13.** $3\frac{2}{3}$ **14.** $2\frac{3}{4}$ **15.** $4\frac{3}{8}$ **16.** $3\frac{4}{5}$

17. $5\frac{1}{3}$ **18.** $2\frac{4}{9}$ **19.** $2\frac{3}{8}$ **20.** $3\frac{5}{7}$ **21.** $5\frac{7}{9}$ **22.** $3\frac{5}{8}$ **23.** $4\frac{2}{3}$ **24.** $3\frac{3}{8}$

25. $6\frac{3}{4}$ **26.** $2\frac{5}{6}$ **27.** $4\frac{7}{8}$ **28.** $5\frac{3}{8}$ **29.** $17\frac{2}{3}$ **30.** $13\frac{5}{9}$ **31.** $15\frac{6}{7}$ **32.** $16\frac{2}{3}$

Write as a whole number or mixed number in simplest form.

33. $\frac{3}{2}$ **34.** $\frac{4}{3}$ **35.** $\frac{6}{2}$ **36.** $\frac{5}{3}$ **37.** $\frac{7}{4}$ **38.** $\frac{8}{3}$ **39.** $\frac{6}{5}$ **40.** $\frac{10}{4}$

41. $\frac{7}{2}$ **42.** $\frac{11}{6}$ **43.** $\frac{12}{5}$ **44.** $\frac{16}{8}$ **45.** $\frac{15}{9}$ **46.** $\frac{14}{3}$ **47.** $\frac{20}{9}$ **48.** $\frac{46}{8}$

49. $\frac{36}{10}$ **50.** $\frac{85}{7}$ **51.** $\frac{63}{9}$ **52.** $\frac{25}{8}$ **53.** $\frac{32}{6}$ **54.** $\frac{18}{5}$ **55.** $\frac{28}{8}$ **56.** $\frac{39}{9}$

57. $\frac{17}{3}$ **58.** $\frac{21}{7}$ **59.** $\frac{30}{4}$ **60.** $\frac{27}{6}$ **61.** $\frac{45}{8}$ **62.** $\frac{39}{5}$ **63.** $\frac{22}{8}$ **64.** $\frac{38}{3}$

MIXED REVIEW

Here are Stephanie's science grades in the second semester.

Test 1: 85 Test 2: 88 Test 3: 92 Test 4: 78 Test 5: 92

Compute.

65. Range = ■ **66.** Median = ■ **67.** Mode = ■ **68.** Mean = ■

Use the list of ingredients to solve.

69. Jon has a $\frac{1}{4}$ cup for measuring. How many $\frac{1}{4}$ cups of wheat flour does he need?

70. How many $\frac{1}{4}$ cups of honey does he need?

71. Which is greater, the amount of water or the amount of honey?

HONEY OATMEAL BREAD

1 c oats	$1\frac{3}{4}$ c sifted wheat flour
1 stick butter	1 t baking soda
$1\frac{1}{4}$ c boiling water	$\frac{3}{4}$ t salt
$1\frac{1}{2}$ c honey	1 t cinnamon
1 t vanilla	$\frac{1}{4}$ t nutmeg
2 eggs	

PROBLEM *Solving* STRATEGY

5.6 INTERPRETING REMAINDERS

Situation:

The first and largest Ferris Wheel was built in Chicago in 1893. It had 36 cars. Each car held 60 people. There were 372 people waiting to go on the Ferris Wheel.

Strategy:

The same division example can be used to answer different questions about a situation. Make sure that you answer the question that is asked.

Applying the Strategy:

This division example can be used to answer different questions about seating people on the Ferris Wheel.

$$\begin{array}{r} 6\text{ R}12 \\ 60\overline{)372} \\ \underline{360} \\ 12 \end{array}$$

Notice the different answers to each of the questions below.

A. How many cars would have been filled by 372 people?
Use only the quotient. — 6 cars would have been filled.

B. How many people would have been in the last car?
Use only the remainder. — 12 people would have been in the last car.

C. How many cars would have been needed to hold 372 people?
Raise the quotient by 1. — **THINK:** 6 full cars + 1 car with 12 people.
7 cars would have been needed.

D. What fraction of the last car would have been filled?
Use a fractional remainder. — **THINK:** **People in last car →** / **People in full car →** $\frac{12}{60} = \frac{1}{5}$

$\frac{1}{5}$ of the last car would have been filled.

PRACTICE EXERCISES

Write the letter of the correct answer.

1. The Whirly Bird ride holds 25 people. If 185 people are taking a ride, how many people will go on the last trip?
a. Use only the quotient.
b. Use only the remainder.
c. Raise the quotient by 1.

2. There are 39 people who want to ride the Safari Train. If 9 people can ride in each car, how many cars are needed?
a. Use only the quotient.
b. Use only the remainder.
c. Raise the quotient by 1.

Solve.

Rose sold sweatshirts at an amusement park. She packed 20 sweatshirts in each carton. She had 185 sweatshirts to pack.

3. How many cartons did Rose fill?

4. How many sweatshirts did she put into the last partially filled carton?

5. How many cartons did Rose use to pack 185 sweatshirts?

6. What fraction of the last carton was filled?

Jeff sold lemonade at an amusement park. He made 1 qt (32 oz) at a time. He served the lemonade in 6-oz cups.

7. How many cups could Jeff fill with 1 qt of lemonade?

8. How many ounces of lemonade were poured into the last cup?

9. How many cups were needed to serve the quart of lemonade?

10. What fraction of the last cup was filled?

11. One paper cup is needed for each of 115 people at the Snack Bar. If a package contains 20 cups, how many packages are needed?

12. Bob has $20 to spend on tickets for rides at the amusement park. How many tickets can he buy if each ticket costs $3?

5.7 FRACTIONS AND DECIMALS

You can write a decimal as a fraction or a mixed number.

EXAMPLE 1 Write as a fraction or mixed number in simplest form.

a. 0.08

THINK: 0.08 means 8 hundredths

$0.08 = \frac{8}{100} = \frac{2}{25}$

b. 2.3

THINK: 2.3 means 2 and 3 tenths

$2.3 = 2\frac{3}{10}$

You can write a fraction as a decimal by dividing the numerator by the denominator.

EXAMPLE 2 Write $\frac{3}{4}$ as a decimal.

$$\frac{3}{4} \rightarrow 4\overline{)3.00} = 0.75$$

```
   0.75
4)3.00
  2 8
    20
    20
     0
```

THINK: Additional zeros are needed to complete the division.

So, $\frac{3}{4} = 0.75$.

You can also write a mixed number as a decimal.

EXAMPLE 3 Write $2\frac{1}{8}$ as a decimal.

1. Write the fraction as a decimal. $\frac{1}{8} \rightarrow 8\overline{)1.000} = 0.125$
2. Add the whole number. $2 + 0.125 = 2.125$

So, $2\frac{1}{8} = 2.125$.

Sometimes when you divide to change a fraction to a decimal, the remainder will never be zero. These are called **nonterminating decimals.**

EXAMPLE 4 Write $\frac{2}{9}$ as a decimal.

```
  0.222...
9)2.000
  1 8
    20
    18
     20
     18
      2
```

Notice that the remainder will never be 0.
This answer can be written in two ways:

$\frac{2}{9} = 0.22$ when rounded to the nearest hundredth.

$\frac{2}{9} = 0.\overline{2}$ which means the digit, 2, repeats.

CHECKPOINT Write the letter of the correct answer.

1. What is 2.82 as a mixed number in simplest form?

a. $2\frac{41}{50}$ **b.** $\frac{41}{50}$ **c.** $\frac{282}{100}$ **d.** $2\frac{82}{100}$

2. What is $\frac{3}{5}$ as a decimal?

a. 1.6 **b.** 0.06 **c.** 0.6 **d.** 0.006

3. What is $\frac{3}{7}$ as a decimal rounded to the nearest hundredth?

a. 0.043 **b.** 4.28 **c.** 42.85 **d.** 0.43

PRACTICE EXERCISES

Write as a fraction or mixed number in simplest form.

1. 0.3 **2.** 0.6 **3.** 0.5 **4.** 0.8 **5.** 0.9 **6.** 2.7

7. 3.4 **8.** 4.2 **9.** 0.12 **10.** 0.25 **11.** 0.35 **12.** 0.56

13. 0.75 **14.** 2.47 **15.** 8.05 **16.** 3.75 **17.** 0.221 **18.** 0.792

19. 0.125 **20.** 0.075 **21.** 0.625 **22.** 4.375 **23.** 5.025 **24.** 2.085

Write the fraction or mixed number as a decimal.

25. $\frac{1}{2}$ **26.** $\frac{3}{4}$ **27.** $\frac{4}{5}$ **28.** $\frac{1}{4}$ **29.** $\frac{2}{5}$ **30.** $\frac{7}{10}$ **31.** $\frac{1}{5}$ **32.** $\frac{1}{20}$

33. $1\frac{3}{5}$ **34.** $2\frac{1}{4}$ **35.** $5\frac{9}{10}$ **36.** $3\frac{1}{5}$ **37.** $8\frac{3}{4}$ **38.** $6\frac{4}{5}$ **39.** $11\frac{1}{10}$ **40.** $6\frac{1}{50}$

41. $\frac{1}{8}$ **42.** $\frac{3}{8}$ **43.** $2\frac{7}{8}$ **44.** $\frac{6}{15}$ **45.** $9\frac{5}{8}$ **46.** $\frac{5}{16}$ **47.** $4\frac{1}{16}$ **48.** $10\frac{5}{16}$

Write the fraction or mixed number as a decimal. Round to the nearest hundredth.

49. $\frac{1}{3}$ **50.** $\frac{5}{6}$ **51.** $\frac{2}{7}$ **52.** $\frac{4}{9}$ **53.** $4\frac{3}{14}$ **54.** $7\frac{5}{9}$ **55.** $6\frac{5}{12}$ **56.** $9\frac{7}{15}$

Write the fraction or mixed number as a decimal. Use the bar notation.

57. $\frac{1}{3}$ **58.** $\frac{7}{9}$ **59.** $\frac{6}{11}$ **60.** $\frac{2}{3}$ **61.** $2\frac{2}{9}$ **62.** $4\frac{6}{11}$ **63.** $5\frac{1}{3}$ **64.** $7\frac{4}{9}$

Solve.

65. Ellen Travers wants to make hamburgers for her party. She needs $2\frac{1}{2}$ pounds of meat. She buys 2.75 pounds. Does she have enough hamburger?

66. Ellen is making fruit salad for the party. She buys $2\frac{1}{2}$ pounds of bananas and and 2.25 pounds of grapes. Does she buy more bananas or grapes?

PROBLEM *Solving* APPLICATION

5.8 CONSUMER: BUDGETING

A **budget** is a plan for using money. The Williams family decided to control their expenses by making and following a budget. This circle graph shows how they decided to allot their family income after taxes.

Use the circle graph to answer Exercises 1–12.

THE WILLIAMS' BUDGET

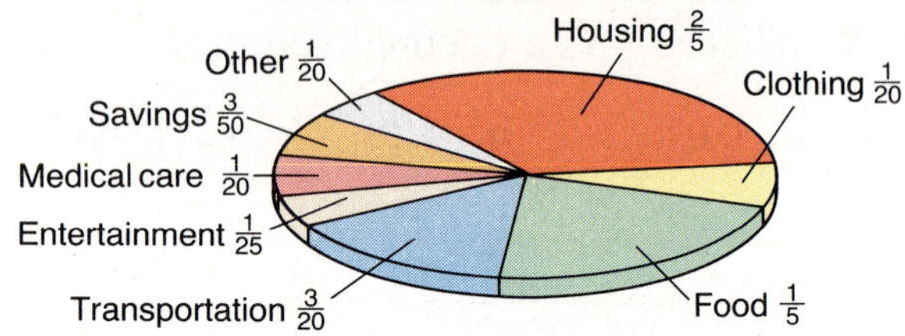

What part of the budget is allotted to:

1. housing?
2. food?
3. clothing?
4. savings?

Did they allot more money to:

5. transportation or clothing?
6. entertainment or medical care?
7. food or clothing?
8. savings or entertainment?
9. For which type of expense did they allot the greatest part of their budget?

Which expense represents:

10. $\frac{40}{100}$ of the budget?
11. $\frac{4}{100}$ of the budget?
12. 0.15 of the budget?

Solve.

13. The Bartner family made a budget. They allotted their income as follows: food, 0.16; housing, 0.30; clothing, 0.06; savings, 0.125; transportation, 0.20; entertainment, 0.05; medical care, 0.05; other, 0.055. For which types of expense did they allot more than 0.15?
14. Rewrite the Bartners' budget using fractions in lowest terms.
15. Suppose you wanted to arrange the Bartner budget items in order from greatest to least. Do you think it would be easier to use decimals or fractions? Why?

PROBLEM Solving APPLICATION

5.9 CAREER: ADVERTISING EXECUTIVE

Michelle Morales works for a small advertising agency. She is the Account Executive for a car manufacturer. It is her job to make and present an advertising campaign budget for the new Aurora sports car.

Michelle used a **divided-bar graph** in her presentation to the client.

Aurora Advertising Campaign

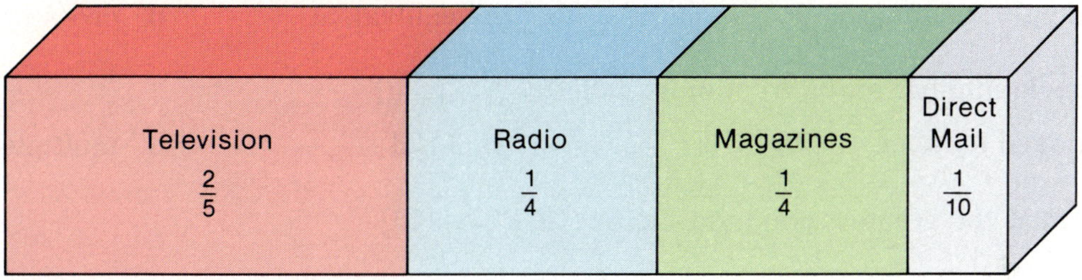

Use the divided-bar graph to answer Exercises 1–10.

What part of the advertising budget is devoted to:

1. radio? **2.** television? **3.** magazines? **4.** direct mail?

5. Which type of advertising received the smallest part of the budget?

6. Which types of advertising were allotted equal parts of the budget?

Solve.

7. Use graph paper to draw a rectangle 20 units long and 5 units wide. Show the Aurora graph using decimals instead of fractions.

8. What is the sum of the decimals in the graph you drew for Exercise 7?

9. The part of the budget allotted for television is how many times as great as the part allotted for direct mail? (Use the decimals in your graph.)

10. Suppose a total of $8,500,000 will be used for the Aurora campaign. How much money will be used for each of the four types of advertising?

11. Why do you think Michelle has chosen to spend more money on television advertising than on any other type of advertising?

12. Which types of magazines would be best for advertising a new sports car? Why?

CHAPTER REVIEW

Vocabulary. Choose the letter of the word(s) that completes each statement.

1. The ■ of the fraction identifies the total number of equal parts. [122]
   a. numerator **b.** denominator **c.** mixed number **d.** multiple

2. Fractions that name the same amount using different terms are ■. [126]
   a. nonterminating **b.** common **c.** equivalent **d.** cross products

3. A whole number and a fraction is called a ■. [132]
   a. mixed number **b.** GCF **c.** LCM **d.** multiple

Skills. Find the greatest common factor (GCF). [124]

4, 10 *2*

4. 3, 9 **5.** 15, 25 **6.** 12, 20 **7.** 18, 30

Find the least common multiple (LCM). [124]

4, 5 *20*

8. 2, 6 **9.** 6, 10 **10.** 8, 12 **11.** 10, 25

Write the fraction in lowest terms. [126]

$\frac{6}{9}$ $\frac{2}{3}$

12. $\frac{12}{15}$ **13.** $\frac{14}{18}$ **14.** $\frac{15}{21}$ **15.** $\frac{25}{35}$

Complete. [126]

$\frac{9}{12} = \frac{■}{4}$ *3*

16. $\frac{1}{4} = \frac{■}{36}$ **17.** $\frac{2}{3} = \frac{■}{12}$ **18.** $\frac{20}{24} = \frac{■}{6}$ **19.** $\frac{12}{16} = \frac{3}{■}$

Compare. Write =, <, or >. [128]

$\frac{2}{3}$ ● $\frac{3}{4}$ *<*

20. $\frac{1}{2}$ ● $\frac{1}{3}$ **21.** $\frac{3}{4}$ ● $\frac{12}{16}$ **22.** $\frac{3}{7}$ ● $\frac{4}{5}$ **23.** $\frac{7}{8}$ ● $\frac{2}{3}$

Write as an improper fraction. [132]

$2\frac{3}{4}$ $\frac{11}{3}$

24. $1\frac{2}{3}$ **25.** $5\frac{2}{7}$ **26.** $3\frac{5}{6}$ **27.** $8\frac{3}{10}$

Write the fraction or mixed number as a decimal. [136]

$1\frac{1}{2}$ *1.5*

28. $\frac{1}{4}$ **29.** $\frac{3}{5}$ **30.** $1\frac{1}{10}$ **31.** $2\frac{4}{5}$ **32.** $3\frac{1}{8}$

CHAPTER TEST

Write a fraction for the shaded part.

1. 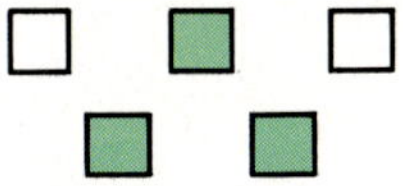

2. 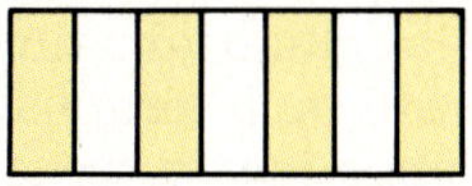

Write a fraction for the point.

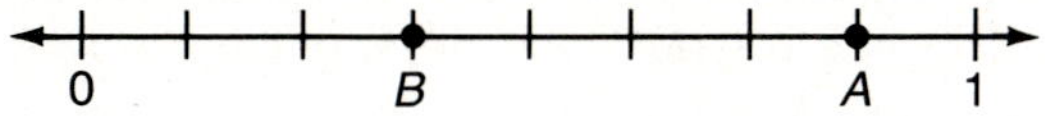

3. A **4.** B

Find the greatest common factor (GCF).

5. 6 and 15 **6.** 6, 12, and 24

Find the least common multiple (LCM).

7. 4 and 10 **8.** 2, 5, and 8

Write the fraction in lowest terms.

9. $\frac{15}{40}$ **10.** $\frac{9}{12}$

Complete.

11. $\frac{1}{5} = \frac{\blacksquare}{15}$ **12.** $\frac{9}{21} = \frac{\blacksquare}{7}$

Compare. Write >, <, or =.

13. $\frac{5}{8}$ ● $\frac{3}{8}$ **14.** $\frac{3}{4}$ ● $\frac{9}{12}$ **15.** $\frac{1}{5}$ ● $\frac{1}{4}$ **16.** $\frac{2}{3}$ ● $\frac{5}{8}$

Write as an improper fraction.

17. $1\frac{1}{4}$ **18.** $3\frac{2}{3}$ **19.** $6\frac{4}{9}$

Write as a whole number or mixed number.

20. $\frac{5}{3}$ **21.** $\frac{15}{6}$ **22.** $\frac{21}{7}$

Write as a fraction or mixed number in simplest form.

23. 0.8 **24.** 0.24 **25.** 5.3 **26.** 10.45

Write as a decimal. Round to the nearest hundredth when necessary.

27. $\frac{1}{4}$ **28.** $\frac{9}{20}$ **29.** $2\frac{1}{2}$ **30.** $5\frac{5}{6}$

Solve.

31. Larry has \$29. How many tee shirts can he buy if each tee shirt costs \$6?

32. Hamburger buns come 8 to a package. If there are 36 hamburgers, how many packages are needed?

33. What fraction of Jim's expenses is spent on books?

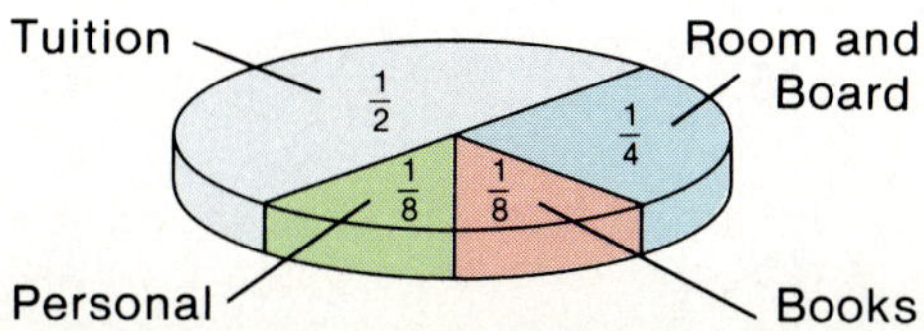

34. Will more money be spent on advertising on TV or radio?

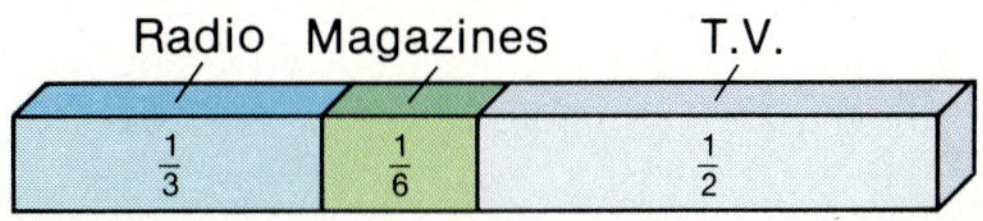

ENRICHMENT TERMINATING DECIMALS

All fractions can be written as **terminating** or **nonterminating decimals.** 0.435 is a terminating decimal. 0.434343 . . . is a nonterminating decimal. A fraction written in simplest form will be a terminating decimal only if all the prime factors of the denominator are 2 or 5.

Examples Will $\frac{28}{50}$ be a terminating decimal?

1. Write $\frac{28}{50}$ in simplest form. $\frac{28}{50} = \frac{14}{25}$
2. Factor the denominator.

   $25 = 5 \times 5$

   THINK: 5 is the only factor.

So, $\frac{28}{50}$ will terminate.

Will $\frac{10}{24}$ be a terminating decimal?

1. Write $\frac{10}{24}$ in simplest form. $\frac{10}{24} = \frac{5}{12}$
2. Factor the denominator.

   $12 = 4 \times 3 = 2 \times 2 \times 3$

   THINK: 3 is a prime factor.

So, $\frac{10}{24}$ will not terminate.

Will the decimal terminate? Write *yes* or *no*.

1. $\frac{1}{2}$	**2.** $\frac{2}{10}$	**3.** $\frac{14}{18}$	**4.** $\frac{6}{20}$
5. $\frac{33}{45}$	**6.** $\frac{13}{16}$	**7.** $\frac{18}{24}$	**8.** $\frac{9}{20}$
9. $\frac{4}{25}$	**10.** $\frac{3}{125}$	**11.** $\frac{9}{21}$	**12.** $\frac{6}{8}$
13. $\frac{3}{32}$	**14.** $\frac{6}{27}$	**15.** $\frac{5}{75}$	**16.** $\frac{121}{625}$
17. $\frac{3}{36}$	**18.** $\frac{8}{128}$	**19.** $\frac{12}{15}$	**20.** $\frac{10}{65}$

CALCULATOR

FRACTIONS ON A CALCULATOR

You have learned that a fraction can be written as a decimal by dividing the numerator by the denominator.

$$\frac{1}{4} \rightarrow 4\overline{)1.00}\ \ 0.25$$

You can also use a calculator to change a fraction or mixed number to a decimal.

Change $\frac{1}{3}$ to a decimal.

Procedure	**Calculator Entry**	**Calculator Display**
Divide the numerator by the denominator.	[1] [÷] [3] [=]	0.33333333

$\frac{1}{3} = 0.333 \leftarrow$ **Repeating decimal rounded to the nearest thousandth**

Change $5\frac{3}{8}$ to a decimal.

Procedure	**Calculator Entry**	**Calculator Display**
Change the fraction to a decimal.	[3] [÷] [8] [=]	0.375
Add the whole number.	[+] [5] [=]	5.375

$5\frac{3}{8} = 5.375$

Use a calculator to change the fraction or mixed number to a decimal. When necessary, round your answer to the nearest thousandth.

1. $\frac{5}{8}$ **2.** $\frac{7}{20}$ **3.** $\frac{13}{25}$ **4.** $\frac{3}{7}$ **5.** $\frac{5}{6}$

6. $4\frac{3}{5}$ **7.** $8\frac{18}{25}$ **8.** $7\frac{11}{16}$ **9.** $6\frac{1}{6}$ **10.** $8\frac{2}{3}$

Use a calculator to change the fraction to a decimal.

11. $\frac{1}{11}$ **12.** $\frac{2}{11}$ **13.** $\frac{3}{11}$ **14.** $\frac{4}{11}$ **15.** $\frac{5}{11}$

What patterns do you see? Use the pattern to write a decimal for the fraction.

16. $\frac{6}{11}$ **17.** $\frac{7}{11}$ **18.** $\frac{8}{11}$ **19.** $\frac{9}{11}$ **20.** $\frac{10}{11}$

PreSkills TEST

Find the greatest common factor (GCF). [124]

1. 8, 6　**2.** 16, 24　**3.** 21, 81　**4.** 54, 18

5. 6, 16　**6.** 12, 16　**7.** 11, 19　**8.** 18, 45

Write the fraction in lowest terms. [126]

9. $\frac{6}{8}$　**10.** $\frac{3}{9}$　**11.** $\frac{4}{10}$　**12.** $\frac{8}{12}$

13. $\frac{16}{24}$　**14.** $\frac{7}{21}$　**15.** $\frac{6}{15}$　**16.** $\frac{12}{21}$

Write as an improper fraction. [132]

17. 3　**18.** $4\frac{1}{5}$　**19.** $7\frac{5}{8}$　**20.** $5\frac{1}{6}$　**21.** $8\frac{4}{9}$

22. $4\frac{1}{9}$　**23.** $6\frac{2}{3}$　**24.** $2\frac{7}{8}$　**25.** $10\frac{1}{2}$　**26.** $9\frac{2}{3}$

Write as a mixed number in simplest form. [132]

27. $\frac{5}{3}$　**28.** $\frac{8}{5}$　**29.** $\frac{23}{3}$　**30.** $\frac{13}{2}$　**31.** $\frac{19}{5}$

32. $\frac{9}{4}$　**33.** $\frac{33}{6}$　**34.** $\frac{20}{2}$　**35.** $\frac{26}{8}$　**36.** $\frac{15}{4}$

Find the least common multiple (LCM) for the pair of numbers. [124]

37. 3, 4　**38.** 6, 8　**39.** 9, 6　**40.** 12, 8　**41.** 12, 16

Compare. Use $<$, $>$, or $=$. [128]

42. $\frac{4}{5}$ ● $\frac{3}{5}$　**43.** $\frac{1}{2}$ ● $\frac{1}{3}$　**44.** $\frac{7}{10}$ ● $\frac{7}{8}$　**45.** $\frac{5}{12}$ ● $\frac{7}{12}$　**46.** $\frac{31}{40}$ ● $\frac{31}{60}$

47. $\frac{3}{8}$ ● $\frac{1}{2}$　**48.** $\frac{2}{4}$ ● $\frac{1}{2}$　**49.** $\frac{5}{6}$ ● $\frac{1}{2}$　**50.** $\frac{7}{12}$ ● $\frac{1}{2}$　**51.** $\frac{7}{16}$ ● $\frac{1}{2}$

Chapter 6

FRACTION COMPUTATION

Often people have to follow the same time schedule each day. What are the advantages or disadvantages?

6.1 MULTIPLYING FRACTIONS

Frank Torelli saves $\frac{1}{5}$ of his income. He deposits $\frac{2}{3}$ of his savings into a savings account. He buys stock with the rest of his savings. What part of his total earnings does Frank deposit into his savings account?

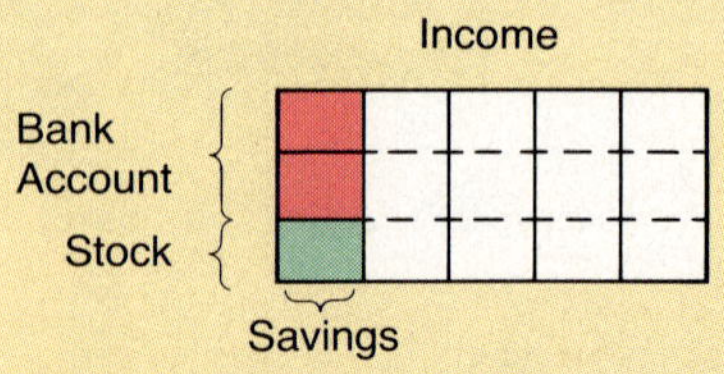

Multiply $\frac{1}{5}$ by $\frac{2}{3}$ to find the part of his total earnings that Frank puts into his savings account.

EXAMPLE 1

Multiply: $\frac{2}{3} \times \frac{1}{5}$

1. Multiply the numerators.

$$\frac{2}{3} \times \frac{1}{5} = 2$$

2. Multiply the denominators.

$$\frac{2}{3} \times \frac{1}{5} = \frac{2}{15}$$

Frank deposits $\frac{2}{15}$ of his income into a savings account.

Sometimes the answer needs to be written in simplest form.

EXAMPLE 2

Multiply: $\frac{3}{4} \times \frac{8}{9}$ Write the answer in simplest form.

1. Multiply.

$$\frac{3}{4} \times \frac{8}{9} = \frac{24}{36}$$

So, $\frac{3}{4} \times \frac{8}{9} = \frac{2}{3}$.

2. Write the product in simplest form.

$$\frac{24}{36} = \frac{24 \div 12}{36 \div 12} = \frac{2}{3}$$

When numerators and denominators have common factors, you can use a shortcut to multiply.

EXAMPLE 3

Multiply: $\frac{4}{9} \times \frac{5}{8} \times \frac{3}{7}$

1. Find common factors. Simplify the fractions.

**THINK: 4 is the GCF of 4 and 8.
3 is the GCF of 3 and 9.**

$$\frac{\overset{1}{\cancel{4}}}{\underset{3}{\cancel{9}}} \times \frac{5}{\underset{2}{\cancel{8}}} \times \frac{\overset{1}{\cancel{3}}}{7}$$

2. Multiply.

$$\frac{\overset{1}{\cancel{4}}}{\underset{3}{\cancel{9}}} \times \frac{5}{\underset{2}{\cancel{8}}} \times \frac{\overset{1}{\cancel{3}}}{7} = \frac{5}{42}$$

CHECKPOINT

Write the letter of the correct answer.

Multiply. What is the product in simplest form?

1. $\frac{3}{8} \times \frac{5}{6}$ **a.** $\frac{8}{14}$ **b.** $\frac{15}{48}$ **c.** $\frac{5}{16}$ **d.** $\frac{35}{86}$

2. $\frac{3}{8} \times \frac{1}{2} \times \frac{4}{9}$ **a.** $\frac{8}{19}$ **b.** $\frac{12}{144}$ **c.** $\frac{1}{24}$ **d.** $\frac{1}{12}$

PRACTICE EXERCISES

Multiply. Write the product in simplest form.

1. $\frac{1}{3} \times \frac{1}{2}$ **2.** $\frac{1}{4} \times \frac{1}{2}$ **3.** $\frac{2}{3} \times \frac{1}{5}$ **4.** $\frac{1}{6} \times \frac{5}{8}$ **5.** $\frac{5}{9} \times \frac{1}{2}$ **6.** $\frac{1}{7} \times \frac{5}{6}$

7. $\frac{2}{3} \times \frac{2}{5}$ **8.** $\frac{3}{4} \times \frac{3}{4}$ **9.** $\frac{3}{4} \times \frac{5}{7}$ **10.** $\frac{5}{9} \times \frac{2}{3}$ **11.** $\frac{2}{3} \times \frac{4}{5}$ **12.** $\frac{2}{5} \times \frac{3}{7}$

13. $\frac{3}{4} \times \frac{4}{5}$ **14.** $\frac{2}{3} \times \frac{5}{6}$ **15.** $\frac{3}{4} \times \frac{3}{9}$ **16.** $\frac{2}{7} \times \frac{3}{8}$ **17.** $\frac{3}{8} \times \frac{4}{6}$ **18.** $\frac{5}{6} \times \frac{3}{5}$

19. $\frac{5}{12} \times \frac{2}{3}$ **20.** $\frac{3}{4} \times \frac{7}{10}$ **21.** $\frac{9}{15} \times \frac{5}{8}$ **22.** $\frac{2}{3} \times \frac{6}{13}$ **23.** $\frac{8}{11} \times \frac{3}{8}$ **24.** $\frac{4}{5} \times \frac{3}{8}$

25. $\frac{4}{15} \times \frac{5}{12}$ **26.** $\frac{3}{14} \times \frac{7}{12}$ **27.** $\frac{9}{10} \times \frac{10}{11}$ **28.** $\frac{11}{20} \times \frac{5}{22}$ **29.** $\frac{10}{21} \times \frac{14}{25}$ **30.** $\frac{5}{6} \times \frac{3}{10}$

31. $\frac{1}{4} \times \frac{8}{9}$ **32.** $\frac{4}{5} \times \frac{2}{3}$ **33.** $\frac{3}{8} \times \frac{1}{6}$ **34.** $\frac{7}{12} \times \frac{3}{14}$ **35.** $\frac{1}{8} \times \frac{1}{4}$ **36.** $\frac{9}{16} \times \frac{2}{3}$

Multiply. Use the shortcut when possible.

37. $\frac{1}{3} \times \frac{2}{5} \times \frac{1}{2}$ **38.** $\frac{2}{3} \times \frac{3}{4} \times \frac{1}{2}$ **39.** $\frac{1}{3} \times \frac{1}{8} \times \frac{4}{5}$ **40.** $\frac{3}{4} \times \frac{2}{3} \times \frac{3}{5}$

41. $\frac{4}{5} \times \frac{5}{8} \times \frac{1}{2}$ **42.** $\frac{6}{9} \times \frac{3}{4} \times \frac{3}{4}$ **43.** $\frac{5}{24} \times \frac{8}{15} \times \frac{9}{10}$ **44.** $\frac{10}{21} \times \frac{6}{15} \times \frac{7}{8}$

MIXED REVIEW

Add or subtract.

45. $347 + 68$ **46.** $1{,}980 + 3{,}268$ **47.** $825{,}971 + 3{,}009$ **48.** $61{,}803 + 2{,}999$ **49.** $572 + 46$

50. $47 - 28$ **51.** $852 - 690$ **52.** $78{,}542 - 12{,}855$ **53.** $623{,}401 - 9{,}426$ **54.** $937 - 486$

Solve.

55. Yoko spends $\frac{1}{8}$ of her total income on entertainment. Of that amount, she spends $\frac{2}{5}$ on movies. What part of her total income does she spend on movies?

56. Erik deducted $\frac{1}{4}$ of his income for taxes. He received $\frac{1}{3}$ of his total deductions as a tax refund. What part of his total income did he receive as a refund?

6.2 MULTIPLYING MIXED NUMBERS

Amir Johnson is building a cabin cruiser. The hull of the cruiser is 48 feet long. The owner wants $\frac{3}{4}$ of the hull painted white. How many feet of the hull should Amir paint white?

THINK: $\frac{3}{4}$ of 48 feet is the same as $\frac{3}{4} \times 48$.

EXAMPLE 1

Multiply: $\frac{3}{4} \times 48$

1. Rename the whole number as an improper fraction.

   $48 = \frac{48}{1}$

2. Multiply.

   $\frac{3}{4} \times \frac{48}{1} = \frac{36}{1} = 36$

Amir should paint 36 feet of the hull white.

To multiply mixed numbers and whole numbers, write each factor as an improper fraction.

EXAMPLE 2

Multiply:

a. $3 \times 2\frac{1}{6}$

$3 \times 2\frac{1}{6} = \frac{3}{1} \times \frac{13}{6} = \frac{13}{2} = 6\frac{1}{2}$

b. $4\frac{2}{3} \times 5\frac{1}{4}$

$4\frac{2}{3} \times 5\frac{1}{4} = \frac{14}{3} \times \frac{21}{4} = \frac{49}{2} = 24\frac{1}{2}$

CHECKPOINT

Write the letter of the correct answer.

Multiply. What is the product in simplest form?

1. $\frac{3}{4} \times 12$ **a.** $\frac{36}{4}$ **b.** $12\frac{3}{4}$ **c.** 16 **d.** 9

2. $3 \times 4\frac{2}{3}$ **a.** $12\frac{2}{3}$ **b.** 14 **c.** $12\frac{6}{3}$ **d.** 6

3. $3\frac{2}{5} \times 1\frac{3}{4}$ **a.** $3\frac{6}{20}$ **b.** $3\frac{3}{10}$ **c.** $5\frac{19}{20}$ **d.** $4\frac{5}{9}$

PRACTICE EXERCISES

Multiply. Write the product in simplest form.

1. $2 \times \frac{1}{2}$ **2.** $3 \times \frac{2}{3}$ **3.** $5 \times \frac{3}{5}$ **4.** $\frac{1}{2} \times 6$ **5.** $\frac{2}{3} \times 6$

6. $8 \times \frac{3}{4}$ **7.** $\frac{1}{3} \times 9$ **8.** $12 \times \frac{5}{6}$ **9.** $5 \times \frac{2}{3}$ **10.** $\frac{4}{5} \times 3$

11. $2 \times 1\frac{1}{2}$ **12.** $3 \times 2\frac{1}{3}$ **13.** $2 \times 3\frac{1}{2}$ **14.** $4 \times 3\frac{1}{4}$ **15.** $3 \times 2\frac{2}{3}$

16. $1\frac{3}{4} \times 8$ **17.** $2 \times 2\frac{5}{8}$ **18.** $3\frac{1}{6} \times 24$ **19.** $3\frac{7}{20} \times 5$ **20.** $4 \times 2\frac{1}{12}$

21. $\frac{7}{8} \times 2\frac{1}{2}$ **22.** $\frac{3}{5} \times 4\frac{1}{3}$ **23.** $7\frac{1}{4} \times \frac{8}{9}$ **24.** $6\frac{2}{3} \times \frac{4}{5}$ **25.** $\frac{5}{6} \times 5\frac{1}{3}$

26. $1\frac{1}{5} \times 2\frac{1}{2}$ **27.** $2\frac{2}{3} \times 1\frac{1}{3}$ **28.** $2\frac{1}{4} \times 2\frac{2}{3}$ **29.** $1\frac{3}{5} \times 2\frac{1}{2}$ **30.** $2\frac{2}{3} \times 3\frac{1}{2}$

31. $4\frac{3}{4} \times 2\frac{1}{3}$ **32.** $3\frac{3}{4} \times 1\frac{3}{4}$ **33.** $4\frac{2}{3} \times 1\frac{2}{7}$ **34.** $4\frac{1}{2} \times 2\frac{2}{3}$ **35.** $1\frac{1}{3} \times 3\frac{2}{7}$

36. $2\frac{1}{4} \times 3$ **37.** $3\frac{2}{3} \times 3\frac{3}{4}$ **38.** $\frac{2}{5} \times 3\frac{1}{2}$ **39.** $5\frac{1}{3} \times 2\frac{3}{8}$ **40.** $\frac{1}{2} \times 9\frac{2}{3}$

41. $5\frac{1}{5} \times 6\frac{1}{4}$ **42.** $7\frac{2}{3} \times 2$ **43.** $\frac{1}{3} \times 3\frac{2}{4}$ **44.** $6\frac{2}{5} \times 10$ **45.** $10\frac{1}{2} \times \frac{4}{7}$

46. $\frac{2}{5} \times \frac{3}{8} \times 10$ **47.** $2\frac{1}{2} \times \frac{1}{3} \times \frac{1}{4}$ **48.** $3\frac{2}{3} \times \frac{1}{4} \times 3$ **49.** $\frac{2}{3} \times \frac{1}{5} \times 2\frac{3}{4}$

50. $1\frac{1}{2} \times 1\frac{1}{4} \times 1\frac{1}{3}$ **51.** $4\frac{2}{5} \times 8\frac{1}{3} \times 2\frac{1}{4}$ **52.** $5\frac{2}{5} \times 7\frac{1}{3} \times 6\frac{1}{11}$ **53.** $9\frac{3}{5} \times 6\frac{1}{8} \times 2\frac{4}{7}$

54. $\left(2 + 1\frac{1}{2}\right) \times \frac{4}{7}$ **55.** $\left(1\frac{1}{4} + 3\frac{3}{8}\right) \times \frac{1}{2}$ **56.** $\left(\frac{3}{4} + \frac{7}{8}\right) \times \frac{2}{3}$ **57.** $\left(\frac{2}{3} + 1\frac{1}{2}\right) \times \frac{3}{4}$

Solve.

58. Maria's boat has a maximum speed of 36 miles per hour. Its cruising speed is $\frac{3}{4}$ of its maximum speed. What is the boat's cruising speed?

59. Latoya earns $18.00 an hour as a carpenter building boats. One week she works 42.5 hours. How much money does she earn that week?

60. Cesar's budget allows him to spend $78,000 on lumber. He has spent $45,963. How much more can he spend without going over budget?

61. A ship can travel at a rate of $18\frac{3}{4}$ miles per hour. How far can it travel in $3\frac{1}{5}$ hours?

TIME OUT

A ship is sailing from New York City to Southhampton, England. The total distance is about 3,600 miles. The ship has traveled $\frac{1}{2}$ the distance to the halfway point. How far has the ship traveled?

6.3 DIVIDING FRACTIONS

If the product of two numbers equals 1, the numbers are called **reciprocals.**

$\frac{3}{4} \times \frac{4}{3} = \frac{12}{12} = 1$ — **Reciprocals**

$6 \times \frac{1}{6} = \frac{6}{1} \times \frac{1}{6} = \frac{6}{6} = 1$ — **Reciprocals**

$2\frac{2}{3} \times \frac{3}{8} = \frac{8}{3} \times \frac{3}{8} = \frac{24}{24} = 1$ — **Reciprocals**

To find the reciprocal of a number, write the number as a fraction and reverse the numerator and denominator.

At the Snapper Creek Scout Camp, the scouts make their own tent stakes. How many stakes $\frac{3}{4}$ ft long can be made out of a piece of lumber that is 12 ft long?

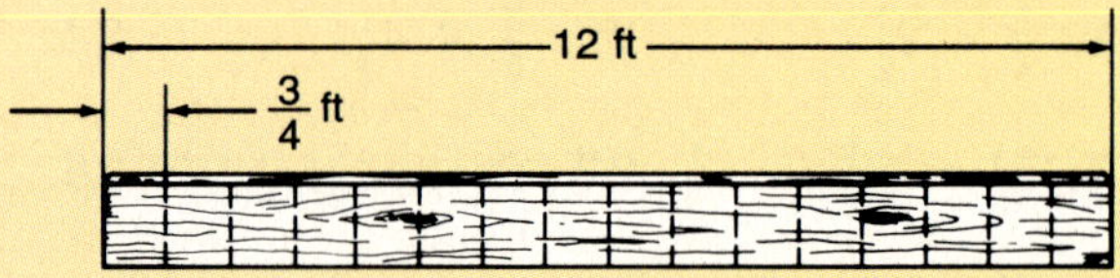

Divide 12 by $\frac{3}{4}$ to find the number of stakes that can be made.

To divide fractions, multiply the dividend by the reciprocal of the divisor.

EXAMPLE 1 Divide: $12 \div \frac{3}{4}$

1. Find the reciprocal of the divisor. The reciprocal of $\frac{3}{4}$ is $\frac{4}{3}$.
2. Multiply. $\frac{12}{1} \times \frac{4}{3} = \frac{\overset{4}{\cancel{12}}}{1} \times \frac{4}{\underset{1}{\cancel{3}}} = \frac{16}{1} = 16$

You can make 16 stakes that are $\frac{3}{4}$ ft long from a 12-ft board.

When the divisor is a whole number, follow the same procedure as when dividing fractions.

EXAMPLE 2 Divide: $\frac{2}{3} \div 8$

1. Find the reciprocal of the divisor. The reciprocal of 8 is $\frac{1}{8}$.
2. Multiply. $\frac{2}{3} \times \frac{1}{8} = \frac{\overset{1}{\cancel{2}}}{3} \times \frac{1}{\underset{4}{\cancel{8}}} = \frac{1}{12}$

CHECKPOINT

Write the letter of the correct answer.

Divide. What is the quotient in simplest form?

1. $\frac{1}{2} \div \frac{2}{3}$ **a.** $\frac{3}{4}$ **b.** $1\frac{1}{3}$ **c.** $\frac{1}{6}$ **d.** $\frac{1}{3}$

2. $\frac{8}{9} \div 4$ **a.** $\frac{8}{36}$ **b.** $\frac{4}{9}$ **c.** $\frac{2}{9}$ **d.** $4\frac{1}{2}$

PRACTICE EXERCISES

Find the reciprocal of each number.

1. 3 **2.** 7 **3.** 15 **4.** 36 **5.** $\frac{1}{4}$ **6.** $\frac{1}{7}$ **7.** $\frac{1}{12}$ **8.** $\frac{1}{40}$

9. $\frac{3}{5}$ **10.** $\frac{4}{9}$ **11.** $\frac{2}{7}$ **12.** $\frac{8}{3}$ **13.** $\frac{3}{11}$ **14.** $\frac{6}{5}$ **15.** $\frac{15}{12}$ **16.** $\frac{25}{3}$

17. $1\frac{1}{3}$ **18.** $2\frac{3}{4}$ **19.** $4\frac{2}{5}$ **20.** $3\frac{1}{6}$ **21.** $5\frac{2}{7}$ **22.** $6\frac{3}{8}$ **23.** $10\frac{1}{2}$ **24.** $12\frac{5}{6}$

Divide. Write the quotient in simplest form.

25. $\frac{2}{3} \div \frac{3}{4}$ **26.** $\frac{1}{2} \div \frac{2}{3}$ **27.** $\frac{3}{8} \div \frac{1}{2}$ **28.** $\frac{3}{4} \div \frac{5}{6}$ **29.** $\frac{2}{9} \div \frac{2}{3}$ **30.** $\frac{1}{8} \div \frac{5}{6}$

31. $\frac{4}{5} \div \frac{5}{6}$ **32.** $\frac{2}{7} \div \frac{7}{8}$ **33.** $\frac{1}{5} \div \frac{3}{4}$ **34.** $\frac{5}{6} \div \frac{8}{9}$ **35.** $\frac{3}{8} \div \frac{6}{7}$ **36.** $\frac{1}{4} \div \frac{7}{8}$

37. $\frac{3}{4} \div \frac{2}{3}$ **38.** $\frac{1}{2} \div \frac{1}{4}$ **39.** $\frac{4}{5} \div \frac{3}{4}$ **40.** $\frac{5}{8} \div \frac{5}{9}$ **41.** $\frac{8}{9} \div \frac{2}{3}$ **42.** $\frac{4}{7} \div \frac{2}{5}$

43. $\frac{3}{4} \div 3$ **44.** $\frac{2}{3} \div 5$ **45.** $\frac{2}{5} \div 4$ **46.** $\frac{3}{8} \div 9$ **47.** $\frac{8}{9} \div 12$ **48.** $\frac{5}{6} \div 10$

49. $2 \div \frac{4}{5}$ **50.** $4 \div \frac{2}{3}$ **51.** $8 \div \frac{8}{9}$ **52.** $10 \div \frac{5}{6}$ **53.** $14 \div \frac{4}{7}$ **54.** $24 \div \frac{3}{8}$

Solve.

55. A hiking trail is $\frac{3}{4}$ mile long. It is marked by stakes $\frac{1}{8}$ mile apart. How many stakes mark the trail? (*Hint:* There is no marker at the beginning of the trail.)

56. It takes 8 stakes to put up 1 tent. How many stakes are needed for 6 tents?

57. Each scout is served two $\frac{1}{4}$-lb hamburgers at lunch. How many pounds of ground beef are needed to make hamburgers for 24 scouts?

58. The scouts wanted to cook frog legs, but they only had $1\frac{1}{2}$ pounds of frog legs. Rewrite the recipe, showing the ingredients they need.

FRIED FROG LEGS

3 lb frog legs
$\frac{1}{2}$ cup bread crumbs
1 teaspoon salt
$\frac{1}{4}$ cup flour
$\frac{1}{4}$ cup Parmesan cheese
2 tablespoons parsley
$\frac{1}{4}$ cup cooking oil
2 eggs

6.4 DIVIDING MIXED NUMBERS

The price of one share of stock in the New Age Electronics Company is $\$2\frac{1}{4}$. Cathy Willis has \$72 to invest. How many shares of stock can she buy?

Divide 72 by $2\frac{1}{4}$ to find how many shares of stock Cathy can buy.

To divide by a mixed number, first rename the mixed number as an improper fraction. Then divide as you would with fractions.

EXAMPLE 1 Divide: $72 \div 2\frac{1}{4}$

$$72 \div 2\frac{1}{4} = \frac{72}{1} \div \frac{9}{4} = \frac{\overset{8}{\cancel{72}}}{1} \times \frac{4}{\underset{1}{\cancel{9}}} = 32$$

Cathy can buy 32 shares of stock in the New Age Electronics Company.

When the dividend is a fraction, divide as you would with fractions.

EXAMPLE 2 Divide: $\frac{5}{6} \div 3\frac{1}{3}$

$$\frac{5}{6} \div 3\frac{1}{3} = \frac{5}{6} \div \frac{10}{3} = \frac{\overset{1}{\cancel{5}}}{\underset{2}{\cancel{6}}} \times \frac{\overset{1}{\cancel{3}}}{\underset{2}{\cancel{10}}} = \frac{1}{4}$$

Remember to rename both mixed numbers as fractions before dividing.

EXAMPLE 3 Divide: $1\frac{1}{2} \div 3\frac{3}{4}$

$$1\frac{1}{2} \div 3\frac{3}{4} = \frac{3}{2} \div \frac{15}{4} = \frac{\overset{1}{\cancel{3}}}{\underset{1}{\cancel{2}}} \times \frac{\overset{2}{\cancel{4}}}{\underset{5}{\cancel{15}}} = \frac{2}{5}$$

CHECKPOINT

Write the letter of the correct answer.

Divide. What is the quotient in the simplest form?

1. $6 \div 1\frac{1}{2}$	**a.** $\frac{12}{3}$	**b.** 9	**c.** 4	**d.** $\frac{1}{4}$
2. $2\frac{2}{3} \div 1\frac{1}{4}$	**a.** $2\frac{2}{15}$	**b.** $\frac{15}{32}$	**c.** $\frac{2}{15}$	**d.** $1\frac{13}{15}$
3. $\frac{4}{7} \div 3\frac{1}{2}$	**a.** 2	**b.** $6\frac{1}{8}$	**c.** $\frac{49}{8}$	**d.** $\frac{8}{49}$
4. $4\frac{1}{5} \div \frac{3}{8}$	**a.** $\frac{15}{168}$	**b.** $11\frac{1}{5}$	**c.** $\frac{63}{40}$	**d.** $1\frac{23}{40}$

PRACTICE EXERCISES

Divide. Write the quotient in simplest form.

1. $5 \div 1\frac{2}{3}$ **2.** $3 \div 2\frac{3}{4}$ **3.** $10 \div 3\frac{1}{3}$ **4.** $15 \div 2\frac{1}{7}$ **5.** $8 \div 2\frac{2}{5}$

6. $2\frac{1}{2} \div 5$ **7.** $1\frac{2}{3} \div 10$ **8.** $4\frac{3}{4} \div 3$ **9.** $5\frac{3}{5} \div 2$ **10.** $3\frac{3}{8} \div 3$

11. $\frac{2}{3} \div 1\frac{1}{2}$ **12.** $\frac{4}{9} \div 2\frac{2}{3}$ **13.** $\frac{2}{5} \div 4\frac{1}{2}$ **14.** $\frac{3}{4} \div 2\frac{1}{4}$ **15.** $\frac{7}{10} \div 4\frac{1}{5}$

16. $2\frac{2}{5} \div \frac{3}{5}$ **17.** $2\frac{1}{4} \div \frac{2}{3}$ **18.** $4\frac{3}{4} \div \frac{5}{6}$ **19.** $2\frac{2}{5} \div \frac{3}{10}$ **20.** $2\frac{5}{6} \div \frac{3}{4}$

21. $1\frac{3}{4} \div 2\frac{1}{3}$ **22.** $1\frac{2}{3} \div 2\frac{1}{2}$ **23.** $2\frac{2}{3} \div 1\frac{3}{5}$ **24.** $3\frac{3}{5} \div 4\frac{1}{4}$ **25.** $4\frac{1}{2} \div 2\frac{5}{8}$

26. $3\frac{3}{4} \div 4\frac{2}{3}$ **27.** $2\frac{3}{4} \div 2\frac{4}{5}$ **28.** $4\frac{2}{3} \div 5\frac{1}{6}$ **29.** $6\frac{1}{8} \div 1\frac{3}{4}$ **30.** $1\frac{2}{3} \div 2\frac{1}{12}$

31. $5\frac{1}{3} \div \frac{3}{4}$ **32.** $3 \div 4\frac{2}{3}$ **33.** $12\frac{1}{4} \div 1\frac{2}{5}$ **34.** $10\frac{5}{8} \div 15\frac{5}{6}$ **35.** $8\frac{2}{3} \div 4$

Solve.

36. The price of Fast Lane Records stock is $\$5\frac{3}{4}$ per share. How many shares can be purchased with $230?

37. 250 shares of Trans Air stock were purchased for $1,200. What was the price per share?

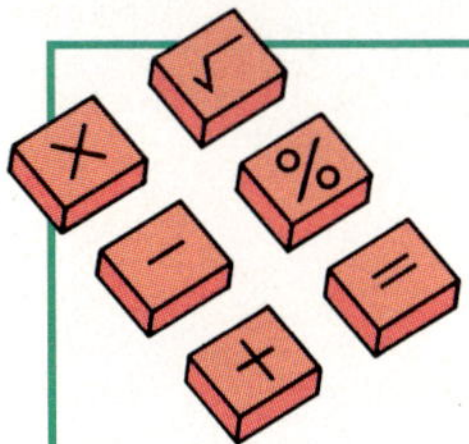

CALCULATOR

You can use what you know about changing a fraction to a decimal to help you multiply fractions and mixed numbers on a calculator.

Ground beef—$1.79 per lb Amount—$24\frac{3}{4}$ lb Cost—■

1. Change $24\frac{3}{4}$ to a decimal.

$24\frac{3}{4} = 24.75$

2. Use your calculator to multiply. Round to the nearest cent.

$24.75 \times \$1.79 = \$44.3025 \rightarrow \$44.30$

Copy and complete the table. Use your calculator to find the cost of each type of meat and the total cost.

	Chicken	Steak	Turkey	Bacon	
Number of pounds	$18\frac{1}{2}$	$21\frac{1}{4}$	$26\frac{1}{2}$	$15\frac{3}{4}$	
Cost per pound	$1.19	$4.99	$0.99	$1.09	Total cost
Cost	**1.** ■	**2.** ■	**3.** ■	**4.** ■	**5.** ■

PROBLEM *Solving* STRATEGY

6.5 WORKING BACKWARD

Situation:

Alonso Ortega has a part-time job at Ernie's Garden Center. This week he worked $\frac{1}{2}$ as many hours as he did last week. He worked 10 hours this week. How many hours did he work last week?

Strategy:

Working backward can help you find a solution to some problems.

Applying the Strategy:

THINK: $\frac{1}{2}$ **of the number of hours worked last week = the number of hours worked this week**

$\frac{1}{2}$ **of ■ hours = 10 hours**

$$10 \div \frac{1}{2} = \frac{10}{1} \times \frac{2}{1} = \frac{20}{1} = 20$$

Alonso worked 20 hours last week.

FOR DISCUSSION

Blanca deposits her paycheck in a savings account. She has $285 in her account. Last week she deposited $100. Two weeks ago she withdrew $40. How much was in Blanca's account before she withdrew the money?

1. What facts are you given?
2. Explain how you can work backward to solve the problem.
3. What is the solution?

PRACTICE EXERCISES

Solve the problem by working backward.

1. During the first week of October, Steve worked $\frac{1}{3}$ as many hours as he did the second week of October. He worked 9 hours the first week.

a. How many times as many hours did Steve work during the second week as compared with the first week?
b. How many hours did he work the second week of October?

2. A customer owes Ernie's Garden Center $75. Last week she charged a $20 plant to her account. Three weeks ago she paid $30 of her bill.

a. How much was the customer's bill before she charged the $20 plant?
b. How much was the customer's bill before she paid $30?

Steve made this sale sign for a weekend special. Find the original price.

3. Begonia

4. Geranium

5. Fern

SALE: 1/2 OFF REGULAR PRICE		
Begonias	—	Now $2.00 each
Geraniums	—	Now $1.75 each
Ferns	—	Now $4.25 each

6. Donna works in the greenhouse. This week she worked $\frac{2}{3}$ as many hours as she did last week. She worked 36 hours this week. How many hours did she work last week?

7. Ernie has $3,000 in his bank account. Yesterday he deposited $500. Last week he withdrew $1,000. How much was in Ernie's account before he withdrew the money?

8. In the garden shop, Fred bought twice as many plants as Juanita. Juanita bought three fewer plants than Ann. Ann bought five plants. How many plants did each person buy?

9. Betty wants radishes by July 12. The seed package states that radishes take 24 days to grow. Steve looks at the calendar and tells Betty there are 30 days in June. What is the last day that Betty can plant radish seeds?

MIDCHAPTER REVIEW

Multiply. Write the product in simplest form. [146, 148]

1. $\frac{1}{2} \times \frac{1}{4}$ **2.** $\frac{1}{8} \times \frac{1}{3}$ **3.** $\frac{1}{6} \times \frac{1}{5}$ **4.** $\frac{2}{3} \times \frac{1}{5}$ **5.** $\frac{5}{8} \times \frac{1}{2}$

6. $\frac{4}{5} \times \frac{1}{2}$ **7.** $\frac{2}{3} \times \frac{2}{5}$ **8.** $\frac{7}{8} \times \frac{3}{4}$ **9.** $\frac{5}{12} \times \frac{2}{3}$ **10.** $\frac{3}{5} \times \frac{3}{8}$

11. $\frac{1}{6} \times \frac{2}{7}$ **12.** $\frac{3}{4} \times \frac{2}{9}$ **13.** $\frac{5}{8} \times \frac{3}{10}$ **14.** $\frac{5}{6} \times \frac{3}{5}$ **15.** $\frac{7}{8} \times \frac{4}{9}$

16. $6 \times \frac{2}{3}$ **17.** $5 \times \frac{7}{10}$ **18.** $\frac{3}{4} \times 12$ **19.** $\frac{5}{6} \times 7$ **20.** $15 \times \frac{4}{5}$

21. $\frac{1}{2} \times 3\frac{1}{4}$ **22.** $\frac{2}{3} \times 6\frac{1}{2}$ **23.** $3\frac{3}{4} \times \frac{5}{8}$ **24.** $4\frac{2}{3} \times \frac{3}{8}$ **25.** $\frac{3}{5} \times 8\frac{1}{2}$

26. $1\frac{1}{3} \times 2\frac{1}{2}$ **27.** $3\frac{1}{4} \times 5\frac{1}{2}$ **28.** $4\frac{2}{5} \times 3\frac{1}{6}$ **29.** $8\frac{3}{8} \times 1\frac{3}{4}$ **30.** $2\frac{1}{6} \times 3\frac{5}{12}$

Find the reciprocal. [150]

31. $\frac{2}{3}$ **32.** $\frac{4}{9}$ **33.** $\frac{7}{15}$ **34.** 6 **35.** 18

36. $2\frac{1}{5}$ **37.** $3\frac{1}{7}$ **38.** $10\frac{1}{2}$ **39.** $6\frac{3}{4}$ **40.** $12\frac{4}{5}$

Divide. Write the quotient in simplest form. [150, 152]

41. $\frac{2}{3} \div \frac{5}{6}$ **42.** $\frac{2}{3} \div \frac{4}{5}$ **43.** $\frac{3}{4} \div \frac{3}{8}$ **44.** $\frac{5}{6} \div \frac{4}{9}$ **45.** $\frac{6}{7} \div \frac{3}{4}$

46. $6 \div \frac{2}{5}$ **47.** $\frac{7}{8} \div 3$ **48.** $9 \div \frac{1}{2}$ **49.** $10 \div \frac{7}{10}$ **50.** $\frac{5}{6} \div 10$

51. $2\frac{3}{4} \div 6$ **52.** $3\frac{2}{3} \div 4$ **53.** $5\frac{3}{5} \div 7$ **54.** $8 \div 2\frac{2}{3}$ **55.** $12 \div 1\frac{1}{5}$

56. $2\frac{1}{2} \div \frac{1}{4}$ **57.** $1\frac{3}{5} \div \frac{7}{10}$ **58.** $\frac{5}{6} \div 4\frac{1}{3}$ **59.** $\frac{7}{12} \div 2\frac{1}{4}$ **60.** $\frac{5}{9} \div 4\frac{2}{3}$

61. $1\frac{1}{3} \div 2\frac{1}{6}$ **62.** $3\frac{1}{2} \div 1\frac{3}{4}$ **63.** $3\frac{2}{5} \div 1\frac{5}{6}$ **64.** $4\frac{1}{2} \div 5\frac{2}{3}$ **65.** $2\frac{7}{8} \div 3\frac{1}{4}$

Solve. [154]

66. Isabel bought twice as many albums as Sam. Sam bought 3 more albums than Becky. Becky bought 2 albums. How many albums did each person buy?

67. Jack works at Disco Record Store. This week he worked $\frac{1}{2}$ as many hours as he did last week. Last week he worked 14 hours. How many hours did he work this week?

PROBLEM Solving APPLICATION

6.6 CAREER: CARPENTER

Kristin Leonard is a carpenter. In her work, she must be able to make careful measurements. Often, she needs to use fractions and mixed numbers.

A board 9 ft long is cut into sections of the given length. How many sections are there?

1. Sections $\frac{3}{4}$ ft long

2. Sections $1\frac{1}{2}$ ft long

A board $7\frac{1}{2}$ ft long is cut into equal parts. How long is each part?

3. 6 equal parts
4. 10 equal parts

Each drawing represents a board marked off into equal parts. Find the total length of the board.

5. 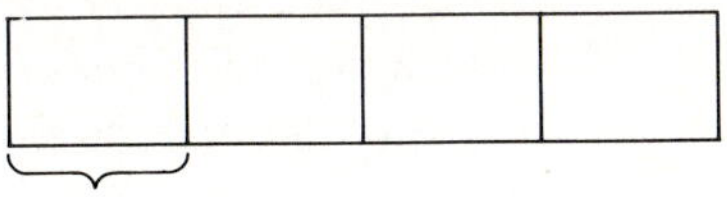

6. 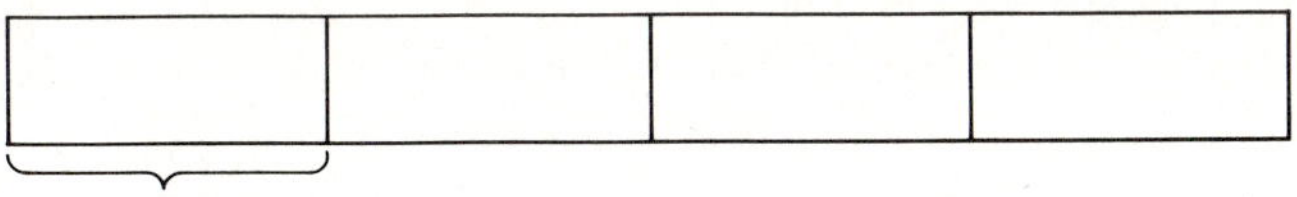

Write the measurement using a fraction or a mixed number in lowest terms.

7. 0.75 yd
8. 2.5 in.
9. 1.125 in.
10. 0.9375 in.

Solve.

11. A board of lumber called a "2-by-4" is actually $1\frac{5}{8}$ in. thick and $3\frac{5}{8}$ in. wide. Kristen stacks four of these boards as shown at the right. What is the total height of the stack?

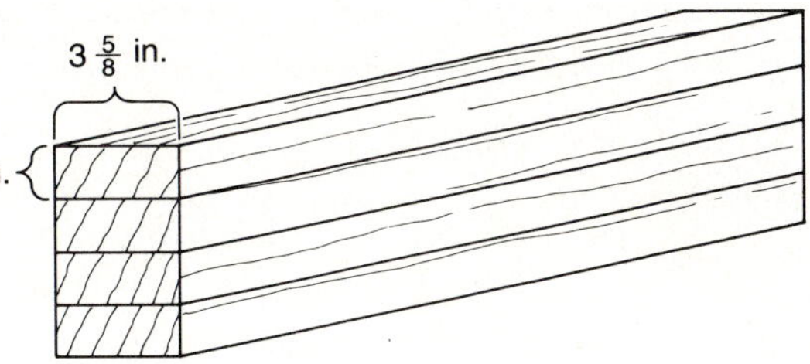

12. Kristen had a board $11\frac{1}{2}$ ft long. She sawed it in half and used one piece. Then she sawed the remaining piece in half and used one of the pieces. How long was the remaining piece?

6.7 ADDING FRACTIONS

Gerry Thomas spends $\frac{3}{8}$ of her income on rent and $\frac{1}{8}$ of her income on utilities. What part of her income is spent on rent and utilities?

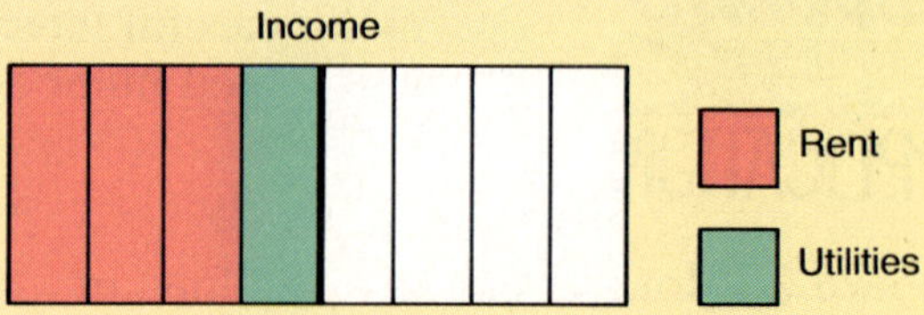

You can add $\frac{3}{8}$ and $\frac{1}{8}$ to find the answer.

EXAMPLE 1 Add: $\frac{3}{8} + \frac{1}{8}$

1. Add the numerators.

$$\begin{array}{r} \frac{3}{8} \\ +\frac{1}{8} \\ \hline 4 \end{array}$$

2. Write the sum over the denominator.

$$\begin{array}{r} \frac{3}{8} \\ +\frac{1}{8} \\ \hline \frac{4}{8} \end{array}$$

3. Write the answer in simplest form.

$$\begin{array}{r} \frac{3}{8} \\ +\frac{1}{8} \\ \hline \frac{4}{8} = \frac{1}{2} \end{array}$$

Gerry spends $\frac{1}{2}$ of her income on rent and utilities.

Sometimes you need to add fractions that have different denominators.

EXAMPLE 2 Add: $\frac{1}{2} + \frac{1}{6}$

1. Write equivalent fractions with a common denominator.

$$\begin{array}{rl} \frac{1}{2} & = \frac{3}{6} \\ +\frac{1}{6} & = \frac{1}{6} \end{array}$$

2. Add.

$$\begin{array}{rl} \frac{1}{2} & = \frac{3}{6} \\ +\frac{1}{6} & = \frac{1}{6} \\ \hline & \frac{4}{6} \end{array}$$

3. Write the answer in simplest form.

$$\begin{array}{rl} \frac{1}{2} & = \frac{3}{6} \\ +\frac{1}{6} & = \frac{1}{6} \\ \hline & \frac{4}{6} = \frac{2}{3} \end{array}$$

When the sum is greater than 1, rename the sum as a mixed number.

EXAMPLE 3 Add:

a. $$\begin{array}{r} \frac{2}{3} \\ +\frac{5}{8} \\ \hline \end{array} \qquad \begin{array}{rl} \frac{2}{3} & = \frac{16}{24} \\ +\frac{5}{8} & = \frac{15}{24} \\ \hline & \frac{31}{24} = 1\frac{7}{24} \end{array}$$

b. $$\begin{array}{r} \frac{1}{3} \\ \frac{1}{4} \\ +\frac{5}{6} \\ \hline \end{array} \qquad \begin{array}{rl} \frac{1}{3} & = \frac{4}{12} \\ \frac{1}{4} & = \frac{3}{12} \\ +\frac{5}{6} & = \frac{10}{12} \\ \hline & \frac{17}{12} = 1\frac{5}{12} \end{array}$$

CHECKPOINT Write the letter of the correct answer.

Add. What is the sum in simplest form?

1. $\frac{1}{8} + \frac{3}{4}$ **a.** $\frac{3}{32}$ **b.** $\frac{4}{8}$ **c.** $\frac{4}{12}$ **d.** $\frac{7}{8}$

2. $\frac{4}{5} + \frac{2}{3}$ **a.** $\frac{6}{15}$ **b.** $\frac{8}{15}$ **c.** $1\frac{7}{15}$ **d.** $\frac{6}{8}$

PRACTICE EXERCISES

Add. Write the sum in simplest form.

1. $\frac{1}{5} + \frac{2}{5}$ **2.** $\frac{3}{7} + \frac{2}{7}$ **3.** $\frac{5}{9} + \frac{2}{9}$ **4.** $\frac{2}{5} + \frac{3}{5}$ **5.** $\frac{3}{8} + \frac{5}{8}$

6. $\frac{3}{4} + \frac{3}{4}$ **7.** $\frac{5}{6} + \frac{2}{6}$ **8.** $\frac{3}{5} + \frac{4}{5}$ **9.** $\frac{5}{12} + \frac{9}{12}$ **10.** $\frac{4}{10} + \frac{8}{10}$

11. $\frac{1}{2} + \frac{2}{5}$ **12.** $\frac{2}{3} + \frac{1}{4}$ **13.** $\frac{2}{5} + \frac{1}{3}$ **14.** $\frac{1}{3} + \frac{3}{5}$ **15.** $\frac{2}{3} + \frac{2}{9}$

16. $\frac{3}{4} + \frac{2}{3}$ **17.** $\frac{1}{3} + \frac{6}{9}$ **18.** $\frac{3}{4} + \frac{3}{8}$ **19.** $\frac{2}{5} + \frac{3}{4}$ **20.** $\frac{3}{4} + \frac{4}{5}$

21. $\frac{1}{8} + \frac{2}{5}$ **22.** $\frac{2}{9} + \frac{4}{9}$ **23.** $\frac{1}{4} + \frac{8}{9}$ **24.** $\frac{3}{4} + \frac{1}{6}$ **25.** $\frac{5}{6} + \frac{3}{8}$

26. $\frac{3}{4} + \frac{5}{12}$ **27.** $\frac{3}{16} + \frac{3}{8}$ **28.** $\frac{7}{16} + \frac{3}{16}$ **29.** $\frac{3}{10} + \frac{5}{8}$ **30.** $\frac{1}{6} + \frac{4}{15}$

31. $\frac{1}{2} + \frac{1}{2} + \frac{1}{3}$ **32.** $\frac{1}{8} + \frac{1}{3} + \frac{1}{4}$ **33.** $\frac{1}{2} + \frac{1}{4} + \frac{5}{6}$ **34.** $\frac{2}{5} + \frac{1}{2} + \frac{3}{4}$

35. $\frac{9}{24} + \frac{3}{8}$ **36.** $\frac{1}{2} + \frac{3}{7}$ **37.** $\frac{3}{4} + \frac{7}{10}$ **38.** $\frac{3}{8} + \frac{7}{8}$ **39.** $\frac{1}{4} + \frac{3}{5}$ **40.** $\frac{3}{5} + \frac{5}{6}$

41. $\frac{1}{3} + \frac{2}{3} + \frac{1}{4}$ **42.** $\frac{1}{2} + \frac{1}{8} + \frac{3}{4}$ **43.** $\frac{4}{9} + \frac{1}{6} + \frac{2}{3}$ **44.** $\frac{1}{8} + \frac{1}{6} + \frac{1}{4}$ **45.** $\frac{3}{5} + \frac{1}{4} + \frac{5}{8}$ **46.** $\frac{2}{3} + \frac{3}{8} + \frac{1}{6}$

Solve.

47. Akira spends $\frac{3}{8}$ of her income on rent and $\frac{1}{12}$ of her income on utilities. What part of her income is spent on housing?

48. Sally spends $\frac{1}{5}$ of her income on food. If she earns \$450 per week, how much money does she spend on food?

49. Carol spends $\frac{1}{3}$ of her pay for food, $\frac{1}{4}$ for rent, and $\frac{1}{8}$ for entertainment. The rest she saves. What part of her pay does Carol spend?

50. Kyle earns \$2,500 per month. He spends \$250 per month on entertainment. What part of his salary does he spend on entertainment?

6.8 ADDING MIXED NUMBERS

Nicole Williams rides her bicycle $8\frac{1}{10}$ miles from Ellisville to Norton, then $3\frac{3}{10}$ miles from Norton to Millstown. How far does she ride?

To find the total distance, add $8\frac{1}{10}$ miles and $3\frac{3}{10}$ miles.

EXAMPLE 1 Add: $8\frac{1}{10} + 3\frac{3}{10}$

1. Add the fractions.

$$\begin{array}{r} 8\frac{1}{10} \\ +3\frac{3}{10} \\ \hline \frac{4}{10} \end{array}$$

2. Add the whole numbers.

$$\begin{array}{r} 8\frac{1}{10} \\ +3\frac{3}{10} \\ \hline 11\frac{4}{10} \end{array}$$

3. Write the answer in simplest form.

$$\begin{array}{r} 8\frac{1}{10} \\ +3\frac{3}{10} \\ \hline 11\frac{4}{10} \end{array} = 11\frac{2}{5}$$

The distance from Ellisville to Millstown is $11\frac{2}{5}$ miles.

Sometimes you need to rename the sum of the fractions.

EXAMPLE 2 Add: $2\frac{3}{8} + 4\frac{7}{8}$

1. Add the fractions. Rename.

$$\begin{array}{r} 2\frac{3}{8} \\ +4\frac{7}{8} \\ \hline \frac{10}{8} \end{array} = 1\frac{2}{8}$$

2. Add the whole numbers.

$$\begin{array}{r} \scriptstyle 1 \\ 2\frac{3}{8} \\ +4\frac{7}{8} \\ \hline 7\frac{2}{8} \end{array}$$

3. Write the answer in simplest form.

$$\begin{array}{r} 2\frac{3}{8} \\ +4\frac{7}{8} \\ \hline 7\frac{2}{8} \end{array} = 7\frac{1}{4}$$

Sometimes the fractions have different denominators.

EXAMPLE 3 Add: $1\frac{3}{5} + 2\frac{1}{10} + 4\frac{3}{4}$

1. Write equivalent fractions with a common denominator.

$$\begin{array}{rcr} 1\frac{3}{5} & = & 1\frac{12}{20} \\ 2\frac{1}{10} & = & 2\frac{2}{20} \\ +4\frac{3}{4} & = & 4\frac{15}{20} \\ \hline \end{array}$$

2. Add the fractions. Rename.

$$\begin{array}{rcr} 1\frac{3}{5} & = & 1\frac{12}{20} \\ 2\frac{1}{10} & = & 2\frac{2}{20} \\ +4\frac{3}{4} & = & 4\frac{15}{20} \\ \hline & & \frac{29}{20} \end{array} = 1\frac{9}{20}$$

3. Add the whole numbers.

$$\begin{array}{rcr} & & \scriptstyle 1 \\ 1\frac{3}{5} & = & 1\frac{12}{20} \\ 2\frac{1}{10} & = & 2\frac{2}{20} \\ +4\frac{3}{4} & = & 4\frac{15}{20} \\ \hline & & 8\frac{9}{20} \end{array}$$

PRACTICE EXERCISES

Add. Write the sum in simplest form.

1. $5\frac{1}{8} + 3\frac{3}{8}$ **2.** $2\frac{2}{5} + 4\frac{2}{5}$ **3.** $4\frac{3}{7} + 2\frac{2}{7}$ **4.** $8\frac{5}{9} + 7\frac{4}{9}$ **5.** $9\frac{3}{4} + 3\frac{3}{4}$

6. $4\frac{3}{8} + 3\frac{1}{4}$ **7.** $6\frac{3}{4} + 7\frac{1}{6}$ **8.** $5\frac{2}{3} + 6\frac{2}{9}$ **9.** $9\frac{2}{3} + 3\frac{5}{6}$ **10.** $7\frac{3}{4} + 7\frac{5}{8}$

11. $7\frac{3}{4} + 4\frac{2}{3}$ **12.** $9\frac{2}{3} + 2\frac{3}{5}$ **13.** $8\frac{1}{2} + 8\frac{2}{3}$ **14.** $6\frac{3}{4} + 3\frac{5}{6}$ **15.** $9\frac{2}{3} + 9\frac{4}{5}$

16. $12\frac{3}{4} + 4\frac{3}{8}$ **17.** $3\frac{11}{12} + 1\frac{1}{8}$ **18.** $23\frac{1}{2} + 5\frac{1}{6}$ **19.** $38\frac{2}{3} + 2\frac{4}{15}$ **20.** $42\frac{7}{12} + 31\frac{5}{16}$

21. $3 + 2\frac{7}{8}$ **22.** $9\frac{3}{5} + 6$ **23.** $18\frac{1}{6} + 7$ **24.** $13 + 12\frac{2}{3}$ **25.** $15\frac{4}{5} + 18$

26. $5\frac{1}{6} + \frac{1}{6}$ **27.** $6\frac{2}{5} + 4\frac{1}{5}$ **28.** $4\frac{1}{2} + 6\frac{1}{3}$ **29.** $\frac{1}{4} + 8\frac{2}{3}$ **30.** $5 + 6\frac{1}{2}$

31. $\begin{array}{r} 7\frac{2}{5} \\ +\ \frac{3}{4} \\ \hline \end{array}$ **32.** $\begin{array}{r} 6\frac{3}{10} \\ +5\frac{4}{10} \\ \hline \end{array}$ **33.** $\begin{array}{r} 8 \\ +6\frac{4}{5} \\ \hline \end{array}$ **34.** $\begin{array}{r} 5\frac{5}{6} \\ +2\frac{2}{5} \\ \hline \end{array}$ **35.** $\begin{array}{r} 9\frac{5}{12} \\ +7\frac{7}{12} \\ \hline \end{array}$

36. $\begin{array}{r} 4\frac{1}{4} \\ 1\frac{3}{4} \\ +8\frac{3}{4} \\ \hline \end{array}$ **37.** $\begin{array}{r} 5\frac{1}{2} \\ 4\frac{1}{3} \\ +3\frac{3}{4} \\ \hline \end{array}$ **38.** $\begin{array}{r} 6\frac{1}{12} \\ 7\frac{2}{3} \\ +18\frac{3}{4} \\ \hline \end{array}$ **39.** $\begin{array}{r} \frac{5}{9} \\ 18 \\ +7\frac{1}{3} \\ \hline \end{array}$ **40.** $\begin{array}{r} 12\frac{3}{8} \\ 54 \\ +\ \frac{7}{12} \\ \hline \end{array}$

Solve. Use the map at the right.

41. Susan lives $3\frac{1}{2}$ miles north of Millstown. Bill lives $2\frac{2}{5}$ miles west of Millstown. How far does Susan live from Bill?

42. To drive from Norton to Sunset Mt., you must pass through Millstown. How far is Sunset Mt. from Norton?

43. The distance from Norton to Brownsville is $1\frac{1}{3}$ times the distance from Norton to Ellisville. How far is Brownsville from Norton?

44. Larry jogged from Millstown to Sunset Mt. in $1\frac{3}{4}$ hours. What was his average rate of speed?

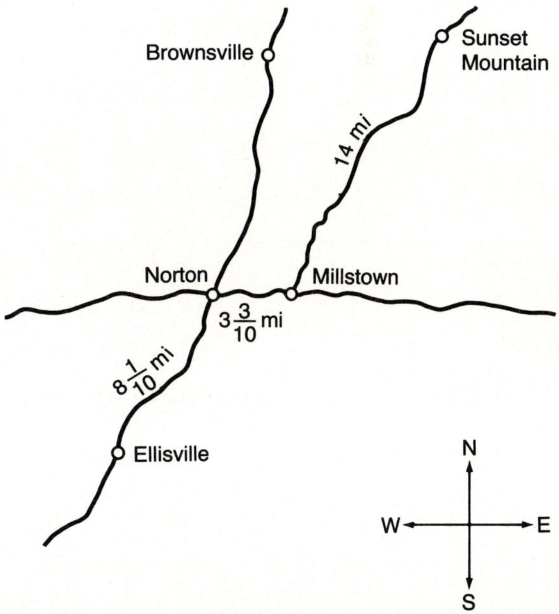

6.9 SUBTRACTING FRACTIONS

Tom Miller worked in his yard for $\frac{3}{4}$ of an hour. He spent $\frac{1}{4}$ of an hour cutting the lawn and the rest of the time raking leaves. How much time did Tom spend raking leaves?

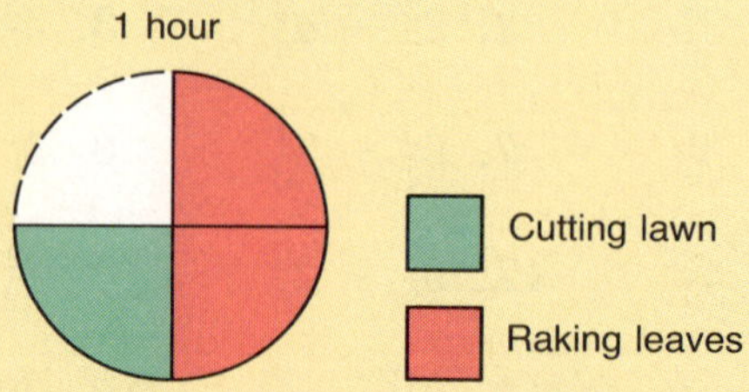

THINK: You know the total and one part. You can subtract $\frac{1}{4}$ from $\frac{3}{4}$ to find the missing part.

EXAMPLE 1

Subtract: $\frac{3}{4} - \frac{1}{4}$

1. Subtract the numerators.

   $\frac{3}{4} - \frac{1}{4} = 2$

2. Write the difference over the denominator.

   $\frac{3}{4} - \frac{1}{4} = \frac{2}{4}$

3. Write the answer in simplest form.

   $\frac{2}{4} = \frac{1}{2}$

Tom spent $\frac{1}{2}$ of an hour raking leaves.

Sometimes you need to subtract fractions with different denominators.

EXAMPLE 2

Subtract: $\frac{11}{15} - \frac{2}{5}$

1. Write equivalent fractions with a common denominator.

   $\frac{11}{15} = \frac{11}{15}$

   $-\frac{2}{5} = \frac{6}{15}$

2. Subtract.

   $\frac{11}{15} = \frac{11}{15}$

   $-\frac{2}{5} = \frac{6}{15}$

   $\frac{5}{15}$

3. Write the answer in simplest form.

   $\frac{5}{15} = \frac{1}{3}$

CHECKPOINT

Write the letter of the correct answer.

Subtract. What is the difference in simplest form?

1. $\frac{5}{6} - \frac{1}{6}$ **a.** $\frac{4}{6}$ **b.** $\frac{4}{12}$ **c.** $\frac{2}{3}$ **d.** $\frac{1}{3}$

2. $\frac{7}{9} - \frac{1}{3}$ **a.** $\frac{6}{27}$ **b.** $\frac{6}{6}$ **c.** $\frac{2}{3}$ **d.** $\frac{4}{9}$

3. $\frac{5}{7} - \frac{2}{3}$ **a.** $\frac{3}{4}$ **b.** $\frac{3}{21}$ **c.** $\frac{3}{10}$ **d.** $\frac{1}{21}$

4. $\frac{3}{5} - \frac{2}{9}$ **a.** $\frac{17}{45}$ **b.** $\frac{17}{35}$ **c.** $\frac{11}{45}$ **d.** $\frac{1}{5}$

PRACTICE EXERCISES

Subtract. Write the difference in simplest form.

1. $\frac{2}{3} - \frac{1}{3}$	**2.** $\frac{4}{5} - \frac{2}{5}$	**3.** $\frac{7}{8} - \frac{4}{8}$	**4.** $\frac{4}{6} - \frac{3}{6}$	**5.** $\frac{5}{7} - \frac{2}{7}$
6. $\frac{7}{9} - \frac{5}{9}$	**7.** $\frac{5}{8} - \frac{3}{8}$	**8.** $\frac{7}{12} - \frac{5}{12}$	**9.** $\frac{9}{16} - \frac{3}{16}$	**10.** $\frac{17}{24} - \frac{11}{24}$
11. $\frac{3}{4} - \frac{1}{2}$	**12.** $\frac{5}{6} - \frac{2}{3}$	**13.** $\frac{3}{4} - \frac{3}{8}$	**14.** $\frac{2}{3} - \frac{2}{9}$	**15.** $\frac{1}{2} - \frac{3}{8}$
16. $\frac{7}{8} - \frac{1}{4}$	**17.** $\frac{11}{12} - \frac{2}{3}$	**18.** $\frac{13}{16} - \frac{3}{4}$	**19.** $\frac{7}{12} - \frac{1}{4}$	**20.** $\frac{9}{10} - \frac{3}{5}$
21. $\frac{2}{3} - \frac{1}{2}$	**22.** $\frac{2}{3} - \frac{2}{5}$	**23.** $\frac{3}{4} - \frac{2}{3}$	**24.** $\frac{1}{2} - \frac{2}{5}$	**25.** $\frac{2}{3} - \frac{5}{8}$
26. $\frac{2}{3} - \frac{3}{8}$	**27.** $\frac{5}{6} - \frac{3}{8}$	**28.** $\frac{8}{9} - \frac{5}{6}$	**29.** $\frac{4}{5} - \frac{4}{7}$	**30.** $\frac{5}{9} - \frac{1}{2}$
31. $\frac{11}{12} - \frac{3}{12}$	**32.** $\frac{2}{3} - \frac{2}{15}$	**33.** $\frac{7}{10} - \frac{7}{12}$	**34.** $\frac{19}{21} - \frac{4}{7}$	**35.** $\frac{9}{10} - \frac{3}{10}$
36. $\frac{5}{12} - \frac{3}{14}$	**37.** $\frac{11}{12} - \frac{5}{18}$	**38.** $\frac{11}{18} - \frac{5}{9}$	**39.** $\frac{15}{16} - \frac{13}{16}$	**40.** $\frac{13}{24} - \frac{12}{36}$

MIXED REVIEW

Multiply or divide.

41. 23×9	**42.** 187×40	**43.** $7{,}301 \times 568$	**44.** $24{,}664 \times 100$	**45.** $32{,}412 \times 372$
46. 3)204	**47.** 33)1,914	**48.** 100)5,869	**49.** 541)8,926.5	**50.** 89)3,675.7

Solve.

51. Gerry mows the lawn in $\frac{2}{3}$ of an hour. Matt mows the lawn in $\frac{3}{4}$ of an hour. How much longer does it take Matt to mow the lawn?

52. It takes Frank $\frac{5}{6}$ of an hour to mow and edge the lawn. He can mow the lawn in $\frac{2}{3}$ of an hour. How long does it take him to edge the lawn?

53. Esther charges $5.25 per hour for doing lawn work. How much would she charge for $3\frac{2}{3}$ hours?

54. Fran earned $13.75 for working $2\frac{1}{2}$ hours mowing lawns. What was her average hourly rate of pay?

6.10 SUBTRACTING MIXED NUMBERS

A mutton snapper weighed $9\frac{7}{8}$ pounds. A red grouper weighed $5\frac{3}{8}$ pounds. How much more did the mutton snapper weigh?

THINK: You are comparing weights. Subtract to find how much more the mutton snapper weighs.

EXAMPLE 1

Subtract: $9\frac{7}{8} - 5\frac{3}{8}$

1. Subtract the fractions.

$$\begin{array}{r} 9\frac{7}{8} \\ -5\frac{3}{8} \\ \hline \frac{4}{8} \end{array}$$

2. Subtract the whole numbers.

$$\begin{array}{r} 9\frac{7}{8} \\ -5\frac{3}{8} \\ \hline 4\frac{4}{8} \end{array}$$

3. Write the answer in simplest form.

$$4\frac{4}{8} = 4\frac{1}{2}$$

The mutton snapper weighs $4\frac{1}{2}$ lb more than the red grouper.

Sometimes you need to rename when subtracting mixed numbers.

EXAMPLE 2

Subtract: $4\frac{1}{4} - 2\frac{3}{4}$

1. Rename. Subtract the fractions.

$$\begin{array}{rcr} 4\frac{1}{4} & = & 3\frac{5}{4} \\ -2\frac{3}{4} & = & 2\frac{3}{4} \\ \hline & & \frac{2}{4} \end{array}$$

2. Subtract the whole numbers.

$$\begin{array}{rcr} 4\frac{1}{4} & = & 3\frac{5}{4} \\ -2\frac{3}{4} & = & 2\frac{3}{4} \\ \hline & & 1\frac{2}{4} \end{array}$$

3. Write the answer in simplest form.

$$1\frac{2}{4} = 1\frac{1}{2}$$

Sometimes you need to rename twice when subtracting mixed numbers.

EXAMPLE 3

Subtract: $8\frac{1}{4} - 5\frac{5}{6}$

1. Write equivalent fractions with a common denominator.

$$\begin{array}{rcr} 8\frac{1}{4} & = & 8\frac{3}{12} \\ -5\frac{5}{6} & = & 5\frac{10}{12} \end{array}$$

2. Rename if necessary. Subtract.

$$\begin{array}{rcr} 8\frac{3}{12} & = & 7\frac{15}{12} \\ -5\frac{10}{12} & = & 5\frac{10}{12} \\ \hline & & 2\frac{5}{12} \end{array}$$

CHECKPOINT

Write the letter of the correct answer.

Subtract. What is the difference in simplest form?

1. $8\frac{3}{4} - 5\frac{1}{4}$ **a.** $3\frac{2}{4}$ **b.** $2\frac{1}{2}$ **c.** $2\frac{2}{4}$ **d.** $3\frac{1}{2}$

2. $7\frac{11}{12} - 3\frac{3}{4}$ **a.** $4\frac{1}{6}$ **b.** $4\frac{8}{12}$ **c.** $4\frac{2}{12}$ **d.** $4\frac{8}{8}$

3. $4\frac{1}{3} - 1\frac{2}{5}$ **a.** $3\frac{1}{2}$ **b.** $3\frac{1}{15}$ **c.** $\frac{14}{15}$ **d.** $2\frac{14}{15}$

PRACTICE EXERCISES

Subtract. Write the difference in simplest form.

1. $4\frac{2}{3} - 2\frac{1}{3}$ **2.** $6\frac{4}{5} - 3\frac{2}{5}$ **3.** $5\frac{5}{8} - 1\frac{3}{8}$ **4.** $7\frac{5}{6} - 4\frac{2}{6}$ **5.** $9\frac{6}{7} - 6\frac{2}{7}$

6. $6\frac{7}{8} - 3\frac{1}{4}$ **7.** $18\frac{5}{6} - 5\frac{2}{3}$ **8.** $13\frac{7}{9} - 11\frac{1}{6}$ **9.** $5\frac{3}{4} - 1\frac{2}{3}$ **10.** $11\frac{2}{3} - 9\frac{1}{2}$

11. $9\frac{2}{3} - 6\frac{1}{5}$ **12.** $10\frac{11}{12} - 3\frac{5}{8}$ **13.** $8\frac{13}{16} - 2\frac{1}{4}$ **14.** $7\frac{4}{5} - 2\frac{1}{4}$ **15.** $16\frac{8}{9} - 7\frac{2}{3}$

16. $8\frac{3}{4} - 3\frac{5}{6}$ **17.** $10\frac{1}{3} - 3\frac{1}{2}$ **18.** $9\frac{2}{3} - 8\frac{5}{6}$ **19.** $13\frac{3}{8} - 1\frac{3}{4}$ **20.** $24\frac{2}{9} - 17\frac{5}{12}$

21. $7\frac{1}{4} - 3\frac{1}{2}$ **22.** $8\frac{2}{3} - 1\frac{4}{5}$ **23.** $10\frac{1}{6} - 4\frac{5}{8}$ **24.** $15\frac{5}{12} - 6\frac{1}{2}$ **25.** $11\frac{1}{6} - 7\frac{3}{5}$

26. $9\frac{3}{4} - 6\frac{1}{2}$ **27.** $7\frac{5}{12} - \frac{5}{6}$ **28.** $12\frac{8}{9} - 6\frac{5}{9}$ **29.** $4\frac{2}{3} - 3$ **30.** $8\frac{4}{5} - \frac{3}{5}$

31. $23\frac{11}{16} - 14\frac{3}{4}$ **32.** $37\frac{3}{8} - \frac{7}{12}$ **33.** $4\frac{3}{8} - 4$ **34.** $17\frac{1}{8} - 6\frac{4}{5}$ **35.** $8 - 3\frac{2}{5}$

Write +, −, ×, or ÷ to make the sentence true.

36. $2\frac{1}{4}$ ● $3\frac{1}{3} = 7\frac{1}{2}$ **37.** $6\frac{3}{8}$ ● $5\frac{1}{6} = 11\frac{13}{24}$ **38.** $5\frac{1}{2}$ ● $2\frac{3}{4} = 2$ **39.** $9\frac{1}{6}$ ● $4\frac{5}{8} = 4\frac{13}{24}$

MIXED REVIEW

Add or subtract.

40. 6.21 + 4.7 **41.** 43.344 + 5.03 **42.** 17.4 + 2.28 **43.** 72.29 + 6.15 **44.** 83.156 + 41.97

45. 6 − 4.2 **46.** 87.3 + 104.39 **47.** 28.3 − 12.5 **48.** 9 + 47.6

Solve.

49. The world record for a green sunfish is $2\frac{1}{8}$ lb. The record for red-ear sunfish is $4\frac{1}{2}$ lb. How much heavier is the red-ear sunfish?

50. Camila caught a bluegill weighing $3\frac{7}{8}$ lb. The world record for a bluegill is $4\frac{3}{4}$ lb. How much less than the world record did Camila's fish weigh?

51. The world record for a white marlin is $181\frac{7}{8}$ lb. Larry caught a white marlin that weighed $\frac{4}{5}$ of the record weight. How much did Larry's marlin weigh?

52. Eric caught 3 mutton snappers that weighed $4\frac{1}{2}$ lb, $6\frac{2}{3}$ lb, and $5\frac{3}{4}$ lb. He sold the fish for \$3 per pound. How much money did he earn?

53. Joyce caught four perch during the tournament. Their weights were: 1 lb; $1\frac{1}{4}$ lb; $1\frac{1}{2}$ lb; $1\frac{3}{4}$ lb. Find the average weight of the fish.

54. List the lengths of these fish in order from the longest to the shortest.

Contestant	Kevin	Sarah	John	Glenda
Length of pike	2 ft	20 in.	$1\frac{3}{4}$ ft	$1\frac{5}{6}$ ft

TIME OUT

You are the bookkeeper at Daisy's Diner.
Every week the employees give you a time sheet.
This week they played a joke.
They gave their times in a puzzle.

Use the clues below to match each employee with the number of hours he or she worked.

Employee	Lonnie	Donna	Bob	Steve	Ken
Hours	$42\frac{1}{4}$	46	$39\frac{1}{2}$	$36\frac{1}{4}$	$38\frac{1}{2}$

- Lonnie worked $2\frac{3}{4}$ fewer hours than Donna.
- Donna worked more hours than Steve but fewer than Ken.
- Bob worked $7\frac{1}{4}$ hours a day for 5 days.
- If you divide the number of hours that Steve worked by 7, the answer would be $5\frac{1}{2}$.
- Ken worked $7\frac{1}{2}$ hours more than Steve.

PROBLEM Solving APPLICATION

6.11 CONSUMER: DETERMINING CAB FARES

You enter a taxicab in Ocean City and see this sign. You know that your destination is about 1 mi away. How much will your fare be?

OCEAN CITY RATES
75¢ first $\frac{1}{9}$ mi
15¢ each additional $\frac{1}{9}$ mi

1. What is the cost of the first $\frac{1}{9}$ mi? $0.75
2. How much longer is the trip?

   THINK: 1 mi = $\frac{9}{9}$ mi $\frac{9}{9} - \frac{1}{9} = \frac{8}{9}$

3. What is the cost of the $\frac{8}{9}$ mi? $8 \times \$0.15 = \$1.20$
4. What is the total cost? $\$0.75 + \$1.20 = \$1.95$

So, the total fare for your trip is $1.95.

Use the sign above to answer Exercises 1–4.
Remember to estimate whenever you use your calculator.

What is the cost of a trip of:

1. $1\frac{4}{9}$ mi? **2.** 2 mi? **3.** $4\frac{1}{9}$ mi? **4.** $3\frac{2}{3}$ mi?

Taxicab rates for New London are posted at the right. Use these rates to answer Exercises 5–12.

NEW LONDON RATES
85¢ first $\frac{1}{5}$ mi
35¢ each additional $\frac{1}{5}$ mi

What is the cost of a trip of:

5. 1 mi?

6. 2 mi?

7. 3 mi?

8. 5 mi?

9. 10 mi?

10. Does a trip of 2 mi cost twice as much as a trip of 1 mi?

11. Does a trip of 1 mi cost exactly $\frac{1}{3}$ as much as a trip of 3 mi?

12. What amount will the meter show for a trip of $4\frac{1}{2}$ mi? Explain.

6.12 ESTIMATING WITH MIXED NUMBERS

Ray Harper is a tailor for the Main Event Circus. He is taking an inventory of his supply of fabrics. Here are two situations in which he estimates with mixed numbers.

One roll has $6\frac{3}{4}$ yd of blue satin. Another roll has $8\frac{1}{3}$ yd of blue satin. About how many yards of blue satin are in stock?

Ray needs $24\frac{1}{2}$ yd of beaded material. He has $10\frac{1}{6}$ yd. About how many more yards should he order?

EXAMPLE 1

Estimate:

a. $6\frac{3}{4} + 8\frac{1}{3}$

1. Round. $6\frac{3}{4} \rightarrow 7$, $+8\frac{1}{3} \rightarrow 8$
2. Add. 15

Ray has about 15 yd of blue satin in stock.

b. $24\frac{1}{2} - 10\frac{1}{6}$

1. Round. $24\frac{1}{2} \rightarrow 25$, $-10\frac{1}{6} \rightarrow 10$
2. Subtract. 15

Ray should order about 15 yd of beaded material.

Use the same method to estimate products and quotients.

EXAMPLE 2

Estimate:

a. $7\frac{3}{4} \times 8\frac{1}{8}$

1. Round. $7\frac{3}{4} \times 8\frac{1}{8}$ → 8×8
2. Multiply. $8 \times 8 = 64$

$7\frac{3}{4} \times 8\frac{1}{8}$ is about 64.

b. $5\frac{2}{3} \div 9\frac{3}{7}$

1. Round. $5\frac{2}{3} \div 9\frac{3}{7}$ → $6 \div 9$
2. Divide. $6 \div 9 = \frac{6}{9} = \frac{2}{3}$

$5\frac{2}{3} \div 9\frac{3}{7}$ is about $\frac{2}{3}$.

FOR DISCUSSION

How will the estimated sum compare with the actual answer:

1. If you round both mixed numbers up?

2. If you round both mixed numbers down?

3. If you round one mixed number up and the other down?

PRACTICE EXERCISES

Round to the nearest whole number.

1. $4\frac{1}{2}$
2. $6\frac{1}{3}$
3. $9\frac{3}{4}$
4. $3\frac{3}{5}$
5. $8\frac{3}{7}$
6. $13\frac{1}{4}$
7. $23\frac{1}{2}$
8. $71\frac{4}{5}$
9. $65\frac{6}{11}$
10. $81\frac{6}{13}$

Estimate.

11. $3\frac{1}{4} + 4\frac{2}{3}$
12. $8\frac{1}{7} + 3\frac{1}{3}$
13. $9\frac{5}{8} + 4\frac{3}{4}$
14. $6\frac{2}{3} + 5\frac{1}{6}$
15. $7\frac{5}{9} + 4\frac{2}{3}$
16. $3\frac{1}{3} - 2\frac{1}{4}$
17. $4\frac{1}{2} - 1\frac{2}{3}$
18. $5\frac{7}{8} - 1\frac{2}{9}$
19. $6\frac{1}{3} - 4\frac{2}{5}$
20. $8\frac{1}{2} - 2\frac{3}{8}$
21. $5\frac{1}{5} \times 1\frac{3}{4}$
22. $6\frac{1}{2} \times \frac{7}{8}$
23. $2\frac{3}{4} \times 9\frac{2}{3}$
24. $4\frac{3}{5} \times 6\frac{5}{6}$
25. $3\frac{1}{2} \times 6\frac{1}{8}$
26. $21\frac{1}{3} \div 6\frac{2}{3}$
27. $5\frac{3}{4} \div \frac{7}{8}$
28. $5\frac{3}{4} \div 2\frac{3}{4}$
29. $4\frac{3}{8} \div 2\frac{1}{8}$
30. $15\frac{5}{6} \div 3\frac{3}{4}$
31. $4\frac{1}{5} - 2\frac{1}{6}$
32. $12\frac{1}{3} \times 3\frac{1}{5}$
33. $14\frac{2}{3} \div 5\frac{1}{4}$
34. $23\frac{4}{5} + 3\frac{1}{6}$
35. $5\frac{5}{6} \times 21\frac{3}{8}$
36. $9\frac{3}{8} \div 2\frac{9}{16}$
37. $9\frac{3}{4} + 14\frac{3}{8}$
38. $18\frac{2}{5} - 15\frac{5}{6}$
39. $8\frac{3}{7} - 7\frac{5}{6}$
40. $8\frac{1}{2} \times 31\frac{2}{3}$

RAY HARPER'S FILE OF ANIMAL SIZES

Animal	Neck size	Sleeve length (ft)	Waist size (ft)
Ollie the Ostrich	$12\frac{3}{4}$ in.	—	—
Sparkle the Monkey	$8\frac{1}{2}$ in.	3	$2\frac{1}{2}$
Teddy the Tiger	$3\frac{1}{4}$ ft	$1\frac{1}{12}$	$4\frac{2}{3}$
Minnie the Elephant	$8\frac{3}{4}$ ft	$2\frac{5}{6}$	$15\frac{7}{12}$

Solve. Use the data from the table to estimate the answer.

41. About how much larger is Minnie's waist size than Sparkle's?

42. About how many times larger is Minnie's neck size than Ollie's?

43. Ray is making a new costume for Teddy. The sleeves will be blue satin. Estimate the amount of blue satin that Ray needs. (*Remember:* A tiger has four legs.)

44. Ray wants to make 6 bow ties for Ollie. Each bow tie needs $1\frac{1}{2}$ ft of fabric. He has 6 ft of fabric. How many bow ties can he make? How much more fabric does he need to make all 6 bow ties?

PROBLEM Solving APPLICATION

6.13 CONSUMER: BUYING ON AN INSTALLMENT PLAN

Many stores offer customers different ways to pay for purchases. If you pay for an item all at once, you pay the **cash price.** If you buy the item using a number of payments over a period of time, you are buying the item in **installments.** In this case, the store is providing the credit, and there is a **finance charge** for this service.

You buy a couch and a chair at Helpful Harry's Furniture Company. The cash price of the furniture is $840. You decide to pay in 12 monthly installments. The installments are $76.30 per month. What is the total amount that you pay for the couch and the chair if you use the installment plan? What is the finance charge?

	NUMBER OF INSTALLMENTS	×	AMOUNT OF EACH INSTALLMENT	=	TOTAL AMOUNT
1. Multiply to find the total amount.	12	×	$76.30	=	$915.60

	TOTAL	–	CASH PRICE	=	FINANCE CHARGE
2. Subtract the cash price from the total to find the finance charge.	$915.60	–	$840.00	=	$75.60

So, the total amount you pay is $915.60. The finance charge is $75.60.

Find the total amount and the finance charge.
Remember to estimate whenever you use your calculator.

	Cash Price	No. of Installments	Amount of Each Installment	Total Amount	Finance Charge
1.	$ 480	12	$ 43.60	■	■
2.	$ 720	12	$ 65.40	■	■
3.	$ 540	12	$ 49.05	■	■
4.	$1,092	12	$ 99.19	■	■
5.	$2,400	24	$120.00	■	■
6.	$3,600	24	$175.00	■	■
7.	$5,000	36	$145.02	■	■
8.	$3,895	48	$109.87	■	■

Solve.

9. The cash price of dining room furniture is $960. Mrs. Pannier buys it on an installment plan. She pays in 12 monthly installments, $87.20 per month. What is the total amount she pays? What is the finance charge?

10. Doreen Washington buys a stereo system on an installment plan. The total amount of money she pays is $786. If each installment is $\frac{1}{6}$ of the total amount, how much is each installment?

11. Pat Chang buys a dining room set on an installment plan. He pays in 12 monthly installments, $71.35 per month. When he has finished paying the installments, the total amount he has paid will be $69.40 more than the cash price. What is the cash price?

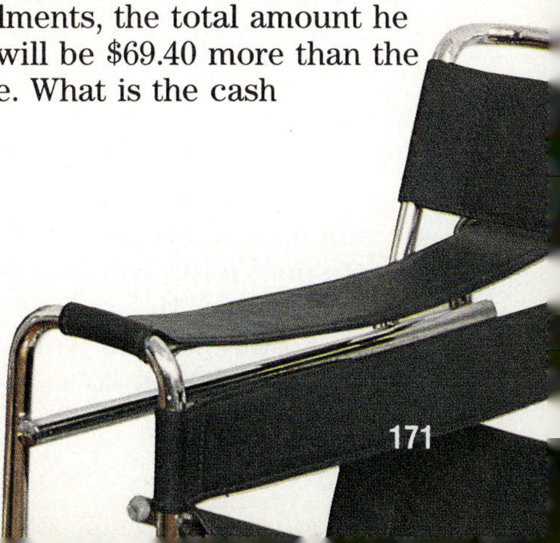

DECISION MAKING

6.14 DO IT YOURSELF

Some household projects may appear too difficult or time consuming to do them yourself. However, estimated costs for a professional job often cause people to reevaluate their options.

PROBLEM

The exterior of the Acevedo family's two-story frame house needs painting and should be done this spring. The estimates from the painting contractors seem high. Therefore, they have decided to develop a plan for painting the house themselves. First they collect all the available data.

Estimates:

Busy Bee Painters
Est. 1986

Paint: $150
Labor: $3,000
Sanding and scraping as necessary: $20 per hour
Time: 5 days

Ace Painters
Est. 1939

Paint: $200 for siding
$ 75 for trim
Labor: $3,500 (includes scraping and sanding)
Time: 3 days

Do It Ourselves

Paint: $225
Brushes: $60
Other supplies: $130
Ladder rental: $50/day (12 days)
Sander rental: $50/day (4 days)
Time: 6 weekends if Victor helps

Mr. and Mrs. Acevedo both have experience with interior painting. Their oldest son, Victor, who is seventeen, painted his bedroom last year. Fifteen-year-old Melissa and ten-year-old Danny have never done any painting. Both parents work full-time, and Victor and Melissa work after school and on Saturdays.

DECISION-MAKING FACTORS

Cost Time Experience

DECISION-MAKING COMPARISONS

Make a chart to compare the three estimates.

	Busy Bee	Ace	Ourselves
Cost	1. ■	2. ■	3. ■
Time	4. ■	5. ■	6. ■
Experience	Limited	7. ■	8. ■

MAKING THE DECISIONS

9. Which of the three estimates is the highest? Lowest?

10. Is the difference in time between the two professional painters significant?

11. Based on experience only, which group would probably do the best job?

12. Based on the given estimates, about how much money would the family save by doing the project themselves?

13. Victor makes about $75 a week, and two-thirds of this is earned on Saturday. About how much pay would he lose if he gave up his job for 6 weekends to help paint?

14. Should the pay Victor would lose be included in the cost of the project?

15. By working on weekends only, what factor must be considered when renting the ladder and sander?

16. Mr. Acevedo could take 1 week of vacation to do the painting, but Mrs. Acevedo would have to take a week off without pay ($750). Since the children would be in school and not available to help, is this a good idea? Why?

17. What differences between interior painting and exterior painting must the family consider?

18. The house is two stories high, and the Acevedos would have to work on long ladders for the first time. What other factor should be considered?

19. What factors could affect the time in any of the plans?

20. Given all the factors, do you think the Acevedos should paint the house themselves? Why?

CHAPTER REVIEW

Vocabulary Choose the letter of the word(s) that completes the statement.

1. If the product of two numbers equals 1, the numbers are ■. [150]

a. improper fractions **b.** reciprocals **c.** mixed numbers **d.** numerators

2. To multiply or divide mixed numbers, first change the mixed numbers to ■. [148, 152]

a. common factors **b.** addends **c.** improper fractions **d.** GCF

Skills Multiply. [146, 148]

$\frac{4}{9} \times \frac{3}{8}$ $\quad$ $\mathit{\frac{1}{6}}$

3. $\frac{2}{3} \times \frac{3}{4}$ **4.** $\frac{1}{6} \times \frac{3}{4}$ **5.** $\frac{7}{8} \times \frac{4}{5}$

6. $2\frac{2}{3} \times 1\frac{1}{6}$ **7.** $4\frac{1}{2} \times 3\frac{2}{3}$ **8.** $3\frac{3}{8} \times 2\frac{3}{4}$ **9.** $\frac{8}{11} \times \frac{2}{5}$

Divide. [150, 152]

$\frac{3}{4} \div \frac{3}{8}$ $\quad$ *2*

10. $\frac{3}{4} \div \frac{1}{2}$ **11.** $\frac{4}{15} \div \frac{3}{5}$ **12.** $3\frac{3}{4} \div 5$

13. $2\frac{2}{5} \div 1\frac{3}{4}$ **14.** $\frac{7}{10} \div \frac{15}{20}$ **15.** $1\frac{9}{15} \div 1\frac{4}{5}$ **16.** $12\frac{1}{2} \div \frac{5}{8}$

Add. [158, 160]

$\frac{2}{3} + \frac{3}{5} = \mathit{1\frac{4}{15}}$

17. $\frac{5}{9} + \frac{5}{6}$ **18.** $\frac{9}{15} + \frac{4}{6}$ **19.** $\frac{7}{16} + \frac{7}{8}$

20. $8\frac{3}{4} + 7\frac{3}{8}$ **21.** $12\frac{5}{8} + 24\frac{3}{5}$ **22.** $4\frac{3}{5} + 2\frac{4}{9}$ **23.** $6 + 5\frac{2}{3}$

24. $6\frac{1}{4} + 8$ **25.** $1\frac{2}{3} + 4\frac{4}{5}$ **26.** $\frac{1}{2} + \frac{1}{6} + \frac{5}{8}$ **27.** $\frac{3}{4} + \frac{1}{5} + \frac{7}{10}$

Subtract. [162, 164]

$\frac{5}{6} - \frac{1}{4} = \mathit{\frac{7}{12}}$

28. $\frac{5}{6} - \frac{3}{8}$ **29.** $\frac{2}{3} - \frac{2}{5}$ **30.** $\frac{15}{16} - \frac{1}{8}$

31. $6\frac{1}{4} - 4\frac{6}{7}$ **32.** $5 - 4\frac{2}{5}$ **33.** $15\frac{4}{7} - 11\frac{3}{8}$ **34.** $4\frac{5}{8} - 1\frac{9}{10}$

CHAPTER TEST

Multiply. Write the product in simplest form.

1. $\frac{1}{3} \times \frac{3}{4}$ **2.** $\frac{4}{5} \times \frac{1}{8}$ **3.** $\frac{1}{2} \times \frac{1}{2}$ **4.** $\frac{7}{11} \times \frac{1}{7}$ **5.** $\frac{6}{7} \times \frac{2}{3}$ **6.** $\frac{5}{6} \times \frac{9}{15}$

7. $9 \times \frac{2}{3}$ **8.** $\frac{1}{2} \times 8$ **9.** $1\frac{1}{2} \times 6$ **10.** $3\frac{3}{4} \times 2\frac{1}{2}$ **11.** $2\frac{1}{3} \times 5\frac{3}{6}$ **12.** $\frac{1}{3} \times 6\frac{1}{2}$

Divide. Write the quotient in simplest form.

13. $\frac{2}{5} \div \frac{3}{4}$ **14.** $\frac{1}{4} \div \frac{1}{2}$ **15.** $\frac{3}{5} \div 5$ **16.** $\frac{3}{4} \div \frac{7}{8}$ **17.** $9 \div \frac{2}{3}$ **18.** $\frac{4}{9} \div \frac{2}{3}$

19. $4 \div 2\frac{1}{2}$ **20.** $9 \div 5\frac{2}{3}$ **21.** $3\frac{3}{4} \div 2$ **22.** $\frac{5}{9} \div 1\frac{2}{3}$ **23.** $2\frac{1}{8} \div \frac{3}{4}$ **24.** $6\frac{4}{5} \div 2$

Add. Write the sum in simplest form.

25. $\frac{4}{7} + \frac{2}{7}$ **26.** $\frac{5}{8} + \frac{5}{8}$ **27.** $\frac{4}{9} + \frac{5}{9}$ **28.** $\frac{2}{5} + \frac{2}{3}$ **29.** $\frac{3}{4} + \frac{7}{12}$ **30.** $\frac{2}{3} + \frac{4}{9}$

31. $4\frac{1}{5} + 1\frac{3}{5}$ **32.** $3\frac{3}{8} + 3\frac{7}{8}$ **33.** $\frac{1}{3} + 5\frac{4}{6}$ **34.** $9\frac{4}{9} + 3\frac{7}{9}$ **35.** $12 + 5\frac{1}{6}$ **36.** $3\frac{1}{4} + 4\frac{3}{4}$

Subtract. Write the difference in simplest form.

37. $\frac{3}{4} - \frac{1}{4}$ **38.** $\frac{8}{9} - \frac{5}{9}$ **39.** $\frac{5}{6} - \frac{1}{3}$ **40.** $\frac{11}{12} - \frac{1}{4}$ **41.** $\frac{2}{3} - \frac{1}{8}$ **42.** $\frac{11}{12} - \frac{5}{6}$

43. $6\frac{4}{5} - 2\frac{1}{5}$ **44.** $8\frac{7}{8} - 3\frac{3}{8}$ **45.** $9\frac{2}{3} - 7\frac{1}{2}$ **46.** $10 - 4\frac{5}{6}$ **47.** $15\frac{7}{9} - 8\frac{2}{3}$ **48.** $11 - 6\frac{7}{9}$

Solve.

49. Karen has $2,500 in her checking account. She deposited $1,000 yesterday. Last week she deposited $500. How much money was in the account before she deposited the $500?

50. Terry Foster bought a couch on an installment plan. The total amount of money she paid is $575. If each installment is $\frac{1}{5}$ of the total amount, how much is each installment?

Use the rate table at the right to answer Exercises 51 and 52.

51. What is the cost of a taxi trip of 1 mi?

52. What is the cost of a taxi trip of 2 mi?

BRADLEY RATES

95¢ first $\frac{1}{5}$ mi

35¢ each additional $\frac{1}{5}$ mi

CUMULATIVE REVIEW

Choose the letter of the correct answer.

1. A(n) ▩ shows how many times a number or base is used as a factor.

a. exponent **b.** average **c.** product **d.** none of these

2. The ▩ of a set of data is the same as the average.

a. median **b.** mode **c.** mean **d.** none of these

3. Which of the following is greater than $\frac{5}{6}$?

a. $\frac{3}{4}$ **b.** $\frac{5}{7}$ **c.** $\frac{19}{24}$ **d.** none of these

4. A ▩ is a bar graph that shows the frequencies of intervals of data.

a. frequency table **b.** pictograph **c.** histogram **d.** none of these

5. A(n) ▩ fraction has a numerator equal to or greater than its denominator.

a. equivalent **b.** mixed number **c.** improper **d.** none of these

Select the best estimated answer.

6. 892×114	**a.** 800,000	**b.** 9,000	**c.** 18,000	**d.** 90,000
7. $7{,}198 \times 342$	**a.** 210,000	**b.** 2,100,000	**c.** 3,800,000	**d.** 280,000
8. $16{,}263 \div 829$	**a.** 20	**b.** 200	**c.** 2,000	**d.** 2
9. $4\frac{5}{8} \div \frac{3}{4}$	**a.** 9	**b.** 1	**c.** 5	**d.** 2
10. $432 + 1{,}983 + 96$	**a.** 2,500	**b.** 1,000	**c.** 600	**d.** 1,500

Compute.

11. $125\overline{)200}$	**a.** 0.16	**b.** 16	**c.** 1.6	**d.** none of these
12. $\frac{5}{8} \times \frac{1}{3}$	**a.** $1\frac{7}{8}$	**b.** $\frac{5}{24}$	**c.** $1\frac{5}{72}$	**d.** none of these
13. $\frac{360{,}000}{9{,}000}$	**a.** 40	**b.** 400	**c.** 4,000	**d.** none of these
14. $8\overline{)3{,}245}$	**a.** 405 R5	**b.** 45 R5	**c.** 4,005 R5	**d.** none of these
15. $75{,}000 \times 4{,}000$	**a.** 300,000	**b.** 3,000,000	**c.** 300,000,000	**d.** none of these

Solve.

16. Sal's Taxi Service charges 75¢ for the first mile and 35¢ for each additional mile. What is the cost of a 3-mi trip?

a. \$1.50 **b.** \$1.45 **c.** \$1.10 **d.** none of these

17. Jane wants to purchase a new car called the Gazelle which has a base price of \$10,826. She would like the optional air conditioning which costs \$552 and the automatic transmission which costs \$823. What is the total price of the car?

a. \$12,201 **b.** \$11,649 **c.** \$17,172 **d.** none of these

18. A board of lumber is $7\frac{7}{8}$ ft long. Into how many $\frac{7}{8}$-ft sections can it be cut?

a. $6\frac{11}{32}$ **b.** 8 **c.** 9 **d.** none of these

19. Dennis buys clothes through a mail order catalog. He orders 3 plaid shirts which cost \$16.75 each, 2 flannel shirts which are \$22.50 each, and 4 white dress shirts at \$24.95 each. Shipping costs are \$2.75. What is the total amount of his order?

a. \$193.05 **b.** \$193.95 **c.** \$183.05 **d.** none of these

20. The Carini family made a budget. They allotted 0.24 for food, 0.28 for rent, 0.36 for daily expenses, and 0.12 for vacation expenses. Which expense is $\frac{3}{25}$ of their total budget?

a. food **b.** vacation **c.** daily expenses **d.** none of these

THINKING ABOUT MATH

1. Mark, José, and Joel each belong to one of three sports teams. Mark and Joel are not on the basketball team. Joel does not play on the tennis team. Who plays on the baseball team? Describe the best method for solving the problem.

2. Tom ran the 100-yd dash several times. To find his average speed, would you calculate his median speed or his mean speed? Explain.

3. You need to make a graph showing the approximate number of cars manufactured in Detroit, Michigan, last year. Which graph would you use: a broken-line graph, a bar graph, or a pictograph? Why?

4. Explain how you can quickly tell that 10^8 is not the exponent form of 10,000,000.

PreSkills Test

Multiply. [68]

1. 100×35 **2.** $1{,}000 \times 22$ **3.** 100×2.4

4. 23.5×100 **5.** 200×10 **6.** $1{,}000 \times 1{,}000$

7. $2.73 \times 1{,}000$ **8.** 0.32×100 **9.** 0.006×100

10. 6.02×100 **11.** 10×0.075 **12.** $10{,}000 \times 46.0041$

Divide. [68]

13. $450 \div 10$ **14.** $3{,}400 \div 100$ **15.** $65 \div 10$

16. $2{,}000 \div 1{,}000$ **17.** $25 \div 1{,}000$ **18.** $64.5 \div 100$

19. $0.35 \div 10$ **20.** $1 \div 1{,}000$ **21.** $0.03 \div 10$

22. $80 \div 100$ **23.** $953 \div 10$ **24.** $645.3 \div 10{,}000$

Multiply. [40, 148]

25. 4×16 **26.** 5×12 **27.** $3 \times 1{,}760$

28. $5{,}280 \times 2$ **29.** $8 \times \frac{3}{4}$ **30.** $\frac{5}{8} \times 16$

31. $24 \times 2\frac{1}{3}$ **32.** $60 \times 1\frac{1}{4}$ **33.** $\frac{1}{2} \times 52$

Divide. Write the remainder as a fraction in simplest form. [46, 48]

34. $48 \div 8$ **35.** $128 \div 16$ **36.** $24 \div 16$

37. $30 \div 12$ **38.** $330 \div 60$ **39.** $70 \div 36$

40. $105 \div 7$ **41.** $108 \div 3$ **42.** $160 \div 32$

Add or subtract. [30, 32]

43. $\begin{array}{r} 1{,}760 \\ +\ \ 567 \\ \hline \end{array}$ **44.** $\begin{array}{r} 1{,}567 \\ +1{,}365 \\ \hline \end{array}$ **45.** $\begin{array}{r} 6{,}630 \\ -2{,}000 \\ \hline \end{array}$ **46.** $\begin{array}{r} 2{,}060 \\ -\ \ 500 \\ \hline \end{array}$

47. $\begin{array}{r} 28{,}159 \\ 306 \\ +\ 5{,}856 \\ \hline \end{array}$ **48.** $\begin{array}{r} 9{,}050 \\ -8{,}962 \\ \hline \end{array}$ **49.** $\begin{array}{r} 16{,}849 \\ 82 \\ +\ \ 106 \\ \hline \end{array}$ **50.** $\begin{array}{r} 20{,}147 \\ -16{,}586 \\ \hline \end{array}$

Chapter 7
MEASUREMENT

Measurement units have not always been standard. What problems may develop if people use different systems of measurement?

7.1 CUSTOMARY UNITS OF LENGTH

One type of measurement used in the United States are the Customary Units. The table shows the basic units of length.

LENGTH
12 inches (in.) = 1 foot (ft)
36 in. = 1 yard (yd)
3 ft = 1 yd
5,280 ft = 1 mile (mi)
1,760 yd = 1 mi

To change from a larger unit to a smaller unit, multiply.

EXAMPLE 1 Complete:

a. 2 mi = ■ yd

THINK: 1 mi = 1,760 yd

2 mi = (2 × 1,760) yd

2 mi = 3,520 yd

b. $6\frac{1}{2}$ ft = ■ in.

THINK: 1 ft = 12 in.

$6\frac{1}{2}$ ft = $\left(6\frac{1}{2} \times 12\right)$ in.

$6\frac{1}{2}$ ft = 78 in.

To change from a smaller unit to a larger unit, divide.

EXAMPLE 2 Complete:

a. 12 ft = ■ yd

THINK: 1 yd = 3 ft

12 ft = (12 ÷ 3) yd

12 ft = 4 yd

b. 54 in. = ■ yd

THINK: 1 yd = 36 in.

54 in. = (54 ÷ 36) yd

54 in. = $1\frac{1}{2}$ yd

When measuring, smaller units give more precise measurements.

EXAMPLE 3 What is the length of the pencil measured to the nearest $\frac{1}{16}$ in.?

1. Put the eraser at the 0 mark of the scale.

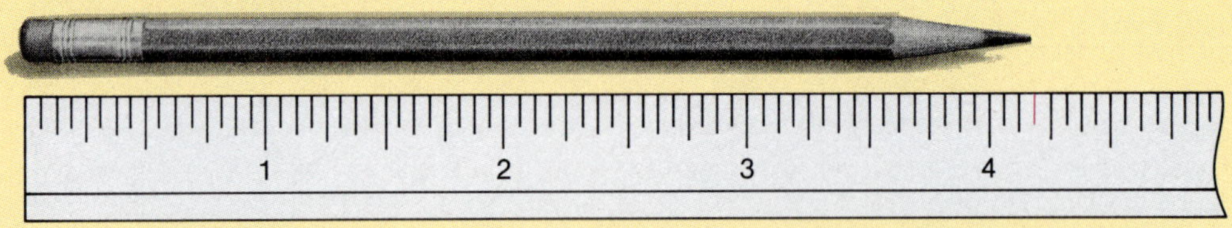

2. Read the mark on the ruler closest to the tip of the pencil.

The pencil is $4\frac{3}{16}$ in. long.

CHECKPOINT

Write the letter of the correct answer.

Complete.

1. $3\frac{1}{2}$ yd = ■ ft   **a.** $1\frac{1}{6}$ **b.** 7 **c.** $10\frac{1}{2}$

2. 48 in. = ■ yd   **a.** $1\frac{1}{3}$ **b.** 4 **c.** 1,728

PRACTICE EXERCISES

Complete.

1. 2 ft = ■ in.	**2.** 3 yd = ■ ft	**3.** 2 mi = ■ yd
4. 20 ft = ■ in.	**5.** 27 yd = ■ ft	**6.** 3 mi = ■ ft
7. 12 ft = ■ in.	**8.** 5 mi = ■ ft	**9.** 43 yd = ■ in.
10. 36 in. = ■ ft	**11.** 27 ft = ■ yd	**12.** 5,280 yd = ■ mi
13. 144 in. = ■ ft	**14.** 2,640 ft = ■ mi	**15.** 432 in. = ■ yd
16. 24 ft = ■ yd	**17.** 2,112 yd = ■ mi	**18.** 440 yd = ■ mi
19. $2\frac{1}{2}$ ft = ■ in.	**20.** $35\frac{2}{3}$ yd = ■ in.	**21.** 66 in. = ■ yd
22. 32 in. = ■ ft	**23.** $\frac{1}{2}$ mi = ■ ft	**24.** 1,760 ft = ■ mi
25. 880 yd = ■ mi	**26.** 27 in. = ■ ft	**27.** $3\frac{2}{3}$ yd = ■ ft

Measure the line segment to the nearest $\frac{1}{16}$ in.

28. ________

29. ________________________

30. ________________

31. ________________

32. ____

Solve.

33. The last event of the meet between Randolph and Fordham is the 440-yd relay. After 30 s, Randolph is at the 220-yd mark. Fordham has 760 ft left to go. Who is ahead?

34. In the 880-yd race, Frankie the Fordham Flash beats Joe "Turtle" James by 57 s. Slick Olivette beats James by 34 s. By how many seconds does Frankie beat Slick?

35. The Randolph High track team water bucket is 52 in. high. The Randolph mascot is 4 ft high. Is the mascot too tall to fit inside the bucket?

36. For his 6-mi workout, Speed Vogel of Randolph runs 24 laps around the school track. How many yards are in each lap?

7.2 CUSTOMARY UNITS OF CAPACITY AND WEIGHT

You can use these relationships to change from one unit to another.

CAPACITY	WEIGHT
8 fluid ounces (fl oz) = 1 cup (c)	16 ounces (oz) = 1 pound (lb)
2 c = 1 pint (pt)	2,000 lb = 1 ton (T)
2 pt = 1 quart (qt)	
4 qt = 1 gallon (gal)	

Dr. Thomas is driving a rebuilt 1964 convertible. She notices that the windshield washer is empty of fluid, so at the next gas station, she pulls over to refill it. She uses 6 c of water to refill the windshield washer. How many fluid ounces does she use?

To change from a larger unit to a smaller unit, multiply.

EXAMPLE 1

Complete: 6 c = ■ fl oz

THINK: 1 c = 8 fl oz 6 c = (6 × 8) fl oz
6 c = 48 fl oz

So, Dr. Thomas uses 48 fl oz of water.

To change from a smaller unit to a larger unit, divide.

EXAMPLE 2

At the gas station, Dr. Thomas remembers to buy a new oil filter for the car. The filter that Dr. Thomas buys weighs 24 oz. How many pounds does the filter weigh?

Complete: 24 oz = ■ lb

THINK: 1 lb = 16 oz 24 oz = (24 ÷ 16) lb
24 oz = $1\frac{1}{2}$ lb

So, the filter weighs $1\frac{1}{2}$ lb.

CHECKPOINT

Write the letter of the correct answer.

Complete.

1. 3 gal = ■ qt	**a.** 12	**b.** 6	**c.** $\frac{3}{4}$
2. $2\frac{3}{4}$ T = ■ lb	**a.** 44	**b.** 2,750	**c.** 5,500
3. 28 oz = ■ lb	**a.** $1\frac{3}{4}$	**b.** $3\frac{1}{2}$	**c.** 448

PRACTICE EXERCISES

Complete.

1. 4 c = ■ fl oz	**2.** 5 gal = ■ qt	**3.** 2 pt = ■ c
4. 8 lb = ■ oz	**5.** 3 c = ■ fl oz	**6.** 2 T = ■ lb
7. 4 gal = ■ qt	**8.** 5 pt = ■ c	**9.** 4 qt = ■ pt
10. $2\frac{1}{2}$ pt = ■ c	**11.** $4\frac{3}{8}$ lb = ■ oz	**12.** $\frac{3}{4}$ gal = ■ qt
13. 32 fl oz = ■ c	**14.** 4 pt = ■ qt	**15.** 8 c = ■ pt
16. 36 oz = ■ lb	**17.** 4,000 lb = ■ T	**18.** 12 qt = ■ gal
19. 20 fl oz = ■ c	**20.** 5 c = ■ pt	**21.** 2,500 lb = ■ T
22. 3,500 lb = ■ T	**23.** 16 qt = ■ gal	**24.** 12 pt = ■ qt
25. $2\frac{1}{4}$ lb = ■ oz	**26.** 3 T = ■ lb	**27.** 128 fl oz = ■ c
28. 1 oz = ■ lb	**29.** $3\frac{3}{4}$ c = ■ fl oz	**30.** 15 qt = ■ gal
31. $6\frac{1}{2}$ gal = ■ qt	**32.** 12,000 lb = ■ T	**33.** 20 qt = ■ pt
34. $3\frac{3}{8}$ lb = ■ oz	**35.** $2\frac{1}{5}$ T = ■ lb	**36.** 4 fl oz = ■ c
37. 1 pt = ■ fl oz	**38.** 1 qt = ■ fl oz	**39.** 1 gal = ■ fl oz
40. 6 c = ■ qt	**41.** 200 fl oz = ■ gal	**42.** 35 c = ■ gal

Solve.

43. A 1964 convertible weighs 3,000 lb. How much does it weigh in tons?

44. The convertible's engine holds 6 qt of oil. How many pints is that?

45. In 1964, gasoline sold for 30.4¢ per gallon. The convertible's tank holds 17 gal. How much did it cost to fill up the car in 1964?

46. In highway driving, the convertible uses up an average of 8 fl oz of gas each mile. How many miles per gallon does the convertible get?

CALCULATOR

Mrs. Hoffman pulls into a gas station for some unleaded gasoline. The gasoline sells for 93.9¢ per gallon. Mrs. Hoffman buys $10 worth. How many gallons did she get? Round to the nearest tenth.

7.3 COMPUTING WITH CUSTOMARY MEASURES

The stage manager for the rock band *Deffening* has to make 2 shelves for guitar cases that are 3 ft 8 in. long. How much shelving does the stage manager need? If wood shelving is sold only in 9-ft lengths, how much wood will be left over?

You can add to find how much shelving the stage manager needs.

EXAMPLE 1

Add: 3 ft 8 in. + 3 ft 8 in.

1. Add the inches. Regroup.

THINK: 8 in. + 8 in. = 16 in.
16 in. = 1 ft 4 in.

```
  1 ft
  3 ft 8 in.
 +3 ft 8 in.
 ----------
       4 in.
```

2. Add the feet.

```
  1 ft
  3 ft 8 in.
 +3 ft 8 in.
 ----------
  7 ft 4 in.
```

So, the stage manager needs 7 ft 4 in. of shelving.

You can subtract to find how much wood will be left over.

EXAMPLE 2

Subtract: 9 ft − 7 ft 4 in.

1. Regroup.

THINK: 9 ft = 8 ft 12 in.

```
  8 ft 12 in.
  9 ft  0 in.
 -7 ft  4 in.
 -----------
```

2. Subtract.

```
  8 ft 12 in.
  9 ft  0 in.
 -7 ft  4 in.
 -----------
  1 ft  8 in.
```

There will be 1 ft 8 in. of wood left over.

CHECKPOINT

Write the letter of the correct answer.

Add or subtract.

1. 5 ft 9 in. + 4 ft 7 in. **a.** 1 ft 2 in. **b.** 10 ft 4 in. **c.** 10 ft 6 in.

2. 4 yd 1 ft − 3 yd 2 ft **a.** 1 yd 2 ft **b.** 1 yd 1 ft **c.** 2 ft

3. 1 lb 12 oz + 12 oz **a.** 1 lb **b.** 2 lb 4 oz **c.** 2 lb 8 oz

PRACTICE EXERCISES

Add or subtract.

1. 3 yd 1 ft +4 yd 1 ft	**2.** 4 lb 3 oz +7 lb 5 oz	**3.** 1 ft 3 in. +1 ft 8 in.	**4.** 6 gal 1 qt +18 gal 2 qt
5. 2 yd 2 ft +1 yd 2 ft	**6.** 6 gal 3 qt +7 gal 2 qt	**7.** 1 pt 1 c +1 pt 1 c	**8.** 3 mi 2,875 ft +6 mi 2,475 ft
9. 6 ft 2 in. −4 ft 1 in.	**10.** 3 yd 2 ft −2 yd 1 ft	**11.** 5 c 7 fl oz −3 c 3 fl oz	**12.** 4 mi 1,300 yd −3 mi 500 yd
13. 7 qt −5 qt 1 pt	**14.** 5 lb 6 oz −3 lb 7 oz	**15.** 2 gal 1 qt −1 gal 2 qt	**16.** 6 T 450 lb −4 T 780 lb
17. 5 T 750 lb −3 T 460 lb	**18.** 3 pt −1 pt 4 fl oz	**19.** 1 pt 5 fl oz +1 pt 3 fl oz	**20.** 8 mi 567 yd +7 mi 56 yd
21. 8 lb 5 oz −3 lb 8 oz	**22.** 7 lb 4 oz +9 lb 8 oz	**23.** 5 T 1,500 lb +4 T 375 lb	**24.** 5 mi 1,350 ft −2 mi 2,000 ft
25. 1 c 3 fl oz + 6 fl oz	**26.** 12 T 1,100 lb − 5 T 1,500 lb	**27.** 7 yd 1 ft −3 yd 2 ft	**28.** 1 ft 9 in. +1 ft 9 in.

Solve.

29. *Deffening* is giving a concert so loud that you can hear it 6 mi away. The Morosco sisters are late. They have been able to hear the band for the last 3 mi 650 yd. How far are they from the arena?

30. Lead guitarist Ogre Krebs starts the show wearing 12 lb 6 oz of gold. During a solo, he takes off a gold chain weighing 5 lb 12 oz. How much gold is he wearing now?

31. When the Morosco sisters arrive, they order refreshments. Rosie buys a 2-qt Big Deal, while Rhonda orders a 72-fl oz Thirst Thrasher. Which sister ordered more?

32. The *Deffening* stage speakers consist of 3 ft 7 in. tweeters stacked up on top of woofers that are twice as high. How tall is each speaker?

TIME OUT

This is *Deffening*'s "My Sink Says Drip" Tour. The stage show features a giant 20-ft tall kitchen sink that drips during the 210-min concert from a 35-gal water tank hidden behind the faucet. The faucet drips at a constant rate and is timed to empty the tank just as the show ends. How much water drips from the faucet each minute? How many gallons are left in the tank after the band has played for 120 min?

7.4 COMPUTING WITH TIME

The standard unit of measurement for time is the day. A **day** is the time it takes the Earth to rotate one full turn on its axis. A **year** is the time it takes the Earth to revolve around the sun.

The 24 h of the day are separated into two 12-h time periods: before noon (A.M.) and after noon (P.M.).

To find elapsed time, subtract. Sometimes it helps to use a 24-h clock.

TIME
60 seconds (s) = 1 minute (min)
60 min = 1 hour (h)
24 h = 1 day (d)
7 d = 1 week (wk)
365 d = 1 year (y)
52 wk = 1 y
12 months (mo) = 1 y

EXAMPLE 1 At the Memphis railway station, Trevor Rivera boards the City of New Orleans at 8:52 A.M. He arrives in Jackson at 1:19 P.M. How long did the trip take?

1. Change both times to hours and minutes.

   THINK: 1:19 P.M. = 1 h 19 min + 12 h = 13 h 19 min

   $$\begin{array}{r} 13 \text{ h } 19 \text{ min} \\ -\ 8 \text{ h } 52 \text{ min} \\ \hline \end{array}$$

2. Regroup. Subtract.

   THINK: 13 h 19 min = 12 h 79 min

   $$\begin{array}{r} \overset{12 \text{ h}}{\cancel{13}} \text{ h } \overset{79 \text{ min}}{\cancel{19}} \text{ min} \\ -\ 8 \text{ h } 52 \text{ min} \\ \hline 4 \text{ h } 27 \text{ min} \end{array}$$

The trip took 4 h 27 min.

To find the end time, add.

EXAMPLE 2 The City of New Orleans pulls into Kankakee at 9:32 A.M. to pick up some freight. The loading takes 4 h 40 min. At what time is the last of the freight loaded aboard?

1. Change the starting time to hours and minutes.

   THINK: 9:32 A.M. = 9 h 32 min

   $$\begin{array}{r} 9 \text{ h } 32 \text{ min} \\ +4 \text{ h } 40 \text{ min} \\ \hline \end{array}$$

2. Add the minutes and regroup.

   72 min = 1 h 12 min

   $$\begin{array}{r} \overset{1 \text{ h}}{9} \text{ h } 32 \text{ min} \\ +4 \text{ h } 40 \text{ min} \\ \hline 12 \text{ min} \end{array}$$

3. Add the hours. Change to clock time.

   14 h 12 min − 12 h = 2 h 12 min = 2:12 P.M.

   $$\begin{array}{r} \overset{1 \text{ h}}{9} \text{ h } 32 \text{ min} \\ +4 \text{ h } 40 \text{ min} \\ \hline 14 \text{ h } 12 \text{ min} \end{array}$$

The last of the freight is loaded aboard at 2:12 P.M.

FOR DISCUSSION

European train schedules all use the 24-h clock. So does the military. What are some of its advantages?

PRACTICE EXERCISES

You get on the train at 8:35 A.M. How long does your trip take if you arrive at:

1. 8:52 A.M.? **2.** 9:17 A.M.? **3.** 9:26 A.M.? **4.** 10:47 A.M.?

You get on the train at 12 noon. How long does it take to get home if you arrive at:

5. 4:29 P.M.? **6.** 7:17 P.M.? **7.** 12 midnight? **8.** 8:39 A.M.?

The time is now 9:45 P.M. What will be the correct time in:

9. $\frac{3}{4}$ hour? **10.** 3 hours 15 minutes? **11.** 5 hours 25 minutes?

The time is now 5:20 A.M. What was the correct time:

12. 2 h 47 min ago? **13.** $8\frac{1}{2}$ h ago? **14.** 33 min ago?

Complete the schedule of the Northeast Corridor Rail Service.

Route	Departure Time	Arrival Time	Length of Trip
15. Boston–New York	7:32 A.M.	11:05 A.M.	▩
16. Boston–Washington	8:29 A.M.	3:47 P.M.	▩
17. Philadelphia–Miami	10:27 A.M.	▩	19 h 27 min
18. New York–Atlanta	1:40 P.M.	4:36 A.M.	▩
19. Savannah–Richmond	▩	12 midnight	8 h 48 min
20. Buffalo–Cleveland	9:52 P.M.	1:06 P.M.	▩

Solve.

21. The City of New Orleans stops at 11:37 A.M. to allow another train to pass. The City of New Orleans starts up again 1 h 6 min later. At what time does the train start up again?

22. There is a jukebox in the dining car. Rivera puts in a quarter and picks 2 songs that are each 2 min 50 s long. How long does the jukebox play Rivera's songs?

23. The City of New Orleans runs on a special summer schedule from May 24 until September 6. How many weeks does the train run on its special schedule?

24. It took Rivera 18 min to get from the train to a taxi and another 24 min to get to his hotel. If the train arrived at 1:19 P.M., when did Rivera get to his hotel?

7.5 CONSUMER: TIME ZONES AND TRAVEL SCHEDULES

When you call a person who lives in another part of the United States or in another country, be sure you know what time it is there. Remember, there are 24 time zones in the world.

You are working for a company in Washington, D.C. You want to call Mr. Jones as soon as he arrives at his office in Los Angeles, California. He arrives at his office at 8 A.M. Pacific Time. At what time should you make the call?

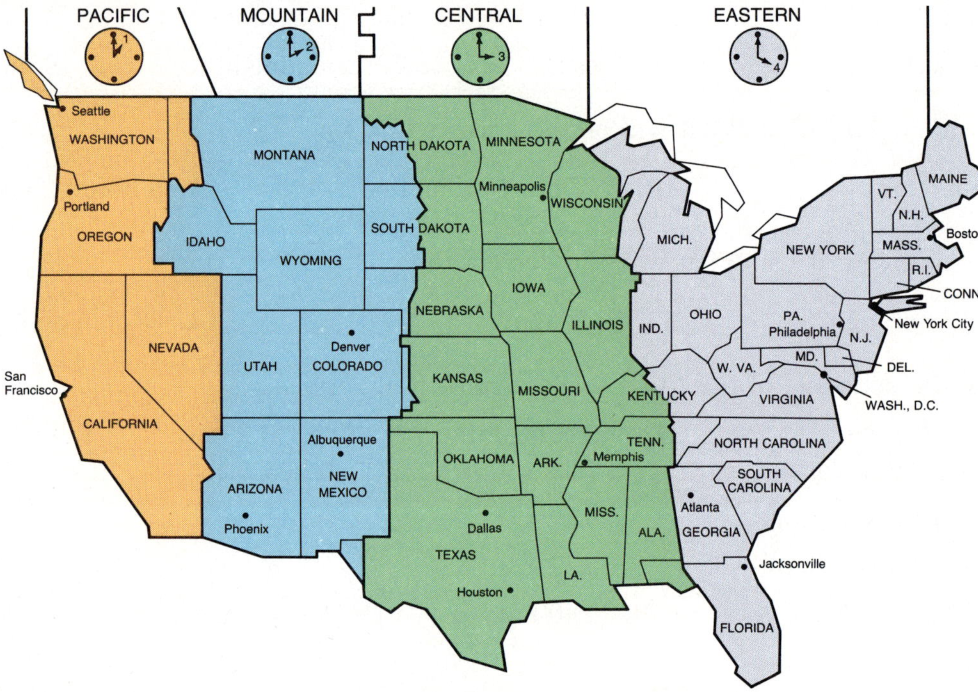

As the map shows, there is a 3-hour time difference between the Eastern zone and the Pacific zone. The time in Washington is 3 hours later than the time in Los Angeles.

So, to call Mr. Jones when it is 8 A.M. in Los Angeles, you should make the call at 11 A.M. Eastern Time.

Use the map above to answer Exercises 1–10.

If the time is 3 P.M. in New York, NY, what time is it in:

1. Minneapolis, MN? **2.** Albuquerque, NM? **3.** Seattle, WA?

If the time is 1 P.M. in Reno, NV, what time is it in:

4. Denver, CO? **5.** Dallas, TX? **6.** Atlanta, GA?

Suppose the time is:	What time is it in:
7. 9:30 A.M. in Boston, MA	Houston, TX?
8. 7:05 P.M. in Phoenix, AZ	Jacksonville, FL?
9. 1:15 P.M. in Philadelphia, PA	San Francisco, CA?
10. 11:08 P.M. in Portland, OR	Memphis, TN?

Often when you plan a trip by plane, you need to be aware of the different time zones. A flight schedule shows departures and arrivals in terms of the local time at each city.

Use the flight schedule at the right to answer Exercises 11–15.

PHILADELPHIA, PA				
Leave	**Arrive**	**Flt.**	**Stops/Meal**	
TO: Chicago, IL				
6:30a	7:56a	181	NS	M
8:00a	9:30a	197	NS	M
10:00a	11:16a	207	NS	S
12:00p	1:15p	219	NS	M
TO: Denver, CO				
6:30a	10:15a	181/317	C	M
10:00a	4:05p	207/365	C	M
a-A.M.		p-P.M.		
NS-Nonstop		C-Connection		
M-Meal		S-Snack		

11. Mrs. Carelli plans to fly from Philadelphia to Denver. Does the schedule show any nonstop flights?

12. Mrs. Carelli takes the 6:30 A.M. flight to Denver. How many hours will it take?

13. Is Flight 181 to Chicago less than 2 hours? If not, how long is it?

14. How long is Flight 219?

15. How much longer is the 10 A.M. flight to Denver than the 6:30 A.M. flight?

MIDCHAPTER REVIEW

Complete. [180, 182, 186]

1. 5 ft = ■ in. **2.** 3 T = ■ lb **3.** 8 fl oz = ■ pt

4. 60 oz = ■ lb **5.** $2\frac{1}{2}$ yd = ■ ft **6.** 2 gal = ■ fl oz

7. 7 qt = ■ c **8.** 5,280 yd = ■ mi **9.** 180 s = ■ min

10. 12 h = ■ d **11.** 5,000 lb = ■ T **12.** 1 wk = ■ h

13. 25 ft = ■ yd **14.** 64 c = ■ gal **15.** 3 h = ■ min

16. $6\frac{3}{4}$ lb = ■ oz **17.** $4\frac{1}{2}$ qt = ■ fl oz **18.** 66 in. = ■ ft

Measure the line segment to the nearest $\frac{1}{16}$ in. [180]

19. ______________________________ **20.** ____

21. ____________ **22.** ____________________

Add or subtract. [184]

23. 7 T 500 lb − 6 T 350 lb

24. 8 pt − 6 pt 1 c

25. 3 gal 3 qt + 1 gal 2 qt

26. 3 mi 855 yd + 3 mi 855 yd

Complete the Cinema 6 schedule. [186]

	Movie	Begins	Ends	Length of Film
27.	"Brown Cars, Silver Bridges"	1:30 P.M.	3:05 P.M.	■
28.	"Don't Worry, Baby"	8:20 P.M.	10:12 P.M.	■
29.	"In Concert: Deffening"	10:45 P.M.	■	2 hours 17 minutes
30.	"Lonely Is as Lonely Does"	12:40 P.M.	3:36 P.M.	■
31.	"Bach to Bach"	■	12 midnight	1 hour 48 minutes

Use the flight schedule at the right to answer Exercises 32–33. [188]

Leave	Arrive	Flt.	Stops/Meal	
TO: Chicago, IL				
8:00a	10:00a	118	NS	M
10:00a	11:40a	121	NS	S
11:00a	12:45p	201	NS	S

32. John Allen takes Flight 118. How long is his flight?

33. Susan Raimes wants to arrive in Chicago as close to 11:00 A.M. as possible. Which flight should she take?

PROBLEM Solving APPLICATION

7.6 CONSUMER: HOURLY WAGES AND OVERTIME PAY

In some jobs, employees are paid an hourly rate for the first 40 hours that they work in a week. For every hour over 40 hours, they are paid an **overtime rate** calculated at "time and a half."

Suppose your rate of pay is \$5.00 per hour and you work 42 hours one week. What is your total pay including overtime?

1. Multiply to find the regular pay. $40 \times \$5.00 = \$200.00$
2. Find the overtime pay.
   - Subtract to find the overtime hours. $42 - 40 = 2$
   - Multiply to find the overtime rate.

     THINK: $1\frac{1}{2} = 1.5$ $1.5 \times \$5.00 = \$7.50$
   - Multiply to find the total overtime pay. $2 \times \$7.50 = \$15.00$
3. Add to find the total pay. $\$200.00 + \$15.00 = \$215.00$

So, your total pay for the week is \$215.00.

Find the total pay for the week including overtime pay. Remember to estimate whenever you use your calculator.

	Number of Hours Worked	Rate per Hour		Number of Hours Worked	Rate per Hour
1.	42	\$4.90	**2.**	44	\$5.20
3.	48	\$5.60	**4.**	40	\$4.80
5.	41	\$4.50	**6.**	45	\$8.70
7.	47	\$5.90	**8.**	43.5	\$6.60

Solve.

9. Your friend earns \$4.80 per hour. One week he works 46 hours. What is his total pay including overtime?

10. Your cousin earns \$485.00 per week for 40 weeks of the year. For the other 12 weeks, she works overtime and earns \$533.50 per week. What is her total pay per year? What is her average total pay per week?

7.7 METRIC UNITS OF LENGTH

The table below shows the most commonly used metric units of length.

kilometer (km) 1,000 m	meter (m) 1 m	centimeter (cm) 0.01 m	millimeter (mm) 0.001 m
	 1 m	 1 cm	 1 mm
A kilometer is about the length of 12 city blocks.	A meter is about the width of a door.	A centimeter is about the width of a bandaid.	A millimeter is about the thickness of a dime.

To choose a reasonable estimate, determine the most likely unit of measure.

EXAMPLE 1 Choose the most reasonable estimate for the width of a shoelace.

a. 4 mm b. 4 cm c. 4 m d. 4 km

THINK: The width of a shoelace is several times thicker than a dime.

Since a dime is 1 mm thick, the most reasonable estimate for the width of a shoelace is 4 mm.

The smaller the unit, the more precise the measurement.

EXAMPLE 2 What is the length of the stamp measured to the nearest millimeter?

1. Put the 0 mark of the scale at the left edge of the stamp.

2. Read the mark closest to the right edge of the stamp.

The length of the stamp is 33 mm.

FOR DISCUSSION

To measure the height of several people, would you use a tape measure marked only in meters or one marked only in centimeters? Explain.

PRACTICE EXERCISES

Choose the most reasonable estimate.

1. length of a golf club	**a.** 1 km	**b.** 1 m	**c.** 1 cm
2. height of an office building	**a.** 100 km	**b.** 100 m	**c.** 100 mm
3. length of your textbook	**a.** 24 m	**b.** 24 cm	**c.** 24 mm
4. thickness of your textbook	**a.** 30 m	**b.** 30 cm	**c.** 30 mm
5. distance from Detroit to Miami	**a.** 2,000 km	**b.** 2,000 m	**c.** 2,000 cm
6. height of a one-story house	**a.** 2 km	**b.** 4 m	**c.** 200 cm
7. height of a ceiling	**a.** 0.5 km	**b.** 3 m	**c.** 122 cm
8. length of a marathon course	**a.** 44 km	**b.** 1,800 m	**c.** 3,500 cm
9. thickness of a wallet	**a.** 2 m	**b.** 5 cm	**c.** 12 mm
10. length of a tennis racket	**a.** 2 m	**b.** 70 cm	**c.** 200 mm

Use a ruler to measure the line segment or object to the nearest centimeter and the nearest millimeter.

11.

12.

13.

14.

15.

16. length of a pen

17. width of your desk

18. length of a new piece of chalk

19. length of a paper clip

Solve.

20. Ralph Plotnick spills some coffee on the blueprint for his shed. Some labels are smudged. The blueprint lists the window size as 175 by 75. To which metric unit of length does the blueprint likely refer?

21. Ralph needs 6 sheets of paper towels to clean up the 8 fl oz of coffee he spilled. On average, how much coffee does each sheet pick up?

22. The waitress gives Ralph a new cup of coffee and they begin to talk. She asks Ralph which metric unit he would be likely to use to measure the height of the letters on the menu. What should Ralph answer?

23. The cook prides himself on his skill in estimating the heights of customers while they are still seated. Ralph is about medium height. The cook estimates Ralph's height to be 245 cm. Is the estimate reasonable?

7.8 METRIC UNITS OF MASS AND CAPACITY

The table below shows the most commonly used metric units of mass. The gram (g) is the basic unit.

kilogram (kg) 1,000 g	gram (g) 1 g	milligram (mg) 0.001 g
	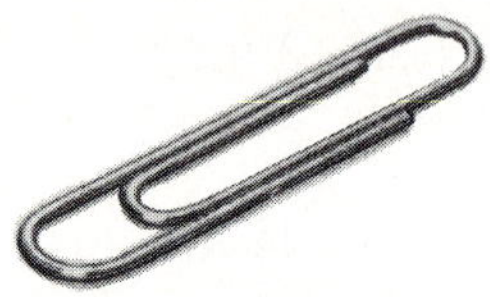	
A kilogram is about the mass of a textbook.	A gram is about the mass of a paper clip.	A milligram is about the mass of a seed.

To choose a reasonable estimate for the mass of an object, determine the most likely unit of measure.

EXAMPLE 1 Choose the most reasonable estimate for the mass of a quarter.

a. 10 kg b. 10 g c. 10 mg

THINK: The mass of a quarter is several times greater than the mass of a paper clip.

Since the mass of a paper clip is 1 g, then the most reasonable estimate for the mass of a quarter is 10 g.

The table shows the most commonly used metric units of capacity. The liter (L) is the basic unit.

kiloliter (kL) 1,000 L	liter (L) 1 L	milliliter (mL) 0.001 L
	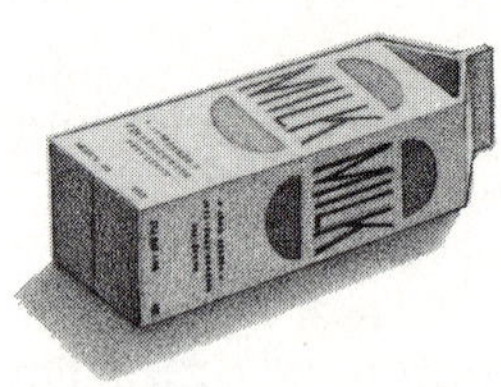	
A kiloliter is about the amount of water in a small pond.	A liter is the amount of liquid in a liter container (or carton).	A milliliter is about the amount of water in a raindrop.

To choose a reasonable estimate for the capacity of an object, determine the most likely unit of measure.

EXAMPLE 2 Choose the most reasonable estimate for the amount of ink in a ballpoint pen. a. 125 kL b. 125 L c. 125 mL

THINK: The amount of ink in a ballpoint pen is greater than that of a raindrop, but less than that of a liter bottle.

Since the capacity of a raindrop is 1 mL, then the most reasonable estimate for the amount of ink in a ballpoint pen is 125 mL.

PRACTICE EXERCISES

Choose the most reasonable estimate.

1.	mass of a light bulb	**a.** 40 kg	**b.** 40 g	**c.** 40 mg
2.	mass of a compact car	**a.** 1,000 kg	**b.** 1,000 g	**c.** 1,000 mg
3.	mass of a chicken leg	**a.** 125 kg	**b.** 125 g	**c.** 125 mg
4.	mass of a tennis ball	**a.** 80 kg	**b.** 80 g	**c.** 80 mg
5.	mass of a television	**a.** 25 kg	**b.** 25 g	**c.** 25 mg
6.	mass of a toothpick	**a.** 15 kg	**b.** 15 g	**c.** 15 mg
7.	capacity of a vase	**a.** 800 kL	**b.** 800 L	**c.** 800 mL
8.	capacity of a swimming pool	**a.** 75 kL	**b.** 75 L	**c.** 75 mL
9.	capacity of a canteen	**a.** 0.6 kL	**b.** 0.6 L	**c.** 0.6 mL
10.	capacity of a sink	**a.** 8 kL	**b.** 8 L	**c.** 8 mL
11.	capacity of a hot-air balloon	**a.** 15 kL	**b.** 15 L	**c.** 15 mL
12.	amount of water in a lake	**a.** 5,000 kL	**b.** 5,000 L	**c.** 5,000 mL
13.	mass of a sandwich	**a.** 2 kg	**b.** 225 g	**c.** 200 mg
14.	mass of a tissue	**a.** 6 kg	**b.** 13 g	**c.** 875 mg
15.	mass of a bowling ball	**a.** 6 kg	**b.** 45 g	**c.** 900 mg
16.	mass of a fork	**a.** 45 kg	**b.** 90 g	**c.** 75 mg
17.	mass of a register receipt	**a.** 25 kg	**b.** 1 g	**c.** 5 mg
18.	amount of water in an ice cube	**a.** 0.3 kL	**b.** 0.8 L	**c.** 35 mL
19.	capacity of a bucket	**a.** 12 kL	**b.** 4 L	**c.** 12 mL
20.	amount of juice in a lemon	**a.** 25 kL	**b.** 9 L	**c.** 50 mL
21.	capacity of a water tower	**a.** 10 kL	**b.** 50 L	**c.** 3 mL
22.	capacity of a washing machine	**a.** 60 kL	**b.** 50 L	**c.** 40 mL

MIXED REVIEW

Multiply or divide.

23. $\frac{1}{4} \times \frac{2}{3}$ **24.** $\frac{4}{7} \times \frac{5}{8}$ **25.** $\frac{7}{8} \div \frac{7}{9}$ **26.** $\frac{6}{7} \div \frac{3}{4}$

27. $\frac{3}{7} \times \frac{7}{8}$ **28.** $\frac{3}{8} \times \frac{5}{6}$ **29.** $\frac{8}{9} \div \frac{2}{3}$ **30.** $\frac{6}{7} \div \frac{7}{8}$

31. $3\frac{1}{2} \times 2\frac{1}{4}$ **32.** $4\frac{3}{8} \times 1\frac{6}{7}$ **33.** $6\frac{7}{8} \div 1\frac{4}{7}$ **34.** $5\frac{2}{3} \div 2\frac{5}{6}$

35. $1\frac{5}{7} \times 4\frac{2}{3}$ **36.** $3\frac{3}{5} \times 2\frac{2}{9}$ **37.** $1\frac{1}{7} \div 1\frac{1}{7}$ **38.** $4\frac{2}{9} \div 2\frac{5}{7}$

Solve.

39. Renaldo Thomas is an apprentice cook at a downtown cafeteria. The head chef tells Renaldo that the stew recipe starts with 45 kL of beef broth. Renaldo doesn't believe him. Why?

40. The head chef tells Renaldo to peel 22 lb of onions for the stew. The onions come in 32-oz bags. How many bags of onions does Renaldo have to peel?

41. Renaldo empties a full bottle of hot sauce into the stew. He estimates that the bottle held 150 mL. Is his estimate reasonable?

42. Today's stew is a lamb stew. Each of the 200 servings includes about 250 g of lamb. How many kilograms of lamb are used in the stew altogether?

TIME OUT

You have three buckets. The largest one is filled with 8 L of water. The other two buckets hold 5 L and 2 L. They are empty.

How can you pour water back and forth, without spilling, so that you end up with exactly 7 L in the largest bucket?

PROBLEM Solving APPLICATION

7.9 CAREER: DRAFTSPERSON

Ellen Jessup is a draftsperson. As part of her job, she makes mechanical drawings that show how to construct parts of machines. The drawings must be very precise. Otherwise, the parts of the machines will not be correct.

Find the measurement to the nearest tenth of a centimeter.

1. 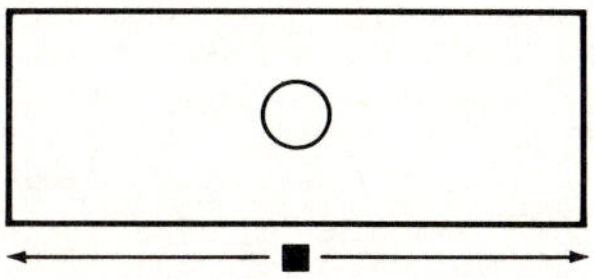

The actual machine part is 2.5 times as long as the drawing. What is the length of the actual part?

2. 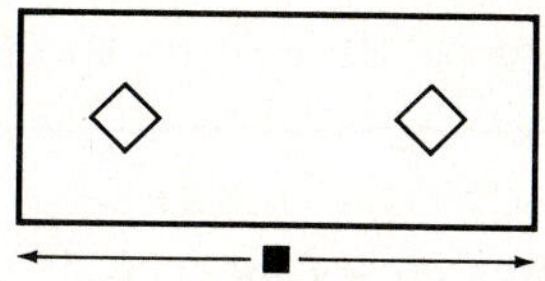

A metal strip is 4.5 times as long as the drawing. What is the length of the actual metal strip?

3. 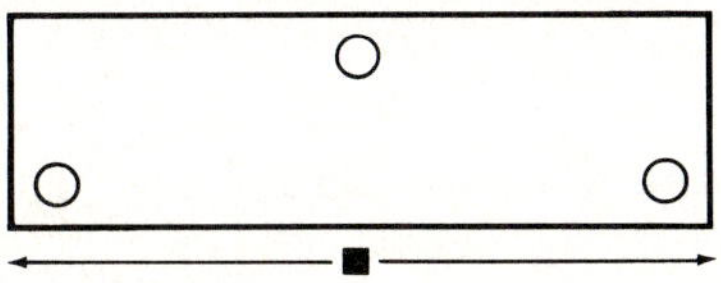

How long is the drawing?

4. 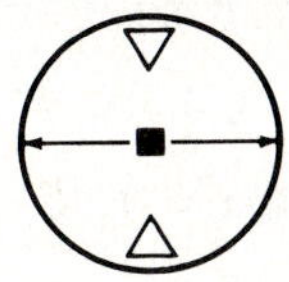

What is the diameter of the washer?

Sometimes, large drawings are made for small machine parts. In Exercises 5–8, the actual part is 0.25 as long as the dimension of the given drawing.

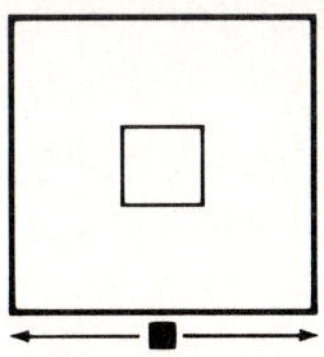

A

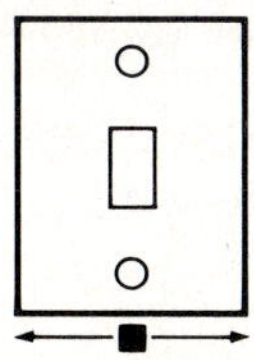

C

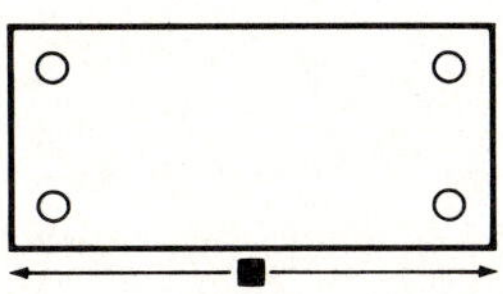

B

D

Find the missing numbers to the nearest tenth of a centimeter. Remember to estimate whenever you use your calculator.

	Drawing	Length of Drawing	Length of Part
5.	A	■	■
6.	B	■	■
7.	C	■	■
8.	D	■	■

7.10 CONVERTING METRIC UNITS

Terry Partridge walked the five blocks from her house to the Surf City swimming pool. The distance is 0.4 km. How many meters did she walk from her house to the pool?

To solve, you need to change 0.4 km to meters.

You can use these relationships to change from one unit to another.

LENGTH	MASS	CAPACITY
1 km = 1,000 m	1 kg = 1,000 g	1 kL = 1,000 L
1 m = 100 cm	1 g = 1,000 mg	1 L = 1,000 mL
1 m = 1,000 mm		

To change from a larger unit to a smaller unit, multiply.

EXAMPLE 1 Complete: 0.4 km = ■ m

THINK: 1 km = 1,000 m

0.4 km = (0.4 × 1,000) m
0.4 km = 400 m

So, Terry walked 400 m from her house to the pool.

To change from a smaller unit to a larger unit, divide.

EXAMPLE 2 After a short swim, Terry comes out of the pool and applies 800 mg of suntan lotion to her arms and legs. How many grams of suntan lotion does Terry apply?

Complete: 800 mg = ■ g

THINK: 1 g = 1,000 mg

800 mg = (800 ÷ 1,000) g
800 mg = 0.8 g

So, Terry applies 0.8 g of suntan lotion.

CHECKPOINT

Write the letter of the correct answer.

Complete.

1. 0.35 m = ■ cm **a.** 350 **b.** 35 **c.** 0.0035

2. 4,500 L = ■ kL **a.** 450 **b.** 45 **c.** 4.5

3. 0.04 g = ■ mg **a.** 400 **b.** 40 **c.** 4

PRACTICE EXERCISES

Complete.

1. 2 m = ■ mm
2. 4 km = ■ m
3. 1 cm = ■ mm
4. 4,000 mm = ■ m
5. 600 cm = ■ m
6. 9,000 m = ■ km
7. 500 mm = ■ m
8. 40 m = ■ km
9. 2,000 cm = ■ m
10. 4 g = ■ mg
11. 7 kg = ■ g
12. 1 kg = ■ mg
13. 9,000 g = ■ kg
14. 12,000 mg = ■ g
15. 21,000 g = ■ kg
16. 80 mg = ■ g
17. 140 g = ■ kg
18. 3 g = ■ kg
19. 9 L = ■ mL
20. 3 kL = ■ L
21. 1 kL = ■ mL
22. 8,000 L = ■ kL
23. 89,000 L = ■ kL
24. 2,000 mL = ■ L
25. 80 L = ■ kL
26. 750 mL = ■ L
27. 15 mL = ■ L
28. 0.01 m = ■ mm
29. 3.75 g = ■ mg
30. 23.45 mL = ■ L
31. 10.5 kL = ■ L
32. 0.125 kg = ■ mg
33. 45.6 mm = ■ cm
34. 43 mg = ■ g
35. 2.225 km = ■ m
36. 13,321 L = ■ kL

MIXED REVIEW

Complete.

Data	Range	Mean	Median	Mode
12, 17, 19, 26, 43, 12, 11	**37.** ■	**38.** ■	**39.** ■	**40.** ■
212, 568, 345, 420, 345	**41.** ■	**42.** ■	**43.** ■	**44.** ■
3.6, 4.2, 7.4, 1.7, 4.2, 5.3, 3.7	**45.** ■	**46.** ■	**47.** ■	**48.** ■
2.5, 0.4, 3.4, 5.2, 6.7, 5.2	**49.** ■	**50.** ■	**51.** ■	**52.** ■

Solve.

53. The Surf City swimming pool holds 75 kL of water. About 600 L a day are lost to evaporation, splashing, and so on. How many kiloliters are left in the pool at the end of the day before the pool is refilled?

54. Felicia, the lifeguard, puts 350 mg of sun block on her nose each day. How many days will it take her to use up a 140-g tube of sun block?

55. The Surf City pool is open daily from 8:30 A.M. to 7:15 P.M. How long is the pool open each day?

56. Barbie is sunning herself on a towel that is 120 cm long. Barbie is 1.63 m tall. How much shorter is the towel than Barbie?

7.11 MEASURING TEMPERATURE

Thermometers are used to measure temperature. Temperature can be measured in **degrees Celsius (°C)** or in **degrees Fahrenheit (°F).** To find the temperature, read the scale.

EXAMPLE 1

What is the temperature?

THINK: 50° + 8° = 58°

The temperature is 58°C.

The table below lists temperatures in both systems.

	°C	°F
Boiling point of water	100	212
Record high air temperature	58	136
Normal body temperature	37	98.6
Comfortable room temperature	20	68
Freezing point of water	0	32
Record low air temperature	−88	−127

Use the standard temperatures you already know to help you choose the most reasonable temperature.

EXAMPLE 2

Choose the most reasonable temperature for a very hot cup of tea.
a. 72°F b. 95°C c. 100°F

THINK: Water boils at 100°C or 212°F.

Compare. 95°C is close to 100°C.

So, the most reasonable temperature for a very hot cup of tea is 95°C.

FOR DISCUSSION

Kim and Jane are talking on the phone. Kim lives in Southern California. She says that the temperature is 30° and that she's going swimming. Jane lives in Vermont and says that the temperature there is 30° but she's going ice skating. How can this be?

PRACTICE EXERCISES

Write the temperature.

1. 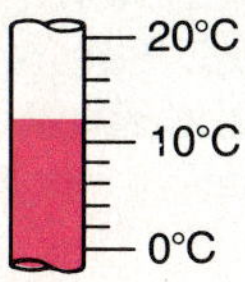

2. 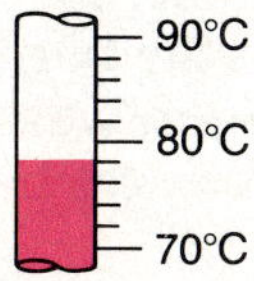

3. 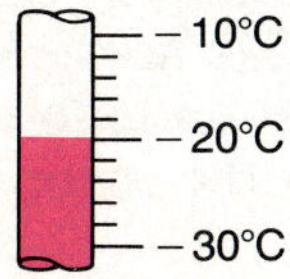

4. 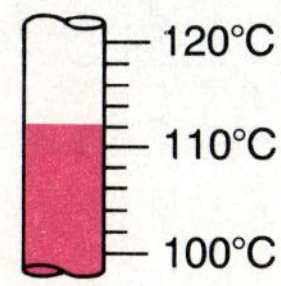

5. 

6. 

7. 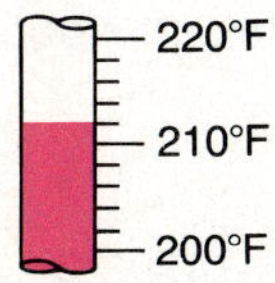

8. 

Choose the most reasonable temperature.

9.	hot oven	**a.** 80°C	**b.** 100°C	**c.** 200°C
10.	sunny summer day	**a.** 10°C	**b.** 30°C	**c.** 50°C
11.	ice cube	**a.** 22°C	**b.** 8°C	**c.** −4°C
12.	refrigerator	**a.** −12°C	**b.** 0°C	**c.** 4°C
13.	snowstorm	**a.** 28°F	**b.** 42°F	**c.** 112°F
14.	chicken roasting	**a.** 100°F	**b.** 212°F	**c.** 400°F
15.	swimming pool	**a.** 72°F	**b.** 32°F	**c.** 15°F
16.	heat wave	**a.** 185°F	**b.** 97°F	**c.** 65°F
17.	cool fall night	**a.** 20°F	**b.** 40°C	**c.** 55°F
18.	hot bathwater	**a.** 25°C	**b.** 100°F	**c.** 100°C
19.	aquarium	**a.** 75°F	**b.** 65°C	**c.** 10°C
20.	sweater weather	**a.** 75°C	**b.** 55°F	**c.** 35°C

Solve.

21. Horace Clark is going fishing. The weather report says that the temperature on the bay will be about 32°C. Does Horace need to wear anything heavier than a T-shirt?

22. The captain of the boat tells Horace that the ocean temperature is 62°. Does he mean degrees Celsius or degrees Fahrenheit?

23. Horace catches a 4-lb flounder. After he cleans the fish, there are 54 oz of fish left. How many ounces of fish did Horace discard?

24. Horace breads the flounder and fries it at about 250°. Is that degrees Celsius or degrees Fahrenheit?

PROBLEM *Solving* STRATEGY

7.12 DRAWING A DIAGRAM

Situation:

In filming a Western movie scene, the director asks an actor to ride his horse 35 m east from the barn and then stop. The next direction is to turn and trot 20 m west and stop again. How far must the actor ride the horse to get back to the barn?

Strategy:

Drawing a diagram can help you solve some problems.

Applying the Strategy:

1. Draw the horse and rider going east. Draw the horse and rider going west.

2. Subtract to find how far the actor must ride the horse to get back to the barn.

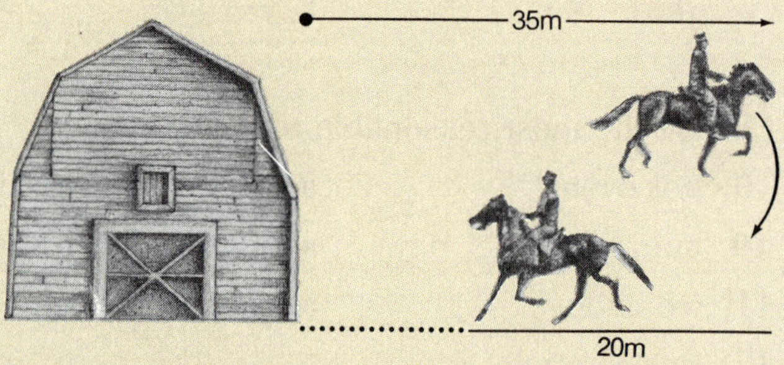

The actor must ride the horse 15 m to get back to the barn.

$$\begin{array}{r} 35 \\ -20 \\ \hline 15 \end{array}$$

FOR DISCUSSION

A flag being used on the movie set has 9 stripes. The first stripe is blue. The second stripe is white. This pattern is repeated on the flag. What color is the seventh stripe?

1. What would you draw to help you solve the problem?

2. What is the solution?

PRACTICE EXERCISES

Solve the problem by drawing a diagram.

1. A set designer hangs a picture on the wall of a model room. The picture is 84 cm wide. The wall is 218 cm wide. If the picture is hung in the center of the wall, how much wall space is on each side of the picture?

2. A camera crew sets up a special light board. There are 25 light bulbs. There are five light bulbs in each row. The center and corner lights are yellow. How many yellow light bulbs are there in all?

Solve each problem by drawing a diagram.

3. A director tells an actor to walk 19 m from the stable to the barn. The next direction is to turn around and walk back 12 m to the well. How far from the stable is the actor?

4. The stage crew sets up 2 shelves on one wall of a movie set. The first shelf is 1.4 m off the ground. The second is 1.2 m above that and 1.8 m from the ceiling. What is the height of the set?

5. A set designer places a rug and a sofa on the living room floor. The rug has 12 stripes. The first stripe is white. The second is blue. The third is green. This pattern is repeated. How many times is the complete pattern repeated?

6. The camera crew moves around the stage for the filming of the Western. They move 6 m north, then 8.2 m east, then 10.5 m south, then 8.2 m west, and finally 4.5 m south. How far is the camera crew from its original location?

7. A sound technician needs wire cut into 1.5-m lengths. The technician has 6 sections of wire. Each section is 5 m long. How many 1.5-m lengths can the technician cut?

CHAPTER REVIEW

Vocabulary Choose the letter of the word(s) that completes each statement.

1. The "A.M." in 7:00 A.M. indicates ■. [186]

a. before noon **b.** noon **c.** after noon **d.** midnight

2. You would likely use centimeters to measure the width of a ■. [192]

a. barn **b.** pin **c.** calculator **d.** canyon

Skills Complete. [180, 182, 186, 198]

7.45 L = *7,450 mL*	**3.** $2\frac{1}{2}$ lb = ■ oz	**4.** $2\frac{1}{3}$ ft = ■ in.
5. 7 yd = ■ ft	**6.** 500 g = ■ kg	**7.** 10 qt = ■ gal
8. 6.5 m = ■ cm	**9.** 120 s = ■ h	**10.** 3,800 mL = ■ L

Choose the most reasonable estimate. [192, 194, 198]

the mass of a cotton ball	*a*	**a.** 750 mg	**b.** 5 g	**c.** 0.01 kg
11. thickness of a scarf		**a.** 3 mm	**b.** 3 cm	**c.** 3 m
12. capacity of a bathroom sink		**a.** 0.08 kL	**b.** 8 L	**c.** 800 mL
13. mass of a credit card		**a.** 12 kg	**b.** 4 g	**c.** 75 mg
14. temperature of boiling water		**a.** 100°F	**b.** 212°C	**c.** 212°F

Add or subtract. [184]

$$\begin{array}{r} 3 \text{ qt } 1 \text{ pt} \\ +2 \text{ qt } 1 \text{ pt} \\ \hline \textit{6 qt} \end{array}$$

15. $\begin{array}{r} 2 \text{ T } 1{,}500 \text{ lb} \\ +1 \text{ T } \quad 800 \text{ lb} \\ \hline \end{array}$ **16.** $\begin{array}{r} 5 \text{ ft } 3 \text{ in.} \\ -3 \text{ ft } 9 \text{ in.} \\ \hline \end{array}$ **17.** $\begin{array}{r} 3 \text{ c } 4 \text{ fl oz} \\ - \quad 6 \text{ fl oz} \\ \hline \end{array}$

18. $\begin{array}{r} 4 \text{ yd } 2 \text{ ft} \\ +3 \text{ yd } 2 \text{ ft} \\ \hline \end{array}$ **19.** $\begin{array}{r} 6 \text{ lb } 3 \text{ oz} \\ -4 \text{ lb } 5 \text{ oz} \\ \hline \end{array}$ **20.** $\begin{array}{r} 9 \text{ gal} \\ -2 \text{ gal } 3 \text{ qt} \\ \hline \end{array}$ **21.** $\begin{array}{r} 5 \text{ mi } 1{,}300 \text{ yd} \\ +3 \text{ mi } \quad 800 \text{ yd} \\ \hline \end{array}$

Find the missing time. [186]

Start time: 12 noon End time: 4:19 P.M. Elapsed time: *4 h 19 min*	**22.**	Start time: 6:28 P.M. End time: ■ Elapsed time: 7 h 54 min	**23.** Start time: 9:32 A.M. End time: 3:17 P.M. Elapsed time: ■

CHAPTER TEST

Complete.

1. 45 s = ■ min
2. 2.3 mm = ■ cm
3. 125 L = ■ kL
4. $9\frac{3}{4}$ gal = ■ qt
5. 21,120 ft = ■ mi
6. 1.05 g = ■ mg
7. 36 in. = ■ ft
8. $3\frac{1}{2}$ T = ■ lb
9. 7 wk = ■ d

Choose the most reasonable estimate.

10. The thickness of a notepad	**a.** 10 mm	**b.** 10 cm	**c.** 10 m
11. The mass of a football player	**a.** 250 g	**b.** 100 kg	**c.** 300 kg
12. The capacity of a bathtub	**a.** 1 kL	**b.** 180 L	**c.** 250 mL
13. The temperature of ice water	**a.** 0°C	**b.** 0°F	**c.** 4°C

Add or subtract.

14.
$$\begin{array}{r} 2 \text{ ft } 9 \text{ in.} \\ +4 \text{ ft } 7 \text{ in.} \\ \hline \end{array}$$

15.
$$\begin{array}{r} 6 \text{ c } 3 \text{ fl oz} \\ -4 \text{ c } 5 \text{ fl oz} \\ \hline \end{array}$$

16.
$$\begin{array}{r} 3 \text{ T } 1{,}500 \text{ lb} \\ +2 \text{ T } \quad 750 \text{ lb} \\ \hline \end{array}$$

Find the missing time.

17. Start time: 10:18 A.M.
End time: 6:32 P.M.
Elapsed time: ■

18. Start time: 3:57 P.M.
End time: ■
Elapsed time: 10 hours 13 minutes

Write the temperature.

19.

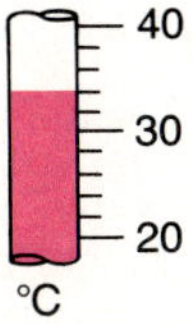

20.

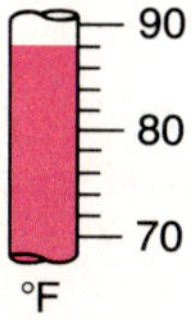

Find the length to the nearest $\frac{1}{16}$ in.

21. ________________

Find the length to the nearest millimeter.

22. ____________

Solve. Draw a diagram if necessary.

23. Isabel designed a scarf to knit. The scarf has 15 blocks. The first block is green, the second is blue, and the third is yellow. How many times is the complete pattern repeated?

24. Jack Cook earns $8.00 per hour. One week he works 48 hours. What is his total pay, including overtime?

ENRICHMENT PRECISION AND GREATEST POSSIBLE ERROR

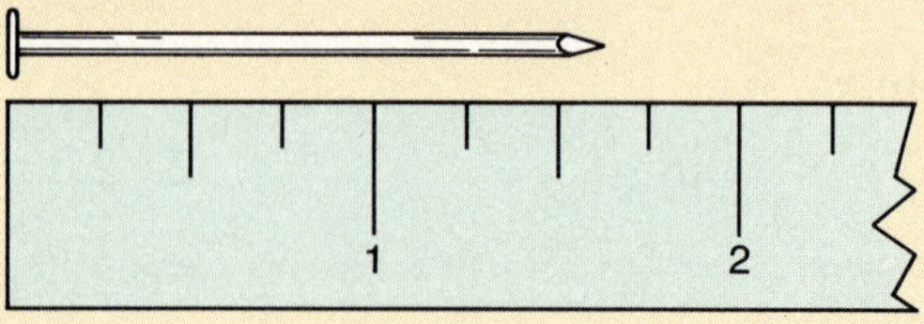

The length of this nail is closer in length to 2 in. than 1 in. So, the length of the nail is 2 in. to the nearest inch.

If you want to measure the nail more precisely, you can measure the nail to the nearest half-inch. The length of the nail is $1\frac{1}{2}$ in. to the nearest half-inch. The smaller the unit of measure, the more precise the measurement will be.

Examples Choose the more precise measurement.

8 cm or 8 mm

THINK: The smaller unit is millimeters.

The more precise measurement is 8 mm.

3 ft or 37 in.

THINK: The smaller unit is inches.

The more precise measurement is 37 in.

Since no measurement is exact, the greatest possible error is equal to one-half of the smallest unit of the measurement.

If your height is measured as $56\frac{1}{2}$ in., the greatest possible error is $\pm\frac{1}{4}$ in. So, the range of your height is between $56\frac{1}{4}\left(56\frac{1}{2} - \frac{1}{4}\right)$ and $56\frac{3}{4}\left(56\frac{1}{2} + \frac{1}{4}\right)$ in.

Examples Find the greatest possible error of measurement. Then find the range of length.

30 in.

1. Find the greatest possible error.

   THINK: The smallest unit is 1 in.

   $\frac{1}{2} \times 1 = \frac{1}{2}$

The greatest possible error is $\pm\frac{1}{2}$ in.

2. Find the range.

   $30 - \frac{1}{2} = 29\frac{1}{2}$ in.

   $30 + \frac{1}{2} = 30\frac{1}{2}$ in.

41.5 km

1. Find the greatest possible error.

   THINK: The smallest unit is 0.1 km.

   $\frac{1}{2} \times 0.1 = 0.05$

The greatest possible error is ± 0.05 km.

2. Find the range.

   $41.5 - 0.05 = 41.45$ km

   $41.5 + 0.05 = 41.55$ km

Find the greatest possible error of measurement. Then find the range of length.

1. 42 in. **2.** 16 mm **3.** 84 km **4.** 93 mi

5. 32.4 m **6.** $36\frac{1}{2}$ in. **7.** 56 ft **8.** 89.2 cm

ESTIMATION SKILLS

ESTIMATING WITH CUSTOMARY MEASURES

When you estimate the sum or difference of measures expressed in two units, round each measure to the greater unit.

Estimate the difference: 15 ft 2 in. − 8 ft 11 in.

1. Round each measure to the nearest number of feet.

   THINK: 2 in. $< \frac{1}{2}$ ft, so 15 ft 2 in. rounds to 15 ft.

   11 in. $> \frac{1}{2}$ ft, so 8 ft 11 in. rounds to 9 ft.

2. Subtract.

   $15 - 9 = 6$

The estimated difference is 6 ft.

Estimate the sum: 1 h 15 min + 5 h 37 min + 3 h 6 min

1. Round each measure to the nearest number of hours.

   THINK: 15 min $< \frac{1}{2}$ h, so 1 h 15 min rounds to 1 h.

   37 min $> \frac{1}{2}$ h, so 5 h 37 min rounds to 6 h.

   6 min $< \frac{1}{2}$ h, so 3 h 6 min rounds to 3 h.

2. Add.

   $1 + 6 + 3 = 10$

The estimated sum is 10 hours.

Estimate the sum or difference.

1. 18 ft 4 in. + 6 ft 9 in.

2. 9 ft 5 in. − 3 ft 3 in.

3. 4 h 8 min + 43 min

4. 16 h 40 min − 5 h 52

5. 9 lb 2 oz + 8 lb 3 oz

6. 24 lb 3 oz − 15 lb 6 oz

7. 2 gal 3 qt + 3 gal 1 qt

8. 7 gal − 2 gal 1 qt

9. 35 min 2 s + 20 min 8 s

10. 10 min 4 s − 5 min 53 s

11. 6 c 7 fl oz + 8 c 2 fl oz

12. 8 c 5 fl oz − 3 c 7 fl oz

13. 2 yd 32 in. + 3 yd 27 in.

14. 10 yd 32 in. − 5 yd 5 in.

15. 1 ft 5 in. + 3 ft 7 in. + 2 ft 9 in.

16. 5 min 16 s + 4 min 42 s + 8 min 9 s

17. 2 lb 9 oz + 1 lb 3 oz + 15 oz

18. 1 c 6 fl oz + 3 c 4 fl oz + 5 fl oz

PreSkills Test

Multiply. [40]

1. 5×27 **2.** 6×53 **3.** 8×96

4. 74×3 **5.** 87×4 **6.** 18×42

7. 39×15 **8.** 63×29 **9.** 84×57

Divide. [46, 48]

10. $4\overline{)72}$ **11.** $7\overline{)91}$ **12.** $5\overline{)85}$ **13.** $8\overline{)128}$ **14.** $9\overline{)306}$

15. $12\overline{)96}$ **16.** $15\overline{)225}$ **17.** $23\overline{)138}$ **18.** $47\overline{)360}$ **19.** $76\overline{)532}$

Write the fraction in simplest form. [126, 132]

20. $\frac{9}{12}$ **21.** $\frac{16}{20}$ **22.** $\frac{18}{24}$ **23.** $\frac{54}{63}$ **24.** $\frac{56}{72}$ **25.** $\frac{35}{42}$

26. $\frac{12}{3}$ **27.** $\frac{26}{22}$ **28.** $\frac{62}{48}$ **29.** $\frac{112}{28}$ **30.** $\frac{224}{178}$ **31.** $\frac{500}{425}$

Use cross products to determine if the fractions are equivalent. Write *yes* or *no*. [126]

32. $\frac{3}{7} \bullet \frac{9}{24}$ **33.** $\frac{5}{8} \bullet \frac{35}{56}$ **34.** $\frac{24}{48} \bullet \frac{4}{8}$ **35.** $\frac{56}{49} \bullet \frac{7}{8}$

36. $\frac{12}{4} \bullet \frac{48}{16}$ **37.** $\frac{56}{8} \bullet \frac{118}{16}$ **38.** $\frac{256}{16} \bullet \frac{32}{2}$ **39.** $\frac{300}{250} \bullet \frac{60}{50}$

Complete. [126]

40. $\frac{7}{9}, \frac{14}{18}, \frac{21}{27}, \frac{\blacksquare}{\blacksquare}, \frac{\blacksquare}{\blacksquare}, \frac{\blacksquare}{\blacksquare}$ **41.** $\frac{5}{8}, \frac{10}{16}, \frac{15}{24}, \frac{\blacksquare}{\blacksquare}, \frac{\blacksquare}{\blacksquare}, \frac{\blacksquare}{\blacksquare}$

42. $\frac{3}{7}, \frac{6}{14}, \frac{9}{21}, \frac{\blacksquare}{\blacksquare}, \frac{\blacksquare}{\blacksquare}, \frac{\blacksquare}{\blacksquare}$ **43.** $\frac{4}{11}, \frac{8}{22}, \frac{12}{33}, \frac{\blacksquare}{\blacksquare}, \frac{\blacksquare}{\blacksquare}, \frac{\blacksquare}{\blacksquare}$

Find the missing term. [126]

44. $\frac{2}{3} = \frac{\blacksquare}{12}$ **45.** $\frac{7}{8} = \frac{\blacksquare}{56}$ **46.** $\frac{5}{9} = \frac{25}{\blacksquare}$ **47.** $\frac{25}{45} = \frac{\blacksquare}{9}$

48. $\frac{75}{15} = \frac{\blacksquare}{3}$ **49.** $\frac{\blacksquare}{31} = \frac{124}{62}$ **50.** $\frac{48}{45} = \frac{144}{\blacksquare}$ **51.** $\frac{153}{33} = \frac{\blacksquare}{11}$

Chapter 8

RATIO AND PROPORTION

Some patterns are easy to see; others are less obvious. How are patterns used? Why are patterns used?

8.1 RATIO

A **ratio** is a comparison of two numbers by division.
It can be written

In words	As a fraction	With a colon
5 to 6	$\frac{5}{6}$	5:6

You read each the same way: "5 to 6."

EXAMPLE 1 At Vic's Video, there are 1,400 videos in stock. The comedy section has 217 videos. What is the ratio of the number of comedy videos to the total number of videos?

THINK: 217 comedy videos out of 1,400 videos

WRITE:

In words	As a fraction	With a colon
217 to 1,400	$\frac{217}{1{,}400}$	217:1,400

So, the ratio is 217 comedy videos to 1,400 total videos.

The numbers in a ratio are called **terms.** When you multiply or divide both terms of a ratio by the same nonzero number, you get an equivalent ratio.

EXAMPLE 2 Complete to form an equivalent ratio.

a. $\frac{7}{8} = \frac{35}{■}$

THINK: 7 × 5 = 35
8 × 5 = ■

$$\frac{7}{8} = \frac{7 \times 5}{8 \times 5} = \frac{35}{40}$$

So, $\frac{7}{8} = \frac{35}{40}$.

b. $\frac{15}{24} = \frac{■}{8}$

THINK: 24 ÷ 3 = 8
15 ÷ 3 = ■

$$\frac{15}{24} = \frac{15 \div 3}{24 \div 3} = \frac{5}{8}$$

So, $\frac{15}{24} = \frac{5}{8}$.

A ratio is in simplest form when the greatest common factor of both terms is 1. To write a ratio in simplest form, divide both terms by the greatest common factor.

EXAMPLE 3 Write the ratio $\frac{24}{40}$ in simplest form.

THINK: The GCF of 24 and 40 is 8.

Divide each term by 8. $\frac{24}{40} = \frac{24 \div 8}{40 \div 8} = \frac{3}{5}$

So, the ratio $\frac{24}{40}$ is $\frac{3}{5}$ in simplest form.

PRACTICE EXERCISES

Complete to show the ratio in three ways.

Words	Fraction	Colon	Words	Fraction	Colon
6 to 1	**1.** $\blacksquare$	**2.** $\blacksquare$	**3.** $\blacksquare$	**4.** $\blacksquare$	5:8
5. $\blacksquare$	$\frac{21}{8}$	**6.** $\blacksquare$	120 to 250	**7.** $\blacksquare$	**8.** $\blacksquare$

Complete to form an equivalent ratio.

9. $\frac{3}{5} = \frac{\blacksquare}{25}$ **10.** $\frac{4}{7} = \frac{\blacksquare}{28}$ **11.** $\frac{3}{10} = \frac{27}{\blacksquare}$ **12.** $\frac{2}{3} = \frac{16}{\blacksquare}$ **13.** $\frac{8}{13} = \frac{24}{\blacksquare}$

14. $\frac{18}{10} = \frac{\blacksquare}{5}$ **15.** $\frac{35}{42} = \frac{\blacksquare}{6}$ **16.** $\frac{27}{15} = \frac{9}{\blacksquare}$ **17.** $\frac{30}{140} = \frac{3}{\blacksquare}$ **18.** $\frac{32}{44} = \frac{8}{\blacksquare}$

19. $\frac{28}{48} = \frac{7}{\blacksquare}$ **20.** $\frac{5}{16} = \frac{\blacksquare}{32}$ **21.** $\frac{6}{13} = \frac{54}{\blacksquare}$ **22.** $\frac{14}{30} = \frac{\blacksquare}{15}$ **23.** $\frac{12}{51} = \frac{4}{\blacksquare}$

Use the denominator to write a ratio equivalent to $\frac{3}{5}$.

24. 15 **25.** 30 **26.** 20 **27.** 45 **28.** 60 **29.** 100 **30.** 500 **31.** 250

Use the numerator to write a ratio equivalent to $\frac{90}{120}$.

32. 9 **33.** 3 **34.** 6 **35.** 30 **36.** 180 **37.** 900 **38.** 450 **39.** 9,000

Write the ratio in simplest form.

40. $\frac{7}{21}$ **41.** $\frac{5}{30}$ **42.** $\frac{6}{18}$ **43.** $\frac{8}{38}$ **44.** $\frac{9}{72}$ **45.** $\frac{4}{36}$ **46.** $\frac{2}{22}$ **47.** $\frac{3}{36}$

48. $\frac{12}{28}$ **49.** $\frac{15}{27}$ **50.** $\frac{24}{36}$ **51.** $\frac{18}{42}$ **52.** $\frac{45}{63}$ **53.** $\frac{24}{48}$ **54.** $\frac{14}{35}$ **55.** $\frac{18}{72}$

Solve.

56. Vic's Video has 420 adventure videos and 140 music videos. Write the ratio in simplest form of adventure videos to music videos.

57. Mike rents 15 videos for every 6 that Sue rents. Write the ratio in simplest form of Sue's rentals to Mike's rentals.

58. Joan rented 7 videos at $1.84 each. How much did she pay?

59. In 16 weeks, Sally has seen 96 videos. On an average, how many did she see a week?

TIME OUT

The ratio of rock music videos to jazz music videos is 3 to 2. If the total number of videos is 85, how many rock music videos are there? How many jazz music videos are there?

8.2 RATE

In 28 minutes, eight telephone calls were placed from Mount Wilson to Bayville. What was the rate of calls per minutes?

A **rate** is a ratio that compares different units. Rates, like ratios, can be written as fractions in simplest form.

EXAMPLE 1 Write the rate of calls per minutes as a fraction in simplest form.

THINK: The rate is $\frac{\text{8 calls}}{\text{28 minutes}}$

$$\frac{\text{8 calls}}{\text{28 minutes}} \begin{matrix}\rightarrow \\ \rightarrow\end{matrix} \frac{8}{28} = \frac{2}{7}$$

You can compare rates using cross products.

EXAMPLE 2 Phone rates in Bowville depend on the phone company. Dash charges \$1.30 for 8 minutes. GME charges \$0.50 for 3 minutes. Which company has the lower rate?

	Dash		**GME**
1. Write the rates.	$\frac{\$1.30}{\text{8 minutes}}$		$\frac{\$0.50}{\text{3 minutes}}$
2. Find the cross products.	\$1.30 × 3		8 × \$0.50
3. Compare.	\$3.90	<	\$4.00

Since \$1.30 per 8 minutes is less than \$0.50 for 3 minutes, Dash has the lower rate.

When the second term of a rate is 1, the rate is called a **unit rate.**

EXAMPLE 3 Write the rate of 272 calls for 8 phones as a unit rate.

THINK: To find a unit rate, divide both terms by the second term.

$$\frac{\text{272 calls}}{\text{8 phones}} = \frac{272 \div 8}{8 \div 8} = \text{34 calls per phone}$$

So, the unit rate is 34 calls per phone.

PRACTICE EXERCISES

Write the rate as a fraction in simplest form.

1. \$10 for 3 records
2. 7 laps in 9 minutes
3. 3 points in 4 seconds
4. 14 hits in 15 games
5. 213 words in 6 minutes
6. 420 km in 6 hours
7. \$45 for 7 tickets
8. 15 ft in 17 minutes
9. 353 km on 2 L
10. \$5.34 for 6 lb
11. 15 m in 10 seconds
12. 20 defects in 500 bulbs
13. 44 books in 14 days
14. 72 pages in 45 minutes
15. \$46.20 for 22 balloons
16. 35 cars in 14 minutes
17. 125 people in 5 buses
18. 266.6 mi in 8.6 gal

Compare the rates. Write $>$, $<$, or $=$.

19. $\frac{4 \text{ cans}}{\$0.88}$ ● $\frac{6 \text{ cans}}{\$1.32}$
20. $\frac{87 \text{ seats}}{3 \text{ rows}}$ ● $\frac{96 \text{ seats}}{4 \text{ rows}}$
21. $\frac{5 \text{ oz}}{\$0.39}$ ● $\frac{8 \text{ oz}}{\$0.57}$
22. $\frac{91 \text{ pens}}{7 \text{ boxes}}$ ● $\frac{84 \text{ pens}}{6 \text{ boxes}}$
23. $\frac{115 \text{ minutes}}{5 \text{ tapes}}$ ● $\frac{180 \text{ minutes}}{8 \text{ tapes}}$
24. $\frac{247 \text{ mi}}{9 \text{ gal}}$ ● $\frac{286 \text{ mi}}{7 \text{ gal}}$
25. $\frac{145 \text{ yd}}{7 \text{ passes}}$ ● $\frac{159 \text{ yd}}{8 \text{ passes}}$
26. $\frac{\$9.06}{11 \text{ gal}}$ ● $\frac{\$12.46}{14 \text{ gal}}$
27. $\frac{37 \text{ plants}}{100 \text{ seeds}}$ ● $\frac{68 \text{ plants}}{150 \text{ seeds}}$

Write the rate as a unit rate.

28. $\frac{32 \text{ pencils}}{4 \text{ packs}}$
29. $\frac{84 \text{ people}}{3 \text{ buses}}$
30. $\frac{108 \text{ mi}}{9 \text{ hours}}$
31. $\frac{\$3.25}{5 \text{ lb}}$
32. $\frac{750 \text{ points}}{30 \text{ games}}$
33. $\frac{\$186}{6 \text{ tickets}}$
34. $\frac{112 \text{ players}}{8 \text{ teams}}$
35. $\frac{126 \text{ calls}}{3 \text{ minutes}}$
36. $\frac{2600 \text{ words}}{13 \text{ pages}}$
37. $\frac{\$367.20}{18 \text{ months}}$

MIXED REVIEW

Multiply or divide.

38. 100×32.8
39. 0.654×10
40. $1{,}000 \times 0.0037$
41. 100×0.0005
42. $986 \div 100$
43. $5.48 \div 1{,}000$
44. $67.3 \div 10$
45. $0.25 \div 100$

Solve.

46. A telephone operator quoted \$25.99 charges for a 23-minute call. How much did the call cost per minute?

47. The telephone company installed 25,640,000,000 feet of cable. Write the word name for this number.

48. BRC had a \$112.74 phone bill in October and a \$181.32 bill in November. How much more was the November bill?

49. Dash charges \$2.38 for a 7-minute call. GME charges \$1.70 for a 5-minute call. Which company offers the lower rate?

PROBLEM Solving APPLICATION

8.3 CONSUMER: COMPARISON SHOPPING

When you shop, you often need to decide which of two items is the better buy. In comparing prices, you need to use the unit price of each item. The **unit price** is the price per standard amount, such as the price per ounce or the price per dozen.

A 12-oz box of Superbran Flakes is selling for \$1.92.
A 16-oz box of Superbran Flakes is selling for \$2.44.
Which is the better buy?

THINK: You can use this rate: $\frac{\text{price of cereal}}{\text{amount of cereal}}$

The unit price is a unit rate, cost per ounce, or $\frac{\text{cost}}{1\text{ oz}}$.

1. Find the unit price for the 12-oz box. $\frac{\$1.92}{12\text{ oz}} = 12\overline{)\$1.92} = \$0.16$ per ounce
2. Find the unit price for the 16-oz box. $\frac{\$2.44}{16\text{ oz}} = 16\overline{)\$2.44} = \$0.1525$ per ounce
3. Compare the unit prices. $\$0.1525 < \$0.16$

So, the 16-oz box is the better buy.

FOR DISCUSSION

When choosing between two different brands of a product, what other considerations are there besides the unit price of each box? Explain.

Find the unit price. Remember to estimate whenever you use your calculator.

1. Fruit: $1.76 for a 16-oz can

2. Twine: $3.25 for 50 ft

3. Cheese: 24 oz for $2.88

4. Soap: 3 bars for $1.26

5. Apples: 6 for $1.50

6. Meat: 2.6 lb for $8.06

Find the unit price to the nearest tenth of a cent.

7. An 8-oz box of Ultrawheat Flakes is selling for $1.25.

8. 5 cans of fruit salad are selling for $3.36.

9. 3 green peppers are selling for $1.07.

10. 4 avocados are selling for $3.50.

11. A 12-oz bottle of fruit juice is selling for $0.89.

Which is the better buy if price is the only consideration?

12. 3 lb of apples for $1.89 or 5 lb of apples for $3.17

13. A 4-oz jar of olives for $1.04 or a 6-oz jar of olives for $1.53

14. 50 sheets of notebook paper for $0.65 or 144 sheets for $2.16

15. A 15-ft extension cord for $3.60 or a 4-yd extension cord for $2.58

16. A 6-oz jar of tomato sauce for $1.49 or a 16-oz jar for $3.89

17. A 6-oz bag of grated mozzarella cheese for $1.79 or a 16-oz piece for $4.59

18. In many supermarkets and groceries, information is provided about the unit prices of various items. The weights of items and the total prices of the items are also indicated. Find such information about five different items. Then check to see that

$$\textit{weight} \times \textit{unit price} = \textit{total price of item.}$$

8.4 SOLVING A PROPORTION

A **proportion** is a statement that two ratios are equivalent.
To determine if two ratios form a proportion, check the cross products.
If the cross products of the ratios are equal, the ratios form a proportion.

EXAMPLE 1 Determine if the ratios form a proportion.

a. $\frac{8}{5} \stackrel{?}{=} \frac{2}{3}$

$\frac{8}{5} \times \frac{2}{3}$

$8 \times 3 \bullet 5 \times 2$

$24 \neq 10$

$\frac{8}{5}$ and $\frac{2}{3}$ do not form a proportion.

b. $\frac{6}{9} \stackrel{?}{=} \frac{5}{7.5}$

$\frac{6}{9} \times \frac{5}{7.5}$

$6 \times 7.5 \bullet 9 \times 5$

$45 = 45$

$\frac{6}{9}$ and $\frac{5}{7.5}$ form a proportion.

Many problems can be solved by writing a proportion and then finding the missing term.

EXAMPLE 2 Eric used a pancake mix that called for 3 eggs to make 16 pancakes. How many eggs would he need to make 48 pancakes?

1. Write a proportion.

   THINK: **3 eggs is to 16 pancakes as n eggs is to 48 pancakes**

   $\frac{3}{16} = \frac{n}{48}$ ← **eggs** / ← **pancakes**

2. Multiply to find cross products.

   $3 \times 48 = 16 \times n$
   $144 = 16 \times n$

3. Find n.

   THINK: To find the missing factor, divide 144 by 16.

   $144 \div 16 = n$
   $9 = n$

So, 9 eggs are needed to make 48 pancakes.

CHECKPOINT

Write the letter of the correct answer.

Find n.

1. $\frac{n}{16} = \frac{4}{8}$ **a.** 2 **b.** 4 **c.** 8 **d.** 32

2. $\frac{6}{5} = \frac{n}{35}$ **a.** 7 **b.** 30 **c.** 36 **d.** 42

3. $\frac{9}{n} = \frac{3}{15}$ **a.** 5 **b.** 15 **c.** 45 **d.** 135

4. $\frac{12}{18} = \frac{2}{n}$ **a.** 3 **b.** 6 **c.** 24 **d.** 36

PRACTICE EXERCISES

Determine if the ratios form a proportion. Write *yes* or *no*.

1. $\frac{6}{7} \stackrel{?}{=} \frac{14}{12}$ **2.** $\frac{8}{3} \stackrel{?}{=} \frac{56}{21}$ **3.** $\frac{5}{9} \stackrel{?}{=} \frac{30}{54}$ **4.** $\frac{72}{35} \stackrel{?}{=} \frac{8}{5}$

5. $\frac{49}{14.7} \stackrel{?}{=} \frac{7}{2.1}$ **6.** $\frac{0.25}{0.3} \stackrel{?}{=} \frac{0.1}{0.12}$ **7.** $\frac{2.1}{30} \stackrel{?}{=} \frac{9}{1.5}$ **8.** $\frac{4.8}{3.2} \stackrel{?}{=} \frac{9.4}{6.4}$

Find n.

9. $\frac{3}{5} = \frac{n}{25}$ **10.** $\frac{7}{4} = \frac{n}{28}$ **11.** $\frac{6}{15} = \frac{n}{45}$ **12.** $\frac{9}{7} = \frac{n}{56}$

13. $\frac{n}{48} = \frac{8}{12}$ **14.** $\frac{n}{35} = \frac{8}{7}$ **15.** $\frac{n}{26} = \frac{10}{13}$ **16.** $\frac{n}{72} = \frac{3}{8}$

17. $\frac{0.5}{0.8} = \frac{n}{3.2}$ **18.** $\frac{n}{6.3} = \frac{0.4}{0.9}$ **19.** $\frac{1.4}{0.3} = \frac{2.8}{n}$ **20.** $\frac{4.5}{n} = \frac{6.3}{0.7}$

21. $\frac{n}{15.9} = \frac{4.2}{12.6}$ **22.** $\frac{n}{3.7} = \frac{18.9}{2.1}$ **23.** $\frac{30.6}{n} = \frac{20.4}{4}$ **24.** $\frac{5.2}{n} = \frac{3.5}{24.5}$

MIXED REVIEW

Write the fraction or mixed number as a decimal.

25. $\frac{25}{100}$ **26.** $\frac{45}{50}$ **27.** $\frac{1}{8}$ **28.** $4\frac{45}{60}$ **29.** $30\frac{10}{16}$ **30.** $7\frac{4}{5}$

Use the table below for Exercises 31–34.

Item	Number	Mix	Milk	Eggs	Oil
Pancakes	25	$2\frac{1}{4}$ cups	2 cups	2	2 tbsp
Muffins	144	8 cups	$6\frac{1}{2}$ cups	4	$\frac{1}{4}$ cup

31. How much milk is needed to make 175 pancakes?

32. How much mix is needed to make 36 muffins?

33. If you had 10 cups of milk, how much would be left after making 144 muffins?

34. To triple the pancake recipe, how many cups of mix would you need?

TIME OUT

There are three plates. On one plate there are three pancakes stacked so that the largest one is on the bottom and the smallest one is on the top. You can only move one pancake at a time from plate to plate, and you can never place a larger pancake on a smaller one. What is the least number of moves that it will take to stack the pancakes in the same order on a different plate?

PROBLEM *Solving* STRATEGY

8.5 USING ESTIMATION

Situation:

At Byrne's Automotive Shop, front floor mats are priced at \$9.29 each. Rear floor mats are priced at \$6.45 each.

Strategy:

Sometimes an estimate is all that is needed to solve a problem.

Applying the Strategy:

A. Chris has \$25. Does he have enough money to buy one front mat and two rear mats?

THINK: When you want to be sure that you have enough money to make a purchase, make an **overestimate.**

1. Estimate by rounding up. $\$9.29 \rightarrow \$10 \quad \$6.45 \rightarrow \$7$

$2 \times \$7 = \$14 \qquad \$10 + \$14 = \$24$

Since \$25 is greater than the overestimated total, Chris will have enough to buy the mats.

B. Last week Byrne's Shop sold 26 front floor mats. Were the total sales of front floor mats at least \$200 during the week?

THINK: When you want to be sure that you have reached a minimum total amount, make an **underestimate.**

Estimate by rounding down. $26 \rightarrow 20 \quad \$9.29 \rightarrow \$9 \quad 20 \times \$9 = \$180$

Since the underestimated total is less than \$200, you cannot be sure that the sales reached \$200.

Find the exact answer. $26 \times \$9.29 = \$241.54$

The total sales were at least \$200 during the week.

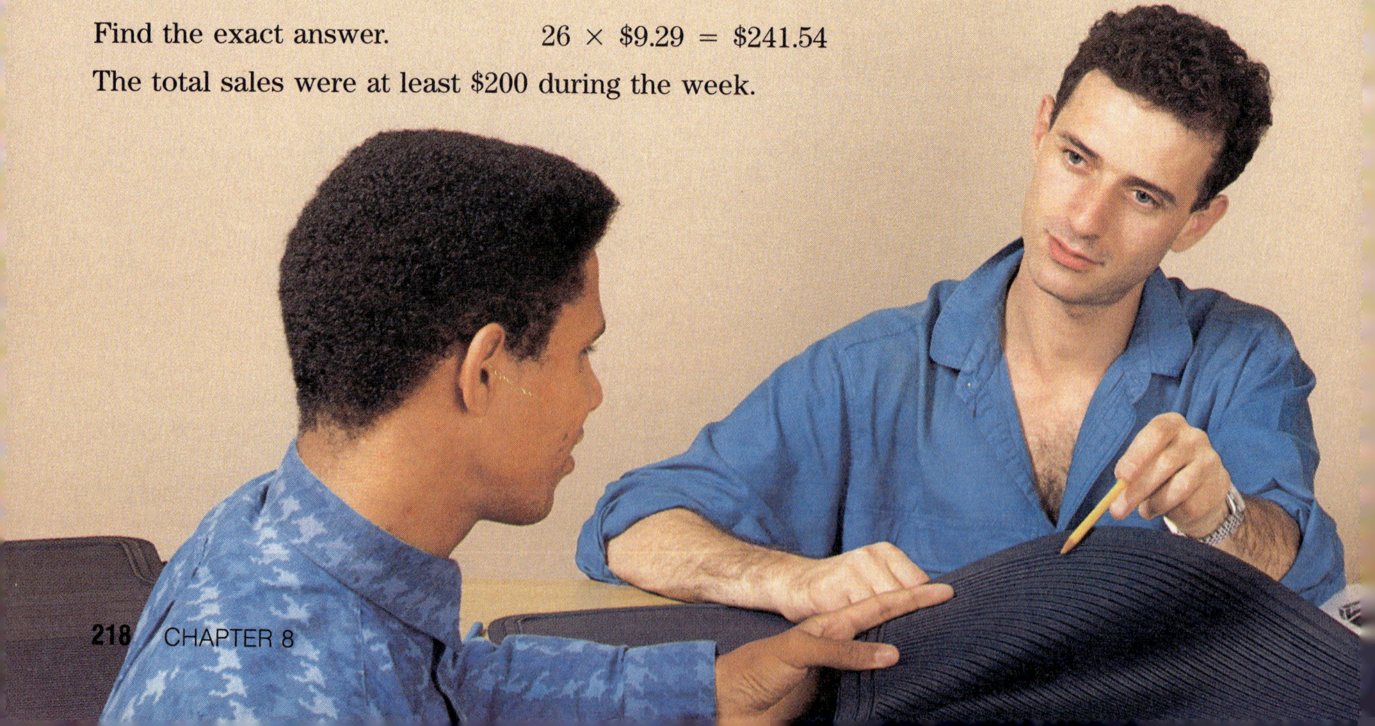

PRACTICE EXERCISES

Which is the better estimate?

1. Amy wants to buy 3 car polishing cloths and a flashlight. Each cloth costs $2.49. A flashlight costs $5.40. Will $12 be enough?

   a. 3 × $2 = $6
   $6 + $5 = $11

   b. 3 × $3 = $9
   $9 + $6 = $15

Use the advertisement to answer Exercises 2–5.

2. Will $6 be enough to buy 2 containers of vinyl cleaner?
3. Will $15 be enough to buy safety flares and paste wax?
4. Will $25 be enough to buy 6 bottles of dry gas and safety flares?
5. Is $20 enough to buy one of each item?

Byrne's Sale

Paste Wax	$6.39
Safety Flares	$8.65
Dry Gas	$1.25
Vinyl Cleaner	$2.79

Use the table of projected profits to answer Exercises 6–9.

Byrne's Automotive Shop can hire another cashier if the projected profits are over $100,000 for this 4-month period.

6. Should you overestimate or underestimate the projected profits to determine if the goal of $100,000 has been reached?
7. What is the projected estimated profit for this 4-month period?
8. What is the projected profit for this 4-month period?
9. Can Byrne's Automotive hire another cashier?

BYRNE'S AUTOMOTIVE
PROJECTED PROFIT RECORD

July	$28,750
August	$36,109
September	$24,930
October	$41,885

PROBLEM Solving APPLICATION

8.6 USING PROPORTIONS TO SOLVE PROBLEMS

Speed, time, and distance can be calculated using proportion. When you travel, it takes a certain amount of time to go a certain number of miles. If you know your rate of speed, it is easy to calculate how far you can travel in a certain amount of time.

A. Your family is on a vacation trip.
In highway driving, you travel 125 miles in 2.5 hours.
If you continue to travel at this rate,
how many miles will you travel in 6 hours?

A proportion can be used to solve the problem.

THINK: Use the rate, $\frac{\text{distance}}{\text{time}}$.

You know that $\frac{125}{2.5}$ is the rate of speed.

1. Write a proportion.

THINK: 125 mi is to 2.5 hours as *n* mi is to 6 hours.

$$\frac{125}{2.5} = \frac{n}{6} \begin{matrix}\leftarrow \textbf{miles} \\ \leftarrow \textbf{hours}\end{matrix}$$

2. Multiply to find the cross products.

$$125 \times 6 = 2.5 \times n$$
$$750 = 2.5 \times n$$

3. Find n.

THINK: To find the missing factor, divide 750 by 2.5.

$$n = 750 \div 2.5$$
$$n = 300$$

You will travel 300 miles in 6 hours.

B. Suppose a car travels 112 miles on 3.5 gal of gasoline.
How many gallons are needed to travel 272 miles?

THINK: Use the rate $\frac{\text{distance}}{\text{gallons}}$ to write a proportion.

1. Write a proportion.

THINK: 112 mi is to 3.5 gal as 272 mi is to *n* gal.

$$\frac{112}{3.5} = \frac{272}{n} \begin{matrix}\leftarrow \textbf{miles} \\ \leftarrow \textbf{gallons}\end{matrix}$$

2. Multiply to find the cross products.

$$112 \times n = 3.5 \times 272$$
$$112 \times n = 952$$

3. Find n.

THINK: To find the missing factor, divide 952 by 112.

$$n = 952 \div 112$$
$$n = 8.5$$

8.5 gal are needed to travel 272 miles.

Solve.

How far can you travel in 5 hours if you travel at the rate of:

1. 174 mi in 3 hours? **2.** 130 mi in 2.5 hours? **3.** 320 mi in 6.4 hours?

How far can you travel in 9.2 hours if you travel at the rate of:

4. 162 mi in 3 hours? **5.** 216 mi in 4.5 hours? **6.** 234 mi in 5.2 hours?

If a car travels:	**It can travel:**
• **7.** 84 mi on 3 gal of gasoline	140 mi on ■ gal.
• **8.** 99 mi on 4.5 gal of gasoline	275 mi on ■ gal.
9. 207 mi on 6.9 gal of gasoline	441 mi on ■ gal.
10. 72 mi on 4 gal of gasoline	■ mi on 10 gal.
11. 156 mi on 6.5 gal of gasoline	■ mi on 8.5 gal.
12. 153 mi on 5.1 gal of gasoline	■ mi on 9.6 gal.

13. One morning, you drive 216 mi in 4 hours. At that rate, how many miles will you drive in 7 hours?

14. The Carruthers family drives 96 mi in 2.4 hours. At that rate, how long will it take them to drive 48 mi?

15. The Carruthers family car travels 221 mi on 8.5 gal of gasoline. At that rate, how many miles will the car travel on 14.5 gal?

16. The Carruthers family paid $13.44 for 12 gal of gas for their car. Their tank holds 16 gal. At that rate, how much would a full tank cost?

MIDCHAPTER REVIEW

Write the ratio in two ways. [210]

		Colon		Fraction
75 oz to 29 oz	**1.**	■	**2.**	■
32 days to 76 days	**3.**	■	**4.**	■
$14.12 for 3 tapes	**5.**	■	**6.**	■
329 mi on 13 gal	**7.**	■	**8.**	■
4.5 cups to 8 oz	**9.**	■	**10.**	■

Write the ratio in simplest form. [210]

11. $\frac{7}{28}$ **12.** $\frac{5}{45}$ **13.** $\frac{20}{32}$ **14.** $\frac{14}{63}$ **15.** $\frac{9}{21}$

Write as a unit rate. [212]

16. $\frac{\text{24 pens}}{\text{8 boxes}}$ **17.** $\frac{\text{35 pages}}{\text{7 chapters}}$ **18.** $\frac{\text{28 people}}{\text{14 seats}}$ **19.** $\frac{\text{30 miles}}{\text{6 gallons}}$ **20.** $\frac{\text{120 calls}}{\text{40 minutes}}$

Determine if the ratios form a proportion. Write *yes* or *no*. [216]

21. $\frac{3}{5} \stackrel{?}{=} \frac{9}{15}$ **22.** $\frac{4}{9} \stackrel{?}{=} \frac{18}{36}$ **23.** $\frac{5}{7} \stackrel{?}{=} \frac{15}{24}$ **24.** $\frac{8}{11} \stackrel{?}{=} \frac{32}{44}$ **25.** $\frac{7}{3} \stackrel{?}{=} \frac{9}{21}$

26. $\frac{18}{21} \stackrel{?}{=} \frac{6}{7}$ **27.** $\frac{45}{15} \stackrel{?}{=} \frac{9}{5}$ **28.** $\frac{30}{20} \stackrel{?}{=} \frac{60}{40}$ **29.** $\frac{7.5}{2.5} \stackrel{?}{=} \frac{1.8}{0.6}$ **30.** $\frac{4.8}{1.6} \stackrel{?}{=} \frac{22.6}{7.2}$

Find n. [216]

31. $\frac{n}{6} = \frac{25}{30}$ **32.** $\frac{14}{n} = \frac{4}{8}$ **33.** $\frac{18}{21} = \frac{n}{7}$ **34.** $\frac{24}{15} = \frac{8}{n}$ **35.** $\frac{54}{72} = \frac{n}{12}$

Solve. [214, 218, 220]

36. Which is the better buy if price is the only consideration: an 8-oz box of spinach pasta for $0.69 or a 12-oz box of pasta for $1.05?

37. Beth wants to buy three cans of soup for $0.89, one loaf of wheat bread for $1.09, and 4 lb of fish at $1.59 pound. Will $10 be enough?

38. How far can you go in 6 hours if you are traveling at a rate of 225 mi in 5 hours?

39. A car goes 256 mi on 16 gal of gas. How far can it go on 10 gal?

PROBLEM Solving APPLICATION

8.7 CAREER: PHOTOGRAPHER

Juan Mendoza is a photographer. He takes photographs for large corporations, magazines, and models. Sometimes Juan enlarges or reduces a photograph he has taken.

The relationships of the dimensions can be written in equivalent ratios.

$$\frac{\text{LENGTH OF ORIGINAL}}{\text{LENGTH OF ENLARGEMENT}} = \frac{\text{WIDTH OF ORIGINAL}}{\text{WIDTH OF ENLARGEMENT}}$$

A photograph is 7 in. long and 5 in. wide. Juan makes an enlargement that is 10.5 in. long. What is the width of the enlargement?

1. Write a proportion.

   THINK: 7 in. is to 10.5 in. as 5 in. is to *n* in.

   $$\frac{7}{10.5} = \frac{5}{n} \begin{array}{l}\leftarrow \textbf{original (in.)} \\ \leftarrow \textbf{enlargement (in.)}\end{array}$$

2. Multiply to find the cross products.

   $7 \times n = 10.5 \times 5$

   $7 \times n = 52.5$

3. Find n.

   THINK: To find the missing factor, divide 52.5 by 7.

   $n = 52.5 \div 7$

   $n = 7.5$

The width of the enlargement is 7.5 in.

Find the missing dimension.

Original photograph		Enlargement		Original photograph		Reduction	
Length	Width	Length	Width	Length	Width	Length	Width
1. 7 in.	5 in.	17.5 in.	■ in.	**2.** 9 in.	6 in.	4.5 in.	■ in.
3. 3 in.	5 in.	■ in.	11 in.	**4.** 10 in.	8 in.	■ in.	6 in.
5. 8 in.	5.6 in.	20 in.	■ in.	**6.** 11 in.	8.5 in.	4.4 in.	■ in.

Solve.

7. A photograph has a length of 6 in. and a width of 4 in. Juan makes an enlargement that has a length of 13.5 in. What is the width of the enlargement?

8. In Exercises 1, 3, and 5, is the ratio of the length to the width in the enlargement equivalent to the ratio of the length to the width in the original photograph?

8.8 USING A SCALE

A **scale drawing** uses a ratio to compare an object's size on a drawing to its actual size. The ratio of the drawing size to the actual size is called the **scale.**

When you know the scale and the actual size, you can write and solve a proportion to find the scale size.

EXAMPLE 1 The Spruce Goose has the largest wingspan of any plane. It measures almost 320 ft tip to tip. For a scale model of the plane, a scale where 3 in. represents 40 ft was used. How long is the wingspan of the scale model of the plane?

1. Write a proportion. $\frac{3}{40} = \frac{n}{320}$ ← **model (in.)** ← **actual (ft)**
2. Multiply to find the cross products. $3 \times 320 = 40 \times n$; $960 = 40 \times n$
3. Find n. $960 \div 40 = n$; $24 = n$

So, the wingspan of the scale model is 24 in.

Scale drawings show something that is too large or too small to be shown easily in its actual size. You can also use a proportion to find the actual size.

EXAMPLE 2 Find the distance from the foul line to the basket.

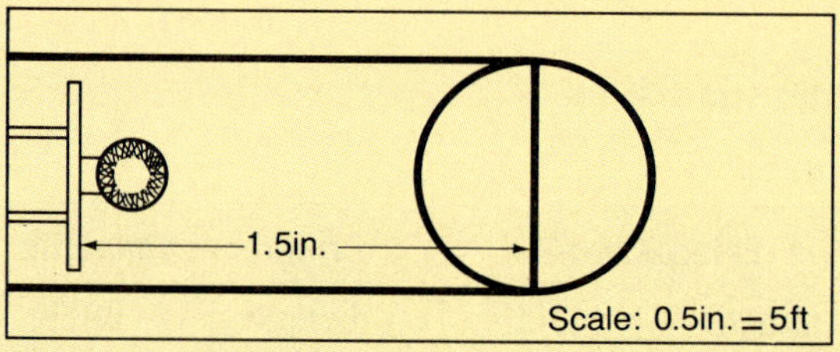

1. Write the proportion. $\frac{0.5}{5} = \frac{1.5}{n}$ ← **scale (in.)** ← **actual (ft)**
2. Multiply to find the cross products. $0.5 \times n = 5 \times 1.5$; $0.5 \times n = 7.5$
3. Find n. $n = 7.5 \div 0.5$; $n = 15$

So, the distance is 15 ft.

PRACTICE EXERCISES

Use the floor plan to complete the table.

	Room	Actual measure
1.	Living room (a)	■
2.	Living room (b)	■
3.	Family room (c)	■
4.	Family room (d)	■
5.	Kitchen (e)	■
6.	Laundry room (f)	■
7.	Laundry room (g)	■
8.	Bath (h)	■
9.	Bath (i)	■
10.	Closet (j)	■
11.	Dining room (k)	■
12.	Dining room (l)	■
13.	Hall (m)	■
14.	Living/family (n)	■
15.	Bath/laundry (o)	■
16.	Laundry/ kitchen/ family (p)	■

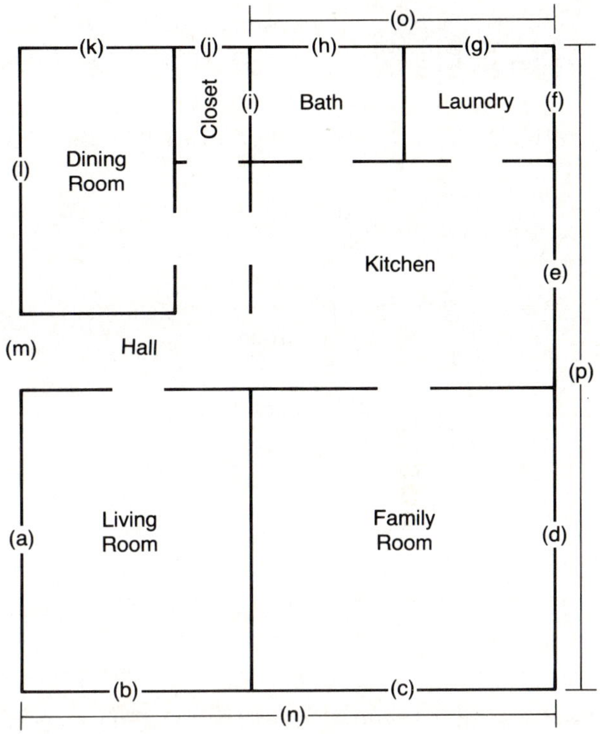

Scale: 0.5 cm = 1 m

A model train was made to the scale of 0.5 in.: 2 ft. Find the actual measure.

Car	Engine	Caboose	Flat car	Freight	Passenger
Scale measure	6 in.	4.5 in.	10 in.	5.5 in.	10.5 in.
Actual measure	**17.** ■	**18.** ■	**19.** ■	**20.** ■	**21.** ■

Solve.

22. A scale drawing of an insect is 4.2 cm long. The scale is 6 cm : 1 cm. How long is the insect?

23. A scale of 2 cm : 5 mm is used to draw a bolt. The bolt is 7.4 mm long. How long is the drawing?

24. On a map, two towns are 5.8 in. apart. The actual distance is 40.6 mi. What is the scale?

25. A scale drawing uses 13 cm to represent 390 km. What scale is being used?

PROBLEM Solving APPLICATION

8.9 CONSUMER: USING MAPS

A map is a scale drawing. You can use a map to determine distances between places. In the map of the Florida panhandle shown below the scale is 1.8 cm = 40 km. Each 1.8 cm on the map represents 40 km of air distance.

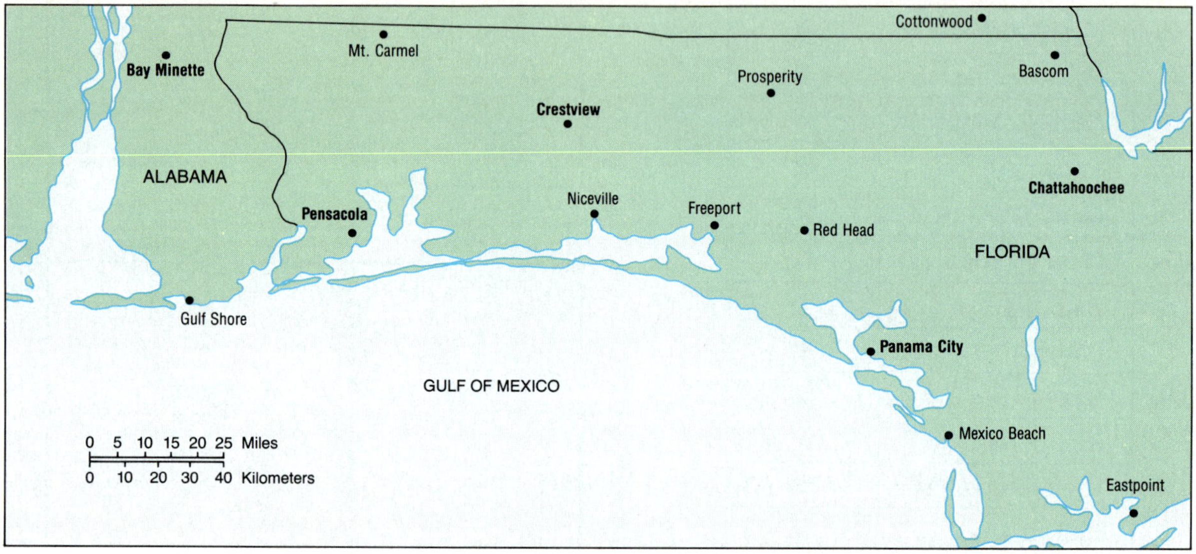

You are on a business trip for the IBQ Corporation. A company plane will fly you from Pensacola to Panama City. What is the air distance between these cities?

1. Measure the distance from Pensacola to Panama City on the map.

   The distance is 7 cm to the nearest tenth of a centimeter.

2. Write a proportion.

   THINK: 1.8 cm is to 40 km as 7 cm is to *n* km

   $$\frac{1.8}{40} = \frac{7}{n} \quad \begin{matrix} \leftarrow \textbf{scale (cm)} \\ \leftarrow \textbf{actual (km)} \end{matrix}$$

3. Solve the proportion.

   $$\begin{aligned} 1.8 \times n &= 40 \times 7 \\ 1.8 \times n &= 280 \\ n &= 280 \div 1.8 \\ n &= 155.6 \end{aligned}$$

So, the air distance from Pensacola to Panama City is about 156 km.

FOR DISCUSSION

Which of the following would be best for the scale of a state road map? Why?

1 in. represents: 100 ft, 100 yd, or 100 mi

Use the map on page 226. Measure the distance to the nearest tenth of a centimeter. Then find the air distance in kilometers between the Florida cities below.

	Distance on Map	Air Distance
Panama City to Crestview	**1.** ■	**2.** ■
Crestview to Mexico Beach	**3.** ■	**4.** ■
Mexico Beach to Bascom	**5.** ■	**6.** ■
Bascom to Niceville	**7.** ■	**8.** ■

Find the air distance in kilometers.

9. Cottonwood, AL, to Freeport, FL

10. Pensacola, FL, to Prosperity, FL

11. Mt. Carmel, FL, to Bay Minette, AL

12. Eastpoint, FL, to Red Head, FL

Solve.

13. What is the total air distance if you fly from Chattahoochee, FL, to Cottonwood, AL?

14. What is the total air distance if you fly from Pensacola, FL, to Crestview, FL, and on to Panama City, FL?

Suppose you fly from Bay Minette, AL, to Prosperity, FL, and on to Eastpoint, FL.

15. What is the total air distance?

16. About how many kilometers longer is the distance than if you fly directly from Bay Minette to Eastpoint?

Suppose you fly from Gulf Shores, AL, to Pensacola, FL, to Mexico Beach, FL.

17. What is the total air distance?

18. About how many kilometers longer is the distance than if you fly directly from Gulf Shores to Mexico Beach?

19. If the plane flies at an average of 140 mph, about how long will it take to fly from Gulf Shores to Mexico Beach?

20. In a reference library, find a road map of the Florida-Alabama border. Compute the road distance of a trip from Gulf Shores, AL, to Pensacola, FL, to Mexico Beach, FL.

CHAPTER REVIEW

Vocabulary Choose the letter of the word that completes each statement.

1. A comparison of two numbers by division is a ■. [210]
   a. quotient **b.** colon **c.** ratio **d.** proportion
2. You can compare rates by comparing ■. [212]
   a. means **b.** extremes **c.** units **d.** cross products
3. In a proportion there are two equivalent ■. [216]
   a. equations **b.** ratios **c.** extremes **d.** means

Skills Write the ratio in simplest form. [210]

$\frac{18}{42}$ *$\frac{3}{7}$*

4. $\frac{10}{25}$ **5.** $\frac{12}{36}$ **6.** $\frac{8}{28}$ **7.** $\frac{21}{30}$ **8.** $\frac{18}{63}$ **9.** $\frac{28}{56}$ **10.** $\frac{32}{48}$

Compare the rates. Write >, <, or =. [212]

$\frac{2 \text{ in.}}{3 \text{ ft}}$ ● $\frac{5 \text{ in.}}{6 \text{ ft}}$ $<$

11. $\frac{6 \text{ lb}}{12 \text{ oz}}$ ● $\frac{5 \text{ lb}}{10 \text{ oz}}$ **12.** $\frac{3 \text{ in.}}{8 \text{ ft}}$ ● $\frac{7 \text{ in.}}{9 \text{ ft}}$

13. $\frac{30 \text{ mi}}{10 \text{ gal}}$ ● $\frac{240 \text{ mi}}{6 \text{ gal}}$ **14.** $\frac{\$0.68}{5 \text{ cans}}$ ● $\frac{\$0.72}{7 \text{ cans}}$ **15.** $\frac{49 \text{ cars}}{20 \text{ minutes}}$ ● $\frac{63 \text{ cars}}{30 \text{ minutes}}$

Write as a unit rate. [212]

$\frac{250 \text{ mi}}{10 \text{ gal}}$ *25 mi per gal*

16. $\frac{91 \text{ chairs}}{7 \text{ rows}}$ **17.** $\frac{132 \text{ points}}{4 \text{ games}}$ **18.** $\frac{\$0.90}{3 \text{ cans}}$

19. $\frac{150 \text{ pencils}}{6 \text{ boxes}}$ **20.** $\frac{235 \text{ coins}}{5 \text{ banks}}$ **21.** $\frac{400 \text{ mi}}{8 \text{ hours}}$ **22.** $\frac{126 \text{ yd}}{9 \text{ rushes}}$

Determine if the ratios form a proportion. Write *yes* or *no*. [216]

$\frac{5}{9} \stackrel{?}{=} \frac{18}{10}$ *no*

23. $\frac{4}{7} \stackrel{?}{=} \frac{18}{28}$ **24.** $\frac{3}{5} \stackrel{?}{=} \frac{15}{9}$ **25.** $\frac{2}{9} \stackrel{?}{=} \frac{12}{54}$

26. $\frac{35}{14} \stackrel{?}{=} \frac{5}{2}$ **27.** $\frac{1.2}{2.1} \stackrel{?}{=} \frac{0.3}{0.5}$ **28.** $\frac{15}{0.24} \stackrel{?}{=} \frac{0.5}{0.8}$ **29.** $\frac{3.6}{4.8} \stackrel{?}{=} \frac{0.6}{0.8}$

Find *n*. [216]

$\frac{n}{8} = \frac{21}{56}$ *3*

30. $\frac{n}{12} = \frac{5}{6}$ **31.** $\frac{13}{n} = \frac{26}{8}$ **32.** $\frac{5}{9} = \frac{n}{63}$

33. $\frac{3}{8} = \frac{21}{n}$ **34.** $\frac{10}{n} = \frac{100}{130}$ **35.** $\frac{15}{24} = \frac{n}{8}$ **36.** $\frac{28}{49} = \frac{4}{n}$

CHAPTER TEST

Write the ratio in two ways.

14 L water to 15 L ammonia

5 cm to 13 m

32 mi to 4 mi

29 hours to 25 hours

Colon	Fraction
1. ■	**2.** ■
3. ■	**4.** ■
5. ■	**6.** ■
7. ■	**8.** ■

Write the ratio in simplest form.

9. $\frac{12}{48}$ **10.** $\frac{16}{24}$ **11.** $\frac{42}{28}$ **12.** $\frac{45}{27}$ **13.** $\frac{33}{72}$

Write as a unit rate.

14. $\frac{\text{84 seeds}}{\text{4 packs}}$ **15.** $\frac{\text{129 candles}}{\text{3 boxes}}$ **16.** $\frac{\text{469 albums}}{\text{7 boxes}}$ **17.** $\frac{\text{835 students}}{\text{5 classes}}$ **18.** $\frac{\text{564 hours}}{\text{6 months}}$

Compare the rates. Write >, <, or =.

19. $\frac{\text{2 in.}}{\text{3 ft}}$ ● $\frac{\text{7 in.}}{\text{8 ft}}$ **20.** $\frac{\$1.52}{\text{7 oz}}$ ● $\frac{\$0.98}{\text{4 oz}}$ **21.** $\frac{\text{280 mi}}{\text{14 gal}}$ ● $\frac{\text{360 mi}}{\text{18 gal}}$

22. $\frac{\text{142 words}}{\text{2 minutes}}$ ● $\frac{\text{201 words}}{\text{3 minutes}}$ **23.** $\frac{\text{220 mi}}{\text{4.5 in.}}$ ● $\frac{\text{305 mi}}{\text{3.2 in.}}$ **24.** $\frac{\text{3 cm}}{\text{125 km}}$ ● $\frac{\text{7 cm}}{\text{342 km}}$

Determine if the ratios form a proportion. Write *yes* or *no*.

25. $\frac{2}{3} \stackrel{?}{=} \frac{8}{9}$ **26.** $\frac{5}{7} \stackrel{?}{=} \frac{25}{35}$ **27.** $\frac{6}{11} \stackrel{?}{=} \frac{42}{48}$ **28.** $\frac{4}{9} \stackrel{?}{=} \frac{32}{72}$

Find *n*.

29. $\frac{n}{5} = \frac{8}{20}$ **30.** $\frac{7}{n} = \frac{21}{33}$ **31.** $\frac{15}{9} = \frac{n}{3}$ **32.** $\frac{42}{18} = \frac{7}{n}$

33. $\frac{n}{6} = \frac{7}{8}$ **34.** $\frac{5}{n} = \frac{4}{9}$ **35.** $\frac{3}{10} = \frac{n}{23}$ **36.** $\frac{6}{7} = \frac{9}{n}$

Solve.

37. On a map, the distance from one city to another measures 3.5 in. According to the scale of the map, 1 in. represents 15 miles. How many actual miles is the distance?

38. Wendy wants to buy two ball-point pens, two tablets of paper, and one notebook. The pens cost $1.19 each, the tablets of paper cost $2.49 each, and the notebook costs $2.79. Will $10 be enough?

ENRICHMENT INVERSE PROPORTIONS

Inverse means *opposite.* In an inverse proportion, as the value of x increases, the value of y decreases. Similarly, as the value of x decreases, the value of y increases.

Inverse proportions can be applied to many real-life situations.

Example A 12-kg stone is 2 meters from the fulcrum. How far will an 8-kg stone be from the fulcrum to balance it?

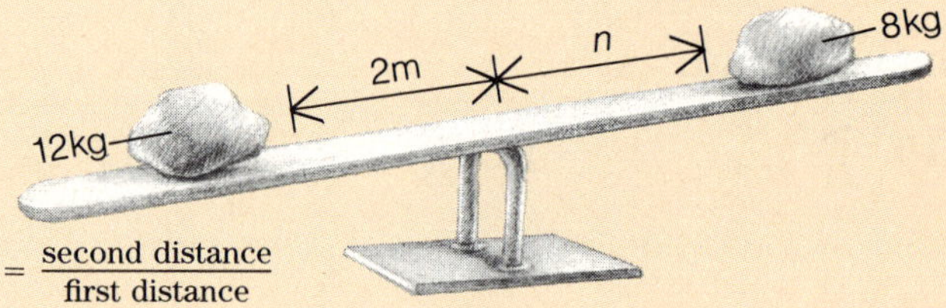

$$\frac{\text{first mass}}{\text{second mass}} = \frac{\text{second distance}}{\text{first distance}}$$

THINK: First mass = 12 kg, second mass = 8 kg
First distance = 2 m, second distance = n meters

$$\frac{12}{8} = \frac{n}{2}$$

$$12 \times 2 = 8 \times n$$

$$24 = 8 \times n$$

$$24 \div 8 = n$$

$$3 = n$$

So, the second stone will be 3 meters from the fulcrum.

Use an inverse proportion to solve.

1. A 32-kg mass is 10 cm from the fulcrum. How far will a 40-kg mass be from the fulcrum to balance it?

2. A 40-kg stone is 9 cm from the fulcrum. To balance it, another stone is 12 cm away from the fulcrum. What is the mass of this stone?

3. A trip takes 2.5 hours at 40 km/hour. How long will the same trip take at 50 km/hour?

4. A trip took 6 hours at 50 mph. If the same trip took 4 hours, what was the rate of speed?

5. Ten word processors can input a manuscript in 6 hours. How long would it take 5 word processors to do the same amount of work?

CALCULATOR SOLVING PROPORTIONS ON A CALCULATOR

You can use a calculator to solve proportions.

Solve for n: $\frac{51}{96} = \frac{n}{80}$

Think	Calculator Entry	Calculator Display
$96n = 51 \times 80$		
$n = \frac{51 \times 80}{96}$	[5] [1] [×] [8] [0] [÷] [9] [6] [=]	42.5
$n = 42.5$		

Solve for n: $\frac{3.5}{n} = \frac{6}{4.9}$ Round the answer to the nearest tenth.

Think	Calculator Entry	Calculator Display
$6n = 3.5 \times 4.9$		
$n = \frac{3.5 \times 4.9}{6}$	[3] [.] [5] [×] [4] [.] [9] [÷] [6] [=]	2.8583333
$n = 2.9$ ← **Rounded to the nearest tenth**		

Solve for n. Use a calculator.
When necessary, round your answer to the nearest tenth.

1. $\frac{n}{95} = \frac{48}{57}$

2. $\frac{63}{91} = \frac{45}{n}$

3. $\frac{385}{n} = \frac{365}{292}$

4. $\frac{23}{n} = \frac{115}{229}$

5. $\frac{153}{594} = \frac{n}{1{,}485}$

6. $\frac{n}{126} = \frac{273}{168}$

7. $\frac{3.4}{10.2} = \frac{2.7}{n}$

8. $\frac{n}{5.2} = \frac{6.8}{20.8}$

9. $\frac{1.5}{0.9} = \frac{n}{19.5}$

10. $\frac{1.9}{n} = \frac{0.2}{1.3}$

11. $\frac{0.7}{0.5} = \frac{2.2}{n}$

12. $\frac{2.5}{11.1} = \frac{n}{3.6}$

Write a proportion to find the answer. Use a calculator to solve the proportion.

13. A carpenter earned $596.80 for a job that took 32 hours. At this rate, how much would he earn on a job that takes 25 hours?

14. A temporary office worker earns $12.65 per hour. Last week she earned $354.20. How many hours did she work last week?

15. A sales representative drove his car 368 mi on 11.5 gal of gasoline. At this rate, how many gallons of gasoline would his car use on a 480-mi trip?

16. A driver drove her school bus 285 mi on 30 gal of gasoline. At this rate, how many miles could she have driven on 12 gal of gasoline?

PreSkills Test

Write as a fraction with a denominator of 100. [126]

1. $\frac{1}{2}$ **2.** $\frac{3}{4}$ **3.** $\frac{7}{10}$ **4.** $\frac{4}{5}$ **5.** $\frac{13}{25}$

6. 0.25 **7.** 0.04 **8.** 0.3 **9.** 3.2 **10.** 5.42

Write as a fraction in simplest form. [126]

11. $\frac{25}{75}$ **12.** $\frac{3}{15}$ **13.** $\frac{5}{100}$ **14.** $\frac{24}{50}$ **15.** $\frac{60}{75}$

16. $\frac{24}{36}$ **17.** $\frac{32}{64}$ **18.** $\frac{35}{100}$ **19.** $\frac{72}{81}$ **20.** $\frac{40}{56}$

Write as a decimal. [136]

21. $\frac{3}{4}$ **22.** $\frac{5}{8}$ **23.** $\frac{9}{100}$ **24.** $\frac{7}{10}$ **25.** $\frac{17}{25}$

26. $\frac{3}{5}$ **27.** $\frac{7}{20}$ **28.** $\frac{9}{40}$ **29.** $\frac{3}{25}$ **30.** $\frac{17}{50}$

Multiply. [74, 146, 148]

31. $0.9 \times 5{,}000$ **32.** 1.3×75 **33.** 0.875×24 **34.** 0.125×72

35. $\frac{3}{4} \times 24$ **36.** $\frac{2}{9} \times 78$ **37.** $1\frac{1}{2} \times 15$ **38.** $2\frac{2}{3} \times 123$

Divide. [80, 150, 152]

39. $4.8 \div 0.32$ **40.** $21 \div 0.7$ **41.** $287 \div 1.4$ **42.** $11 \div 0.125$

43. $300 \div \frac{2}{3}$ **44.** $150 \div \frac{3}{8}$ **45.** $525 \div 3\frac{1}{2}$ **46.** $289 \div 2\frac{3}{7}$

Solve for *n*. [216]

47. $\frac{3}{n} = \frac{21}{63}$ **48.** $\frac{7}{9} = \frac{56}{n}$ **49.** $\frac{5}{6} = \frac{n}{72}$ **50.** $\frac{n}{12} = \frac{144}{24}$

51. $\frac{2}{3} = \frac{n}{24}$ **52.** $\frac{n}{100} = \frac{4}{5}$ **53.** $\frac{3}{n} = \frac{48}{32}$ **54.** $\frac{6}{10} = \frac{54}{n}$

Chapter 9
PERCENT

Many schools sponsor class trips for students. How could you plan a trip that all your classmates would enjoy?

9.1 DECIMALS AND PERCENTS

Here are some examples of percents used in daily life:

Write the three percents as decimals.

THINK: **Percent (%)** means per hundred or hundredths.
15% means 15 per 100, or 15 hundredths. So, 15% = 0.15.
135% means 135 per 100, or 135 hundredths. So, 135% = 1.35.
0.74% means 0.74 per 100, or 0.74 hundredths. So, 0.74% = 0.0074.

You can use a shortcut to write a percent as a decimal. Move the decimal point two places to the left and drop the percent sign.

EXAMPLE 1 Write as a decimal.

a. 15%

15% = 15%
= 0.15

b. 135%

135% = 135%
= 1.35

c. 0.74%

0.74% = 000.74%
= 0.0074

Sometimes you need to write a fractional or mixed-number percent as a decimal. Rename the fraction or mixed number as a decimal.

EXAMPLE 2 Write as a decimal.

a. $\frac{1}{4}\%$

$\frac{1}{4}\%$ = 0.25% = 000.25%
= 0.0025

b. $6\frac{1}{2}\%$

$6\frac{1}{2}\%$ = 6.5% = 06.5%
= 0.065

To write a decimal as a percent, move the decimal point two places to the right and include the percent sign.

EXAMPLE 3 Write as a percent.

a. 0.54

0.54 = 0.54
= 54%

b. 1.7

1.7 = 1.70
= 170%

c. $0.11\frac{1}{9}$

$0.11\frac{1}{9} = 0.11\frac{1}{9}$
= $11\frac{1}{9}\%$

CHECKPOINT Write the letter of the correct answer.

Complete.

1. $12\% = ■$ **a.** 0.012 **b.** 0.12 **c.** 1.2 **d.** 1,200
2. $\frac{1}{2}\% = ■$ **a.** 0.005 **b.** 0.05 **c.** 5 **d.** 50
3. $0.29 = ■$ **a.** 0.29% **b.** 2.9% **c.** 29% **d.** 290%

PRACTICE EXERCISES

Write as a decimal.

1. 28%	**2.** 65%	**3.** 42%	**4.** 10%	**5.** 31%
6. 8%	**7.** 4%	**8.** 6%	**9.** 3%	**10.** 9%
11. 130%	**12.** 200%	**13.** 350%	**14.** 245%	**15.** 157%
16. 0.35%	**17.** 0.17%	**18.** 0.51%	**19.** 0.8%	**20.** 0.5%
21. 7.2%	**22.** 3.09%	**23.** 90.4%	**24.** 11.15%	**25.** 200.6%
26. $\frac{1}{4}\%$	**27.** $\frac{1}{5}\%$	**28.** $\frac{3}{4}\%$	**29.** $\frac{4}{5}\%$	**30.** $\frac{3}{8}\%$
31. $3\frac{1}{2}\%$	**32.** $2\frac{3}{4}\%$	**33.** $8\frac{2}{5}\%$	**34.** $10\frac{1}{4}\%$	**35.** $9\frac{3}{8}\%$

Write as a percent.

36. 0.23	**37.** 0.45	**38.** 0.15	**39.** 0.37	**40.** 0.25
41. 0.07	**42.** 0.03	**43.** 0.092	**44.** 0.013	**45.** 0.055
46. 3.7	**47.** 2.1	**48.** 0.9	**49.** 4.2	**50.** 0.8
51. $0.66\frac{2}{3}$	**52.** $0.16\frac{2}{3}$	**53.** $0.37\frac{1}{2}$	**54.** $0.83\frac{1}{3}$	**55.** $0.22\frac{2}{9}$

Solve.

56. Carol Heilman plans to finance her new car. The dealer is offering an annual interest rate of 8.9%. The bank's offer is an annual interest rate of $8\frac{9}{11}\%$. Which is the lower rate?

57. Carol's new car gets 1.7 times more miles to the gallon than her old car. Write this decimal as a percent.

58. Carol must pay a sales tax of $8\frac{1}{4}\%$ on her purchase. Write this percent as a decimal.

9.2 WRITING PERCENTS AS FRACTIONS

The Kennesaw Gazette conducted a poll to find out what its readers thought of their congressional representative. Of the 97% who had an opinion, 67% supported Representative Sherman and 30% opposed him. What fraction of those polled supported Sherman? What fraction opposed him?

To change a percent to a fraction, write the percent as a fraction with a denominator of 100. Then, simplify the fraction.

EXAMPLE 1 Write as a fraction.

a. 67%

$$67\% = \frac{67}{100}$$

Of those polled, $\frac{67}{100}$ supported Sherman.

b. 30%

$$30\% = \frac{\overset{3}{\cancel{30}}}{\underset{10}{\cancel{100}}} = \frac{3}{10}$$

Of those responding, $\frac{3}{10}$ opposed Sherman.

Percents greater than or equal to 100 can be written either as whole numbers or as mixed numbers.

EXAMPLE 2 Write as a whole number or a mixed number.

a. 300%

$$300\% = \frac{\overset{3}{\cancel{300}}}{\underset{1}{\cancel{100}}} = 3$$

b. 215%

$$215\% = 2\frac{\overset{3}{\cancel{15}}}{\underset{20}{\cancel{100}}} = 2\frac{3}{20}$$

CHECKPOINT

Write the letter of the correct answer.

Write as a fraction, whole number, or mixed number.

1. 23% **a.** $\frac{23}{100}$ **b.** $\frac{23}{10}$ **c.** $\frac{100}{23}$ **d.** 2,300
2. 50% **a.** $\frac{5}{100}$ **b.** $\frac{1}{2}$ **c.** $\frac{5}{10}$ **d.** 5,000
3. 6% **a.** $\frac{60}{100}$ **b.** $\frac{6}{10}$ **c.** $\frac{3}{50}$ **d.** $\frac{0.06}{100}$
4. 350% **a.** 35,000 **b.** $35\frac{1}{100}$ **c.** $3\frac{1}{2}$ **d.** $\frac{35}{100}$

PRACTICE EXERCISES

Write as a fraction, whole number, or mixed number in simplest form.

1. 3%	**2.** 11%	**3.** 17%	**4.** 29%	**5.** 61%
6. 53%	**7.** 33%	**8.** 97%	**9.** 59%	**10.** 83%
11. 8%	**12.** 10%	**13.** 15%	**14.** 25%	**15.** 50%
16. 85%	**17.** 36%	**18.** 75%	**19.** 40%	**20.** 95%
21. 200%	**22.** 500%	**23.** 300%	**24.** 100%	**25.** 700%
26. 400%	**27.** 600%	**28.** 900%	**29.** 800%	**30.** 1,000%
31. 111%	**32.** 247%	**33.** 323%	**34.** 421%	**35.** 559%
36. 125%	**37.** 250%	**38.** 340%	**39.** 475%	**40.** 515%
41. 26%	**42.** 240%	**43.** 9%	**44.** 45%	**45.** 98%
46. 112%	**47.** 1%	**48.** 49%	**49.** 51%	**50.** 2,000%
51. 30%	**52.** 99%	**53.** 280%	**54.** 444%	**55.** 1,500%

MIXED REVIEW

Solve the proportion.

56. $\frac{5}{100} = \frac{n}{67}$ **57.** $\frac{6}{100} = \frac{n}{36}$ **58.** $\frac{75}{100} = \frac{93}{n}$ **59.** $\frac{n}{100} = \frac{16}{128}$

60. $\frac{n}{100} = \frac{64}{24}$ **61.** $\frac{250}{100} = \frac{n}{44}$ **62.** $\frac{125}{100} = \frac{110}{n}$ **63.** $\frac{25}{n} = \frac{78}{312}$

Solve.

64. The Kennesaw Gazette polled 327 voters by phone. Each call took an average of 4 minutes. How long did it take to conduct the poll?

65. Representative Sherman has served in Congress for 9 years. Each term is 2 years long. How many times has she been re-elected?

66. Included in the results of the poll were 3% who had no opinion. What fractional part of those polled had no opinion?

67. For its poll, the Gazette contacted 16% of Kennesaw's voters. What fractional part of Kennesaw's voters were contacted for the Gazette poll?

TIME OUT In the Kennesaw mayoral election, Joan Bigelow beat Philip Oakley by 212 votes. A total of 21,628 ballots were cast. How many votes did each candidate receive?

9.3 WRITING FRACTIONS AS PERCENTS

In the state finals of the high school basketball tournament, Visalia lost because of poor free-throw shooting. The team made only 11 of 20 attempts. What percent of its free throws did Visalia make?

THINK: The ratio of 11 out of 20 can be written as the fraction $\frac{11}{20}$.

EXAMPLE 1 Write $\frac{11}{20}$ as a percent.

1. Write the fraction as a decimal. $\frac{11}{20} \rightarrow 20\overline{)11.00}$ = 0.55

   THINK: Divide the numerator by the denominator.

2. Write the decimal as a percent. $0.55 = 55\%$

Visalia made 55% of its free throws.

Some fractions are written as mixed-number percents.

EXAMPLE 2 Write as a percent.

a. $\frac{1}{9}$

$\frac{1}{9} \rightarrow 9\overline{)1.00}$, $0.11\frac{1}{9} = 11\frac{1}{9}\%$

b. $\frac{5}{6}$

$\frac{5}{6} \rightarrow 6\overline{)5.00}$, $0.83\frac{1}{3} = 83\frac{1}{3}\%$

To write a whole number or mixed number as a percent, change the number to a decimal. Then, write the decimal as a percent.

EXAMPLE 3 Write as a percent.

a. 5

$5 = 5.00 = 500\%$

b. $2\frac{1}{2}$

$2\frac{1}{2} = 2.50 = 250\%$

CHECKPOINT

Write the letter of the correct answer.

Write as a percent.

1. $\frac{3}{4}$ **a.** 175% **b.** $100\frac{3}{4}\%$ **c.** 75% **d.** 0.75%

2. $\frac{1}{3}$ **a.** 30% **b.** 33% **c.** $33\frac{1}{3}\%$ **d.** $133\frac{1}{3}\%$

3. $2\frac{4}{5}$ **a.** 2.8% **b.** 280% **c.** 24% **d.** 240%

PRACTICE EXERCISES

Write as a percent.

1. $\frac{1}{2}$ **2.** $\frac{2}{5}$ **3.** $\frac{3}{4}$ **4.** $\frac{3}{8}$ **5.** $\frac{4}{5}$ **6.** $\frac{5}{8}$

7. $\frac{9}{10}$ **8.** $\frac{3}{20}$ **9.** $\frac{6}{25}$ **10.** $\frac{7}{50}$ **11.** $\frac{2}{25}$ **12.** $\frac{3}{40}$

13. $\frac{1}{3}$ **14.** $\frac{4}{7}$ **15.** $\frac{2}{9}$ **16.** $\frac{5}{6}$ **17.** $\frac{2}{3}$ **18.** $\frac{8}{9}$

19. $\frac{4}{9}$ **20.** $\frac{1}{7}$ **21.** $\frac{5}{9}$ **22.** $\frac{3}{7}$ **23.** $\frac{2}{7}$ **24.** $\frac{5}{7}$

25. 3 **26.** 1 **27.** 9 **28.** 8 **29.** 6 **30.** 10

31. 7 **32.** 2 **33.** 4 **34.** 11 **35.** 12 **36.** 15

37. $3\frac{3}{4}$ **38.** $5\frac{1}{2}$ **39.** $8\frac{3}{4}$ **40.** $1\frac{4}{5}$ **41.** $6\frac{3}{8}$ **42.** $5\frac{3}{5}$

43. $4\frac{2}{5}$ **44.** $3\frac{1}{6}$ **45.** $6\frac{2}{3}$ **46.** $2\frac{4}{9}$ **47.** $9\frac{1}{3}$ **48.** $12\frac{5}{8}$

49. $\frac{1}{8}$ **50.** $\frac{7}{9}$ **51.** $1\frac{3}{5}$ **52.** 23 **53.** $\frac{7}{8}$ **54.** $1\frac{1}{6}$

Solve.

55. Attendance at the finals was 9,124. The semifinals drew 6,348. How many more people attended the finals?

56. Visalia made 4 of 10 three-point shots. What percent of its three-point shots did Visalia make?

57. Steve Jerome was the high scorer for Athens, the team that beat Visalia. He made 8 of 15 field goals. What percent of his field goals did Jerome make?

58. Athens made 7 of 11 three-point shots and 23 two-point shots. How many points did Athens score?

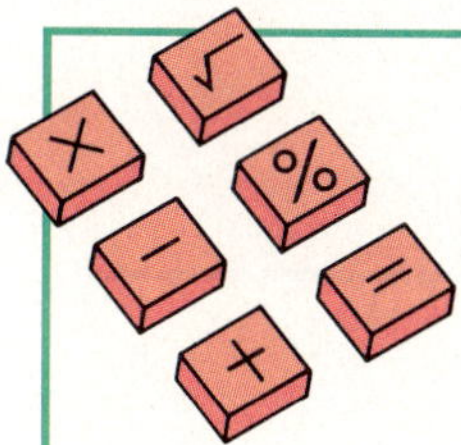

CALCULATOR

Copy and complete. Round to the nearest percent.

ATHENS TOURNAMENT FREE-THROW SHOOTING

	Free throws attempted	Free throws made	Percent made
Round of 32	25	21	**1.** ■
Round of 16	11	9	**2.** ■
Round of 8	17	15	**3.** ■
Semifinals	32	26	**4.** ■
Finals	12	7	**5.** ■

9.4 FINDING A PERCENT OF A NUMBER

Audio Magazine's subscription price is 80% of its newsstand price. The newsstand price is $30 per year. What is the subscription price?

To solve, you need to find what number is 80% of $30.

EXAMPLE 1 What is 80% of 30?

1. Write a number sentence.

   What is 80% of 30?

   $n = 0.8 \times 30$

   THINK: 80% = 0.8

2. Find n.

   $n = 24$

The subscription rate is $24 per year.

Sometimes it is easier to use a fraction to find the percent of a number.

TABLE OF FRACTIONAL EQUIVALENTS

$25\% = \frac{1}{4}$	$50\% = \frac{1}{2}$	$75\% = \frac{3}{4}$	
$20\% = \frac{1}{5}$	$40\% = \frac{2}{5}$	$60\% = \frac{3}{5}$	$80\% = \frac{4}{5}$
$16\frac{2}{3}\% = \frac{1}{6}$	$33\frac{1}{3}\% = \frac{1}{3}$	$66\frac{2}{3}\% = \frac{2}{3}$	$83\frac{1}{3}\% = \frac{5}{6}$
$12\frac{1}{2}\% = \frac{1}{8}$	$37\frac{1}{2}\% = \frac{3}{8}$	$62\frac{1}{2}\% = \frac{5}{8}$	$87\frac{1}{2}\% = \frac{7}{8}$

EXAMPLE 2 What is $37\frac{1}{2}\%$ of 240?

1. Write a number sentence.

   What is $37\frac{1}{2}\%$ of 240?

   $n = \frac{3}{8} \times 240$

   THINK: $37\frac{1}{2}\% = \frac{3}{8}$

2. Find n.

   $n = 90$

So, $37\frac{1}{2}\%$ of 240 is 90.

You can estimate the percent of a number.

EXAMPLE 3 Estimate 47.5% of 48.

THINK: 47.5% is about 50% or $\frac{1}{2}$. 48 is about 50.

$n = \frac{1}{2} \times 50$

$n = 25$

So, 47.5% of 48 is about 25.

FOR DISCUSSION

When will the percent of a number be greater than the number itself?

PRACTICE EXERCISES

Find the percent of the number.

1. 2% of 60	**2.** 10% of 30	**3.** 20% of 40	**4.** 46% of 10
5. 25% of 120	**6.** 7.5% of 60	**7.** 2.5% of 200	**8.** 75% of 200
9. $12\frac{1}{2}\%$ of 32	**10.** $37\frac{1}{2}\%$ of 64	**11.** $16\frac{2}{3}\%$ of 36	**12.** $33\frac{1}{3}\%$ of 90
13. $66\frac{2}{3}\%$ of 27	**14.** $87\frac{1}{2}\%$ of 56	**15.** $62\frac{1}{2}\%$ of 40	**16.** $83\frac{1}{3}\%$ of 18

Compute.

17. What is 2% of 200?
18. What is $12\frac{1}{2}\%$ of 480?
19. What is $16\frac{2}{3}\%$ of 360?
20. What is 3% of 100?
21. What is 5% of 50?
22. What is 12.5% of 8?
23. Find 75% of 648.
24. Find $33\frac{1}{3}\%$ of 99.
25. Find 40% of 200.
26. Find $37\frac{1}{2}\%$ of 72.
27. Find 250% of 12.
28. Find 650% of 6.
29. What number is $83\frac{1}{3}\%$ of 42?
30. 9.5% of 20 is what number?

Estimate the percent of the number.

31. 24.7% of 79	**32.** 9.8% of 42	**33.** 52.1% of 95	**34.** 75.7% of 396
35. 19.6% of 107	**36.** 0.9% of 100	**37.** 47.4% of 801	**38.** 97% of 312

Solve.

39. *Audio Magazine*'s cover photo is reprinted at $83\frac{1}{3}\%$ of its original size. Write this percent as a fraction.

40. A full-page ad in *Audio Magazine* costs $87\frac{1}{2}\%$ more than a half-page ad. A half-page ad costs \$500. How much more does a full-page ad cost?

41. For every 3 copies sold on the newsstand, 2 copies are sold to subscribers. Circulation in June was 24,795 copies. How many went to subscribers?

42. One of the tape decks advertised in *Audio Magazine* costs \$175. Sales tax is 6%. What is the final cost of the tape deck?

PROBLEM Solving APPLICATION

9.5 CONSUMER: FEDERAL INCOME TAXES

Federal income tax is withheld from the paychecks of most people. However, when you prepare your tax return, you need to determine whether the amount that has been withheld from January 1 through December 31 is correct.

Marcie Graham is single and her taxable income is $23,850. During the year, she had $3,956.78 withheld from her paychecks. How much in additional taxes does Marcie need to pay?

Single Taxpayers

If taxable income is: over--	but not over--	The tax is:	of the amount over--
$ 0	$17,850	... 15%	$ 0
17,850	43,150	$2,677.50 + 28%	17,850
43,150	89,560	9,761.50 + 33%	43,150
89,560	----	... 28%	0

1. Read the table instructions.

THINK: $23,850 is greater than $17,850 and less than $43,150.

2. Find the total tax.
   - Subtract to find out how much greater $23,850 is than $17,850. $\quad \$23{,}850 - \$17{,}850 = \$6{,}000$
   - Multiply to find 28% of $6,000.
     THINK: 28% = 0.28 $\quad 0.28 \times \$6{,}000 = \$1{,}680$
   - Add to find the total tax. $\quad \$2{,}677.50 + \$1{,}680.00 = \$4{,}357.50$
3. Subtract to find the amount owed. $\quad \$4{,}357.50 - \$3{,}956.78 = \$400.72$

So, Marcie needs to pay $400.72 in additional taxes.

Use the table on page 242. Find the federal income tax for a single taxpayer. Remember to estimate whenever you use your calculator.

Taxable income	Federal income tax	Taxable income	Federal income tax
1. $3,000	■	**2.** $7,000	■
3. $15,000	■	**4.** $24,850	■
5. $98,500	■	**6.** $49,150	■
7. $19,350	■	**8.** $42,950	■
9. $46,050	■	**10.** $88,950	■
11. $17,890	■	**12.** $43,153	■

Find the tax for a single taxpayer. Then determine whether an amount is still owed or is to be refunded.

Taxable income	Federal income tax	Amount withheld	Amount owed	Amount to be refunded
13. $6,575	■	$856	■	■
14. $43,100	■	$5,850	■	■
15. $55,930	■	$25,036	■	■
16. $108,426	■	$42,865	■	■

Solve.

17. Ann Caldor is a single taxpayer. Her taxable income is $32,450. What is her federal income tax?

18. Ben Tyson's taxable income is $20,000. Paul Wingate's is $40,000. Both are single. Is Paul's taxable income twice as great as Ben's? Is Paul's federal income tax twice as great as Ben's?

19. Jessie Pelletier is single. She pays $8,250 in federal income tax. In which category is her taxable income?

MIDCHAPTER REVIEW

Write as a decimal. [234]

1. 6% **2.** 15% **3.** 56% **4.** 150% **5.** 14.7%

6. 0.1% **7.** $16\frac{2}{3}\%$ **8.** $37\frac{1}{2}\%$ **9.** $166\frac{2}{3}\%$ **10.** $\frac{1}{2}\%$

Write as a fraction, whole number, or mixed number. [236]

11. 1% **12.** 4% **13.** 8% **14.** 24% **15.** 50%

16. 80% **17.** 33% **18.** 480% **19.** 300% **20.** 2,000%

Write as a percent. [234, 238]

21. 0.34 **22.** 0.58 **23.** 0.04 **24.** 0.09 **25.** 0.3

26. 0.2 **27.** 0.004 **28.** 0.001 **29.** 0.072 **30.** 1.4

31. 5.82 **32.** $0.12\frac{1}{2}$ **33.** $0.18\frac{2}{3}$ **34.** $0.33\frac{1}{3}$ **35.** $0.44\frac{4}{9}$

36. $\frac{1}{5}$ **37.** $\frac{1}{8}$ **38.** $\frac{3}{4}$ **39.** $\frac{5}{6}$ **40.** $\frac{2}{3}$

41. $\frac{3}{10}$ **42.** $\frac{4}{25}$ **43.** $\frac{17}{50}$ **44.** $\frac{13}{20}$ **45.** $\frac{8}{25}$

46. 8 **47.** 12 **48.** $6\frac{3}{8}$ **49.** $1\frac{2}{3}$ **50.** $7\frac{5}{6}$

Find the percent of the number. [240]

51. 4% of 20 **52.** 25% of 16 **53.** 50% of 34 **54.** 200% of 125

55. $12\frac{1}{2}\%$ of 80 **56.** $16\frac{2}{3}\%$ of 12 **57.** $87\frac{1}{2}\%$ of 96 **58.** $66\frac{2}{3}\%$ of 93

Solve. Use the table on page 242 if necessary. [238, 240, 242]

59. Garth Tilson earned $24,000 last year. He saved 7% of his earnings. How much did Tilson save last year?

60. Garth Tilson spends $\frac{3}{8}$ of his monthly income on rent. What percent of his monthly income does Tilson spend on rent?

61. Tilson's taxable income last year was $21,650. He filed as a single taxpayer. What was Tilson's federal income tax last year?

62. This year, Tilson earned $29,740 in taxable income. He had $6,303.96 withheld from his paychecks. Is Tilson entitled to a refund? If so, how much? If not, how much is owed?

PROBLEM *Solving* APPLICATION

9.6 CAREER: STORE OWNER

Rosa Aquino owns a neighborhood clothing store. She buys merchandise at a certain cost and then sells it to people in small quantities.

Rosa must pay certain expenses such as rent, telephone, and electricity. To meet these expenses and make a profit, she must sell her merchandise for a greater amount than its cost.

To do this, Rosa marks up the merchandise at a rate of 50%. That is, the **markup** is 50% of the cost. The **selling price** is found by adding the markup to the cost. What would Rosa sell a pair of slacks for if they cost her $18?

1. Find the amount of the markup.

   THINK: 50% = 0.50 $0.50 \times \$18 = \$9$

2. Add to the find the selling price. $\$18 + \$9 = \$27$

So, Rosa would sell the pair of slacks for $27.

Find the markup and the selling price. Remember to estimate whenever you use your calculator.

Item	Cost of merchandise	Percent of markup	Amount of markup	Selling price
1. A	$24	50%	■	■
2. B	$15	30%	■	■
3. C	$25	40%	■	■
4. D	$60	45%	■	■
5. E	$8.50	43%	■	■
6. F	$19.75	36%	■	■
7. G	$25.20	$33\frac{1}{3}\%$	■	■
8. H	$18.50	37.5%	■	■

Solve. Rosa buys one type of shirt at $192 per dozen.

9. What is the cost per shirt?

10. What is the selling price per shirt if the markup is 50%?

9.7 FINDING WHAT PERCENT ONE NUMBER IS OF ANOTHER

Of the first 40 Presidents, 22 came from one of the original 13 colonies. What percent of the first 40 Presidents came from one of the original 13 colonies?

To solve, you need to find what percent of 40 is 22.

EXAMPLE 1 What percent of 40 is 22?

1. Write a number sentence.

What percent of 40 is 22?

$$\underbrace{\text{What percent}}_{n} \times 40 = 22$$

2. Find n.

THINK: To find the missing factor, divide 22 by 40.

$$n \times 40 = 22$$
$$n = 22 \div 40$$
$$n = 0.55$$

3. Write the decimal as a percent. $0.55 = 55\%$

Of the first 40 Presidents, 55% came from one of the original 13 colonies.

Sometimes the missing percent is greater than 100%.

EXAMPLE 2 48 is what percent of 16?

1. Write a number sentence.

48 is what percent of 16?

$$48 = \underbrace{\text{what percent}}_{n} \times 16$$

2. Find n.

THINK: To find the missing factor, divide 48 by 16.

$$n \times 16 = 48$$
$$n = 48 \div 16$$
$$n = 3$$

3. Write the whole number as a percent. $3 = 300\%$

So, 48 is 300% of 16.

You can estimate the percent one number is of another.

EXAMPLE 3 Estimate the percent 3 is of 4.8.

THINK: 4.8 is about 5.

$$n \times 5 = 3$$
$$n = 3 \div 5$$
$$n = 0.6\text{, or } 60\%$$

So, 3 is about 60% of 4.8.

FOR DISCUSSION

Which is the only reasonable answer to the following problem? Explain.

62 is what percent of 38? **a.** 6% **b.** 63% **c.** 163%

PRACTICE EXERCISES

Find the missing number.

1. What percent of 5 is 2?
2. What percent of 10 is 3?
3. What percent of 25 is 7?
4. What percent of 20 is 10?
5. 3 is what percent of 4?
6. 8 is what percent of 32?
7. 16 is what percent of 32?
8. 36 is what percent of 48?
9. What percent of 5 is 10?
10. What percent of 5 is 15?
11. What percent of 7 is 35?
12. What percent of 2 is 5?
13. 10 is what percent of 4?
14. 25 is what percent of 10?
15. 24 is what percent of 18?
16. 18 is what percent of 12?
17. What percent of 500 is 2.5?
18. What percent of 200 is 3.5?
19. What percent of 25 is 120?
20. What percent of 125 is 25?
21. What percent of 140 is 70?
22. What percent of 48 is 43.2?
23. 67.2 is what percent of 24?
24. 19.8 is what percent of 99?
25. 37.8 is what percent of 14?
26. 147 is what percent of 245?
27. 148.8 is what percent of 124?
28. 218.4 is what percent of 260?

Estimate.

29. What percent of 13 is 3.25?
30. What percent of 6 is 1.5?
31. What percent of 45 is 4.5?
32. What percent of 12 is 3.8?
33. 26.8 is what percent of 50?
34. 14.5 is what percent of 50?
35. 21.6 is what percent of 10?
36. 45.2 is what percent of 30?

Solve.

37. Of the first 28 Presidents, 8 were born in Virginia. What percent of the first 28 Presidents were born in Virginia?
38. The youngest President ever, John F. Kennedy, was inaugurated on January 20, 1961. Kennedy was born on May 29, 1917. How old was Kennedy when he was inaugurated?
39. Franklin D. Roosevelt's 635 vetoes were the most by any President. About how many vetoes did Roosevelt average each year over his 12 years in office?
40. Of the first 40 Presidents, 25 were elected with 50% or more of the popular vote. What percent of the first 40 Presidents were elected with less than 50% of the popular vote?

9.8 FINDING A NUMBER WHEN A PERCENT OF IT IS KNOWN

Barbara Samuels is a freelance journalist who uses a computer for her writing. Her Letters diskette is 65% full. It has 120.9 kilobytes (K) of information written onto it. How many kilobytes does the Letters diskette hold?

To solve, you need to find the number of which 65% is 120.9.

EXAMPLE 1 65% of what number is 120.9?

1. Write a number sentence.

   65% of what number is 120.9?

   THINK: 65% = 0.65

   $0.65 \times n = 120.9$

2. Find n.

   THINK: To find the missing factor, divide 120.9 by 0.65.

   $0.65 \times n = 120.9$

   $n = 120.9 \div 0.65$

   $n = 186$

The Letters diskette holds 186K.

Sometimes it helps to change the percent to a fraction.

EXAMPLE 2 8 is $33\frac{1}{3}\%$ of what number?

1. Write a number sentence.

   8 is $33\frac{1}{3}\%$ of what number?

   THINK: $33\frac{1}{3}\% = \frac{1}{3}$

   $8 = \frac{1}{3} \times n$

2. Find n.

   THINK: To find the missing factor, divide 8 by $\frac{1}{3}$.

   $\frac{1}{3} \times n = 8$

   $n = 8 \div \frac{1}{3}$

   $n = 24$

8 is $33\frac{1}{3}\%$ of 24.

CHECKPOINT

Write the letter of the correct answer.

Find the number.

1. 4% of what number is 8? **a.** 0.32 **b.** 2 **c.** 20 **d.** 200

2. 3 is 20% of what number? **a.** 0.6 **b.** 1.5 **c.** 6 **d.** 15

3. $87\frac{1}{2}\%$ of what number is 56? **a.** 0.49 **b.** 6.4 **c.** 49 **d.** 64

PRACTICE EXERCISES

Find the number. Change the percent to a decimal.

1. 8% of what number is 4?
2. 1 is 5% of what number?
3. 10% of what number is 6?
4. 10 is 4% of what number?
5. 1% of what number is 10?
6. 18 is 50% of what number?
7. 24% of what number is 48?
8. 102 is 17% of what number?
9. 27% of what number is 81?
10. 9.6 is 8% of what number?
11. 12% of what number is 38.4?
12. 7.5 is 25% of what number?

Find the number. Change the percent to a fraction.

13. 10% of what number is 23?
14. 30 is 15% of what number?
15. 20% of what number is 10?
16. 18 is 3% of what number?
17. 8% of what number is 28?
18. 4 is 25% of what number?
19. 12% of what number is 12?
20. 12 is 40% of what number?
21. $33\frac{1}{3}\%$ of what number is 7?
22. 24 is $37\frac{1}{2}\%$ of what number?
23. $62\frac{1}{2}\%$ of what number is 10?
24. 21 is $87\frac{1}{2}\%$ of what number?

Find the number.

25. 100% of what number is 4.5?
26. 36 is 300% of what number?
27. 200% of what number is 10?
28. 140 is 350% of what number?
29. 4.5% of what number is 4.5?
30. 15 is 7.5% of what number?
31. $12\frac{1}{2}\%$ of what number is 6?
32. 16 is $66\frac{2}{3}\%$ of what number?
33. 87.5% of what number is 56?
34. 28.5 is 9.5% of what number?
35. 6.25% of what number is 250?
36. 3 is 0.5% of what number?
37. 50% of what number is 36?
38. 180 is 60% of what number?

Solve.

39. There are 8 bits in each byte and 1,000 bytes in each kilobyte. How many bits are there in a kilobyte?
40. Barbara paid \$31.60 in sales tax for her printer. Sales tax is 8%. How much was the printer itself?
41. The wholesale price for Barbara's computer is \$1,050. Barbara paid 115% of the wholesale price. How much did she pay?
42. Barbara is printing out a 40-page article. So far, 14 pages have been printed. What percent of the article is left to be printed?

9.9 USING PROPORTIONS TO SOLVE PERCENT PROBLEMS

Deborah Rabinowitz, the art director of an advertising agency, uses a copying machine to reduce the size of a piece of artwork. The machine is set to reduce the artwork to 72% of its original size. The artwork is 25 in. long. How long is the reduced copy of the artwork?

You can use a proportion to find the percent of a number.

EXAMPLE 1 What number is 72% of 25?

1. Write a proportion.

   THINK: 72% is to 100% as *n* inches are to 25 in. $\frac{72}{100} = \frac{n}{25}$

2. Multiply to find the cross products. $100 \times n = 72 \times 25$

   $100 \times n = 1{,}800$

3. Find n. $n = 1{,}800 \div 100$

   $n = 18$

The reduced copy of the artwork is 18 in. long.

You can use a proportion to find the percent one number is of another.

EXAMPLE 2 24 is what percent of 60?

1. Write a proportion.

   THINK: *n*% is to 100% as 24 is to 60. $\frac{n}{100} = \frac{24}{60}$

2. Multiply to find the cross products. $60 \times n = 24 \times 100$

   $60 \times n = 2{,}400$

3. Find n. $n = 2{,}400 \div 60$

   $n = 40\%$

You can use a proportion to find a number when a percent of it is known.

EXAMPLE 3 85% of what number is 17?

1. Write a proportion.

   THINK: 85% is to 100% as 17 is to *n*. $\frac{85}{100} = \frac{17}{n}$

2. Multiply to find the cross products. $85 \times n = 17 \times 100$

   $85 \times n = 1{,}700$

3. Find n. $n = 1{,}700 \div 85$

   $n = 20$

PRACTICE EXERCISES

Use a proportion to solve.

1. What number is 18% of 40?

2. 25% of 16 is what number?

3. What number is 3% of 48?

4. 58% of 200 is what number?

5. What number is 120% of 50?

6. 250% of 18 is what number?

7. 36 is what percent of 48?

8. What percent of 60 is 15?

9. 56 is what percent of 64?

10. What percent of 180 is 45?

11. 21 is what percent of 12?

12. What percent of 50 is 250?

13. 20% of what number is 8?

14. 18 is 36% of what number?

15. 25% of what number is 6?

16. 1 is 50% of what number?

17. 130% of what number is 39?

18. 90 is 225% of what number?

19. 200% of 68 is what number?

20. 150% of what number is 96?

21. 22.4 is what percent of 64?

22. 86 is what percent of 43?

23. 64% of what number is 384?

24. 100% of what number is 2?

25. 240 is what percent of 80?

26. 70% of 490 is what number?

27. 12% of what number is 30?

28. 98 is what percent of 294?

29. 84 is what percent of 36?

30. 345% of what number is 138?

Solve.

31. Bruce Kelvin, the assistant art director, just had his salary raised 11% to $19,425. What was Kelvin's salary before the raise?

32. Deborah and Bruce prepare 15 different logos for a dog food campaign. Of these, 3 will be used. What percent will be used?

33. The agency has 135 employees, 60% of whom receive benefits. What fraction of the employees receive benefits?

34. The agency gets a 12% discount on colored pencils. The pencils list for $1.29. How much does the agency save on an order of 2,000?

TIME OUT

You are the owner of a store that sells artists' supplies. The wholesale price you pay for a drafting table is $150. Before you sell the table, you mark up that price by 60%. One week you put the table on sale for 25% off the marked-up price. In addition, you offer professional artists a 15% discount. If a professional artist buys the drafting table on sale, do you make or lose money? How much?

PROBLEM *Solving* STRATEGY

9.10 SUPPLYING MISSING INFORMATION

Situation:

Captain's Ridge Ski Lodge is open from December through March. Guests are shuttled to the lodge in minibuses that hold one dozen guests each.

Strategy:

Sometimes you have to use information you know or can find to help you solve the problem.

Applying the Strategy:

A. For how many days is Captain's Ridge Ski Lodge open?

Which months is the lodge open?	December, January, February, March.
What information do you need to know?	The number of days in each of the months.
Where could you find this information?	In a calendar.

DECEMBER						
				1	2	3
4	5	6	7	8	9	10
11	12	13	14	15	16	17
18	19	20	21	22	23	24
25	26	27	28	29	30	31

JANUARY						
1	2	3	4	5	6	7
8	9	10	11	12	13	14
15	16	17	18	19	20	21
22	23	24	25	26	27	28
29	30	31				

FEBRUARY						
			1	2	3	4
5	6	7	8	9	10	11
12	13	14	15	16	17	18
19	20	21	22	23	24	25
26	27	28				

MARCH						
			1	2	3	4
5	6	7	8	9	10	11
12	13	14	15	16	17	18
19	20	21	22	23	24	25
26	27	28	29	30	31	

Add to find the total number of days. $31 + 31 + 28 + 31 = 121$

Captain's Ridge Ski Lodge is open for 121 days.

B. If 6 full minibuses arrive, how many guests are the minibuses carrying?

What information are you given?	6 minibuses arrive, a dozen guests in each.
What information do you need to know?	There are 12 in a dozen.
Multiply to find the total number of guests.	$6 \times 12 = 72$

The minibuses are carrying 72 passengers.

PRACTICE EXERCISES

What information do you need to solve the problem?

1. The lodge makes sandwiches for the skiers to eat on the slopes. Each sandwich includes 3 oz of cheese. The cook made 96 sandwiches today. How many pounds of cheese did he use?

a. 8 fl oz = 1 cup
b. 16 oz = 1 lb
c. 2,000 lb = 1 ton

2. It takes 20 minutes to go up the mountain on the chair lift and then ski down again. How many times can a skier ski down the mountain in 2 hours?

a. 60 seconds = 1 minute
b. 60 minutes = 1 hour
c. 24 hours = 1 day

Solve by supplying the missing information.

3. The Captain ordered 9 dozen new ski poles for the rental shop. How many ski poles did he order?

4. The waiter at the lodge is pouring hot apple cider from a 4-qt pitcher into mugs that hold 8 fl oz each. How many mugs can he fill?

5. The ski instructor skis the two expert trails, Beech and Echo, each morning before her lessons begin. Beech Trail is 875 yd long. Echo Trail is 2,500 ft long. How many feet longer is Beech Trail?

6. Joan Galvin wanted to rent skis that were $5\frac{1}{2}$ ft long. The only skis available were 62 in. long. How many inches shorter were these skis than the ones Joan wanted to rent?

7. Every Saturday, Ranger Frederickson is assigned to the ski patrol. He skis the length of the cross-country trail 8 times. The cross-country trail is 1,320 yd long. How many miles does the ranger ski each Saturday?

PROBLEM *Solving* APPLICATION

9.11 CONSUMER: COMMISSION

You have a job as a real estate salesperson. You do not receive a weekly salary. Instead, you receive a **commission** on each house you sell. The **rate of commission** is a percent of the selling price of the house.

A. If your rate of commission is 3%, what is your commission on a selling price of $90,000?

You need to find the number that is 3% of 90,000.

1. Write a number sentence.

   THINK: 3% = 0.03

   What number is 3% of 90,000?

   $$n = 0.03 \times 90{,}000$$

2. Find n

   $$n = 2{,}700$$

So, you receive a commission of $2,700 on a selling price of $90,000.

B. Your friend is a salesperson for another real estate company. She sells a home for $150,000 and receives a $6,000 commission. What is her rate (percent) of commission?

You need to find what percent of 150,000 is 6,000.

1. Write a number sentence.

   Let n represent the rate of commission.

   What percent of 150,000 is 6,000?

   $$n \times 150{,}000 = 6{,}000$$

2. Find n.

   $$n = 6{,}000 \div 150{,}000$$

   $$n = 0.04, \text{ or } 4\%$$

So, your friend's rate of commission is 4%.

Find the commission. Remember to estimate whenever you use your calculator.

	Selling price	Rate of commission	Commission
1.	$80,000	3%	■
2.	$110,000	4%	■
3.	$176,000	4%	■
4.	$158,000	3%	■
5.	$190,000	4%	■
6.	$232,500	3.5%	■

Find the *rate* of commission.

	Selling price	Rate of commission	Amount of commission
7.	$100,000	■	$3,000
8.	$120,000	■	$4,800
9.	$160,000	■	$5,600
10.	$400,000	■	$18,000

Solve.

11. Carl Marder sells boats. He receives a salary and a commission. His rate of commission is 2%. How much commission does he receive for a month in which he makes sales totaling $38,000?

12. Maria Gomez sells commercial real estate. Her rate of commission on sales is 5%. How much commission does she receive if she sells a building for $650,000?

13. Eddie Chang sells cars. On sales of $180,000, he receives commissions totaling $4,500. What is his rate of commission?

14. One month Kelly Mandera receives $2,500 in commissions. Her rate of commission is 4%. What is the total amount of her sales that month?

PROBLEM Solving APPLICATION

9.12 CONSUMER: DISCOUNT

A **discount** is the amount that is subtracted from the list price of an item. The sign below shows the **rate of discount** or **percent of discount** for a sale at Eager Eddie's Electronics Mart.

A. You want to buy a VCR with a list price of $310. What is the amount of discount? What is the sale price?

THINK: List price = $310 **Percent of discount = 20% = 0.2**

LIST PRICE	×	PERCENT OF DISCOUNT	=	DISCOUNT
$310	×	0.20	=	$62

The discount is $62.

Subtract to find the sales price. $310 − $62 = $248

The sale price is $248.

B. Your friend received a discount of $60 on an item with a list price of $400. What was the percent of discount?

THINK: Discount = $60 **List price = $400**

DISCOUNT	÷	LIST PRICE	=	PERCENT OF DISCOUNT
$60	÷	$400	=	0.15, or 15%

The percent of discount was 15%.

FOR DISCUSSION

If two stores sell different brands of equipment, is it always a good idea to buy items in the store that has the greater percent of discount? Why?

Find the discount. Remember to estimate whenever you use your calculator.

	List price	Percent of discount	Amount of discount		List price	Percent of discount	Amount of discount
1.	$80	50%	■	**2.**	$90	10%	■
3.	$120	25%	■	**4.**	$160	30%	■
5.	$140	15%	■	**6.**	$200	35%	■
7.	$28.50	10%	■	**8.**	$39.90	20%	■
9.	$48.72	$12\frac{1}{2}\%$	■	**10.**	$59.97	$33\frac{1}{3}\%$	■

Find the discount and the sale price.

List price		Percent of discount	Amount of discount	Sale price
11.	$90	15%	■	■
12.	$130	20%	■	■
13.	$180	35%	■	■
14.	$45.60	$37\frac{1}{2}\%$	■	■
15.	$90.00	$66\frac{2}{3}\%$	■	■

Solve.

16. Your cousin buys a camera. The list price is $80. The percent of discount is 45%. What is the sale price?

17. Cary James says that he received a discount of $77 on an item listed at $220. What is the percent of discount?

18. Your friend says that if the percent of discount is 20%, then the sale price must be 80% of the list price. Do you agree? Use two methods to find the sale price for this $360 refrigerator.

19. If the percent of discount is 35%, then the sale price is what percent of the list price?

20. The sale price is 55% of the list price. What percent of the list price is the discount?

9.13 PERCENTS OF INCREASE AND DECREASE

At the Griffin Book Store, Claire Martin makes \$100 a week working part time. She gets a raise to \$110 a week. What is the percent of increase in Claire Martin's salary?

To find the **percent of increase,** you first have to find the amount of increase.

EXAMPLE 1

Find the percent of increase in Claire Martin's salary.

1. Subtract to find the amount of increase. — \$110 − \$100 = \$10
2. Write a ratio to show the amount of increase to the original amount. — $\frac{10}{100}$ ← **amount of increase** / ← **original amount**
3. Rename the fraction as a percent. — $\frac{10}{100} = 10\%$

The percent of increase in Claire Martin's salary is 10%.

To find the **percent of decrease,** you first have to find the amount of decrease.

EXAMPLE 2

In the fall, Claire Martin returns to school and cuts back on her hours from 40 to 28. What is the percent of decrease in the number of hours Claire Martin works?

1. Subtract to find the amount of decrease. — 40 − 28 = 12
2. Write a ratio to show the amount of decrease to the original amount. — $\frac{12}{40}$ ← **amount of decrease** / ← **original amount**
3. Rename the fraction as a percent. — $\frac{12}{40} = 30\%$

The number of hours Claire works is decreased by 30%.

FOR DISCUSSION

Yesterday, Amco stock rose 10%. Today, it fell 10%. Is today's closing price for Amco stock the same as yesterday's opening price? Explain.

PRACTICE EXERCISES

Copy and complete the table.

	Original Amount	New Amount	Percent Change		Original Amount	New Amount	Percent Change
1.	200	300	■	**2.**	36	48	■
3.	12	60	■	**4.**	52	65	■
5.	60	150	■	**6.**	85	119	■
7.	1,000	600	■	**8.**	60	42	■
9.	24	16	■	**10.**	72	63	■
11.	27	9	■	**12.**	120	18	■
13.	85	73.1	■	**14.**	45	59.4	■
15.	112	151.2	■	**16.**	65	132.6	■
17.	2,500	2,400	■	**18.**	350	1,554	■
19.	72	1.44	■	**20.**	35	19.6	■
21.	50	■	20% increase	**22.**	132	■	25% increase
23.	200	■	8% decrease	**24.**	250	■	6% decrease
25.	600	■	23% decrease	**26.**	275	■	44% increase
27.	■	305	75% decrease	**28.**	■	804	0.5% increase
29.	■	72	$12\frac{1}{2}\%$ increase	**30.**	■	35	$83\frac{1}{3}\%$ decrease
31.	■	252	110% increase	**32.**	79	■	100% decrease

MIXED REVIEW

Complete.

33. 5 d = ■ h

34. 4 wk = ■ d

35. 90 s = ■ min

36. 30 min = ■ h

37. 42 mo = ■ y

38. 6 min = ■ sec

Solve.

39. Sales of art books at the Griffin totaled $245.50 in March. In April, sales were $279.87. What was the percent of increase in sales?

40. On Labor Day, the Griffin had 3,260 textbooks in stock. By October, that stock had dwindled to 1,141. By how many copies did the stock fall?

41. A customer buys a boxed set of reference books for $13.95. Sales tax is 6.5%. What is the final total cost?

42. March sales at the Griffin totaled 4,500 books. In April, sales were down 6%, but in May they were up 10%. What were sales in May?

PROBLEM *Solving* APPLICATION

9.14 CONSUMER: INTEREST

A charge for the use of money is called **interest.** If you borrow money from a bank, you will need to pay interest for the use of the money. On the other hand, when you have money in a savings account, the bank pays interest for the use of your money.

One type of interest is called **simple interest.** To compute simple interest, use this formula:

$$\text{INTEREST} = \text{PRINCIPAL} \times \text{RATE} \times \text{TIME}$$
$$I = p \times r \times t$$

I is the amount of interest

p is the principal, which is the amount of money that is borrowed or is in the savings account

r is the *yearly* rate of interest

t is the time in years

How much interest do you have to pay if you borrow $500.00 for 6 months at a rate of 12% per year?

THINK: Principal = $500
rate = 12%, or 0.12
time = 6 mo = $\frac{6}{12}$ **y, or** $\frac{1}{2}$ **y**

$I = \$500 \times 0.12 \times \frac{1}{2}$

$I = \$60.00 \times \frac{1}{2}$

$I = \$30.00$

So, you would have to pay $30.00 in interest.

FOR DISCUSSION

Some banks advertise their interest rates in newspapers. They offer free gifts to people who open accounts. Why should banks want additional money on which they will have to pay interest?

Find the simple interest. Remember to estimate whenever you use your calculator.

	Interest	Principal	Rate	Time		Interest	Principal	Rate	Time
1.	■	$1,000	6%	1 y	**2.**	■	$800	7%	2 y
3.	■	$1,200	9%	6 mo	**4.**	■	$2,000	10%	1 y 3 mo
5.	■	$5,000	$7\frac{1}{2}$%	18 mo	**6.**	■	$3,000	5%	2 y 4 mo
7.	■	$4,500	11%	9 mo	**8.**	■	$250	8%	3 mo
9.	■	$8,000	$6\frac{1}{4}$%	4 y	**10.**	■	$1,750	5.5%	1 y
11.	■	$10,000	12%	60 mo	**12.**	■	$9,800	11%	4 y 6 mo
13.	■	$500	15%	9 mo	**14.**	■	$5,000	15%	9 mo
15.	■	$50,000	15%	9 mo	**16.**	■	$500,000	15%	9 mo
17.	■	$12,800	$9\frac{3}{4}$%	3 y	**18.**	■	$975	$4\frac{1}{4}$%	3 mo
19.	■	$15,225	8.75%	16 mo	**20.**	■	$3,788	23%	90 days

Most banks pay **compound interest** on savings accounts. This means the interest is computed two or more times a year. If you do not withdraw your interest, it is credited to your account.

How much is the principal in your account after 6 months if you deposit $500 in a savings account that pays 6% interest compounded quarterly?

THINK: Compounded quarterly means four times a year, or every 3 months.

First quarter earnings

$I = \$500 \times 0.06 \times \frac{1}{4}$

$I = \$30 \times \frac{1}{4}$

$I = \$7.50$

So, the new principal is $500 + $7.50, or $507.50.

Second quarter earnings

$I = \$507.50 \times 0.06 \times \frac{1}{4}$

$I = \$30.45 \times \frac{1}{4}$

$I = \$7.61$ (rounded to nearest cent)

So, the new principal is $507.50 + $7.61, or $515.11.

Use the compound-interest information given above.
Find the amount of money in your account:

21. after 9 months.

22. after 12 months.

23. after 15 months.

24. after 18 months.

25. Find the principal after 1 year if $5,000 is deposited in a savings account that pays an annual interest rate of 8% compounded quarterly.

26. How much interest will you earn in 9 months if you deposit $10,000 into a savings account that pays an annual interest rate of 12% compounded quarterly?

CHAPTER REVIEW

Vocabulary Choose the letter of the word(s) that completes the sentence.

1. Percent is another name for ■. [234]
a. hundreds **b.** hundredths **c.** proportion

2. To find the percent of a number, you would *not* change the percent to a ■. [240]
a. fraction **b.** decimal **c.** proportion

3. The ratio of the amount of increase to the original amount is the ■. [258]
a. proportion **b.** percent of increase **c.** percent of decrease

Skills Write as a percent. [234, 238]

0.37 = *37%*

4. 0.34 **5.** 0.7 **6.** 0.002 **7.** $0.87\frac{1}{2}$

8. $\frac{3}{100}$ **9.** $\frac{4}{5}$ **10.** $\frac{5}{6}$ **11.** 6 **12.** $3\frac{1}{3}$

Find the number. [240, 246, 248, 250]

What number is 25% of 16? *4*

13. $83\frac{1}{3}\%$ of 36 is what number?

14. What percent of 40 is 30?

15. 3 is what percent of 9?

16. 75% of what number is 18?

17. 12 is $37\frac{1}{2}\%$ of what number?

18. 160 is 25% of what number?

19. 338 is what percent of 845?

Estimate. [240, 246]

What is 9.8% of 30? *3*

20. 11.6 is what percent of 48.4?

21. What is 9.75% of 200?

22. 11.3 is what percent of 51?

23. 49.7% of 59.9 is what number?

24. 19.8 is 79.3% of what number?

Find the percent of increase or decrease. [258]

Original amount: \$4.50
New amount: \$1.62
Percent of decrease: *64%*

Original amount: \$225
New amount: \$486
Percent of increase: *116%*

25. Original amount: \$312.00
New amount: \$265.20

26. Original amount: \$65.00
New amount: \$74.10

CHAPTER TEST

Write as a decimal.

1. 12% **2.** 24.5% **3.** 33% **4.** 0.6%

Write as a fraction, whole number, or mixed number.

5. 5% **6.** 1,200% **7.** $12\frac{1}{2}\%$ **8.** 140%

Write as a percent.

9. 0.35 **10.** 2.5 **11.** 15 **12.** $\frac{3}{5}$ **13.** $2\frac{2}{3}$

Find the number.

14. 30% of 45 is what number?

15. What number is 3.5% of 20?

16. 12 is what percent of 18?

17. What percent of 8 is 4?

18. 300% of what number is 16.2?

19. 20 is 4% of what number?

20. What percent of 140 is 7?

21. What number is 150% of 30?

22. 60% of what number is 63?

23. What percent of 50 is 35?

Estimate.

24. 21.65 is 25.3% of what number?

25. 39 is what percent of 49?

26. 31.08 is 74% of what number?

27. What percent of 24.8 is 9.7?

Find the percent of increase or decrease.

28. Original amount: $84.00
New amount: $31.50

29. Original amount: $125
New amount: $165

Solve.

30. Gruen Shoes sells its own brand of black wingtips for $81. The store's markup is 35%. How much profit does Gruen make on each pair of black wing tips sold?

31. Each Gruen salesperson makes a base salary of $425 a week plus a commission bonus equal to 4% of his or her sales. If a salesperson sells $5,000 worth of shoes in a week, how much is that week's paycheck?

32. Gruen Shoes is expecting an order of 28 dozen pairs of sandals. How many pairs of sandals are to be shipped?

33. On Saturdays, a customer walks into Gruen Shoes an average of every 4 minutes. About how many customers enter the store during a 12-hour day?

CUMULATIVE REVIEW

1. To determine if any two fractions are equivalent, you can compare the ■.

   a. numerators **b.** GCF **c.** cross products **d.** none of these

2. A ■ shows the amount of change over a period of time.

   a. bar graph **b.** broken-line graph **c.** pictograph **d.** none of these

3. A ■ means per hundred or hundredths.

   a. percent **b.** decimal **c.** ratio **d.** none of these

4. A(n) ■ is a ratio that compares different units.

   a. range **b.** estimate **c.** rate **d.** none of these

5. To change a mixed number to an improper fraction, you must first multiply the whole number by the ■.

   a. numerator **b.** fraction **c.** denominator **d.** none of these

Choose the most reasonable estimate for each of the following.

6. The temperature of dry ice **a.** $-80°C$ **b.** $0°C$ **c.** $140°C$ **d.** $-5°C$
7. The capacity of a fish bowl **a.** 15 kL **b.** 5 L **c.** 15 mL **d.** 300 L
8. The weight of 4 apples **a.** 2 oz **b.** 10 lb **c.** 18 lb **d.** 32 oz
9. $19.76 - 4.13$ **a.** 16 **b.** 10 **c.** 6 **d.** 1
10. $42.873 + 12.41$ **a.** 20 **b.** 40 **c.** 50 **d.** 60

Find the answer.

11. $\frac{■}{18} = \frac{3}{6}$ **a.** 12 **b.** 6 **c.** 9 **d.** none of these
12. $0.45 = ■$ **a.** $\frac{9}{20}$ **b.** $\frac{5}{9}$ **c.** $\frac{45}{1{,}000}$ **d.** none of these
13. $6\frac{2}{3} \times 5\frac{2}{5}$ **a.** 12 **b.** 36 **c.** $\frac{4}{9}$ **d.** none of these
14. $\frac{19}{40} = ■\%$ **a.** 4.75 **b.** 21 **c.** 47.5 **d.** none of these
15. $65\% = ■$ **a.** $\frac{3}{5}$ **b.** $\frac{11}{20}$ **c.** $\frac{2}{3}$ **d.** none of these

Solve.

16. Sally has a drawing that is 5 in. wide and 7 in. long. She wants the drawing enlarged. The photocopy she has made is 14 in. long. How wide is the photocopy?

a. 19.6 in. **b.** 14 in. **c.** 12 in. **d.** none of these

17. A chicken dinner at the Bunkhouse Cafe costs $9.50. Cole slaw, chicken, corn, pie, and a beverage are included. Jim doesn't want dessert so he orders chicken for $7.25 plus side orders of corn and cole slaw, which cost $0.85 each. A glass of milk costs $0.90. How much would Jim have saved by ordering the complete dinner?

a. $0.35 **b.** $1.25 **c.** $0.05 **d.** none of these

18. The Lucca family went to Sound-A-Rama. They bought a jazz album for $4.79. They gave the cashier a $10 bill. What is the least number of coins they could have received?

a. 4 coins **b.** 3 coins **c.** 5 coins **d.** none of these

19. Last year Erik spent $4,200 on rent. This year his rent was increased to $367.50 per month. How much more will he spend this year on rent?

a. $330.50 **b.** $200.00 **c.** $210.00 **d.** none of these

20. Carmen earns $5.50 per hour. She gets time and a half for every hour over 40 she works in a week. Last week her gross earnings were $277.75. How many hours overtime did she work?

a. 6 hours **b.** 7 hours **c.** 10.5 hours **d.** none of these

THINKING ABOUT MATH

1. A store has a special on paper towels. One roll costs $0.89. If you buy a case of 20 rolls, each roll costs $0.65. Is buying a case of paper towels a better buy? Explain.

2. You need to measure the mass of an elephant. Which metric unit of measure would be the most appropriate? Why?

3. A photocopying machine can enlarge a picture by 100%. How much larger is the new picture compared to the original?

4. Explain why you would have a decimal quotient if you divided a 2-digit number by a 3-digit number.

5. Describe the method you would use to quickly determine if the ratios form a proportion.

PreSkills TEST

Write the ratio as a fraction in simplest form. [210]

1. 2 blue socks to 18 socks

2. 12 color TVs to 5 TVs

3. 4 oriental rugs to 13 rugs

4. 14 tulip bulbs to 35 bulbs

5. 5 tee shirts to 8 shirts

6. 15 jets to 27 planes

Write the fraction in simplest form. [126]

7. $\frac{6}{9}$ **8.** $\frac{4}{12}$ **9.** $\frac{7}{14}$ **10.** $\frac{10}{18}$ **11.** $\frac{25}{30}$

12. $\frac{25}{50}$ **13.** $\frac{42}{49}$ **14.** $\frac{15}{30}$ **15.** $\frac{56}{63}$ **16.** $\frac{54}{81}$

Add. Write the answer in simplest form. [158]

17. $\frac{1}{6} + \frac{5}{6}$ **18.** $\frac{3}{7} + \frac{2}{7}$ **19.** $\frac{5}{8} + \frac{3}{8}$ **20.** $\frac{6}{11} + \frac{4}{11}$

21. $\frac{1}{2} + \frac{1}{3}$ **22.** $\frac{4}{5} + \frac{2}{7}$ **23.** $\frac{7}{8} + \frac{5}{12}$ **24.** $\frac{4}{9} + \frac{2}{3}$

Subtract. Write the answer in simplest form. [162, 164]

25. $\frac{4}{5} - \frac{2}{5}$ **26.** $\frac{7}{8} - \frac{3}{8}$ **27.** $\frac{6}{7} - \frac{4}{7}$ **28.** $\frac{9}{11} - \frac{5}{11}$

29. $\frac{5}{8} - \frac{1}{2}$ **30.** $\frac{6}{7} - \frac{2}{3}$ **31.** $\frac{7}{8} - \frac{3}{4}$ **32.** $\frac{1}{2} - \frac{2}{5}$

33. $1 - \frac{3}{4}$ **34.** $1 - \frac{3}{8}$ **35.** $1 - \frac{5}{9}$ **36.** $1 - \frac{6}{13}$

Multiply. Write the answer in simplest form. [146]

37. $\frac{2}{5} \times \frac{4}{7}$ **38.** $\frac{6}{11} \times \frac{3}{7}$ **39.** $\frac{3}{8} \times \frac{2}{9}$ **40.** $\frac{3}{4} \times \frac{8}{9}$

41. $\frac{21}{25} \times \frac{40}{49}$ **42.** $\frac{18}{45} \times \frac{20}{36}$ **43.** $\frac{15}{63} \times \frac{28}{45}$ **44.** $\frac{32}{54} \times \frac{27}{56}$

Find the percent of the number. [240]

45. 35% of 1,020 **46.** 72% of 830 **47.** 69% of 2,500

48. 84% of 5,000 **49.** 60% of 3,180 **50.** 125% of 7,450

Chapter 10

PROBABILITY

Weather forecasters depend on interpreting data to make their predictions. Do you think they are usually right or wrong?

10.1 PROBABILITY

Suppose you turn all the cards shown at the right face down. If you pick a card at random, it is equally likely that you will pick any one of them. Each card represents a **possible outcome.**

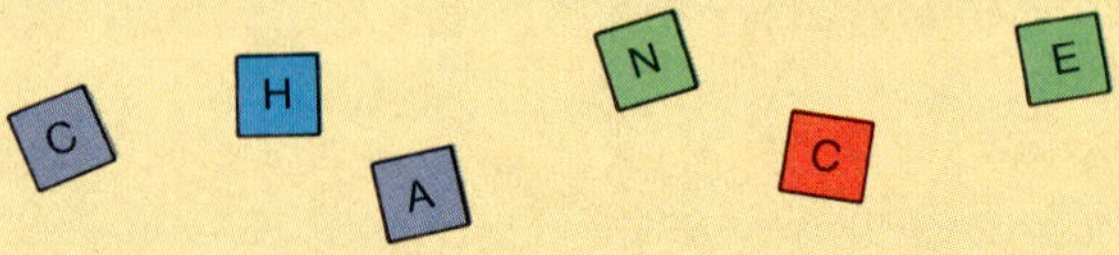

For events that occur at random, **probability** measures the likelihood of a particular outcome. The probability of that event is a ratio:

$$P(\text{event}) = \frac{\text{number of favorable outcomes}}{\text{number of possible outcomes}}$$

EXAMPLE 1 Use the CHANCE cards to find the probability. Write the probability as a ratio in simplest form.

a. $P(\text{N}) =$ ■

**THINK: There is 1 letter N.
There are 6 letters in all.**

$\frac{1}{6}$ **← favorable outcomes**
← possible outcomes

So, $P(\text{N}) = \frac{1}{6}$.

b. $P(\text{H or E}) =$ ■

**THINK: There is 1 letter H
and 1 letter E.
There are 6 letters in all.**

$\frac{2}{6}$ **← favorable outcomes**
← possible outcomes

So, $P(\text{H or E}) = \frac{2}{6}$ or $\frac{1}{3}$.

When no outcome is favorable, the event is **impossible.** The probability of an impossible event is 0.

When every outcome is favorable, the event is **certain.** The probability of a certain event is 1.

EXAMPLE 2 Use the CHANCE cards to find the probability. Write the probability as a ratio in simplest form.

a. $P(\text{a digit}) =$ ■

**THINK: None of the cards shows a digit.
Picking a digit is
an impossible event.**

$P(\text{a digit}) = \frac{0}{6}$ or 0.

b. $P(\text{a letter}) =$ ■

**THINK: All the cards show a letter.
Picking a letter is
a certain event.**

$P(\text{a letter}) = \frac{6}{6}$ or 1.

The **odds** of an event occurring is the ratio of the number of favorable outcomes to the number of unfavorable outcomes.

EXAMPLE 3 Use the CHANCE cards to find the odds of:

a. picking a C.

THINK: $\frac{\text{number of favorable outcomes}}{\text{number of unfavorable outcomes}} = \frac{2}{4}$ $\begin{matrix}\text{(two C's)} \\ \text{(not C's)}\end{matrix}$

So, the odds of picking a C are 2 to 4 or 1:2.

b. not picking a C.

THINK: $\frac{\text{Number of favorable outcomes}}{\text{Number of unfavorable outcomes}} = \frac{4}{2}$ $\begin{matrix}\text{(not C's)} \\ \text{(two C's)}\end{matrix}$

So, the odds of not picking a C are 4 to 2 or 2:1.

FOR DISCUSSION

Why can P(event) never be greater than 1?

PRACTICE EXERCISES

A box contains the 12 counters shown at the right. Each counter is identified by color and by number. A counter is picked at random.

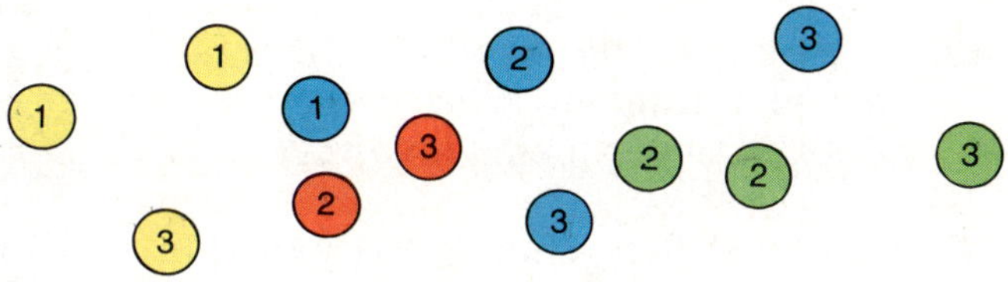

Find the probability.

1. P(green)
2. P(red even)
3. P(yellow or blue)
4. P(odd)
5. P(less than 4)
6. P(not orange)
7. P(red or green)
8. P(yellow prime)
9. P(red, yellow, or green)
10. P(black even)
11. P(white or blue)
12. P(blue less than 3)

Look at the spinner shown at the right. It is spun once.

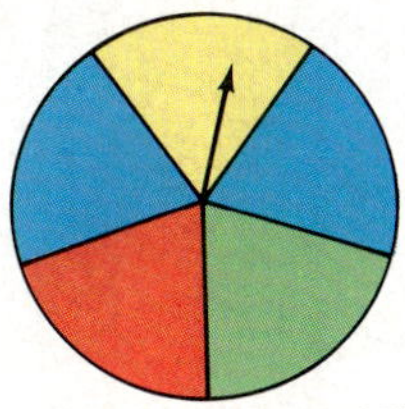

Find the odds of spinning:

13. green.
14. red or blue.
15. red, yellow, or green.
16. red.
17. yellow.
18. red, blue, or green.
19. not red.
20. not green.
21. not blue and not green.
22. blue.
23. not yellow.
24. green, blue, or yellow.

MIXED REVIEW

Complete.

25. 18 L = ■ mL

26. 38,000 g = ■ kg

27. 6,500 km = ■ m

28. 4,200 mg = ■ g

29. 7.9 cm = ■ mm

30. 1,000.3 mL = ■ L

31. 4,500 cm = ■ m

32. 3.1 kg = ■ g

33. 30 L = ■ kL

34. 4 km = ■ cm

35. 0.4 kL = ■ mL

36. 4,500 mg = ■

Solve.

37. Cards with the vowels A, E, I, O, and U are removed from a set of alphabet cards. A consonant card is then picked at random. What is the probability of picking a letter found in the word CHANCE?

38. How many sets of CHANCE cards can be made from a package of 50 index cards? How many index cards will be left over?

39. A letter from the word CHANCE is picked at random. What is the probability of picking a vowel?

40. Earl Llewellyn removes the letters in his name from a set of alphabet cards. How many cards are left?

TIME OUT

The 6 letters in the word CHANCE are written on the alphabet block shown below. Figures A, B, C, and D are all different views of the same block. What is the missing letter in Figure D?

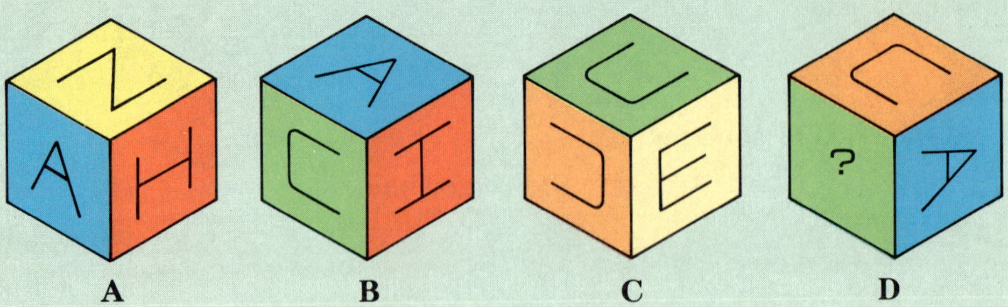

A **B** **C** **D**

PROBLEM *Solving* APPLICATION

10.2 CONSUMER: LIFE INSURANCE

Life insurance companies use probability in determining the rates they charge. When people want to buy life insurance, a company needs to be able to estimate how many years the person is likely to live.

There are many types of tables that provide estimates about life expectancy. Below is a simplified table for a recent year. For example, the table indicates that a person at age 20 is likely to live 56.0 more years. In that case, the total life span would be 76.0 years.

EXPECTATION OF LIFE

At Age	Expectation of Life in Years
5	70.6
10	65.7
15	60.8
20	56.0
25	51.3
30	46.6
35	41.9
40	37.2
45	32.7
50	28.3
55	24.2
60	20.4
65	16.8
70	13.7
75	10.8
80	8.3

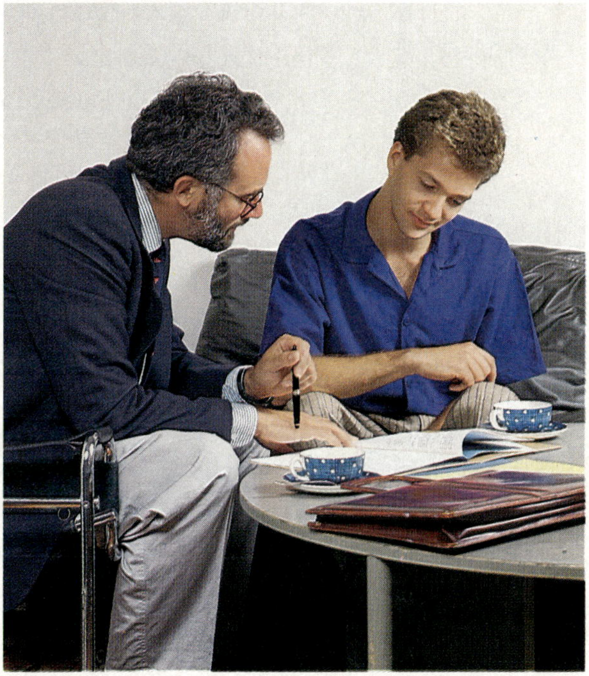

Use the table above. Find the life expectancy in years for a person at age:

1. 15. **2.** 25. **3.** 40. **4.** 55.

Find the estimated *total* life span for a person of:

5. 30. **6.** 45. **7.** 60. **8.** 80.

9. Jackie is likely to live 41.9 more years. How old is she?

10. Abe is 10. His father is 35. Abe's grandfather is 70. What are their life expectancies?

11. What are some of the factors that have caused an increase in life expectancy over the past 60 years?

10.3 PROBABILITY AND PREDICTIONS

A bag contains the 10 marbles shown at the right. A marble is picked at random and then replaced. This is repeated 500 times.

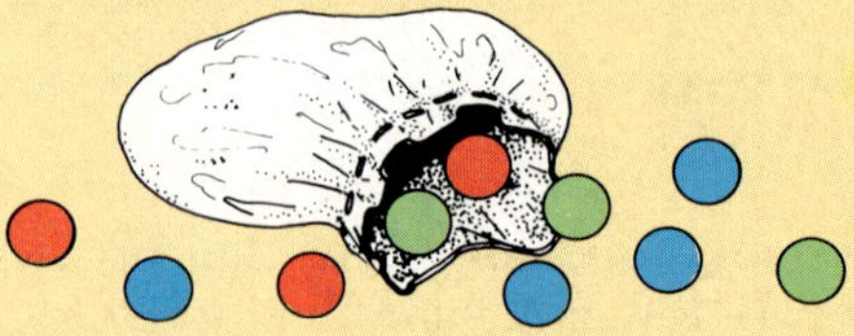

You can use probability to **predict** the number of times a particular event will occur.

EXAMPLE 1 Predict how many times you can expect a red marble to be picked.

1. Find the probability.

   THINK: $\frac{3}{10}$ **← favorable outcomes** / **← possible outcomes** $\qquad P(\text{red}) = \frac{3}{10} = 0.3$

2. Multiply the probability by the number of picks. $\qquad 0.3 \times 500 = 150$

So, you can expect a red marble to be picked 150 times.

The owners of the Amazing Adventure Amusement Park took a random survey of 100 visitors to the park. They summarized the results in the following table:

Type of pass		Favorite ride		Age	
One-day	20	Coast-to-Coaster	40	3–11	23
Two-day	27	Mud Slide	30	12–20	35
Three-day	15	Loop-A-Doo	18	21–59	26
Seasonal	38	Other	12	60 and over	16

Sometimes you can make predictions using data from past events.

EXAMPLE 2 A crowd of 5,000 visitors is expected at Amazing Adventure over the weekend. Predict how many of these visitors will be under 21.

1. Find the probability.

   THINK: Age of visitors 3–11: 23
   Age of visitors 12–20: 35
   58

   $P(\text{under } 21) = \frac{58}{100}$ or 0.58

2. Multiply the probability by the number of expected visitors. $\qquad 0.58 \times 5{,}000 = 2{,}900$

So, Amazing Adventure can expect 2,900 of these visitors to be under 21.

PRACTICE EXERCISES

Complete.

1. A spinner is divided into four equal sections numbered 1 through 4. It is spun 100 times. Predict how many times a 3 will be spun.
2. A coin is flipped 80 times. Predict how many flips will be tails.
3. A spinner is divided into five equal sections numbered 1 through 5. Predict how many times out of 250 an odd number will be spun.

A drawer contains the 15 socks shown at the right. One sock is picked at random and then replaced. This is repeated 300 times. Predict how many times you can expect the color of the sock to be:

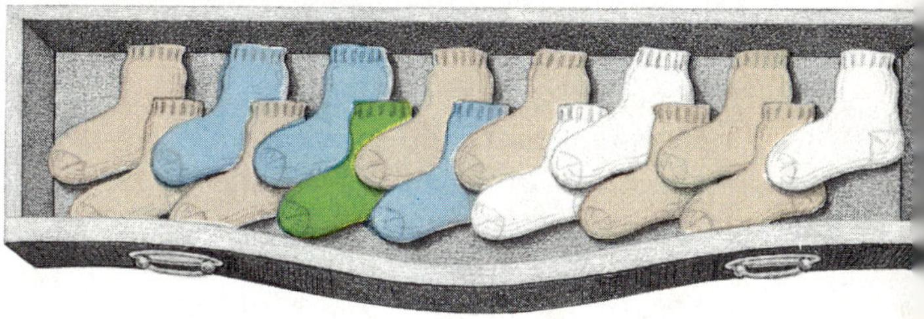

4. brown.
5. white.
6. not white.
7. blue or black.

Use the results of the survey on page 272 for Exercises 8–29.

For each attendance figure, predict how many of the visitors will choose each of the rides listed below as their favorite ride.

Attendance	Coast-to-Coaster	Mud Slide	Loop-A-Doo	Other
10,000	8. ■	9. ■	10. ■	11. ■
7,500	12. ■	13. ■	14. ■	15. ■
3,800	16. ■	17. ■	18. ■	19. ■

Over Labor Day weekend, 25,000 visitors are expected at Amazing Adventure. Predict how many will:

20. be using 3-day passes?
21. be 12–20 years old?
22. be older than 20?
23. be using seasonal passes?
24. be younger than 60?
25. be using a one- or two-day pass?

Solve.

26. What is the probability that the favorite ride of the next visitor to Amazing Adventure will be either the Loop-A-Doo or the Mud Slide?
27. A busload of 250 high school students arrives at Amazing Adventure. Predict how many of them are between 12 and 20 years old.
28. Of those whose favorite ride was not on the survey, half picked the Log Jam. What percent of those surveyed preferred the Log Jam?
29. One Tuesday in May attendance at Amazing Adventure is 5,500. Predict how many of these people are using 1- or 2-day passes?

PROBLEM *Solving* APPLICATION

10.4 CAREER: SCIENTIST

Beth McCaffrey is a scientist. She performs many scientific experiments. One of her experiments is described below.

- A number of plants with white flowers are crossbred with a number of plants with red flowers, and seeds are produced. When these seeds are planted, the resulting plants have pink flowers.
- The plants with pink flowers are bred with each other, and *second-generation* seeds are produced. When these second-generation seeds are planted, some of the resulting plants have flowers that are white; others have flowers that are red; and others have flowers that are pink.

The probability that a plant will have flowers of a given color is as follows:

Probability of white: $\frac{1}{4}$

Probability of red: $\frac{1}{4}$

Probability of pink: $\frac{1}{2}$

Solve.

Suppose 1,000 of the second-generation seeds are planted. About how many of the plants will have flowers that are:

1. red? **2.** pink? **3.** white?

Suppose 2,500 of the second-generation seeds are planted. About how many of the plants will have flowers that are:

4. not red? **5.** not pink? **6.** neither white nor pink?

Write the probability as a decimal. Use the probabilities for second-generation seeds.

7. Probability of a plant with red flowers

8. Probability of a plant that does not have pink flowers

9. Probability of a plant that does not have pink flowers, red flowers, or white flowers

MIDCHAPTER REVIEW

A bag contains the 20 marbles shown.
One marble is picked at random.

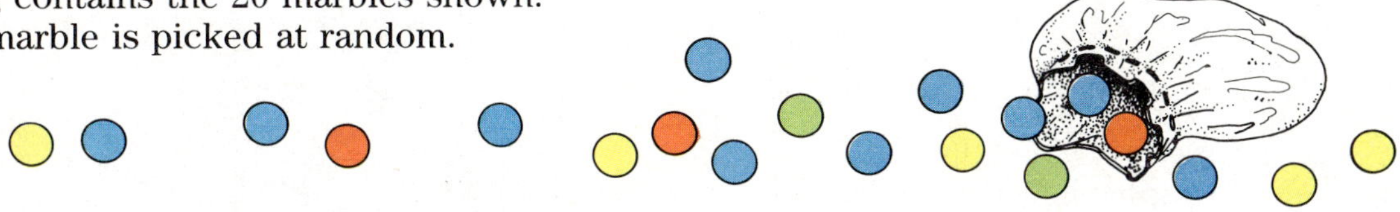

Find the probability. [268]

1. P(yellow)
2. P(blue)
3. P(green)
4. P(red)
5. P(orange)
6. P(green or blue)
7. P(not yellow)
8. P(red or yellow)
9. P(not yellow and not blue)

Find the odds of picking a marble that is: [268]

10. yellow.
11. blue.
12. red.
13. not green.
14. red or yellow.
15. not black and not blue.
16. black or blue.
17. not red.
18. green, red, or yellow.

The spinner shown at the right is spun 300 times. Predict how many times you can expect to spin: [272]

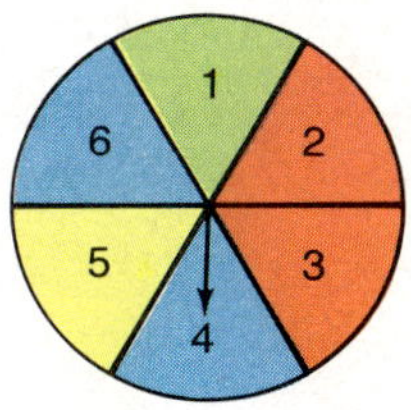

19. green.
20. red.
21 odd.
22. not even.
23. even or yellow.
24. blue or odd.
25. more than 6.
26. green or red.
27. 3 or 4.

Solve. Use the table on page 271 if necessary. [268, 271, 272]

28. Harold Evans is 35 years old. How many more years can he reasonably expect to live?
29. A recent study shows that out of 100 monkeys tested, 34 have the gene for a long tail. What is the probability of having this gene?
30. Jane Kramer is 50 years old. What is her estimated total life span?
31. The probability that a monkey will have the gene for a short tail is $\frac{12}{25}$. Out of 500 monkeys, predict how many will have this gene.

10.5 POSSIBLE OUTCOMES

Berkeley Airlines offers its flight attendants a variety of uniform styles from which to choose. The jackets come in both red and blue. The blouses come in red, white, and blue. How many uniforms are possible?

A **sample space** is a listing of all the possible outcomes in a given situation. You can use a **tree diagram** to show all the possible outcomes. Each "branch" represents a choice or possible outcome.

EXAMPLE 1 Make a tree diagram to show all the possible uniforms.

	Jacket	**Blouse**	**Outcome**
Uniform	Red	Red	Red jacket, red blouse
		White	Red jacket, white blouse
		Blue	Red jacket, blue blouse
	Blue	Red	Blue jacket, red blouse
		White	Blue jacket, white blouse
		Blue	Blue jacket, blue blouse

So, there are 6 possible uniforms.

Sometimes when you have a series of events, you need to find only the number of possible outcomes and not what each of them is.

To find the total number of possible outcomes of a series of events, multiply the number of ways each event can occur.

EXAMPLE 2 The Berkeley Airlines baggage tags use 3-letter city codes. The first and third letters of each code are always consonants and the second letter is always a vowel. How many different codes are possible?

THINK: 21 consonants, 5 vowels.

consonant ↓ 21 × **vowel** ↓ 5 × **consonant** ↓ 21 = 2,205

So, 2,205 different codes are possible.

PRACTICE EXERCISES

Complete the following tree diagram.

1.

BERKELEY AIRLINES SEATING CHOICES IN COACH

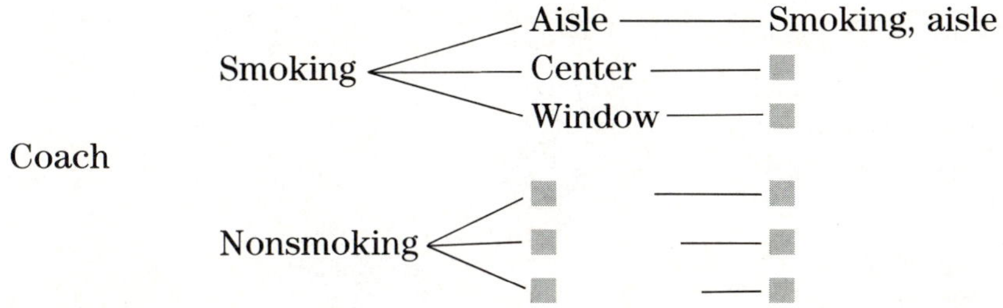

Use the diagram above to answer Exercises 2–7.
How many different types of the seat are available?

2. window **3.** smoking **4.** nonsmoking and center

5. aisle and smoking **6.** coach **7.** nonsmoking

Berkeley Airlines tickets have 4-digit codes. How many codes are possible if the digits are:

8. any 4 digits? **9.** only odd digits? **10.** 2, 3, 5, or 7?

11. all multiples of 3? **12.** less than 3? **13.** 2 even then 2 odd?

Berkeley Airlines routing codes are made up of digits and letters.
How many codes are possible if each code is made up of:

14. 3 digits? **15.** 3 letters?

16. 3 consonants? **17.** 2 odd digits then 2 letters?

18. 3 even digits then 1 letter? **19.** 3 digits then 2 vowels?

Solve.

20. On Flight 1106, 60 passengers paid first-class fares, 277 paid coach fares, and 63 paid excursion fares. What is the probability that a passenger picked at random did not pay a first-class fare?

21. Captain Max Gomez is filing a route plan for a Miami-to-Memphis flight stopping in Atlanta. There are 3 routes from Miami to Atlanta and 4 routes from Atlanta to Memphis. How many route plans are possible?

22. Jon Galvin's ticket originally cost \$373. It carried with it a 25% penalty for change or cancellation. Jon changed his return flight. What was the ticket's final cost?

23. The airlines' fashionable baggage handlers wear pants, shirts, and jackets that each come in the same 4 colors. Pants and jacket must match. How many uniforms are possible?

10.6 INDEPENDENT AND DEPENDENT EVENTS

At the annual Royce High School Fair, Max Lasher is in charge of the Spin-to-Win booth. The wheel is shown at the right. What is the probability of spinning a 1 and then a 4?

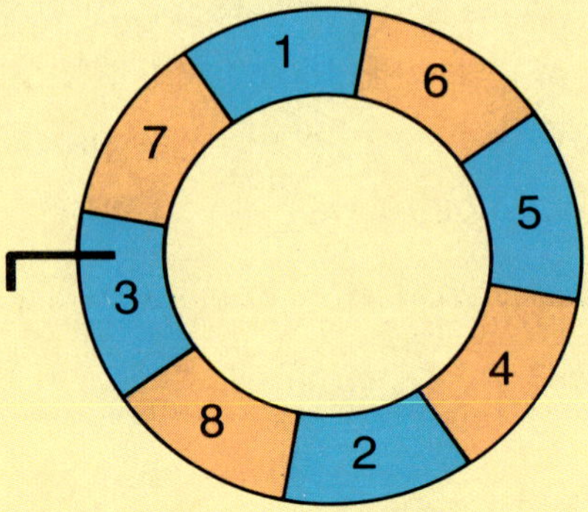

The outcome of the first spin does not affect the outcome of the second spin. These are **independent events.**

To find the probability of a series of independent events, multiply the probabilities of the individual events.

EXAMPLE 1 Find $P(1, \text{then } 4)$

THINK: $P(1, \text{then } 4) = P(1) \times P(\text{then } 4)$

$$P(1, \text{then } 4) = \frac{1}{8} \times \frac{1}{8} = \frac{1}{64}$$

So, the possibility of spinning a 1 and then a 4 is $\frac{1}{64}$.

When the outcome of the first event affects the outcome of the second event, the events are **dependent events.**

EXAMPLE 2 Sally Teicher runs the Match Game. Twelve cards are mixed and placed face down. The object is to pick the two red cards.

One card is picked, left face up, and then another card is picked. What is the probability of winning?

1. Find the probability of the first event. $P(\text{red}) = \frac{2}{12}$, or $\frac{1}{6}$.

2. Find the probability of the second event.

   THINK: The first card is face up. There is one less possible outcome. $P(\text{then red}) = \frac{1}{11}$

3. Multiply the probabilities. $\frac{1}{6} \times \frac{1}{11} = \frac{1}{66}$

So, the probability of winning is $\frac{1}{66}$.

PRACTICE EXERCISES

Write *independent* or *dependent* to describe the series of events.

1. Spinning a spinner 3 times

2. Picking 3 cards without replacing any of them

3. Spinning a spinner, covering that number, and spinning again

Wheel *A* is spun.

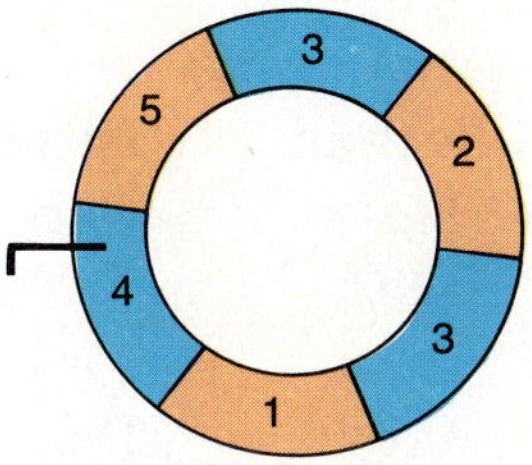

Then Wheel *B* is spun.

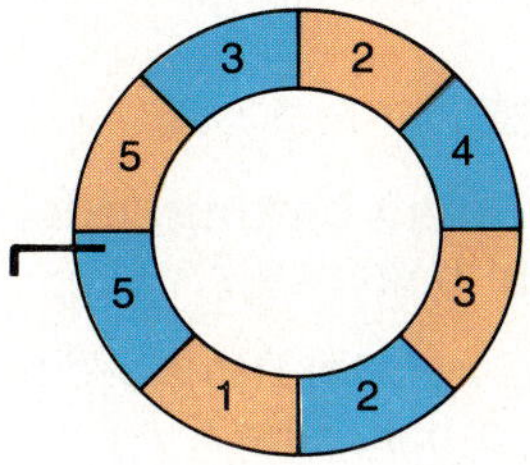

Find the probability.

4. *P*(1, then 4)
5. *P*(3, then 3)
6. *P*(less than 4, then 4)
7. *P*(even, then 6)
8. *P*(even, then even)
9. *P*(a factor of 6, then odd)
10. *P*(5, then 1)
11. *P*(2, then even)
12. *P*(less than 1, then 3)

The cards shown below are placed face down. One at a time, cards are picked at random and then not replaced.

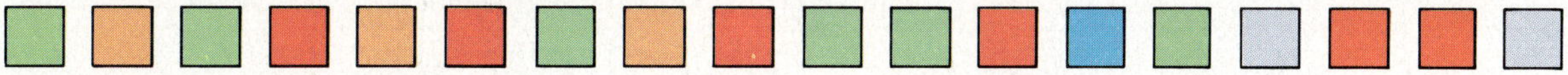

Find the probability.

13. *P*(green, then red)
14. *P*(orange, then blue)
15. *P*(orange, then orange)
16. *P*(blue, then red or green)
17. *P*(purple, then blue or green)
18. *P*(red, then red or blue)
19. *P*(red, then green, then green)
20. *P*(orange, then white, then green)
21. *P*(blue, then green, then red or blue)
22. *P*(red, then red, then red, then red)

MIXED REVIEW

For the set data, find the mean, median, mode, and range.

	Mean	Median	Mode	Range
24, 36, 27, 24, 26, 28, 24, 27	**23.** ■	**24.** ■	**25.** ■	**26.** ■
1.4, 1.8, 2.1, 1.7, 1.8, 1.5, 1.6	**27.** ■	**28.** ■	**29.** ■	**30.** ■
110, 104, 120, 100, 100, 100, 110, 120	**31.** ■	**32.** ■	**33.** ■	**34.** ■

PROBLEM *Solving* STRATEGY

10.7 MAKING AN ORGANIZED LIST

Situation:

Dale, Keva, and Cory are players on a softball team. The coach plans to have them bat in the first 3 positions in the batting order. In how many different ways can the coach arrange these 3 players?

Strategy:

Making an organized list can help solve some problems.

Applying the Strategy:

		Batting Order	
	First	**Second**	**Third**
THINK: How many different batting orders are possible if Dale bats first?	• Dale	Keva	Cory
	• Dale	Cory	Keva
How many different batting orders are possible if Keva bats first?	• Keva	Dale	Cory
	• Keva	Cory	Dale
How many different batting orders are possible if Cory bats first?	• Cory	Dale	Keva
	• Cory	Keva	Dale
How many different batting orders are possible in all?	6		

The coach can arrange these 3 players in 6 different ways.

FOR DISCUSSION

Beth, Helen, and Tom all want to run in a race. A team consists of two runners. Which list shows all the possible combinations of runners?

a. Beth and Helen
Beth and Tom
Helen and Tom

b. Beth and Helen, Helen and Beth
Beth and Tom, Tom and Beth
Helen and Tom, Tom and Helen

PRACTICE EXERCISES

Make an organized list to solve each problem.

1. Jed is picking a 2-digit number for his football jersey. He must choose from the digits 3, 5, and 7. List all 6 possible arrangements.

2. The coach plans to use Joe, Al, and Sam in the outfield. List all 6 ways the coach can arrange them in left field, right field, and center field.

3. Phil is picking a 3-digit number for a license plate. He must choose from the digits 2, 4, 6, and 8. List all the possible arrangements. How many are there?

4. Ben, Ari, Joan, and Sarah go bowling. List all the possible orders in which they can bowl. How many are there?

5. In a round-robin singles tennis tournament, Cathy, Ann, Wes, and Glen all play each other. List all 6 of the matches they play against one another.

6. A basketball team is practicing defense. Each of the 5 players practices with each of the other players. List all 10 possible practice pairings. (*Hint:* You can label the players 1, 2, 3, 4, and 5.)

7. There are 6 sprinters on the track team. List all the possible 4-runner relay teams the coach can form. How many are there?

8. Rudi won 4 medals at the track championships. At least 2 of his medals were silver. What are the possible arrangements of gold, silver, and bronze medals Rudi could have won?

PROBLEM Solving APPLICATION

10.8 SAMPLING AND QUALITY CONTROL

Manufacturers frequently monitor the quality of their products by checking a random **sample** of the items produced. They determine the ratio of defective items to the total number of sample items. They then predict that in the larger group, there is an equivalent ratio of defective items to the total number of items.

The Alpha Corporation produces electrical appliances. In one month it manufactures 5,000 toasters. A random sample of 200 toasters is checked, and 6 are found to be defective. Predict the total number of defective toasters likely to be manufactured during the month.

THINK: In the sample, 6 out of 200 items are defective.
So, the ratio is $\frac{6}{200}$, or $\frac{3}{100}$.

Use a proportion to solve the problem.

1. Write a proportion.
   Let n represent the total number of defective items.

$$\frac{3}{100} = \frac{n}{5{,}000}$$

2. Find n.

$$\begin{aligned} 3 \times 5{,}000 &= 100 \times n \\ 15{,}000 &= 100 \times n \\ 15{,}000 \div 100 &= n \\ 150 &= n \end{aligned}$$

So, it is likely that 150 defective toasters will be manufactured during the month.

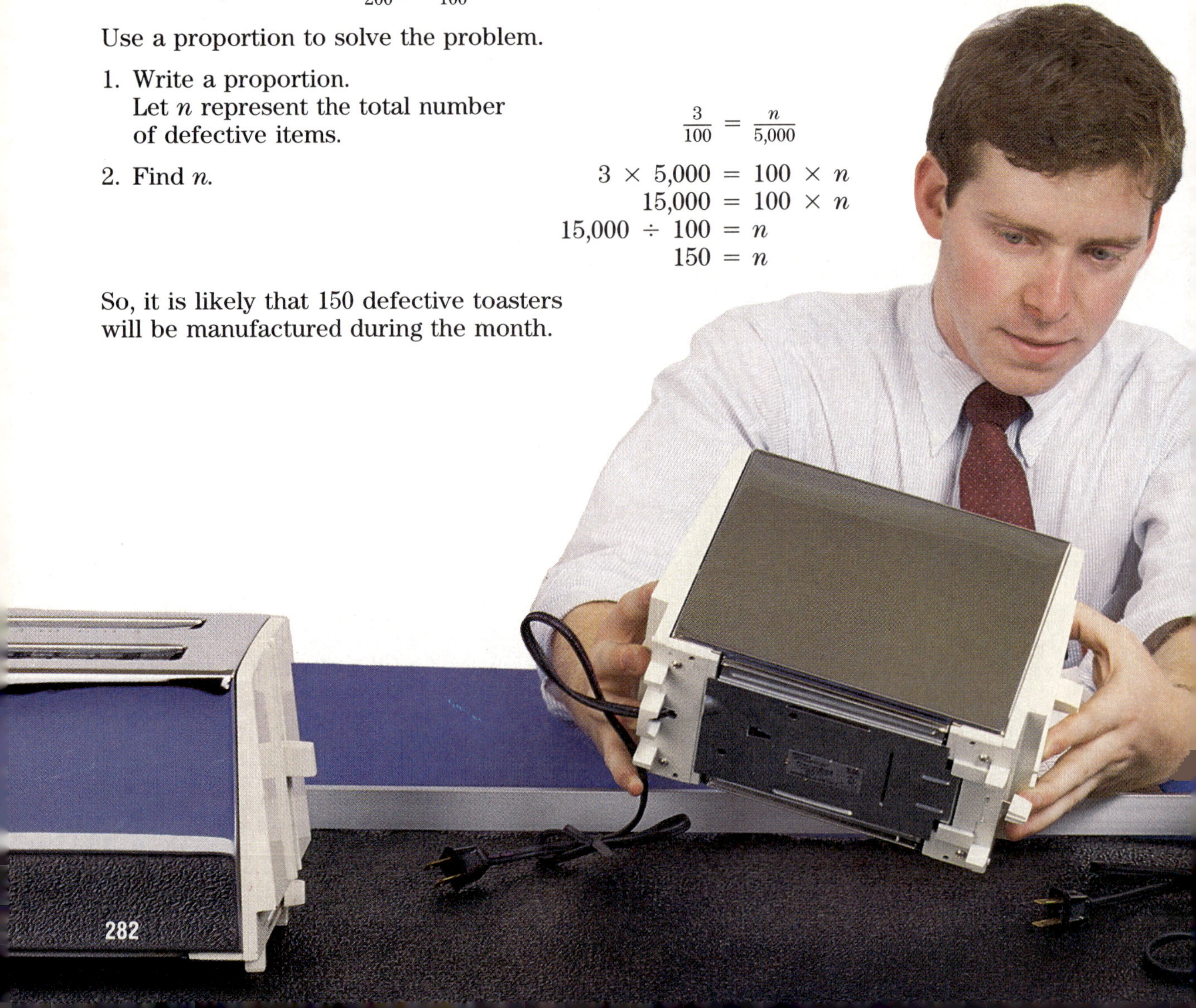

Predict the total number of defective items manufactured each month.

Item	No. in Sample	No. Defective in Sample	Total No. Produced	No. Defective
1. Blender	100	2	3,000	▒
2. Electric knife	500	25	12,500	▒
3. Radio	400	16	8,400	▒
4. Cassette player	300	11	6,300	▒

Solve. Remember to estimate whenever you use your calculator.

5. The Omega Corporation manufactures 16,000 Model A calculators in one month. A random sample of 900 calculators is checked. 20 are defective. Predict the total number of defective Model A calculators likely to be manufactured during the month.

6. In one month the Omega Corporation manufactures 4,500 Model B calculators. A random sample of these calculators is checked. 2% are found to be defective. Predict the total number of Model B calculators that are not likely to be defective.

7. The Gamma Company manufactures 4,000 Super-robots in one month. A random sample of these toys is checked, and 3% are found to be defective. Predict the total number of Super-robots that are likely to be defective.

8. In one month the Gamma Company manufactures 5,680 Mini-robots. A random sample of these robots is checked, and 2.5% are found to be defective. Predict the total number of Mini-robots that are likely to be defective.

9. A quality control technician examines 900 electronic toys. Of these, 18 are defective. About what percent are defective?

10. A quality control technician examines 550 learning devices. Of these, 15 are defective. About what percent are defective?

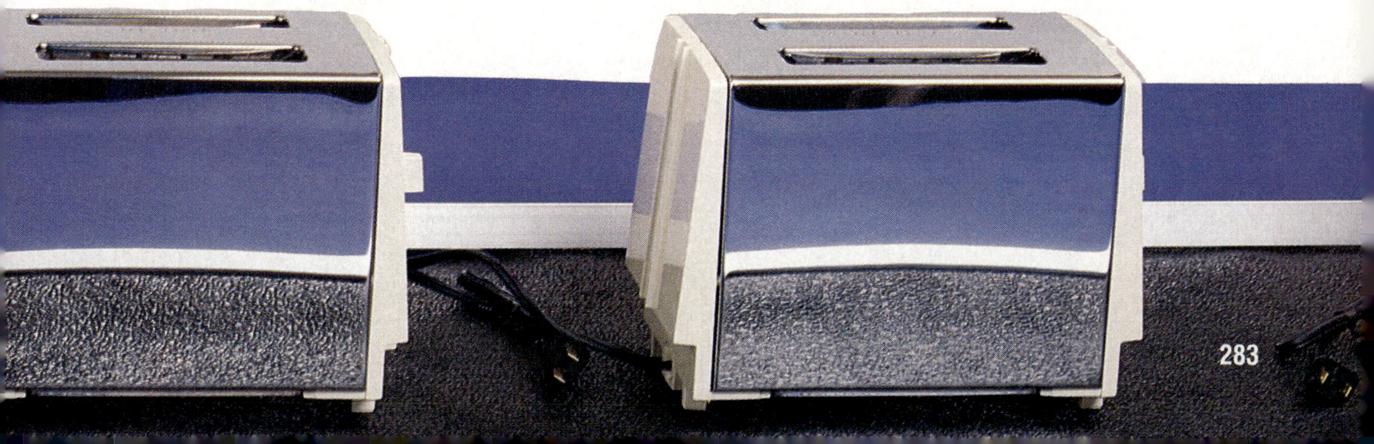

DECISION MAKING

10.9 CHOOSING TRANSPORTATION

Choosing a vacation spot can be a fun decision. Deciding how to get there requires research and some careful considerations.

PROBLEM

Karen, her parents, and her grandmother will spend their vacation at her cousin's house near Orlando, Florida. The adults have decided to allow 10 days for the vacation, and have asked Karen to help them choose the mode of transportation for the trip from New York. Their options are to fly, drive, or take a bus.

Karen made the following notes.

Plane: $2\frac{1}{2}$-hour flight each way
$298 round-trip air fare per person
Taxi to and from airport: $23 each way

Car: 3 days each way
One-way expenses:
Gasoline and tolls: about $100
Motel: 2 nights, $48 per room, per night
(2 rooms required)
Meals: 7, averaging $30 a meal for all 4 people
(Car is a compact model; 4 people can drive)

Bus: 2 days each way
$ 149 one way, half price for senior citizens
(motel for one night included)
Meals: 5, averaging $30 a meal for all 4 people

DECISION-MAKING FACTORS

Cost   Time   Comfort (fatigue)

DECISION-MAKING COMPARISONS

Compare the three modes of transportation.

	Plane	Car	Bus
Cost (round trip for all 4 people)			
Transportation	1. ■	2. ■	3. ■
Motel	4. ■	5. ■	Included
Food	6. ■	7. ■	8. ■
Total cost	9. ■	10. ■	11. ■
Time			
Traveling	12. ■	13. ■	14. ■
In Orlando	10 days	15. ■	16. ■
Comfort	17. ■	18. ■	Could be tiring

MAKING THE DECISIONS

Which mode of transportation should Karen's family use:

19. if cost were the only factor?

20. if time were the only factor?

21. if comfort were the only factor?

22. How much could the family save by driving instead of flying?

23. If the family decided to drive but bring food for the trip, about how much would they save on meals?

24. If the aim of the trip was to see part of the East coast, which mode of transportation would be best?

25. If the family had never been to Florida, which would be the best way to travel? Why?

26. What if Karen's cousin did not have a car the family could use. How could that affect their decision?

27. In terms of meals, how could driving be better than taking the bus?

28. If you were making the decision, which mode of transportation would you choose? Why?

CHAPTER REVIEW

Vocabulary Choose the letter of the word(s) that completes the statement.

1. The probability of an event is a ■. [268]
a. possible outcome **b.** favorable outcome **c.** ratio

2. A tree diagram lists all the ■. [276]
a. possible outcomes **b.** favorable outcomes **c.** odds

3. When one event has no effect on a second, the events are ■. [278]
a. dependent **b.** independent **c.** impossible

Skills A letter is picked at random from the word MATHEMATICS. Find the probability. [268]

P(M) $\frac{2}{11}$

4. *P*(A) **5.** *P*(a vowel) **6.** *P*(a consonant)
7. *P*(M or T) **8.** *P*(A, B, or C) **9.** *P*(K)

A box contains 12 counters. Of these, 2 are white, 3 are green, 3 are blue, and 4 are red. Find the odds of picking at random one that is: [268]

red. *1:2*

10. white. **11.** blue. **12.** green.
13. red or blue. **14.** not green. **15.** red or green. **16.** not white.

A recent survey showed that $\frac{13}{100}$ of those questioned were born on a Saturday. For each sample, predict the number born on a Saturday. [272]

5,000 *650*

17. 12,000 **18.** 20,000 **19.** 3,500

A bag contains 18 marbles. Of these, 2 are orange, 3 are green, 5 are blue, and 8 are red. A marble is picked at random and then replaced. Find the probability. [278]

P(green, then red) $\frac{2}{27}$

20. *P*(orange, then blue) **21.** *P*(purple, then green)
22. *P*(red, then red) **23.** *P*(blue, then green)

Find the probability if the marble is not replaced. [278]

P(green, then red) $\frac{4}{51}$

24. *P*(blue, then red) **25.** *P*(orange, then green)
26. *P*(blue, then blue) **27.** *P*(red, then orange)

CHAPTER TEST

Jurors in Hibbing are selected at random from the voting rolls. The distribution by political party is shown at the right. Find the probability that the next juror will be:

SURVEY RESULTS
Party Membership of 200 Voters

Republican:	Democratic:
Conservative:	Liberal:
Independent:	

1. Conservative. **2.** Liberal. **3.** not a Democrat.

Find the odds that the next juror will be:

4. Republican. **5.** not a Liberal. **6.** Conservative or Democrat.

The results of a recent survey of 75 visitors to Graceland, Elvis Presley's home in Memphis, Tennessee, are shown in the table below.

Number of Elvis albums owned	0	1–3	4–7	8 or more
Number of visitors	9	39	12	15

If 1,200 people visited Graceland yesterday, predict how many own:

7. no albums. **8.** less than 8 albums. **9.** from 1 to 3 albums.

10. A coin is tossed 3 times. How many possible outcomes are there for this series of tosses?

A bag contains 12 marbles. Of these, 1 is white, 2 are blue, 3 are red, and 6 are green. A marble is picked at random and then replaced. A second marble is picked. Find the probability.

11. *P*(green, then blue) **12.** *P*(red, then red) **13.** *P*(white, then not white)

Find the probability if the first marble is not replaced.

14. *P*(red, then green) **15.** *P*(blue, then blue) **16.** *P*(white, then white)

Solve.

17. There are 4 books on a shelf. Make a list to show the different ways they can be ordered.

18. Delta pricing codes are made up of 3 even digits. Make a list to show all the possible codes.

19. Of a random sample of 900 fans, 15 do not work. AMX Inc. orders 30,000. Predict how many do not work.

20. A random sample of 750 mugs shows 25 to be cracked. Out of 18,000 mugs in stock, predict how many are cracked.

ENRICHMENT PERMUTATIONS

The swimming team at Eisenhower High School is practicing for the league championship. Coach Barker has chosen his four best swimmers for the 400-m freestyle relay. Each swimmer will swim a 100-m leg of the relay. In how many different ways can the coach arrange the order of the swimmers?

An arrangement of objects in which order is important is a **permutation.** You can use the following notation to find the number of permutations in the problem above.

$${}_4P_4$$

Number of choices for first position ↗ ↖ **Number of positions**

Multiply to find the answer. $\quad {}_4P_4 = 4 \times 3 \times 2 \times 1 = 24$

THINK: Number of choices for the first leg → 4

Number of choices for the second leg → 3

Number of choices for the third leg. → 2

Number of choices for the fourth leg → 1

So, Coach Barker can arrange the order of the swimmers in 24 different ways.

Examples

Symbol	**Meaning**	**Solution**
${}_5P_3$	The number of permutations of 5 objects taken 3 at a time.	${}_5P_3 = 5 \times 4 \times 3 = 60$
${}_8P_5$	The number of permutations of 8 objects taken 5 at a time.	${}_8P_5 = 8 \times 7 \times 6 \times 5 \times 4 = 6{,}720$

Complete.

1. ${}_5P_4 = 5 \times 4 \times ■ \times ■ = ■$ **2.** ${}_6P_3 = 6 \times ■ \times ■ = ■$

3. ${}_3P_3 = ■ \times ■ \times ■ = ■$ **4.** ${}_7P_4 = ■ \times ■ \times ■ \times ■ = ■$

Find the number of permutations.

5. ${}_4P_2$ **6.** ${}_5P_3$ **7.** ${}_6P_4$ **8.** ${}_7P_3$

9. ${}_7P_5$ **10.** ${}_8P_2$ **11.** ${}_9P_3$ **12.** ${}_{10}P_2$

Solve.

13. In how many ways can 4 flags be ordered on 2 flagpoles?

14. In how many ways can 9 players be ordered in 6 positions?

MENTAL MATH PROBABILITIES AS FRACTIONS

When a probability is expressed as a simple fraction, you may be able to do calculations mentally.

One marble is chosen at random from a bag containing 200 marbles. The probability that the marble is red is $\frac{1}{4}$. How many marbles are red?

Find $\frac{1}{4}$ of 200. **THINK:** $\frac{1}{4}$ of 200 is $\frac{1}{4} \times 200$.

To multiply by $\frac{1}{4}$, divide by 4. $\frac{1}{4} \times 200 = 200 \div 4 = 50$

There are 50 red marbles.

The probability that the marble is blue is $\frac{2}{5}$. How many marbles are blue?

Find $\frac{2}{5}$ of 200. **THINK:** $\frac{2}{5}$ of 200 is $\frac{2}{5} \times 200$.

To multiply by $\frac{2}{5}$:

1. Divide by 5. $200 \div 5 = 40$
2. Multiply by 2. $2 \times 40 = 80$

There are 80 blue marbles.

Compute mentally.

1. $\frac{1}{4}$ of 40 **2.** $\frac{1}{10}$ of 120 **3.** $\frac{1}{8}$ of 240 **4.** $\frac{1}{3}$ of 165

5. $\frac{3}{5}$ of 50 **6.** $\frac{3}{4}$ of 80 **7.** $\frac{2}{3}$ of 210 **8.** $\frac{5}{8}$ of 320

Solve the problems mentally.

One card is chosen at random from a group of 400 marked cards.
The probability that the card shows a book title is $\frac{1}{5}$.
The probability that it shows a song title is $\frac{3}{4}$.

9. How many cards show a book title? **10.** How many show a song title?

One of 240 envelopes is chosen at random.
The probability that the envelope is empty is $\frac{1}{8}$.
The probability that it contains a \$1 bill is $\frac{2}{3}$.

11. How many envelopes are empty? **12.** How many contain a \$1 bill?

PreSkills TEST

Add. [30]

1. $35 + 55$ **2.** $47 + 86$ **3.** $98 + 82$ **4.** $98 + 39$

5. $124 + 35$ **6.** $119 + 61$ **7.** $151 + 13$ **8.** $45 + 123$

Subtract. [32]

9. $90 - 45$ **10.** $180 - 46$ **11.** $180 - 87$ **12.** $90 - 67$

13. $90 - 76$ **14.** $180 - 32$ **15.** $90 - 69$ **16.** $180 - 117$

Measure to the nearest half-inch. [180]

17. 

18.

Measure to the nearest centimeter. [192]

19. 

20. 

Multiply. [40, 74, 146, 148]

21. 2×9 **22.** 2×5.8 **23.** $2 \times 1\frac{3}{4}$ **24.** $2 \times 3\frac{1}{2}$

25. $\frac{1}{2} \times 12$ **26.** $\frac{1}{2} \times 3$ **27.** $\frac{1}{2} \times 4.6$ **28.** $\frac{1}{2} \times 2\frac{3}{4}$

Find the product. [18]

29. 3^2 **30.** 8^2 **31.** 12^2 **32.** 15^2

33. 13^2 **34.** 20^2 **35.** 25^2 **36.** 28^2

Find n. [216]

37. $\frac{3}{5} = \frac{n}{15}$ **38.** $\frac{2}{7} = \frac{8}{n}$ **39.** $\frac{n}{6} = \frac{20}{24}$ **40.** $\frac{3}{n} = \frac{9}{21}$

41. $\frac{5}{8} = \frac{n}{24}$ **42.** $\frac{4}{9} = \frac{16}{n}$ **43.** $\frac{n}{4} = \frac{9}{12}$ **44.** $\frac{4}{n} = \frac{24}{30}$

Chapter 11

GEOMETRY

All builders use geometric shapes in their constructions. What factors determine the size and shape of a building?

11.1 POINTS, LINES, AND PLANES

Geometric figures are made up of sets of points. You can use a drawing or a symbol to represent these figures.

Definition	Drawing	Symbol
A **point** is a location in space with no size.		A
A **plane** is a flat surface extending in all directions without end.	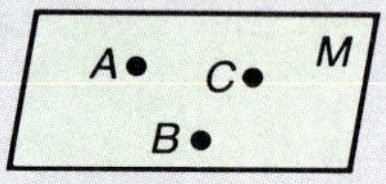	plane ABC or plane M
A **line** is a set of points that forms a straight path extending in opposite directions without end.	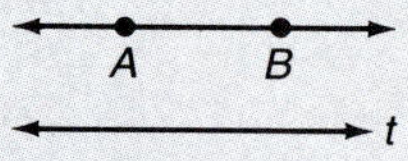	$\overleftrightarrow{AB}$ or $\overleftrightarrow{BA}$ line t
A **line segment** is a part of a line with two endpoints.	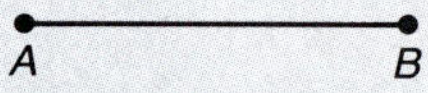	$\overline{AB}$ or $\overline{BA}$
A **ray** is part of a line with one endpoint.	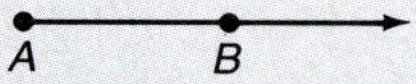	$\overrightarrow{AB}$
An **angle** is made up of two rays with the same endpoint. The endpoint is called the **vertex.**	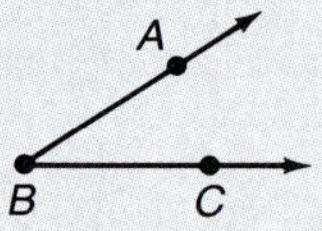	∠ABC

EXAMPLE 1 Draw and label the figure.

a. $\overleftrightarrow{CD}$

THINK: line

b. $\overrightarrow{CD}$

THINK: ray

Some geometric figures can be named in different ways. Look at the following angle.

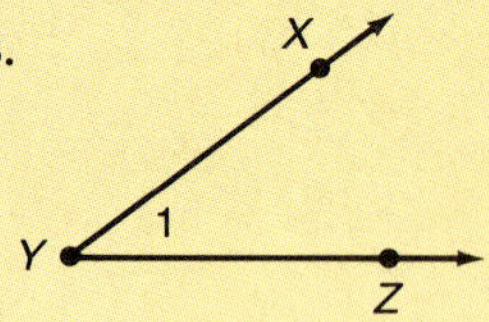

It can be named as $\angle XYZ$, $\angle ZYX$, $\angle Y$, or $\angle 1$.

EXAMPLE 2 Name the figure.

a.

line segment EF, or $\overline{EF}$

b.

ray EF, or $\overrightarrow{EF}$

FOR DISCUSSION

1. How many lines can pass through points *A* and *B*?

2. Explain why some angles must be named by three points and not just the vertex point.

PRACTICE EXERCISES

Draw and label the figure.

1. Plane *RST* **2.** $\overleftrightarrow{RS}$ **3.** $\angle RST$ **4.** $\overline{RS}$ **5.** $\overrightarrow{RS}$ **6.** $\angle SRT$

Identify the geometric figure.

7. $\angle T$ **8.** $\overline{BC}$ **9.** $\overleftrightarrow{UV}$ **10.** $\overrightarrow{MP}$ **11.** *Z* **12.** $\angle RST$

Name the figure.

13.
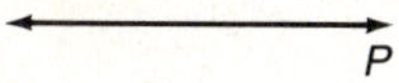

14.
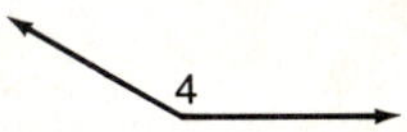

15.

16.

17.
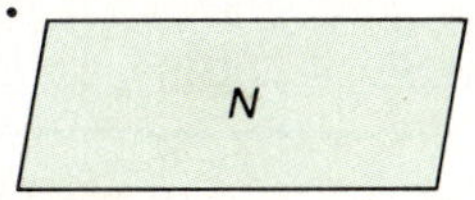

18.
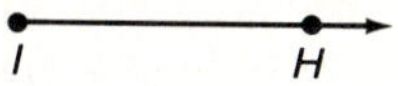

19.
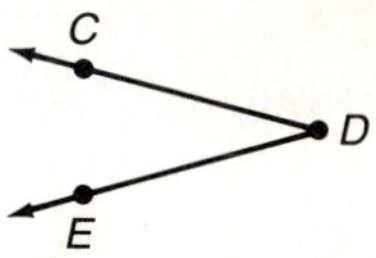

20.
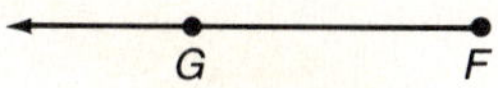

Name the ray in three ways.

21.
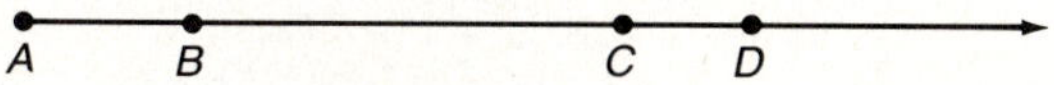

22.
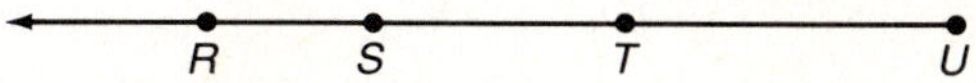

23. Name six line segments.

24. Name the angle in four ways.

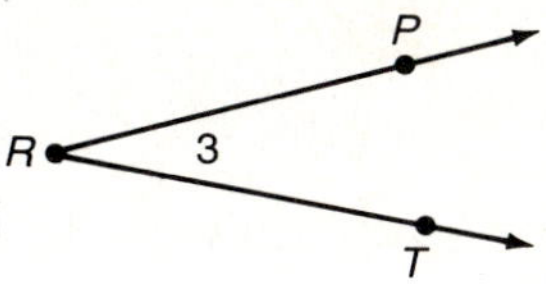

Write *true* or *false*.

25. A line can be measured.

26. A line segment can be measured.

27. A ray extends in two directions without end.

28. Three points will always lie in the same plane.

11.2 MEASURING AND DRAWING ANGLES

You can use a protractor to measure angles. The **degree** (°) is the standard unit of measure.

EXAMPLE 1 Find the measure of $\angle ABC$.

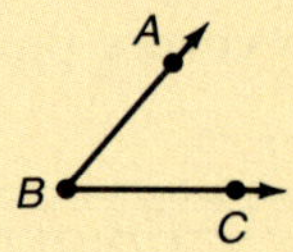

1. Place the center of the protractor on B. Then move the protractor so that the 0° mark is along $\overrightarrow{BC}$.

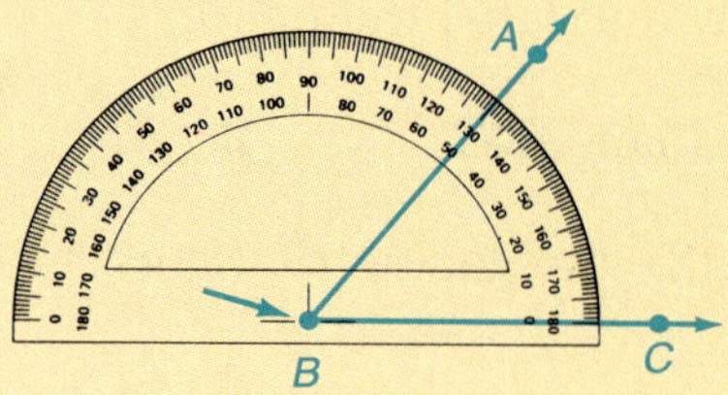

2. Read the measure where $\overrightarrow{BA}$ crosses the same scale as $\overrightarrow{BC}$.

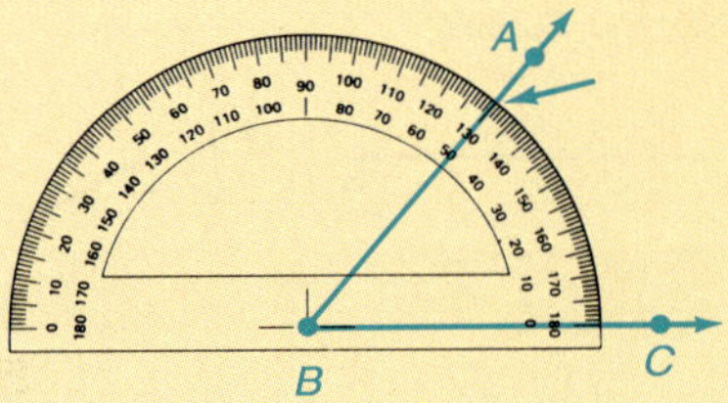

So, the measure of $\angle ABC$ is 50°. This is written $m\angle ABC = 50°$.

A protractor may also be used to draw an angle.

EXAMPLE 2 Draw an angle that measures 120°.

1. Draw $\overrightarrow{AB}$. Place the center of the protractor on A so that the 0° mark is along $\overrightarrow{AB}$.

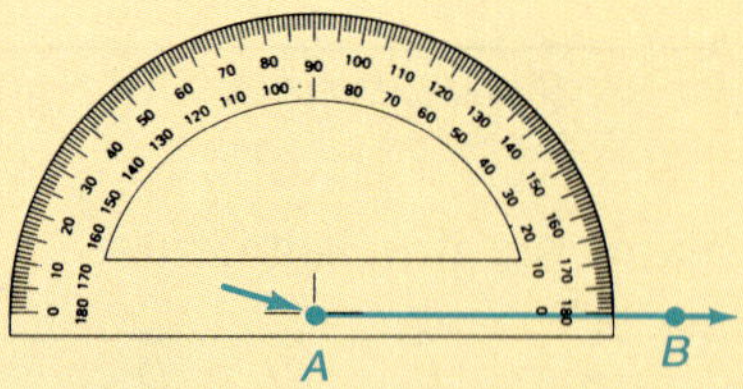

2. Find the 120° mark on the same scale as $\overrightarrow{AB}$. Draw point C. Then draw $\overrightarrow{AC}$.

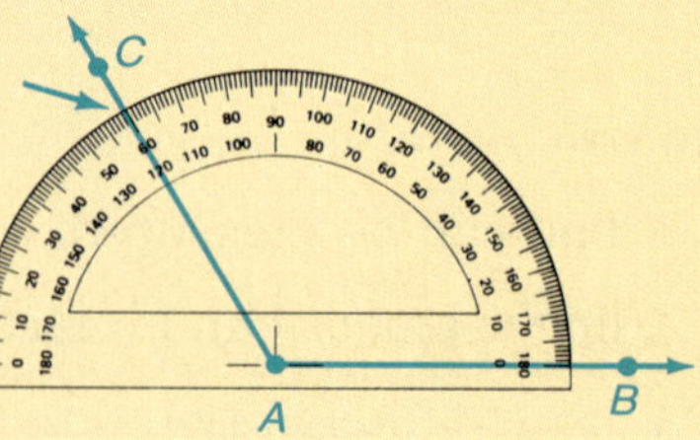

PRACTICE EXERCISES

Use the figure at the right for Exercises 1–9.
Measure the angle.

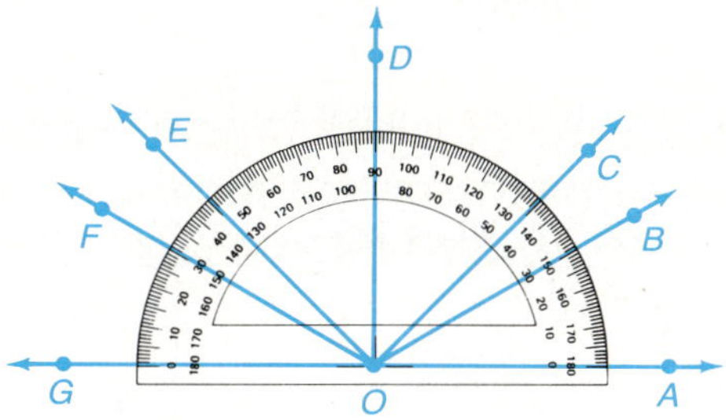

1. $\angle AOB$ **2.** $\angle AOC$ **3.** $\angle AOD$

4. $\angle AOE$ **5.** $\angle AOF$ **6.** $\angle AOG$

7. $\angle GOF$ **8.** $\angle GOE$ **9.** $\angle GOC$

Measure the angle.

10. 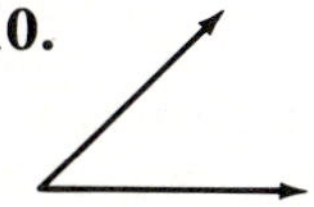

11. 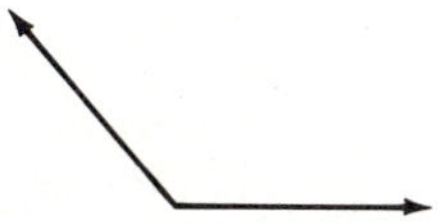

12. 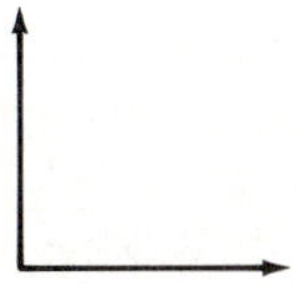

13. 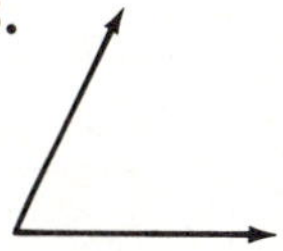

14. 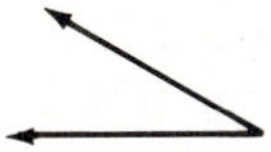

15. 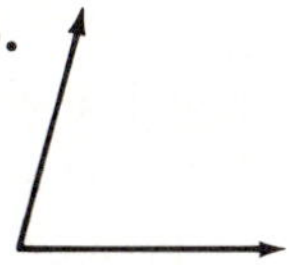

16. 

17. 

Draw an angle with the following measure.

18. 20° **19.** 130° **20.** 90° **21.** 55° **22.** 145° **23.** 70°

Measure the angle.

24. $\angle BAC$

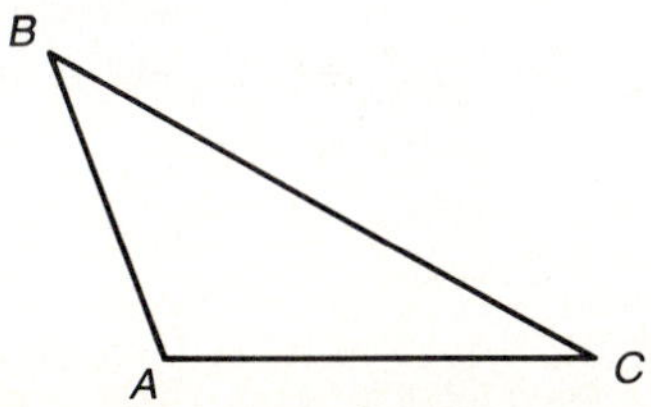

25. $\angle ADC$

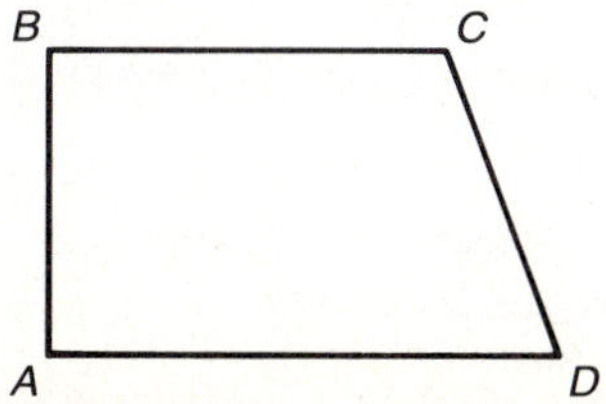

26. $\angle BCD$

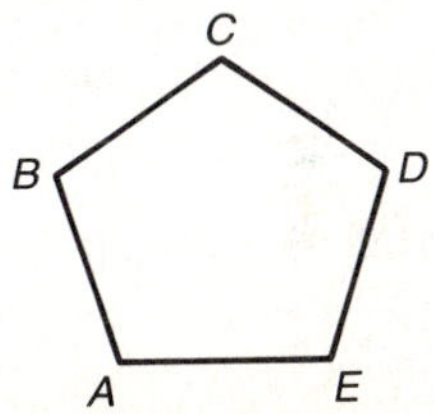

11.3 CLASSIFYING ANGLES

You can classify angles by their measures.

Angle or angle pairs	Name	Definition
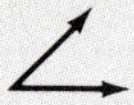	Acute angle	An **acute angle** measures greater than 0° and less than 90°.
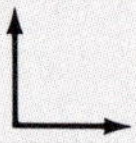	Right angle	A **right angle** measures 90°.
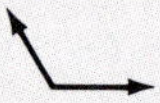	Obtuse angle	An **obtuse angle** measures greater than 90° and less than 180°.
	Straight angle	A **straight angle** measures 180°.
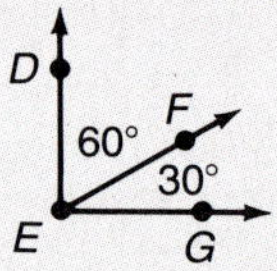 m∠*DEF* + m∠*FEG* = 90°	Complementary angles	Angles are **complementary** when the sum of their measures is 90°.
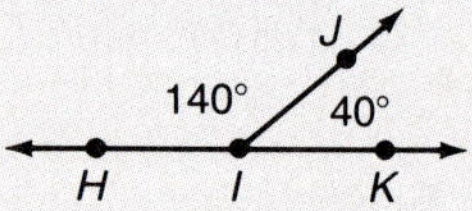 m∠*HIJ* + m∠*JIK* = 180°	Supplementary angles	Angles are **supplementary** when the sum of their measures is 180°.

EXAMPLE 1 Classify ∠*ABC*.

THINK: m∠*ABC* = 135°
135° is greater than 90°
and less than 180°.

∠*ABC* is an obtuse angle.

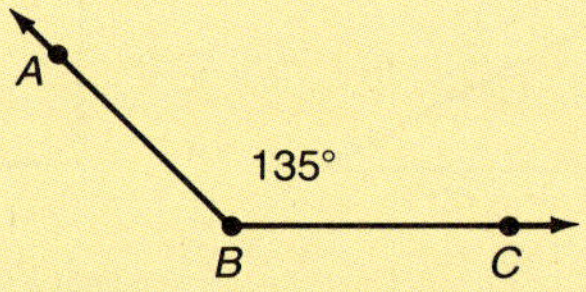

You can classify pairs of angles by the sum of their measures.
The measures (m) of complementary angles add up to 90°.
The measures (m) of supplementary angles add up to 180°.

EXAMPLE 2 a. Which pair of angles is complementary?
b. Which pair of angles is supplementary?

m∠1 = 40° m∠2 = 130° m∠3 = 50°

a. **THINK: m∠1 + m∠3 = 40° + 50°**
= 90°

So, ∠1 and ∠3 are complementary angles.

b. **THINK: m∠2 + m∠3 = 130° + 50°**
= 180°

So, ∠2 and ∠3 are supplementary angles.

You can find the complement or supplement of an angle by subtracting.

EXAMPLE 3 $\angle A$ is 60°.

a. Find the complement of $\angle A$.

THINK: ▩ + 60° = 90°
▩ = 90° − 60°
▩ = 30°

So, the complement is 30°.

b. Find the supplement of $\angle A$.

THINK: ▩ + 60° = 180°
▩ = 180° − 60°
▩ = 120°

So, the supplement is 120°.

PRACTICE EXERCISES

Classify the angle as *acute, right, obtuse,* or *straight.*

1. 

2. 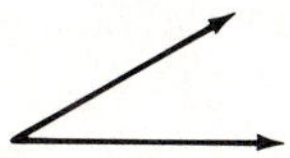

3. 

4. 

5. 28° **6.** 45° **7.** 95° **8.** 180° **9.** 5° **10.** 90°

Tell whether the pair of angles are *complementary, supplementary,* or *neither.*

11. 40°, 30° **12.** 10°, 80° **13.** 35°, 55° **14.** 110°, 80° **15.** 145°, 35°

Use the figure to the right for Exercises 16–19. Tell whether the pair of angles are *complementary, supplementary,* or *neither.*

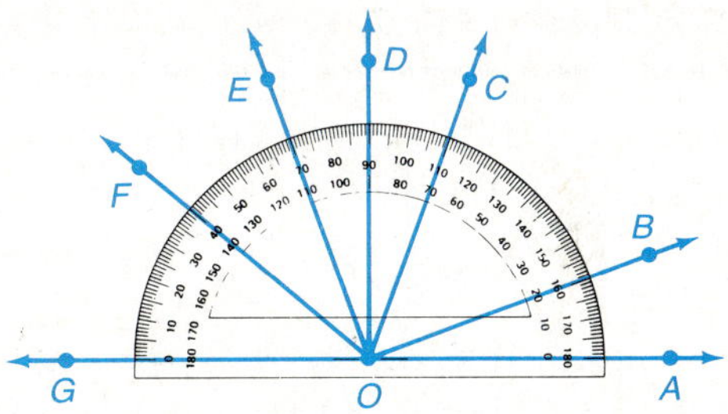

16. $\angle AOB$ and $\angle BOD$

17. $\angle AOE$ and $\angle EOG$

18. $\angle AOC$ and $\angle COG$

19. $\angle AOF$ and $\angle EOG$

Find the complement of the angle.

20. 20° **21.** 55° **22.** 18° **23.** 67° **24.** 5° **25.** 38°

Find the supplement of the angle.

26. 60° **27.** 140° **28.** 95° **29.** 15° **30.** 134° **31.** 87°

Solve.

32. Two angles are complementary and have the same measure. What is the measure of each angle?

33. $\angle AOD$ and $\angle DOE$ are supplementary angles. $\angle DOE$ and $\angle COD$ are complementary angles. If $m\angle AOD = 145°$, what are the measures of $\angle DOE$ and $\angle COD$?

11.4 CONSTRUCTING CONGRUENT LINE SEGMENTS AND ANGLES

Geometric figures that have the same shape and size are **congruent.**

$\overline{ST}$ is congruent to $\overline{UV}$

$\overline{ST} \cong \overline{UV}$

$\angle EFG$ is congruent to $\angle JKL$

$\angle EFG \cong \angle JKL$

You can construct congruent line segments or congruent angles using a straightedge and a compass.

EXAMPLE 1

Construct a line segment congruent to $\overline{MN}$.

1. Draw a ray. Label the endpoint X.

2. Open the compass to the length of $\overline{MN}$. Using the same compass opening, place the compass point on X and draw a small arc that crosses the ray. Label the point of intersection Y.

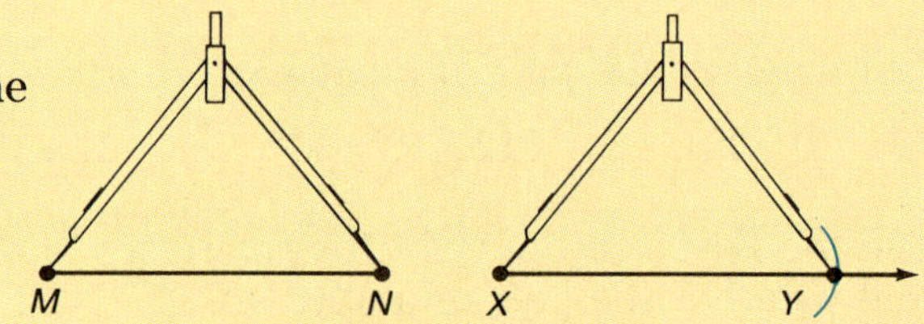

$\overline{XY}$ has the same length as $\overline{MN}$. $\overline{XY} \cong \overline{MN}$

EXAMPLE 2

Construct an angle congruent to $\angle BAC$.

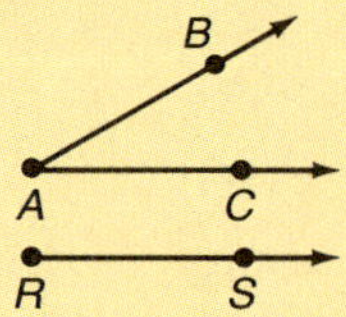

1. Draw $\overrightarrow{RS}$.
2. Draw an arc that intersects both sides of $\angle BAC$. Label these points J and K. Using the same compass opening, place the compass point on R and draw an arc that intersects $\overrightarrow{RS}$. Label the point U.

3. Open the compass to measure the distance between J and K. Using the same compass opening, place the compass point on U and draw an arc that intersects the arc through U. Label this point T.

4. Draw $\overrightarrow{RT}$.

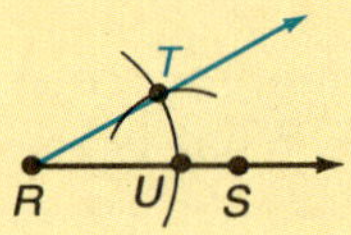

$\angle BAC$ has the same measure as $\angle TRS$. $\angle BAC \cong \angle TRS$

A geometric figure is **bisected** if it is separated into two congruent parts.

$m\angle XYW = 30°$ $m\angle WYZ = 30°$

$\angle XYW \cong \angle WYZ$

$\overrightarrow{YW}$ bisects $\angle XYZ$.

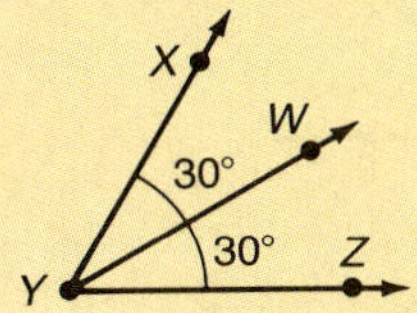

EXAMPLE 3

Construct the **angle bisector** of $\angle ABC$.

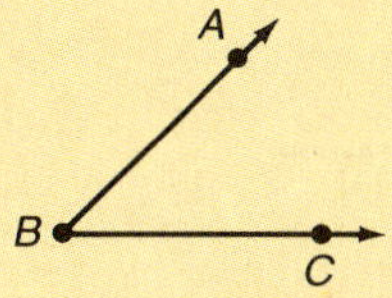

1. Draw an arc that intersects both sides of $\angle ABC$. Label these points D and E.

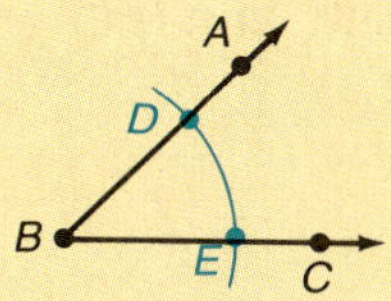

2. Place the compass point on D. Draw an arc in the interior of the angle. Use the same compass opening. Place the compass point on E and draw an arc that intersects the first arc. Label the intersection F.

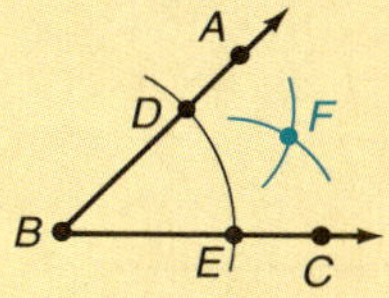

3. Draw $\overrightarrow{BF}$.

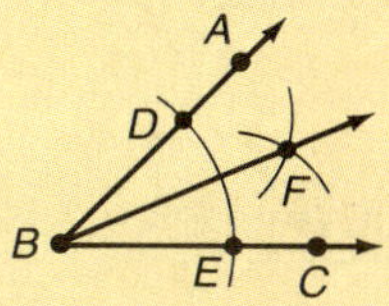

$\angle ABF \cong \angle FBC$. So, $\overrightarrow{BF}$ is the angle bisector of $\angle ABC$.

FOR DISCUSSION

1. Can two angles be congruent even though the drawn sides of one angle are longer than the drawn sides of the other angle? Explain.

2. Is it possible for an angle to have two different bisectors? Explain.

PRACTICE EXERCISES

Trace the line segment and then construct a congruent line segment.

1.

2.

3.

4.

Trace the angle and then construct a congruent angle.

5. 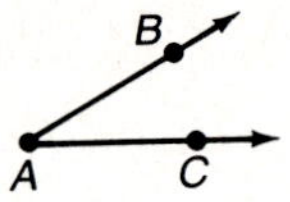

6. 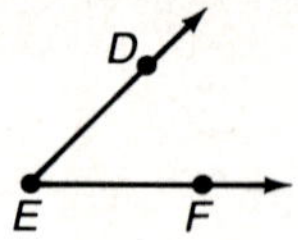

7. 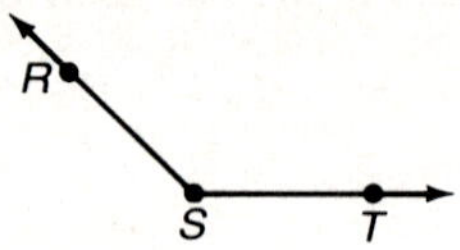

8. 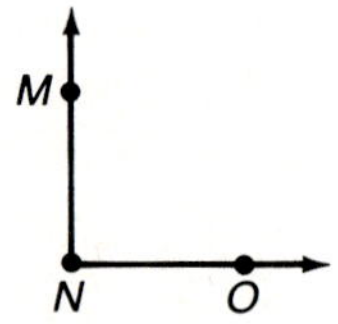

9. 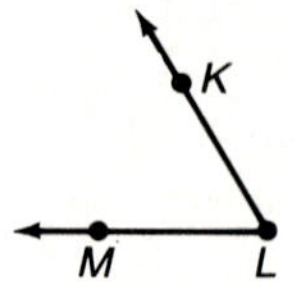

10. 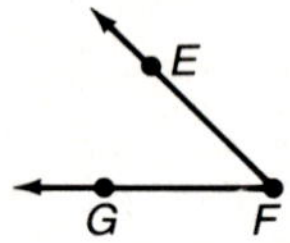

Trace the angle and then construct the angle bisector.

11. 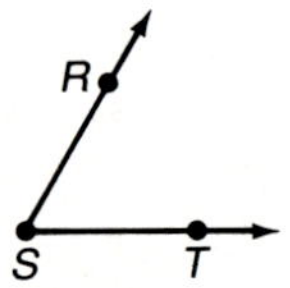

12. 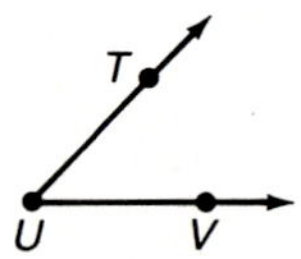

13. 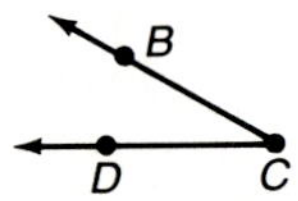

14. 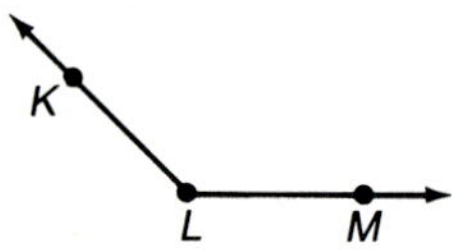

15. 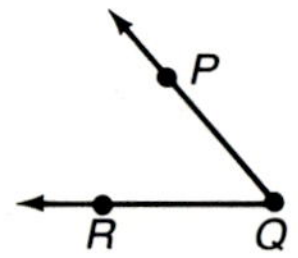

16. 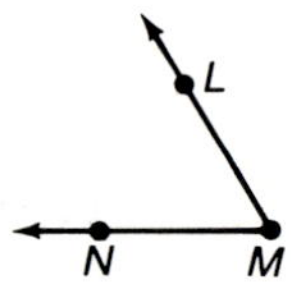

MIXED REVIEW

Complete.

17. 3 ft = ■ in. **18.** 7 qt = ■ pt **19.** 32 oz = ■ lb **20.** 28 qt = ■ gal

21. $8\frac{1}{3}$ yd = ■ ft **22.** 144 in. = ■ yd **23.** $3\frac{1}{2}$ h = ■ min **24.** 75 in. = ■ ft

25. 4.2 m = ■ cm **26.** 0.8 L = ■ mL **27.** 243 g = ■ kg **28.** 72 mm = ■ cm

29. 9.1 mm = ■ m **30.** 83.4 kg = ■ g **31.** 800 m = ■ km **32.** 750 mL = ■ L

Solve.

33. Use a ruler to draw a 4-cm line segment. Label it $\overline{AB}$. Construct $\overline{CD}$ so that $\overline{CD}$ is twice as long as $\overline{AB}$.

34. Use a protractor to draw a 40° angle. Label it $\angle MNO$. Construct $\angle XYZ$ so that m$\angle XYZ$ is twice the m$\angle MNO$.

TIME OUT

Draw a 3-in. line segment and label it $\overline{AB}$. Draw a 30° angle and label it $\angle A$. Draw a 45° angle and label it $\angle B$. Draw a triangle using the two angles and the line segment.

11.5 CONSTRUCTING PERPENDICULAR AND PARALLEL LINES

Two lines in a plane may or may not intersect. If the lines do not intersect, they are **parallel.** If the lines intersect to form right angles, they are **perpendicular.**

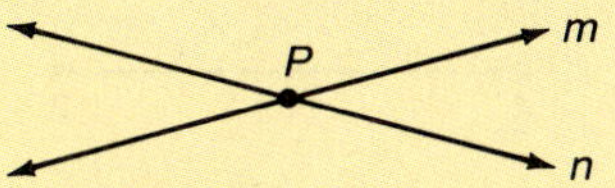

Intersecting lines
m and n intersect at P

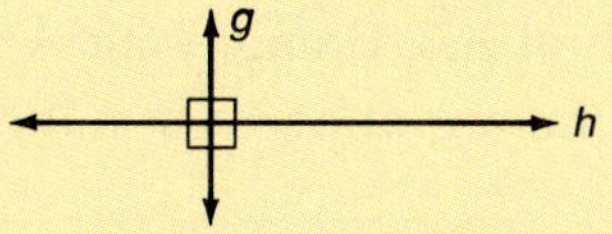

Perpendicular lines
$g \perp h$ and $h \perp g$

Parallel lines
$r \parallel s$ and $s \parallel r$

EXAMPLE 1 Construct a line perpendicular to $\overleftrightarrow{AB}$ through point P on the line.

1. Using the point P as a center, draw an arc that intersects $\overleftrightarrow{AB}$ at points R and S.
2. Open the compass to a distance greater than half the length of $\overline{RS}$. Using points R and S as centers, draw two arcs. Label the point of intersection C.
3. Draw $\overleftrightarrow{CP}$. $\overleftrightarrow{CP}$ is perpendicular to $\overleftrightarrow{AB}$ at P. This is written as $\overleftrightarrow{CP} \perp \overleftrightarrow{AB}$ at P.

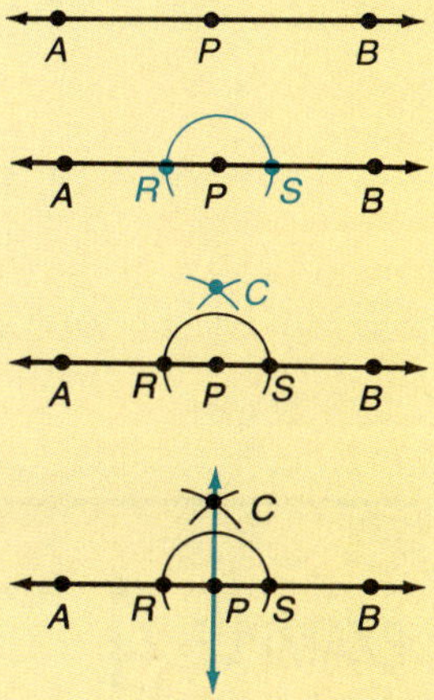

EXAMPLE 2 Construct a line perpendicular to $\overleftrightarrow{AB}$ through point P not on the line.

1. Using point P as a center, draw an arc that intersects $\overleftrightarrow{AB}$ at points R and S.
2. Open the compass to a distance greater than half the length of $\overline{RS}$. Using points R and S as centers, draw two arcs. Label the point of intersection C.
3. Draw $\overleftrightarrow{PC}$. Label the point of intersection D. $\overleftrightarrow{PC}$ is perpendicular to $\overleftrightarrow{AB}$ at D. This is written as $\overleftrightarrow{PC} \perp \overleftrightarrow{AB}$ at D.

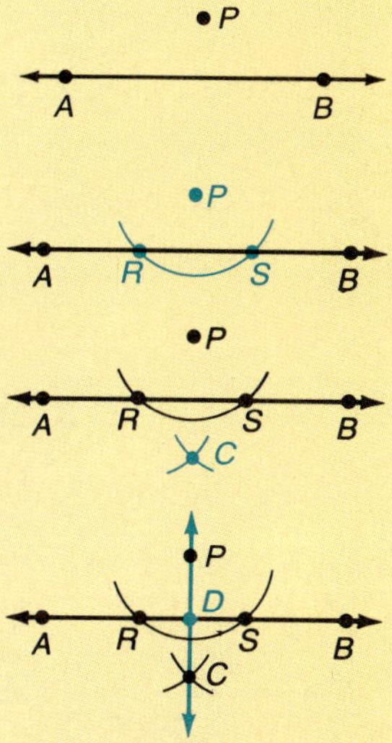

The midpoint of a line segment divides it into two congruent parts. A **perpendicular bisector** is a line perpendicular to a line segment that intersects it at the midpoint.

EXAMPLE 3

Construct the perpendicular bisector of $\overline{AB}$.

1. Open the compass to a distance greater than half the length of $\overline{AB}$. Using point A as a center, draw one arc above $\overline{AB}$ and one arc below $\overline{AB}$.

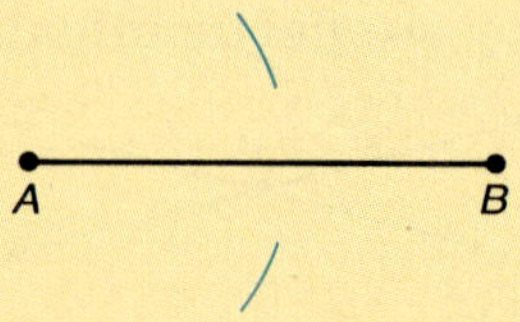

2. Use the same compass opening. Using point B as a center, draw one arc above $\overline{AB}$ and one arc below $\overline{AB}$. Label the points of intersection C and D.

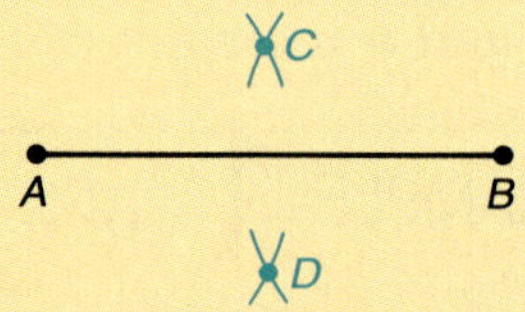

3. Draw $\overleftrightarrow{CD}$. Label the point where $\overleftrightarrow{CD}$ intersects $\overline{AB}$ as M, the midpoint. This is written as $\overleftrightarrow{CD} \perp \overline{AB}$ at M.

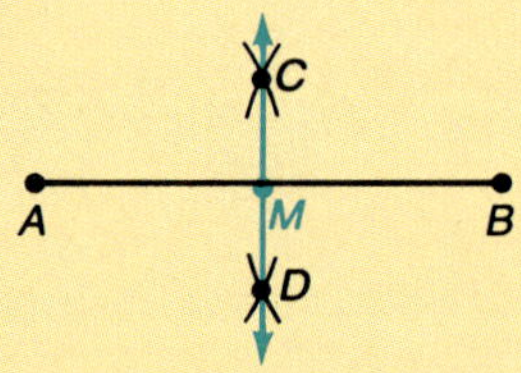

EXAMPLE 4

Given $\overleftrightarrow{AB}$, construct a line parallel to $\overleftrightarrow{AB}$.

1. Mark a point C on $\overleftrightarrow{AB}$.

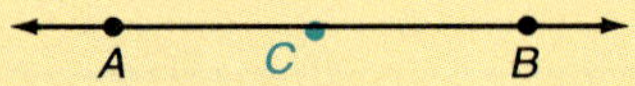

2. Construct $\overleftrightarrow{DC} \perp \overleftrightarrow{AB}$.

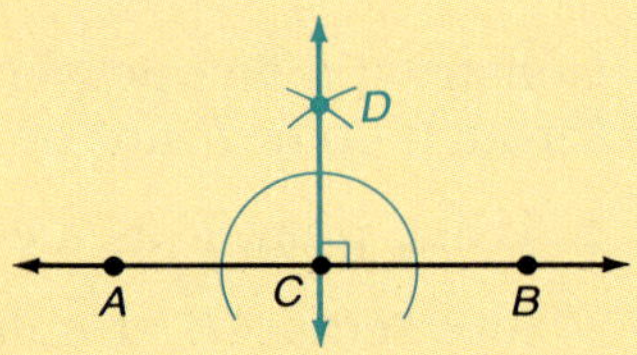

3. Mark a point E on $\overleftrightarrow{DC}$. Construct $\overleftrightarrow{EF} \perp \overleftrightarrow{DC}$. $\overleftrightarrow{EF}$ is parallel to $\overleftrightarrow{AB}$. This is written as $\overleftrightarrow{EF} \parallel \overleftrightarrow{AB}$.

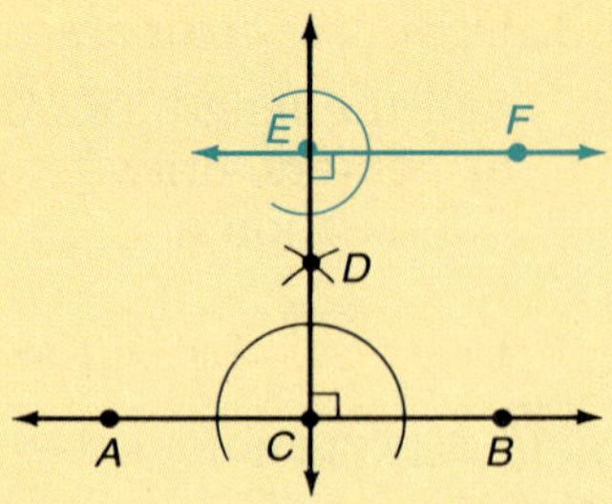

FOR DISCUSSION

1. Can a perpendicular bisector of a line be constructed? Explain.

2. How can you divide a line segment into eight congruent line segments?

PRACTICE EXERCISES

Trace the line and then construct a line perpendicular to it through the point on the line.

1. 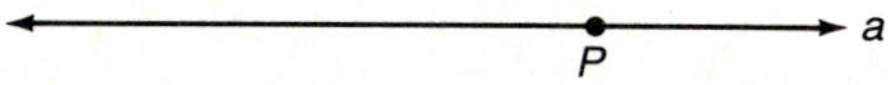

2. 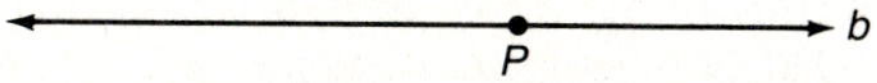

3. 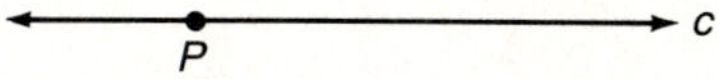

4. 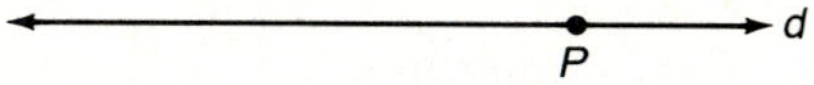

5. Draw $\overleftrightarrow{AB}$. Select a point C not on the line. Construct a line perpendicular to $\overleftrightarrow{AB}$ through point C.

Trace the line segment and then construct the perpendicular bisector.

6. 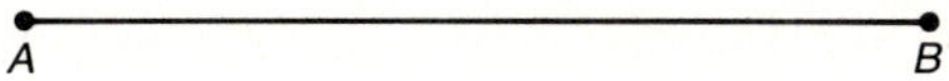

7. 

8. 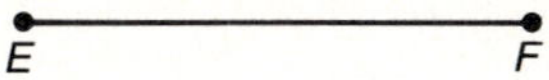

9. 

Trace the line and construct a line parallel to it.

10. 

11. 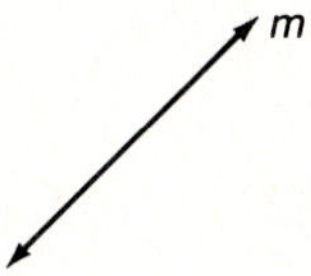

12. 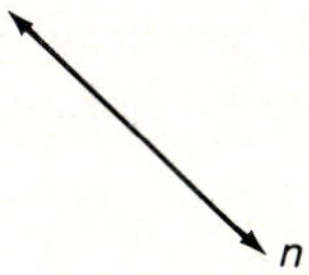

Solve.

13. Draw a 4-in. line segment. Label it $\overline{AB}$. Use a compass and straightedge to bisect $\overline{AB}$ at point C. Bisect $\overline{AC}$ and $\overline{CB}$ at points E and F. The length of $\overline{AE}$ is what fractional part of the length of $\overline{AB}$?

14. Draw a 6-in. line segment. Label it $\overline{XY}$. Use a compass and straightedge to bisect $\overline{XY}$ at point Z. Bisect $\overline{ZY}$ at point P. What is the length in inches of $\overline{XP}$?

11.6 POLYGONS AND CIRCLES

A **polygon** is a simple closed curve consisting entirely of line segments. This is polygon *ABCD*. It is named by its four vertices. $\overline{AC}$ is a diagonal.

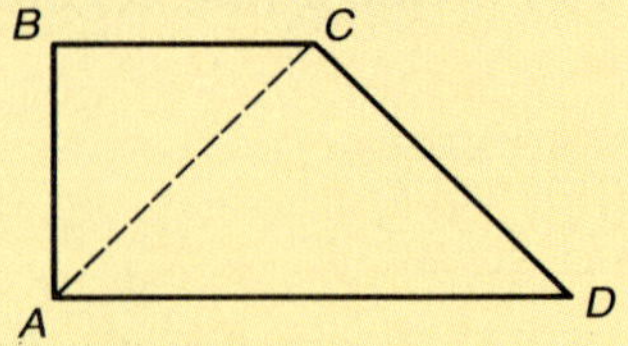

A **regular polygon** has congruent sides and congruent angles. Polygon *EFGHIJ* is a regular polygon.

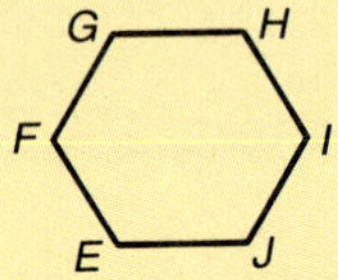

Here are some polygons.

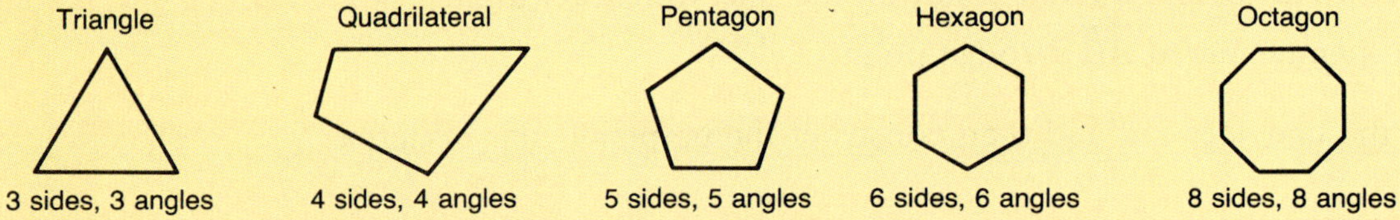

EXAMPLE 1

Name the polygon.

a.

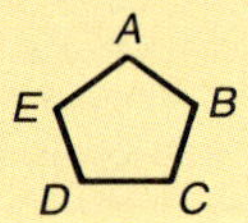

THINK: 5 sides
Vertices: *A, B, C, D, E*

Pentagon *ABCDE*

b.

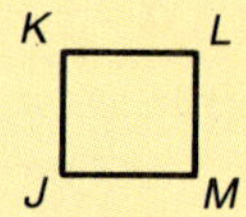

THINK: 4 sides
Vertices: *J, K, L, M*

Quadrilateral *JKLM*

A **circle** is the set of all points in a plane that are the same distance from the center. In this circle:

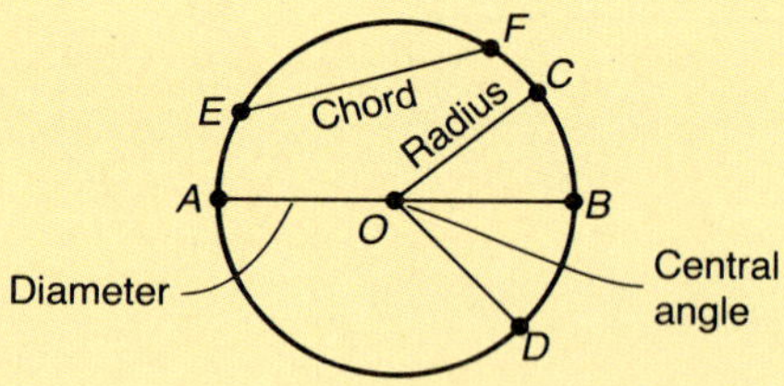

O is the **center.**

$\overline{EF}$ is a chord. A **chord** is a line segment that has endpoints on the circle.

$\overline{AB}$ is a diameter. A **diameter** is a chord that passes through the center of a circle.

$\overline{OC}$ is a radius. A **radius** is a line segment that goes from the center of the circle to a point on it.

$\angle BOD$ is a central angle. A **central angle** is an angle whose vertex is at the center of the circle.

You can find the length of either the diameter or radius of a circle if you know the length of the other.

EXAMPLE 2 Find the missing length.

a. $r = 5.2$ cm, $d = ■$

THINK: $d = 2 \times r$
$= 2 \times 5.2$
$= 10.4$

The diameter is 10.4 cm.

b. $d = 3$ in., $r = ■$

THINK: $r = \frac{1}{2} \times d$
$= \frac{1}{2} \times 3$
$= 1\frac{1}{2}$

The radius is $1\frac{1}{2}$ in.

PRACTICE EXERCISES

Name the polygon.

1. 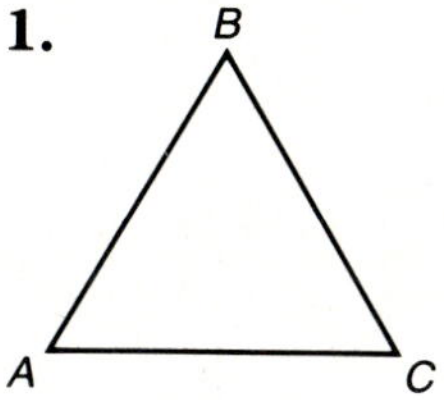

2. 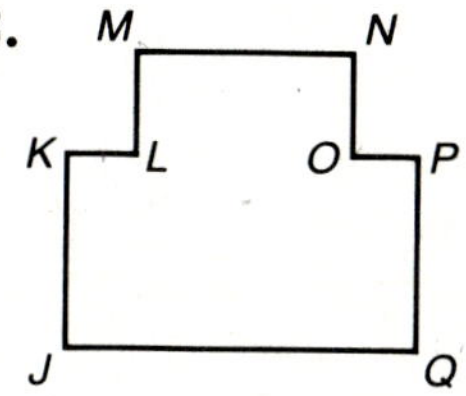

3. 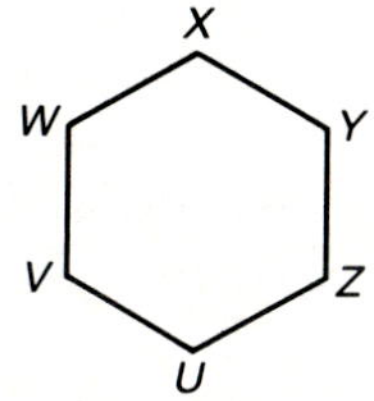

4. 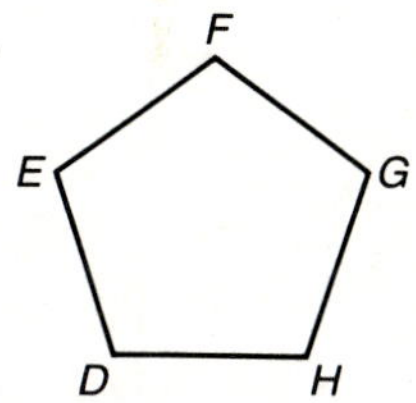

Use the figure to name the part of the circle.

5. $\overline{BD}$ **6.** $\overline{AB}$ **7.** $\overline{OC}$ **8.** $\angle COD$

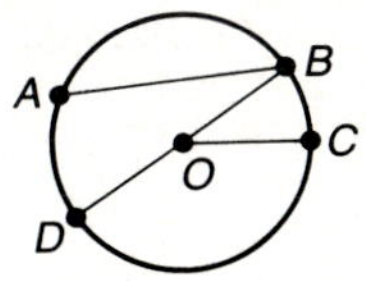

Find the missing length.

9. $r = 5$ cm, $d = ■$
10. $r = 15$ in., $d = ■$
11. $r = 3.5$ cm, $d = ■$
12. $r = 12.4$ cm, $d = ■$
13. $r = 8\frac{1}{2}$ ft, $d = ■$
14. $r = 4\frac{1}{4}$ in., $d = ■$
15. $d = 8$ ft, $r = ■$
16. $d = 58$ mm, $r = ■$
17. $d = 4.8$ cm, $r = ■$
18. $d = 5$ in., $r = ■$
19. $d = 1.3$ m, $r = ■$
20. $d = 12\frac{1}{2}$ ft, $r = ■$

Solve.

21. The schoolyard is in the shape of a quadrilateral. How many sides does the school yard have?

22. The diameter of a circle is 18 in. What is the length of the radius?

23. One side of a regular octagon measures 2 cm. What is the measure of each of the other sides?

24. The radius of circle A is 3 ft. The radius of circle B is twice as long. Find the length of the diameter of circle B.

11.7 TRIANGLES

Triangles may be classified by the number of congruent sides.

Equilateral triangle	Isosceles triangle	Scalene triangle
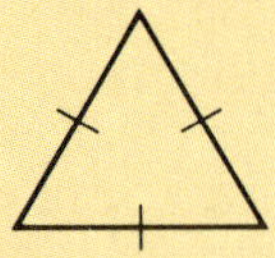	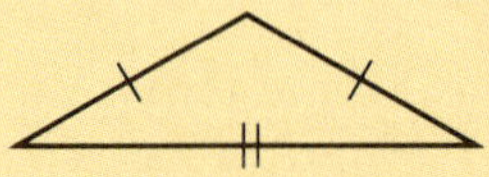	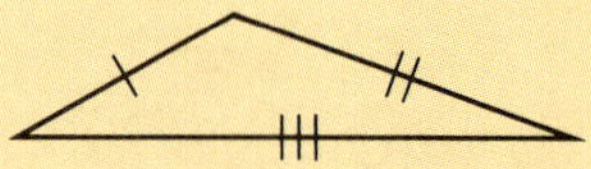
3 congruent sides, 3 congruent angles	2 congruent sides, 2 congruent angles	no congruent sides, no congruent angles

EXAMPLE 1 Identify the triangle as *equilateral, isosceles,* or *scalene.*

a.

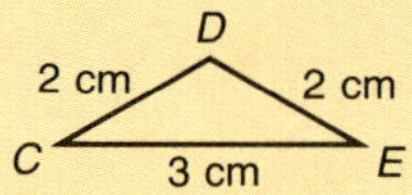

THINK: 2 sides congruent

So, $\triangle CDE$ is an isosceles triangle.

b.

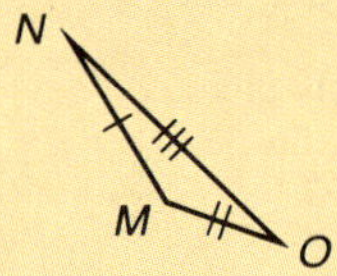

THINK: no sides congruent

So, $\triangle MNO$ is a scalene triangle.

Triangles may also be classified by their angles.

Equiangular triangle	Acute triangle	Obtuse triangle	Right triangle
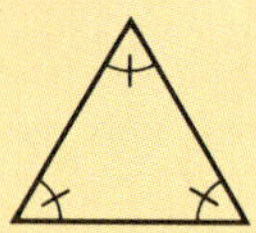	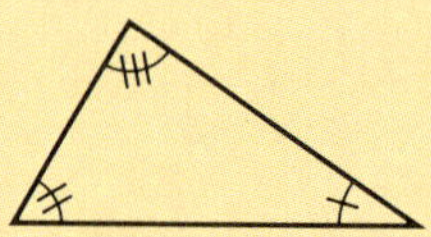	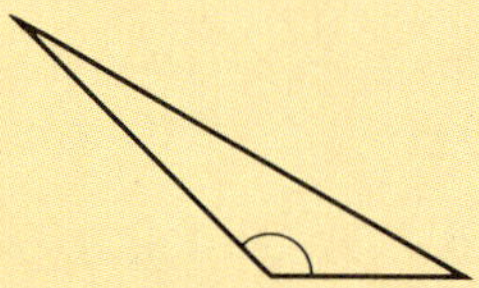	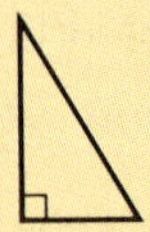
all congruent angles	all acute angles	1 obtuse angle	1 right angle

EXAMPLE 2 Identify the triangle as *equiangular, acute, obtuse,* or *right.*

a.

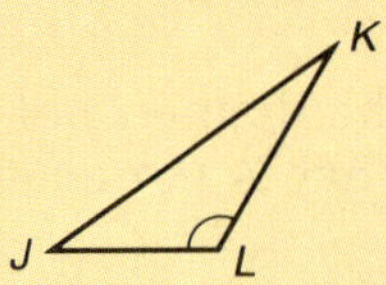

THINK: 1 obtuse angle

So, $\triangle JKL$ is an obtuse triangle.

b.

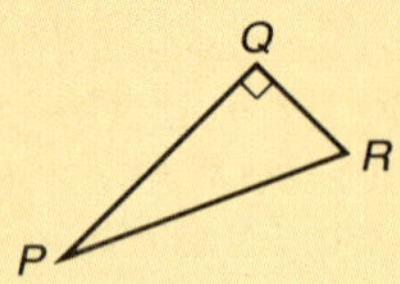

THINK: 1 right angle

So, $\triangle PQR$ is a right triangle.

In all triangles, the sum of the measures of the three angles is 180°. If you know the measure of two angles of a triangle, you can find the measure of the third angle.

EXAMPLE 3 In $\triangle ABC$, $m\angle A = 65°$ and $m\angle B = 35°$. Find $m\angle C$.

1. Add the measures of the two known angles. $65° + 35° = 100°$
2. Subtract the sum from 180°. $180° - 100° = 80°$

So, $m\angle C = 80°$.

PRACTICE EXERCISES

Identify the triangle as *equilateral*, *isosceles*, or *scalene*.

1.
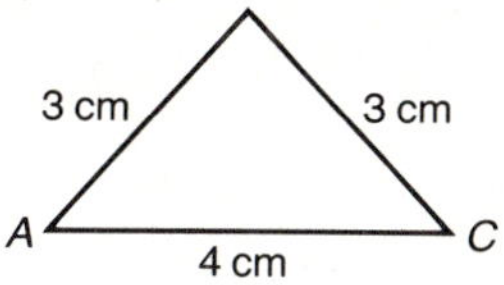

2.
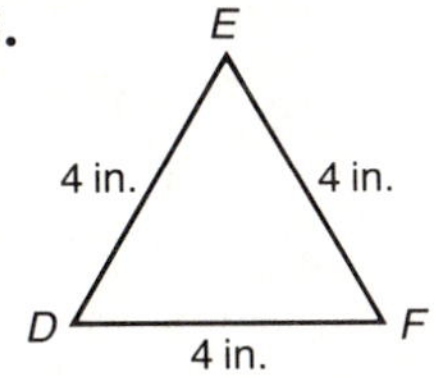

3.
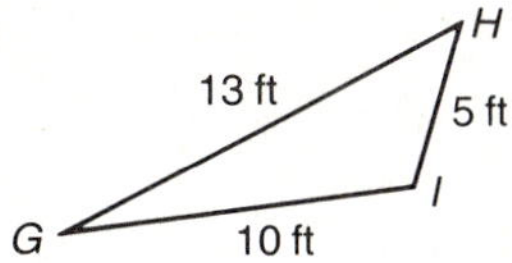

4.
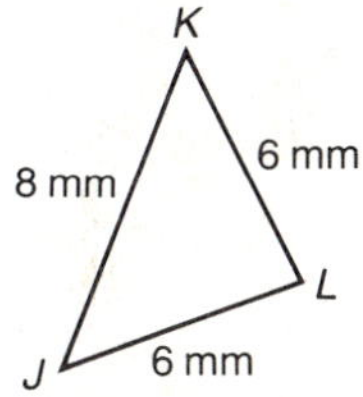

5.
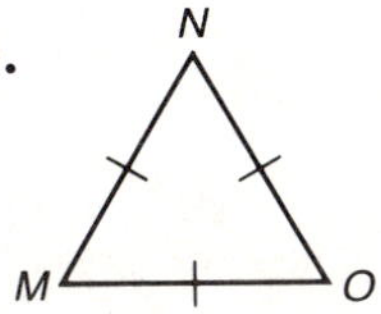

6.
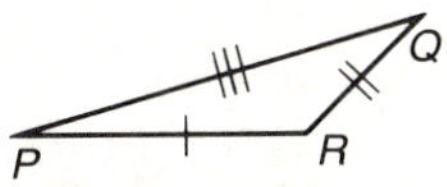

7.
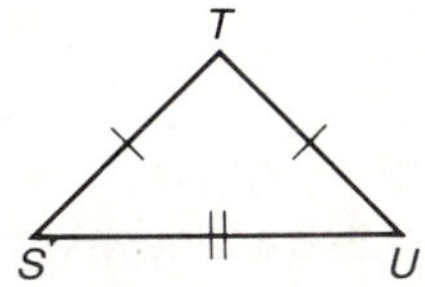

8.
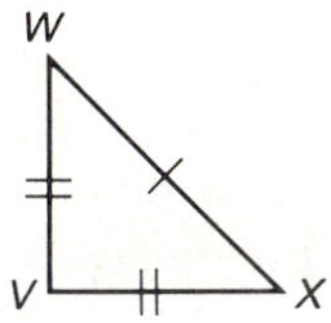

Identify the triangle as equiangular, acute, obtuse, or right.

9.
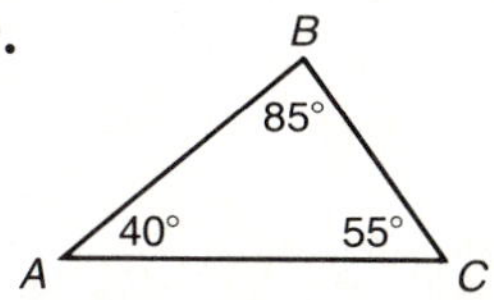

10.
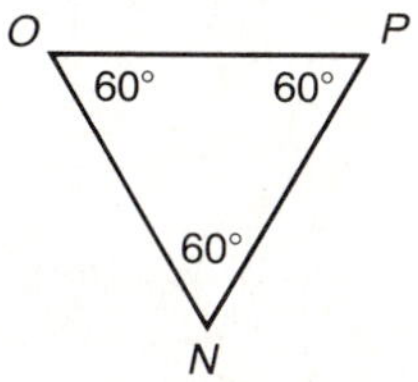

11.
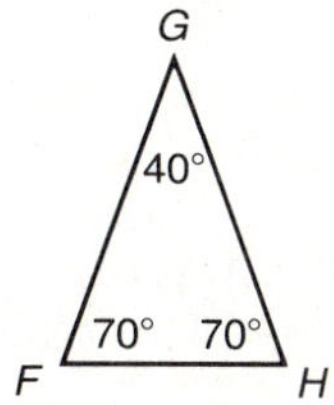

12.
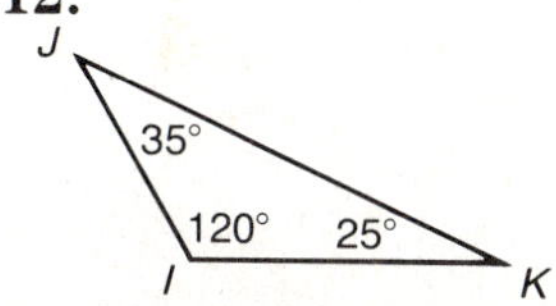

13.
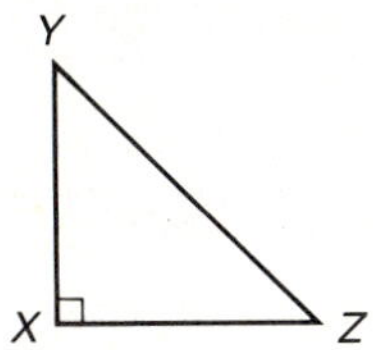

14.
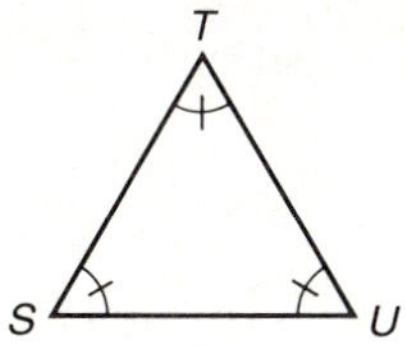

15.
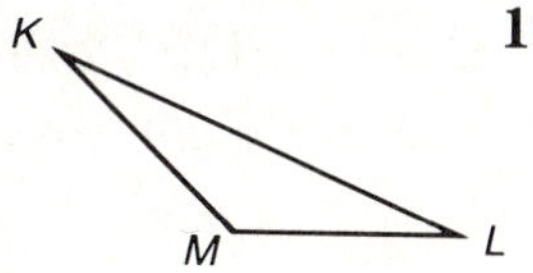

16.
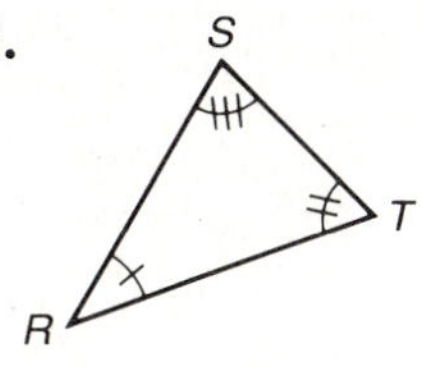

Draw the triangle.

17. Equilateral triangle

18. Acute isosceles triangle

19. Obtuse isosceles triangle

20. Right isosceles triangle

21. Obtuse scalene triangle

22. Acute scalene triangle

The measures of two angles of a triangle are given.
Find the measure of the third angle.

23.

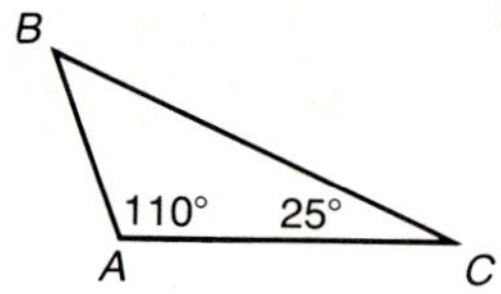

24.

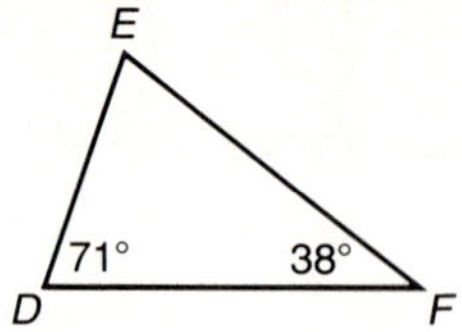

25.

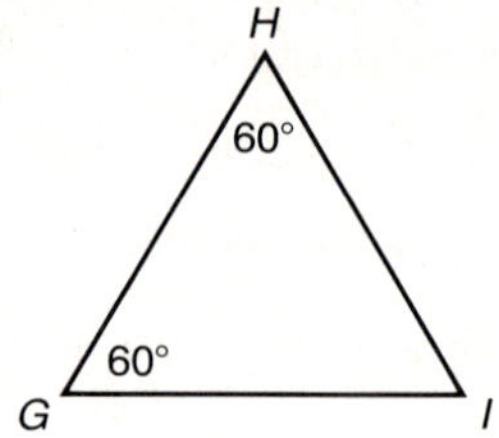

26.

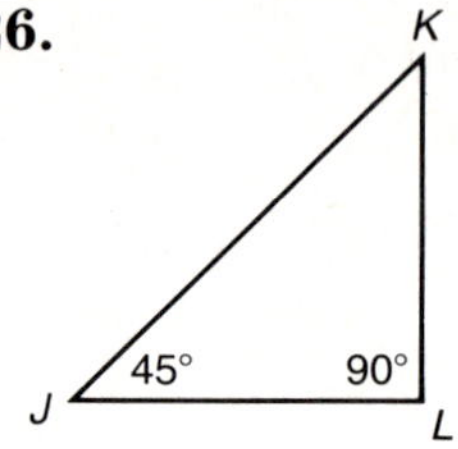

27. 40°, 75° **28.** 120, 45° **29.** 45°, 45° **30.** 110°, 23°

31. 81°, 92° **32.** 100°, 13° **33.** 51°, 113° **34.** 90°, 10°

MIXED REVIEW

Find the number.

35. 28% of 15 **36.** 15% of 50 **37.** 32% of 90 **38.** 68% of 64

39. 8 is what percent of 32?

40. 44 is what percent of 80?

41. 25 is 12.5% of what number?

42. 9.9 is 10% of what number?

Solve.

43. One of the congruent angles of an isosceles triangle is 35°. Find the measures of the other two angles.

44. One angle of an isosceles triangle is 120°. Find the measures of the other two congruent angles.

45. The radius of a circle is 3.2 cm. What is the length of the diameter?

46. How many angles and sides does a regular pentagon have?

TIME OUT

The following activity can be performed to find the sum of the measures of the triangle.

Draw any triangle. Cut out the triangle, and tear off the three angles.

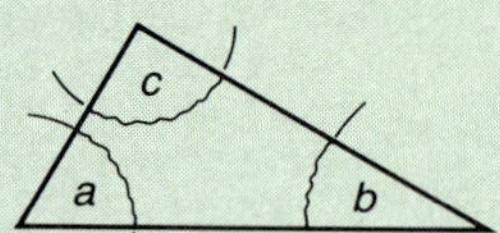

Put the angles together to form a straight angle as in the figure. Remember, a straight angle measures 180°.

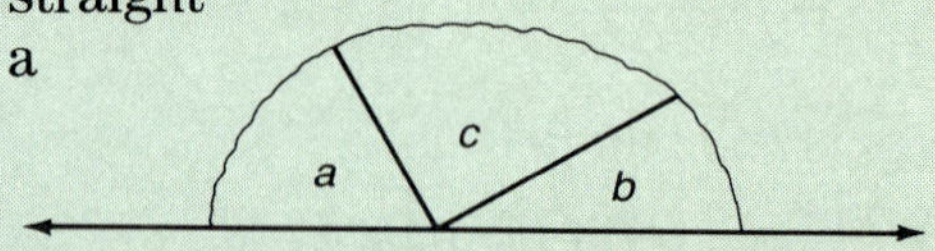

So, the sum of the measures of the angles of a triangle is 180°.

MIDCHAPTER REVIEW

Name the geometric figure. [292]

1. $\angle T$ **2.** $\overline{BC}$ **3.** $\overrightarrow{AB}$ **4.** $.P$ **5.** $\overleftrightarrow{EF}$

Use the figure at the right for Exercises 6–19.

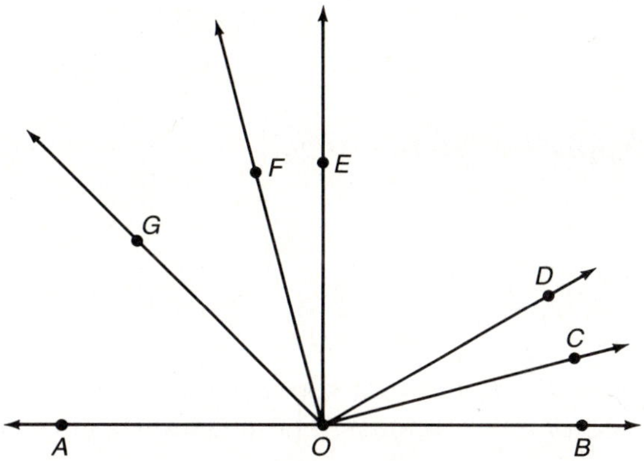

Measure the angle. [294]

6. $\angle AOC$ **7.** $\angle AOE$ **8.** $\angle AOG$

9. $\angle GOF$ **10.** $\angle GOC$ **11.** $\angle GOB$

Identify the angle as *acute, right, obtuse,* or *straight.* [296]

12. $\angle AOC$ **13.** $\angle GOE$ **14.** $\angle AOD$

15. $\angle AOE$ **16.** $\angle BOA$ **17.** $\angle GOC$

Tell whether the pair of angles are *complementary, supplementary,* or *neither.* [296]

18. $\angle AOB$ and $\angle BOD$ **19.** $\angle AOC$ and $\angle COB$

Trace the line segment.

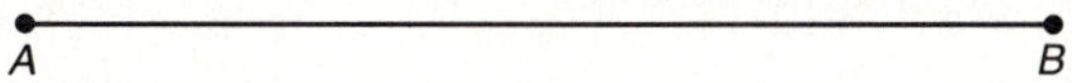

20. Construct a line segment congruent to $\overline{AB}$. [298]

21. Construct a line parallel to $\overline{AB}$. [301]

Identify the triangle as *equilateral, isosceles,* or *scalene.* [306]

22.

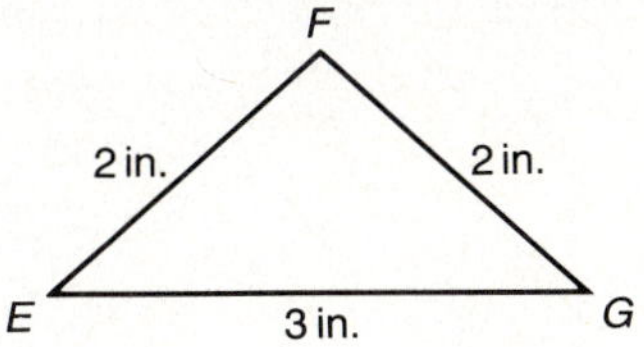

23.

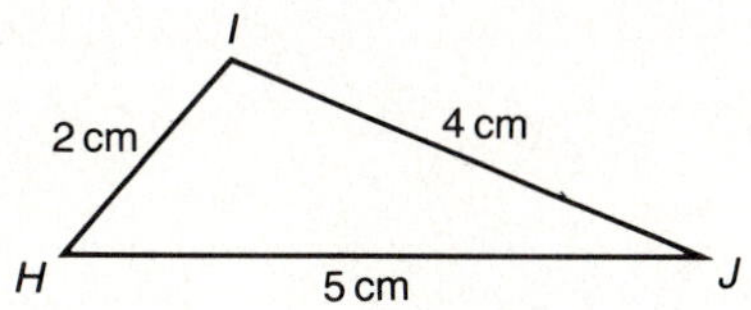

24.

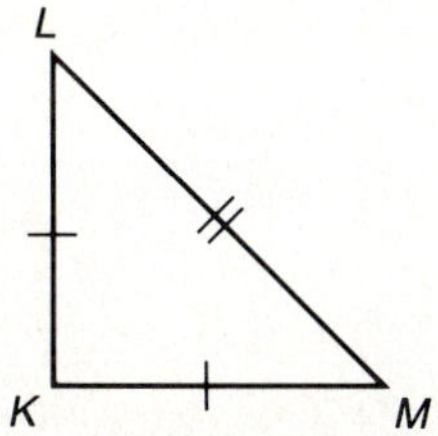

25.

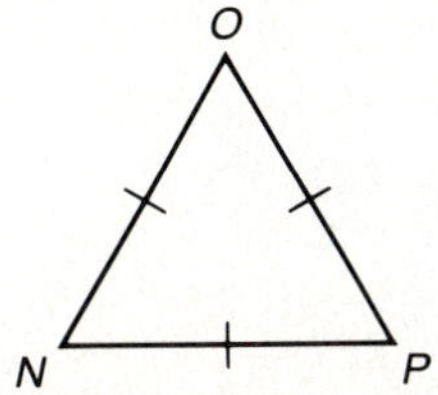

PROBLEM Solving STRATEGY

11.8 LOOKING FOR A PATTERN

Situation:

Fred and Lynne Shaw own a landscape company. They have planned a border of shrubs. The first shrub is planted 4 ft from the road. The second shrub is 9 ft from the road. The third shrub is 16 ft from the road. The next plant is 20 ft from the road, followed by another plant 25 ft from the road. How many feet from the road are the next two shrubs?

Strategy:

Finding a pattern can often help you to solve a problem.

Applying the Strategy:

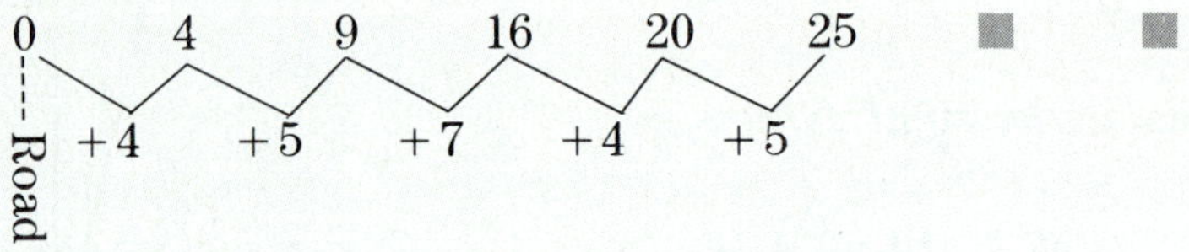

By adding, you can see a pattern: $+4$, $+5$, $+7$, $+4$, $+5$.
The pattern repeats. The next shrub is $+7$ ft away from the last plant.

So, the next shrub is $25 + 7 = 32$ ft from the road.
The following shrub is $32 + 4 = 36$ ft from the road.

PRACTICE EXERCISES

Alex Cook has planned a display around a fountain. The first ring has 16 flowers. The second ring has 32 flowers. The third ring has 64 flowers. How many flowers should be planted in the next two rings?

1. What is the pattern?

2. Complete the pattern:
16, 32, 64, ■, ■

Solve the problem by finding the pattern.

3. Jane doodles as she plans her display. One day this was the sketch on her drawing pad.

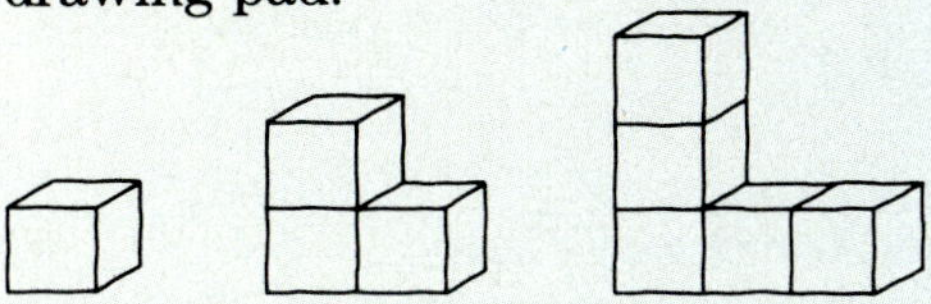

How many blocks should there be in the next picture to continue the pattern? Draw a possible configuration.

4. One day Fred found this drawing on Jane's drawing pad.

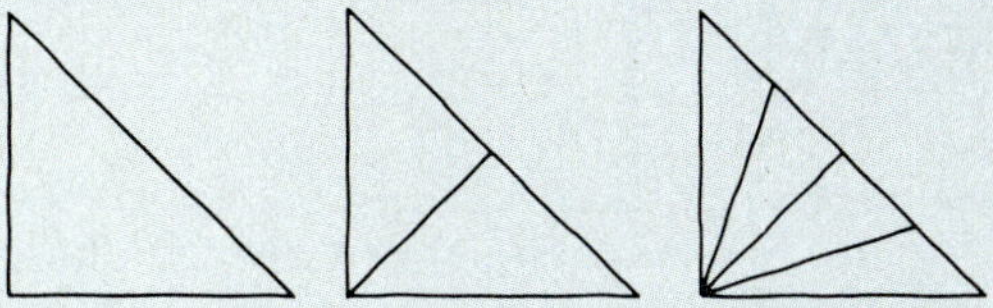

How many small triangles should be drawn in the next triangle to continue the pattern?

5. Marie was planting bushes. The first bush was 12 ft from the street. The second bush was twice the distance from the street. The third bush was 48 ft away from the street. How far from the street was the fourth bush?

6. James was placing rocks in a rock garden. The distances between the rocks were: 3 ft, 6 ft, 16 ft, 19 ft, 29 ft. What is the distance between the fifth and sixth rock? The sixth and seventh rock?

11.9 SQUARE ROOT

For the graduation ceremony, 400 chairs are arranged so that the number of chairs in each row is the same as the number of rows. How many chairs are in each row and how many rows are there?

To solve this problem you need to find the number that when multiplied by itself equals 400. That number is the **square root** of 400. The symbol $\sqrt{}$ means square root.

EXAMPLE 1 Find the square root of 400.

THINK: $2 \times 2 = 4$, so $20 \times 20 = 400$.

$\sqrt{400} = 20$

So, there are 20 chairs per row and 20 rows of chairs.

Sometimes the square root is not a whole number. You may need to use a table of squares and square roots like the one on page 313.

EXAMPLE 2 Find $\sqrt{15}$.

Read the table on page 313.

$\sqrt{15}$ is approximately equal to 3.873. $3.873 \approx 3.9$

You can use the column marked n^2 to find some square roots.

EXAMPLE 3 Find $\sqrt{529}$.

Read the table on page 313.

Look for 529 in the n^2 column and then find n.

THINK:

$$n^2 = 529$$
$$n = \sqrt{529}$$
$$n = 23$$

So, $\sqrt{529} = 23$.

TABLE OF SQUARES AND SQUARE ROOTS

n	n^2	$\sqrt{n}$	n	n^2	$\sqrt{n}$	n	n^2	$\sqrt{n}$	n	n^2	$\sqrt{n}$
1	1	1	26	676	5.099	51	2,601	7.141	76	5,776	8.718
2	4	1.414	27	729	5.196	52	2,704	7.211	77	5,929	8.775
3	9	1.732	28	784	5.292	53	2,809	7.280	78	6,084	8.832
4	16	2	29	841	5.385	54	2,916	7.348	79	6,241	8.888
5	25	2.236	30	900	5.477	55	3,025	7.416	80	6,400	8.944
6	36	2.449	31	961	5.568	56	3,136	7.483	81	6,561	9
7	49	2.646	32	1,024	5.657	57	3,249	7.550	82	6,724	9.055
8	64	2.828	33	1,089	5.745	58	3,364	7.616	83	6,889	9.110
9	81	3	34	1,156	5.831	59	3,481	7.681	84	7,056	9.165
10	100	3.162	35	1,225	5.916	60	3,600	7.746	85	7,225	9.220
11	121	3.317	36	1,296	6	61	3,721	7.810	86	7,396	9.274
12	144	3.464	37	1,369	6.083	62	3,844	7.874	87	7,569	9.327
13	169	3.606	38	1,444	6.164	63	3,969	7.937	88	7,744	9.381
14	196	3.742	39	1,521	6.245	64	4,096	8	89	7,921	9.434
15	225	3.873	40	1,600	6.325	65	4,225	8.062	90	8,100	9.487
16	256	4	41	1,681	6.403	66	4,356	8.124	91	8,281	9.539
17	289	4.123	42	1,764	6.481	67	4,489	8.185	92	8,464	9.592
18	324	4.243	43	1,849	6.557	68	4,624	8.246	93	8,649	9.644
19	361	4.359	44	1,936	6.633	69	4,761	8.307	94	8,836	9.695
20	400	4.472	45	2,025	6.708	70	4,900	8.367	95	9,025	9.747
21	441	4.583	46	2,116	6.782	71	5,041	8.426	96	9,216	9.798
22	484	4.690	47	2,209	6.856	72	5,184	8.485	97	9,409	9.849
23	529	4.796	48	2,304	6.928	73	5,329	8.544	98	9,604	9.899
24	576	4.899	49	2,401	7	74	5,476	8.602	99	9,801	9.950
25	625	5	50	2,500	7.071	75	5,625	8.660	100	10,000	10

PRACTICE EXERCISES

Find the square root.

1. $\sqrt{4}$ **2.** $\sqrt{9}$ **3.** $\sqrt{25}$ **4.** $\sqrt{64}$ **5.** $\sqrt{81}$ **6.** $\sqrt{100}$

7. $\sqrt{10}$ **8.** $\sqrt{37}$ **9.** $\sqrt{58}$ **10.** $\sqrt{72}$ **11.** $\sqrt{87}$ **12.** $\sqrt{95}$

Find the square root. Round to the nearest tenth.

13. $\sqrt{15}$ **14.** $\sqrt{18}$ **15.** $\sqrt{29}$ **16.** $\sqrt{35}$ **17.** $\sqrt{37}$ **18.** $\sqrt{45}$

19. $\sqrt{50}$ **20.** $\sqrt{55}$ **21.** $\sqrt{63}$ **22.** $\sqrt{78}$ **23.** $\sqrt{89}$ **24.** $\sqrt{94}$

Find the square root. Use the table.

25. $\sqrt{196}$ **26.** $\sqrt{576}$ **27.** $\sqrt{1,225}$ **28.** $\sqrt{1,521}$ **29.** $\sqrt{1,681}$

30. $\sqrt{2,304}$ **31.** $\sqrt{3,249}$ **32.** $\sqrt{4,225}$ **33.** $\sqrt{6,241}$ **34.** $\sqrt{9,216}$

11.10 THE PYTHAGOREAN THEOREM

Triangle ABC is a right triangle.

$\angle C$ is the right angle.

The side opposite the right angle is called the **hypotenuse.**

In this triangle, side c is the hypotenuse. Sides a and b are **legs.**

There is a relationship between the hypotenuse and the legs of a right triangle.

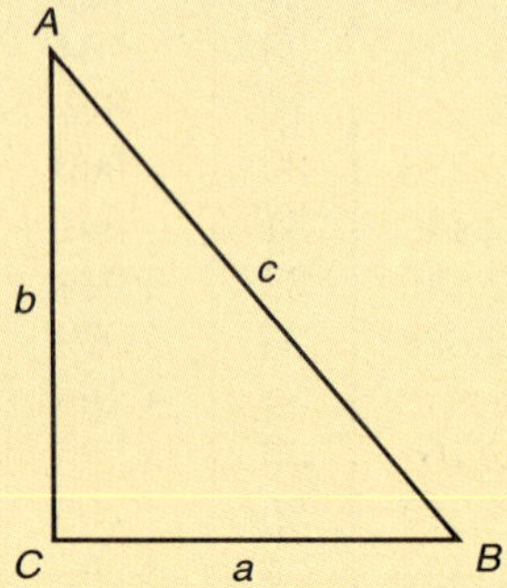

Square the length of each side.

$a^2 = 3 \times 3 = 9$

$b^2 = 4 \times 4 = 16$

$c^2 = 5 \times 5 = 25$

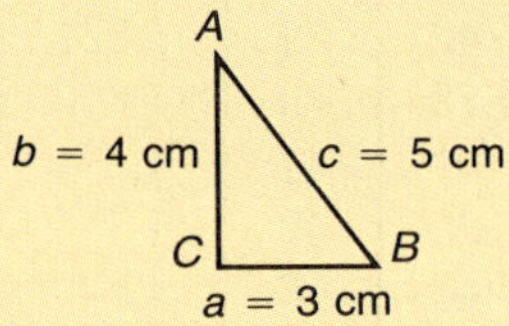

Notice that the square of the length of the hypotenuse is equal to the sum of the squares of the lengths of the legs.

$$\begin{array}{ccccc} 25 & = & 9 & + & 16 \\ \downarrow & & \downarrow & & \downarrow \\ c^2 & = & a^2 & + & b^2 \end{array}$$

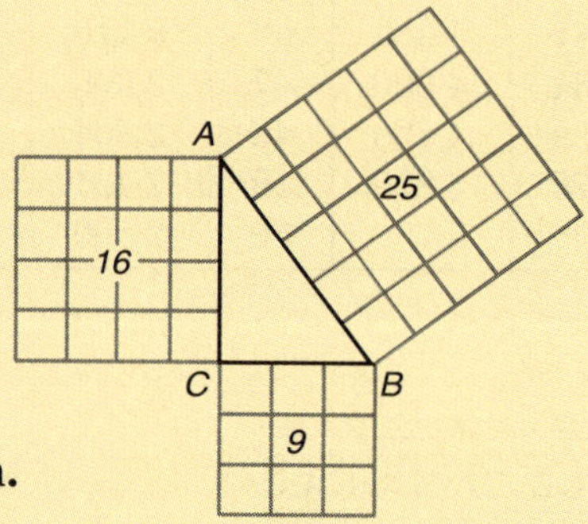

This relationship is called the **Pythagorean theorem.**
It is true for every right triangle.
For right triangle ABC,

$$c^2 = a^2 + b^2 \text{ or } a^2 + b^2 = c^2.$$

Use the Pythagorean theorem to determine if a triangle is a right triangle.

EXAMPLE 1 In $\triangle ABC$, $a = 10$, $b = 12$, and $c = 15$.
Is $\triangle ABC$ a right triangle?

THINK: $c^2 = a^2 + b^2$

$15^2 \stackrel{?}{=} 10^2 + 12^2$

$225 \stackrel{?}{=} 100 + 144$

$225 \neq 244$

The triangle is not a right triangle.

You can also use the Pythagorean theorem to find the length of the hypotenuse or a leg of a right triangle.

EXAMPLE 2 $\triangle ABC$ is a right triangle. Find the missing length.

a. $a = 6, b = 8, c = ■$

THINK: $c^2 = a^2 + b^2$

$c^2 = 6^2 + 8^2$

$c^2 = 36 + 64$

$c^2 = 100$

$c = \sqrt{100}$, or 10

So, the hypotenuse c is equal to 10.

b. $c = 13, b = 12, a = ■$

THINK: $a^2 + b^2 = c^2$

$a^2 + 12^2 = 13^2$

$a^2 + 144 = 169$

$a^2 = 169 - 144$

$a^2 = 25$

$a = \sqrt{25}$, or 5

So, the leg a is equal to 5.

PRACTICE EXERCISES

Do sides of the following lengths form a right triangle? Write *yes* or *no*.

1. $a = 6$, $b = 8$, $c = 10$ **2.** $a = 0$, $b = 12$, $c = 15$ **3.** $a = 24$, $b = 15$, $c = 17$ **4.** $a = 5$, $b = 12$, $c = 13$ **5.** $a = 7$, $b = 10$, $c = 12$

Find the missing length.

6.
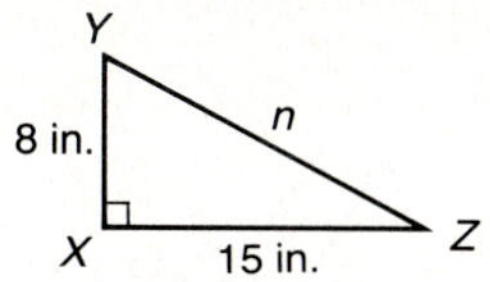

7.
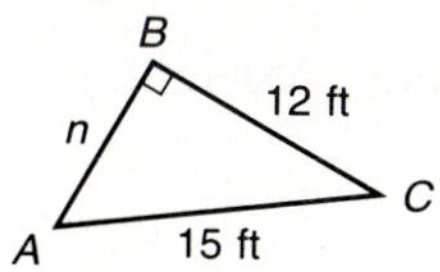

8.
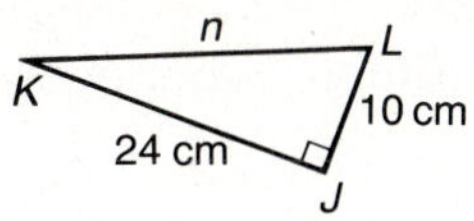

9.
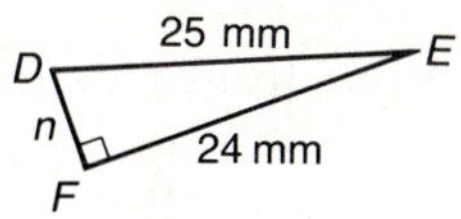

10. $a = 15$, $b = 20$, $c = ■$ **11.** $a = 9$, $b = ■$, $c = 41$ **12.** $a = ■$, $b = 48$, $c = 52$ **13.** $a = 28$, $b = 45$, $c = ■$ **14.** $a = 16$, $b = ■$, $c = 34$

MIXED REVIEW

Multiply.

15. $\frac{1}{2} \times 4$ **16.** $\frac{1}{4} \times 16$ **17.** $4\frac{1}{2} \times \frac{1}{3}$ **18.** $3\frac{5}{6} \times \frac{2}{5}$

19. $\frac{1}{2} \times 4\frac{3}{8}$ **20.** $\frac{1}{15} \times \frac{30}{48}$ **21.** $\frac{1}{8} \times 6\frac{2}{3}$ **22.** $\frac{1}{7} \times 5\frac{4}{9}$

Solve.

23. The foot of a ladder is 10 ft from the base of a building. The top of the ladder is resting against the building, 24 ft above the ground. How long is the ladder?

PROBLEM Solving APPLICATION

11.11 CAREER: SURVEYOR

Janet Soldana is a surveyor. By measuring angles and certain distances that can be conveniently measured, she is able to calculate distances that cannot be directly measured.

Janet uses an instrument called a **transit** to measure angles. A transit is a telescope set up on a three-legged stand. The transit also includes two protractors—one for measuring angles in a horizontal plane and the other for measuring angles in a vertical plane.

There are two facts about right triangles that help Janet perform her job:

- In any right triangle with a 30° angle, the side opposite that angle is $\frac{1}{2}$ of the length of the side opposite the 90° angle.
- In any right triangle with a 37° angle, the side opposite that angle is about $\frac{3}{5}$ of the length of the side opposite the 90° angle.

Find *n*. Remember to estimate whenever you use your calculator.

1.

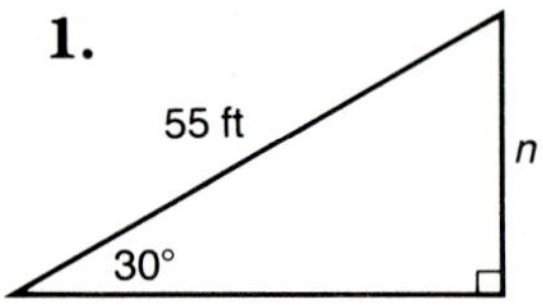

•2.

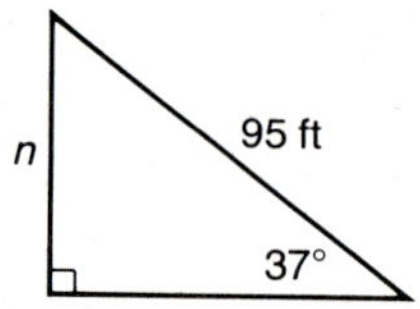

3.

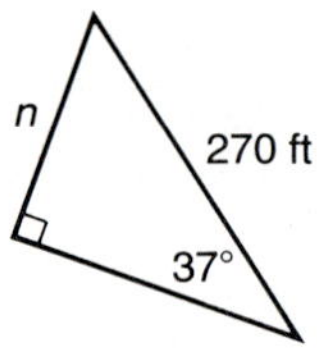

4.

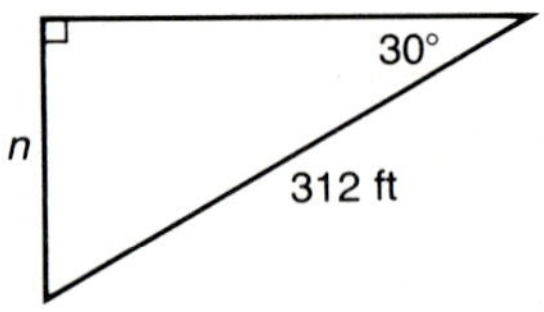

5.

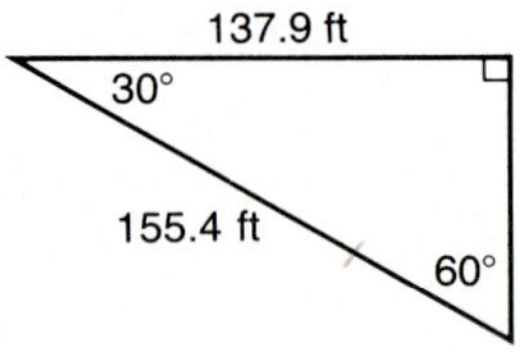

6.

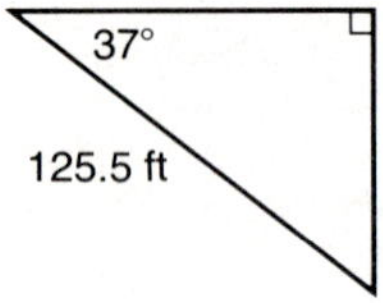

7.

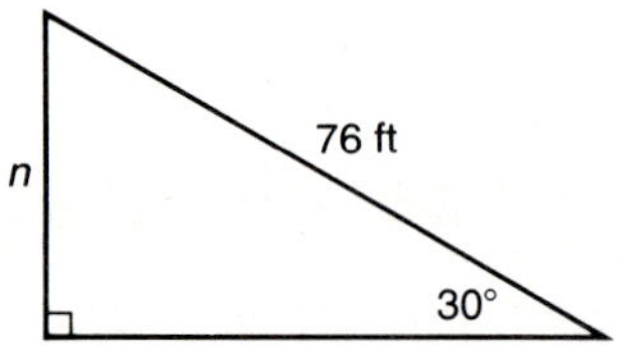

8.

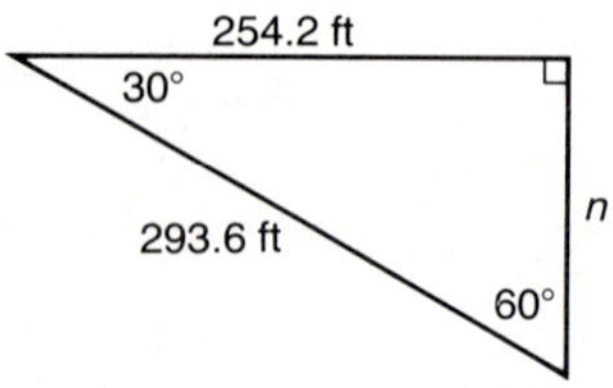

PROBLEM Solving APPLICATION

11.12 CONSTRUCTING CIRCLE GRAPHS

You plan to discuss this year's budget at the next meeting of the Community Center Youth Group. You decide to display the data in a circle graph.

YOUTH GROUP BUDGET

Newsletter	15%	Special Activities	55%
Supplies	20%	Other Expenses	10%

To make a circle graph,

1. Find the measure of each central angle.

THINK: There are 360° in a circle.

Newsletter: 15% of 360° = 54°
Supplies: 20% of 360° = 72°
Special Activities: 55% of 360° = 198°
Other Expenses: 10% of 360° = 36°
Total: = 360°

YOUTH GROUP BUDGET

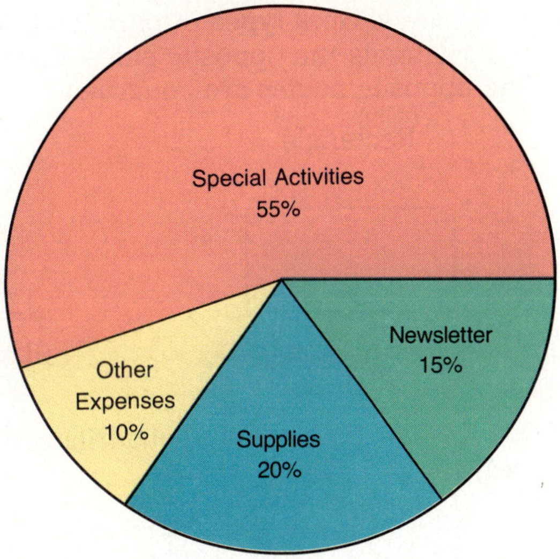

2. Draw a circle. Then draw central angles with the measures you found in Step 1.

3. Label each part of the graph. Remember to title the graph.

Use the circle graph to answer Exercises 1 and 2. If the total Youth Group budget is $575, how much will be spent on:

1. the newsletter? $86.25

2. special activities? $316.25

3. Your own personal budget includes the following categories and percents: school supplies, 25%; snacks, 5%; recreation, 60%; savings, 10%. Make a circle graph for this data. If necessary, round angle measures to the nearest degree.

4. If your yearly budget is $800, how much less per month is spent on snacks than on school supplies? Use the information in Exercise 3.

5. Lucy decides to make a yearly budget for $1,050. She decides to spend $\frac{1}{5}$ on clothes, $\frac{1}{4}$ on food, $\frac{3}{10}$ on sports events, $\frac{1}{20}$ on movies, and $\frac{1}{5}$ on savings. Determine the percents and make a circle graph for this data.

11.13 QUADRILATERALS

Quadrilaterals can be classified according to their sides and angles.

Trapezoid

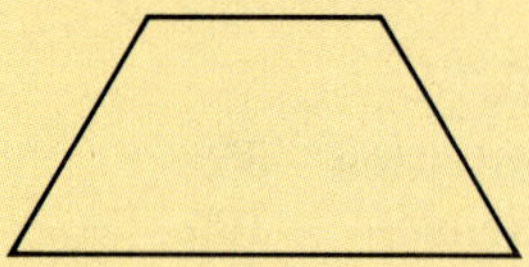

1 pair of parallel sides

Parallelogram

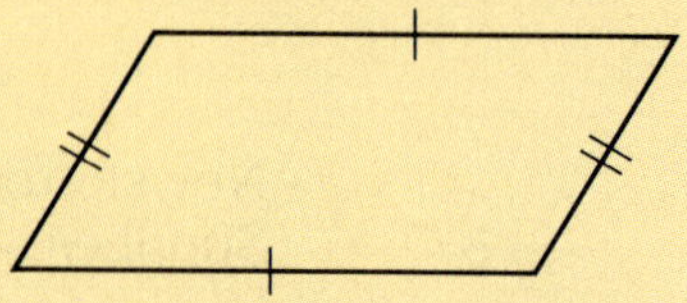

opposite sides parallel,
opposite sides congruent,
opposite angles congruent

There are several types of parallelograms. In each of these quadrilaterals the opposite sides are parallel and congruent. The opposite angles are congruent.

Rectangle

all right angles

Rhombus

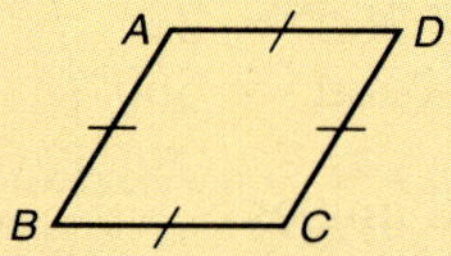

all sides congruent

Square

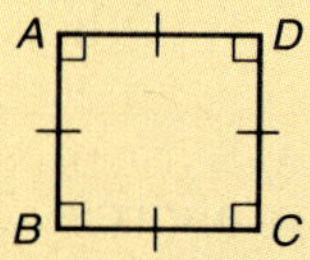

all right angles,
all sides congruent

EXAMPLE

Name the quadrilateral.

a. 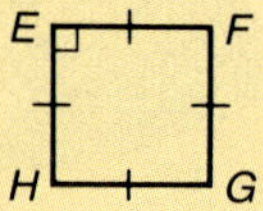

THINK: opposite sides parallel, all sides congruent, all right angles

So, quadrilateral *EFGH* is a square.

b. 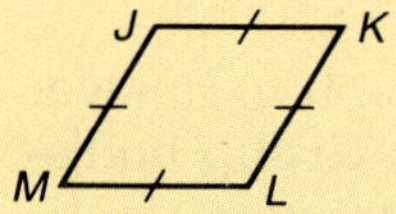

THINK: opposite sides parallel, all sides congruent, opposite angles congruent

So, quadrilateral *JKLM* is a rhombus.

FOR DISCUSSION

In parallelogram *ABCD*, tell how you can determine the measure of $\angle ACB$.

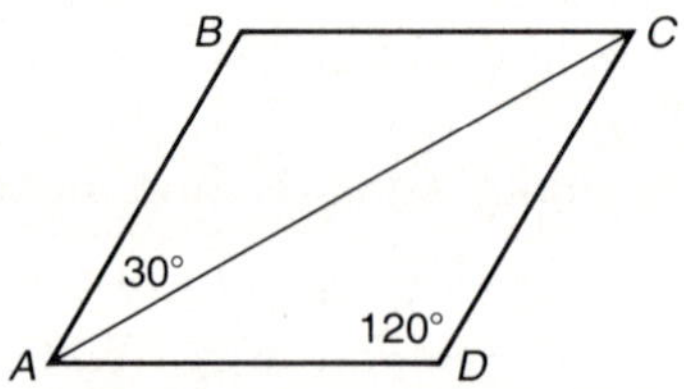

PRACTICE EXERCISES

Name the quadrilateral.

1. 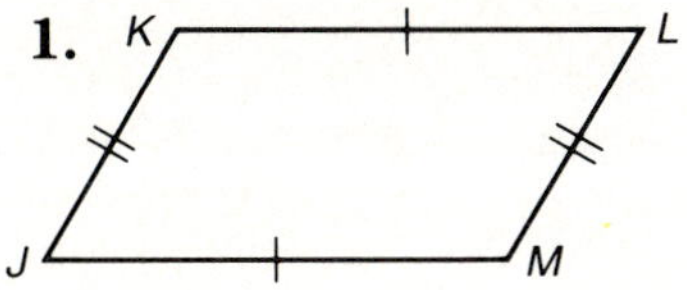

2. 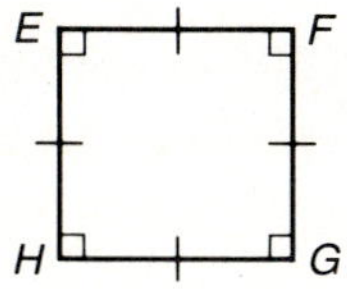

3. 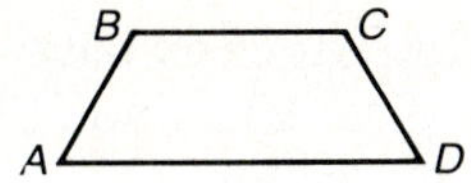

4. 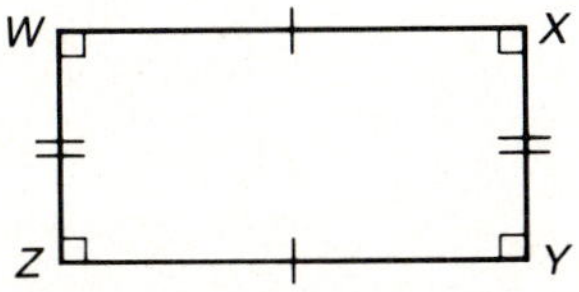

5. 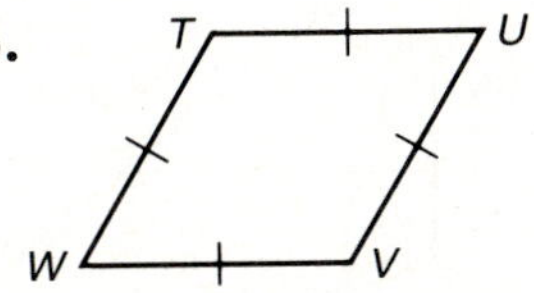

6. 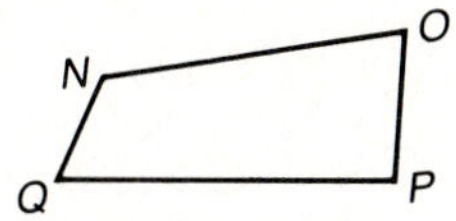

Tell which quadrilateral(s) have the following properties.

7. Four sides
8. Only one pair of parallel sides
9. Opposite sides parallel
10. All right angles
11. All sides congruent
12. All right angles and all sides congruent

Solve.

13. $\angle A$ and $\angle B$ are opposite angles of a parallelogram. If $m\angle A = 60°$, find $m\angle B$.

14. Find the sum of all the angles of a square.

15. The sum of the measures of adjacent angles in a parallelogram is 180°. In a rhombus *ABCD*, $m\angle A = 60°$. Find the measures of the other three angles.

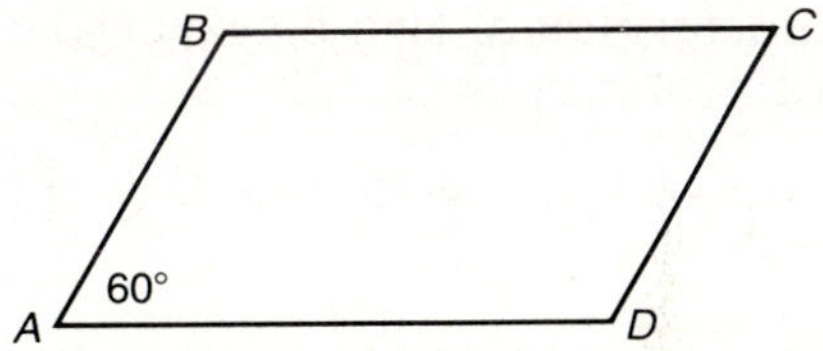

TIME OUT

This diagram shows the relationship between the different types of quadrilaterals.

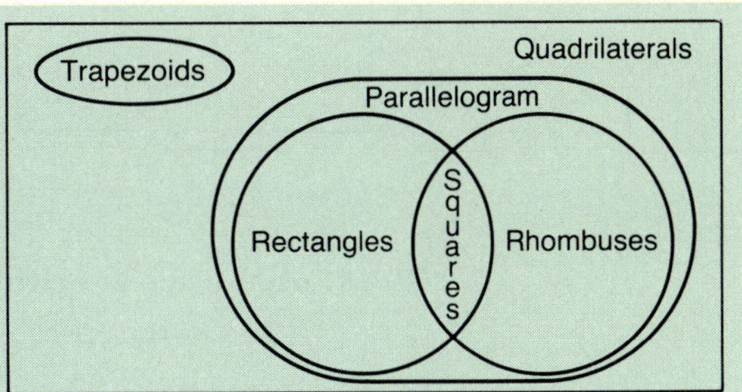

Tell whether the statement is true or false.

1. If a figure is a trapezoid, then it must be a quadrilateral.
2. If a figure is a parallelogram, then it must be a rectangle.
3. If a figure is a rhombus, then it must be a square.
4. If a figure is a rectangle and a rhombus, then it must be a square.

11.14 SIMILAR FIGURES

Similar figures have the same shape but not the same size.
Triangles *ABC* and *DEF* are similar figures.
This is written as $\triangle ABC \sim \triangle DEF$.

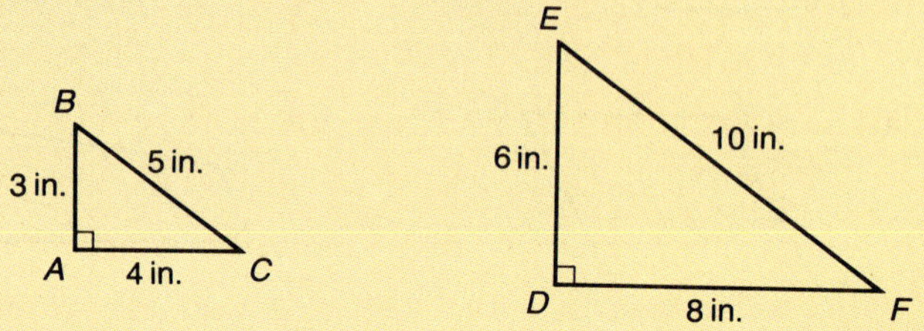

$\angle A \cong \angle D$ $\qquad$ $\angle B \cong \angle E$ $\qquad$ $\angle C \cong \angle F$

$\frac{\overline{AB}}{\overline{DE}} = \frac{3}{6} = \frac{1}{2}$ $\qquad$ $\frac{\overline{BC}}{\overline{EF}} = \frac{5}{10} = \frac{1}{2}$ $\qquad$ $\frac{\overline{AC}}{\overline{DF}} = \frac{4}{8} = \frac{1}{2}$

If two figures are similar:

1. The corresponding angles are congruent.
2. The ratios of the lengths of the corresponding sides are equal.

You can determine if two figures are similar by checking to see that they meet both conditions.

EXAMPLE 1 Are rectangles *ABCD* and *EFGH* similar?

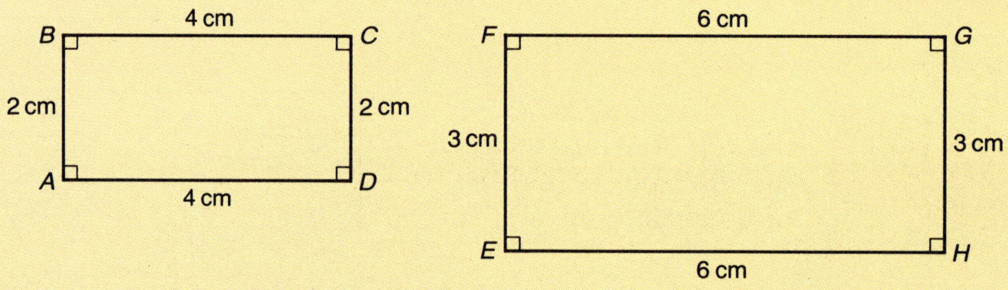

THINK: All angles are right angles.

$\angle \mathbf{A} \cong \angle \mathbf{E}$, $\angle \mathbf{B} \cong \angle \mathbf{F}$, $\angle \mathbf{C} \cong \angle \mathbf{G}$, $\angle \mathbf{D} \cong \angle \mathbf{H}$

All corresponding angles are congruent.

THINK: $\frac{\overline{AB}}{\overline{EF}} = \frac{2}{3}$ $\qquad$ $\frac{\overline{BC}}{\overline{FG}} = \frac{4}{6} = \frac{2}{3}$ $\qquad$ $\frac{\overline{CD}}{\overline{GH}} = \frac{2}{3}$ $\qquad$ $\frac{\overline{DA}}{\overline{HE}} = \frac{4}{6} = \frac{2}{3}$

The ratios of the lengths of the corresponding sides are equal.
So, $ABCD \sim EFGH$.

You can find the length of the missing side of a similar figure using a proportion.

EXAMPLE 2 $\triangle RST \sim \triangle UVW$. Find the length of $\overline{RT}$.

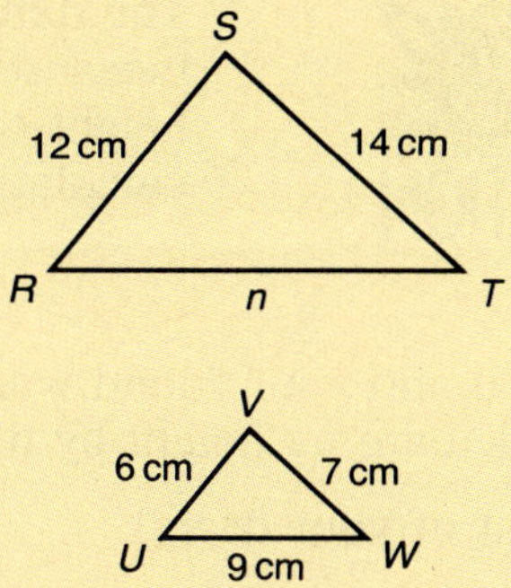

1. Write a proportion.

$$\frac{\text{length of } \overline{RT} \rightarrow n}{\text{length of } \overline{UW} \rightarrow 9} = \frac{12 \leftarrow \text{length of } \overline{RS}}{6 \leftarrow \text{length of } \overline{UV}}$$

2. Solve.

$$n \times 6 = 12 \times 9$$
$$n \times 6 = 108$$
$$n = 108 \div 6$$
$$n = 18$$

So, the length of $\overline{RT}$ is 18 cm.

PRACTICE EXERCISES

Assume that $\triangle RST \sim \triangle XYZ$. Complete the statement.

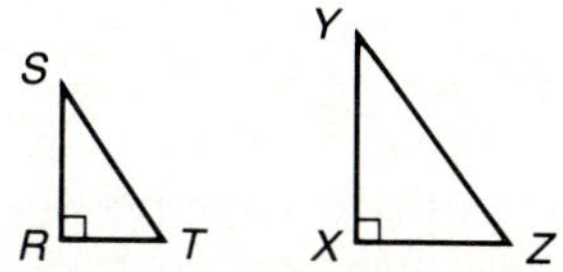

1. $\angle R \cong$ ■ **2.** $\angle S \cong$ ■ **3.** $\angle T \cong$ ■

4. $\frac{\overline{RT}}{\overline{XZ}} = \frac{■}{\overline{YZ}}$ **5.** $\frac{\overline{RS}}{\overline{XY}} = \frac{\overline{ST}}{■}$ **6.** $\frac{\overline{XY}}{■} = \frac{\overline{YZ}}{\overline{ST}}$

State why the figures are, or are not, similar.

7.

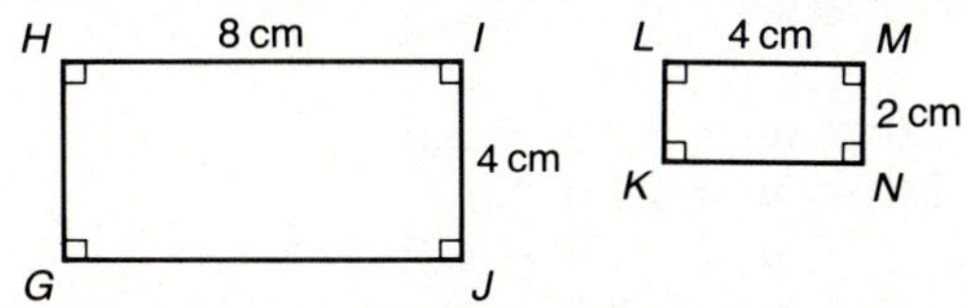

8.

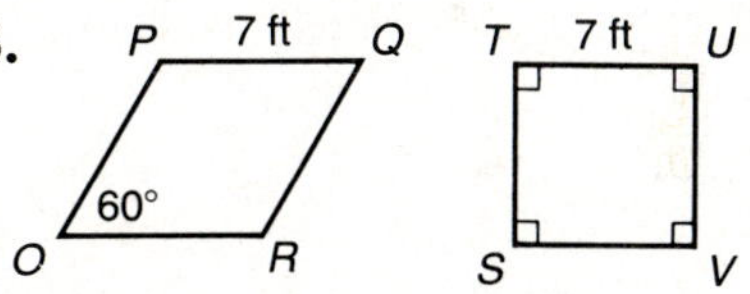

Find the length of the missing side for the pair of similar figures.

9.

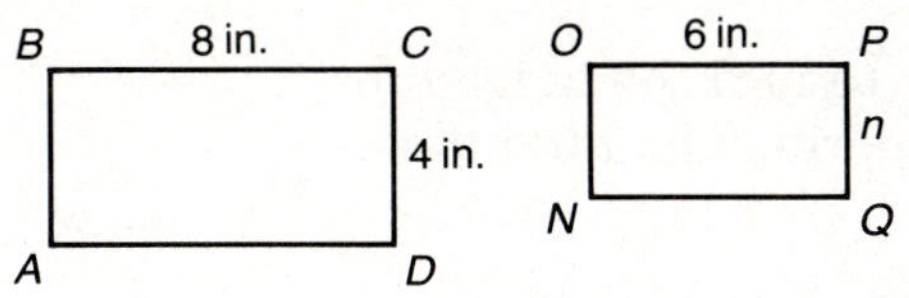

10.

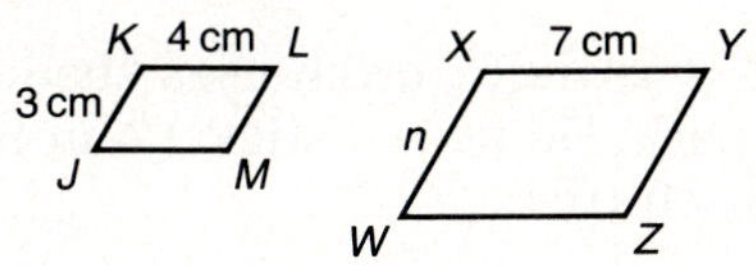

11.

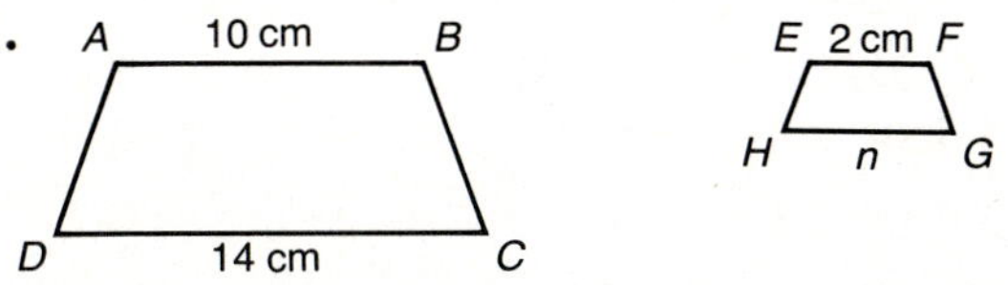

12.

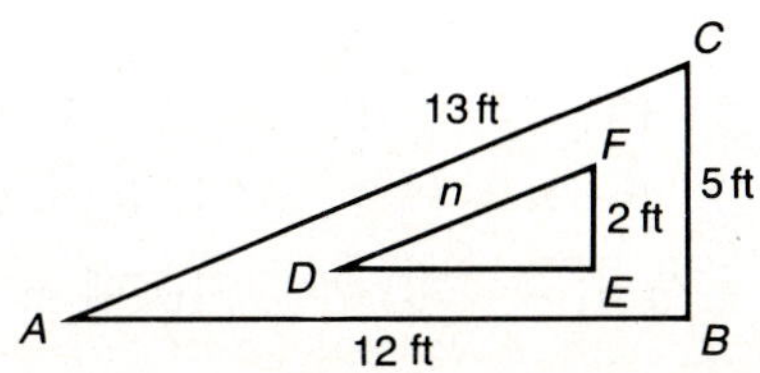

PROBLEM *Solving* APPLICATION

11.15 INDIRECT MEASUREMENT

The dimensions of an object can be found by direct measurement. Indirect measurement can also be used to find the height or length of an object by measuring other objects and then calculating the measurement.

Suppose you and your friend want to know the height of a flagpole. You can calculate its height by measuring:

- The height of your friend.
- The shadows of your friend and the flagpole.

This method, called **shadow reckoning,** uses what you have learned about similar triangles.

A flagpole casts a shadow 13.5 m long. Your friend, who is 1.6 m tall, casts a shadow that is 2.0 m long. What is the flagpole's height?

This drawing shows the two right triangles with the measurements that were made.

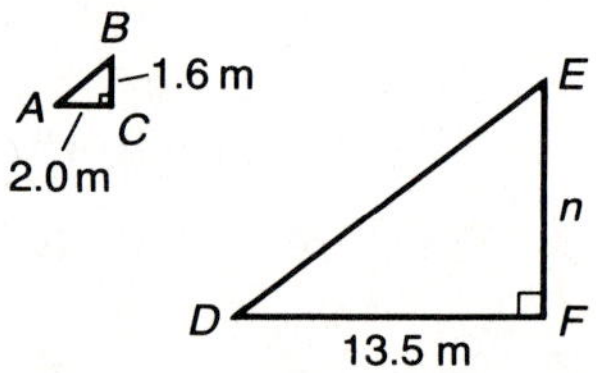

Use a proportion to solve the problem.

1. Write a proportion. Let n represent the height of the flagpole.

$$\frac{2.0}{13.5} = \frac{1.6}{n}$$

2. Find n.

$$2.0 \times n = 13.5 \times 1.6$$
$$2.0 \times n = 21.6$$
$$n = 21.6 \div 2.0$$
$$n = 10.8$$

So, the height of the flagpole is 10.8 m.

On several days Jeff Corbin uses shadow reckoning to find the heights of trees in a park. He uses a stick 1.4 m long in his experiments. Find the height of each tree.

	Length of Shadow of Stick	Length of Shadow of Tree	Height of Stick	Height of Tree
1.	1.2 m	7.8 m	1.4 m	■
2.	1.6 m	11.2 m	1.4 m	■
3.	2.1 m	15.0 m	1.4 m	■
4.	1.75 m	10.50 m	1.4 m	■

Solve.

5. A building casts a 24-m shadow at the same time that a 2-m street sign casts a 4-m shadow. What is the height of the building?

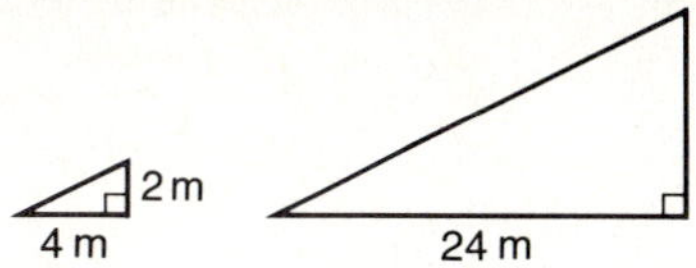

6. A flagpole casts a 16.8-m shadow at the same time that a 2.4-m shadow is cast by a person who is 1.8 m tall. What is the height of the flagpole?

7. Some campers wanted to find the distance across a pond. They laid out two similar triangles by pacing the distances from a tree at point T. (The measurements in the drawing are given in paces rather than meters.) In the two triangles, $\overline{TY}$ and $\overline{TQ}$ are corresponding sides. What is the distance across the pond?

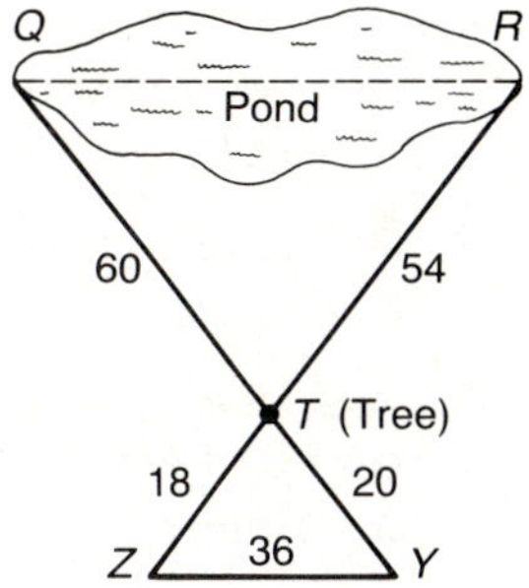

8. Work in teams of two students for this experiment. Use shadow reckoning to find the heights of three trees or other objects. Compare the results of your team with the results of another team.

CHAPTER REVIEW

Vocabulary Choose the letter of the word(s) that correctly completes each statement.

1. A(n) ■ is part of a line with one endpoint. [292]
a. ray **b.** line segment **c.** angle **d.** plane

2. A(n) ■ angle measures greater than 0° and less than 90°. [296]
a. acute **b.** right **c.** obtuse **d.** straight

3. ■ angles are two angles the sum of whose measure is 180°. [296]
a. Straight **b.** Complementary **c.** Supplementary **d.** Similar

4. A ■ is a line segment that goes from the center of a circle to a point on it. [304]
a. diameter **b.** radius **c.** chord **d.** central angle

Skills Find the complement of the angle. [296]

40° *50°*

5. 30° **6.** 55° **7.** 60° **8.** 70°

Find the supplement of the angle. [296]

40° *140°*

9. 35° **10.** 75° **11.** 110° **12.** 130°

Use the figure to name the part of the circle. [304]

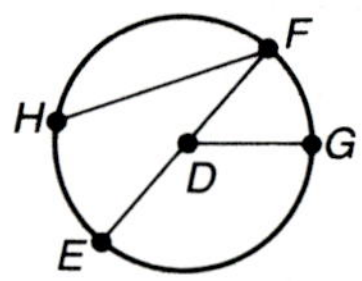

$\overline{EF}$ *diameter*

13. $\overline{HF}$

14. $\angle EDG$

15. $\overline{DE}$

Find the missing length. [304]

$r = 3.2$ cm, $d =$ ■
$d = 6.4$ cm

16. $r = 12.5$ in., $d =$ ■

17. $d = 7$ ft, $r =$ ■

Find the missing length in the right triangle. [314]

$a = 6, b = 8, c =$ ■
$c = 10$

18. $a = 12, b =$ ■, $c = 15$

19. $a =$ ■, $b = 20, c = 25$

20. Trace *AB*. Construct a line segment congruent to *AB*. [298]

21. Trace *AB*. Construct the perpendicular bisector. [301]

Chapter Test

Draw and label the figure.

1. $\overrightarrow{AB}$ **2.** $\overleftrightarrow{CD}$ **3.** $\angle EFG$ **4.** $\overline{HI}$

Use $\angle ABC$ for Exercises 5–10.

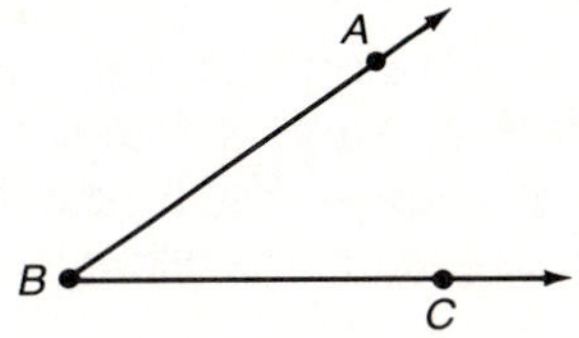

5. Find the measure of $\angle ABC$.

6. Classify $\angle ABC$.

7. Find the complement of $\angle ABC$.

8. Find the supplement of $\angle ABC$.

9. Construct $\angle RST \cong \angle ABC$.

10. Construct $\overrightarrow{BD}$, the bisector of $\angle ABC$.

Find the missing length.

11. $r = 7$ in., $d = ■$ **12.** $r = 5.2$ cm, $d = ■$ **13.** $d = 15$ ft, $r = ■$

Identify the triangle as *equilateral*, *isosceles*, or *scalene*.

14.

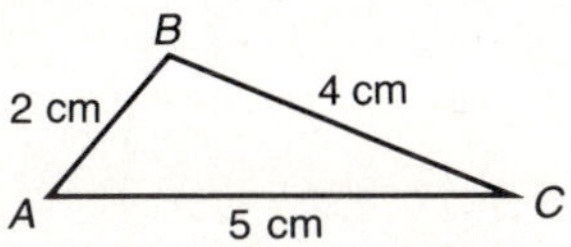

15.

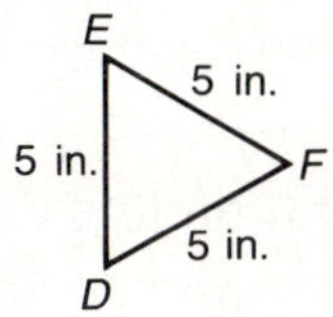

16.

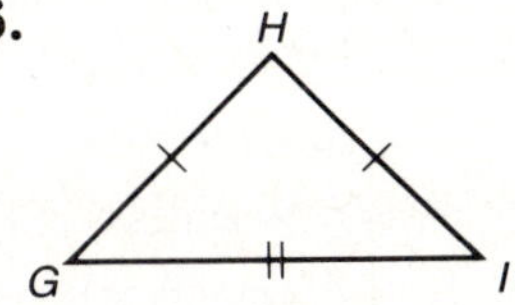

Find the measure of the third angle.

17. 35°, 110° **18.** 70°, 90°

Find the missing length in the right triangle.

19. $a = 12$, $b = ■$, $c = 20$

20. $a = 5$, $b = 12$, $c = ■$

21. $a = 7$, $b = 24$, $c = ■$

Find the length of the missing side for the pair of similar triangles.

22.

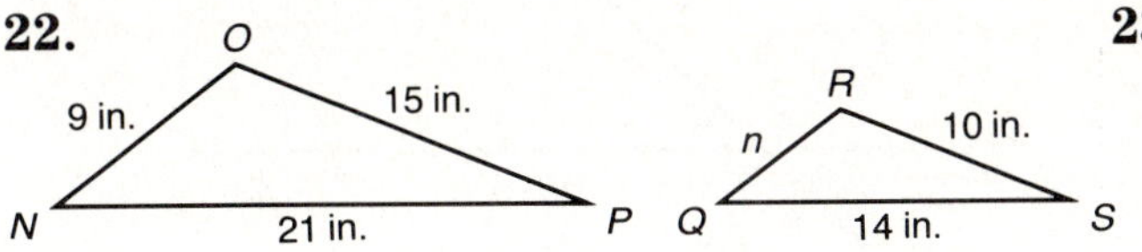

23.

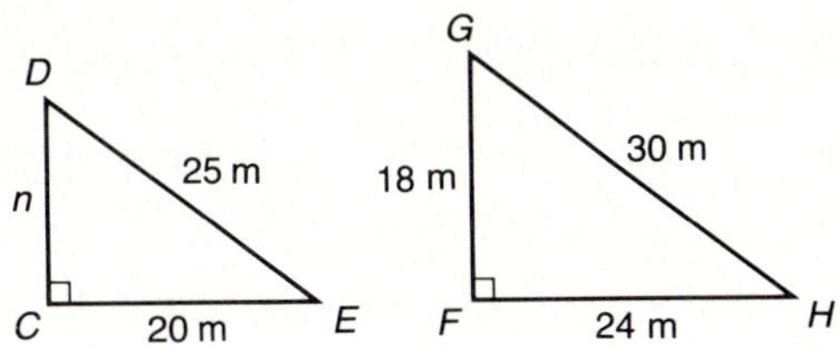

Solve.

24. A square table seats 4 people. Two tables placed end to end seat 6 people. How many people can be seated when 3 tables are placed end to end?

25. A building casts a 45-m shadow at the same time that a 3-m pole casts a 9-m shadow. What is the height of the building?

ENRICHMENT CORRESPONDING ANGLES

In the diagram, $\overleftrightarrow{EF}$ is a **transversal.** It intersects the parallel lines $\overleftrightarrow{AB}$ and $\overleftrightarrow{CD}$.

Eight angles are formed. ∠*1* and ∠*5*, ∠*2* and ∠*6*, ∠*3* and ∠*7*, and ∠*4* and ∠*8* are **corresponding angles.** Each pair of corresponding angles is congruent.

∠*1* and ∠*2* and ∠*1* and ∠*3* are supplementary angles. The sum of supplementary angles always equals 180°.

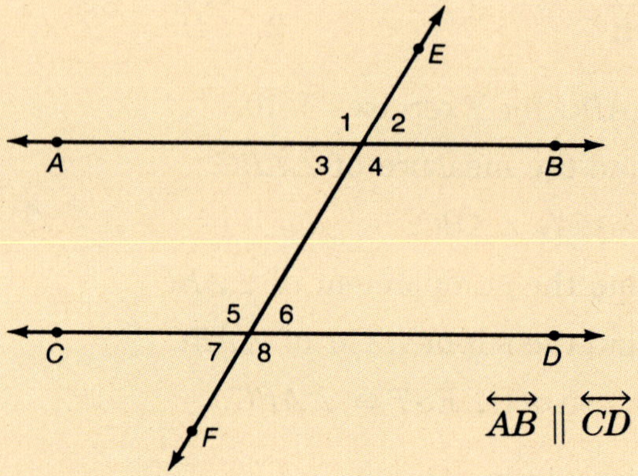

Example In the diagram above, if m∠*1* = 120°, what are the measures of all the other angles?

m∠*2* = 60° because ∠*1* and ∠*2* are supplementary angles.
m∠*3* = 60° because ∠*1* and ∠*3* are supplementary angles.
m∠*4* = 120° because ∠*2* and ∠*4* are supplementary angles.
m∠*5* = 120° because ∠*1* and ∠*5* are corresponding angles.
m∠*6* = 60° because ∠*2* and ∠*6* are corresponding angles.
m∠*7* = 60° because ∠*3* and ∠*7* are corresponding angles.
m∠*8* = 120° because ∠*4* and ∠*8* are corresponding angles.

Using the diagram to the right, identify the pairs of angles as *corresponding* or *supplementary.*

1. ∠*1* and ∠*6* **2.** ∠*2* and ∠*3*

3. ∠*3* and ∠*8* **4.** ∠*1* and ∠*4*

5. ∠*5* and ∠*6* **6.** ∠*2* and ∠*7*

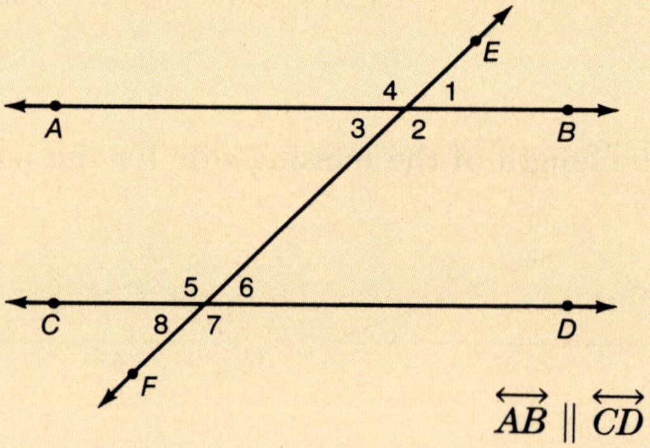

In the diagram to the right, if m∠*1* = 45°, what is:

7. m∠*2*? **8.** m∠*6*?

9. m∠*7*? **10.** m∠*3*?

11. m∠*5*? **12.** m∠*8*?

ESTIMATION SKILLS ESTIMATING ANGLE MEASURES

To estimate angle measures, make mental comparisons with right angles.

Estimate the measure of $\angle ABC$.

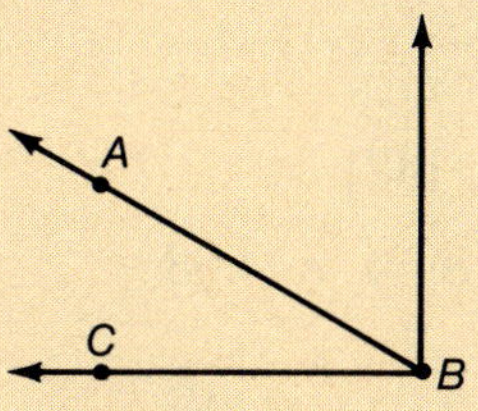

THINK: **$\angle ABC$ is about $\frac{1}{3}$ the size of a right angle.**

$\frac{1}{3} \times 90 = 30$

$\angle ABC$ measures about 30°.

Estimate the measure of $\angle DEF$.

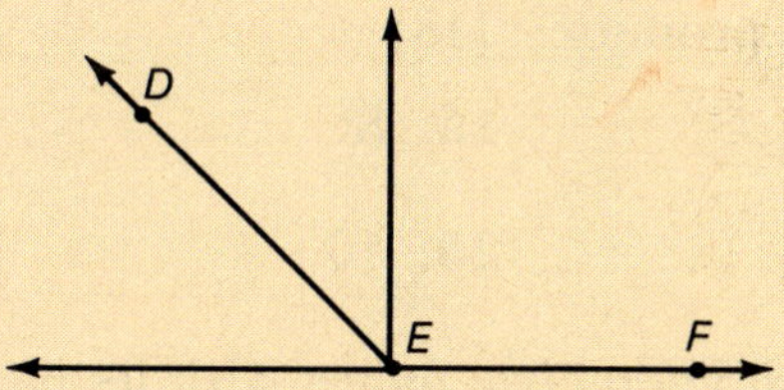

THINK: **$\angle DEF$ is the size of a right angle, plus about $\frac{1}{2}$ of a right angle.**

$\frac{1}{2} \times 90 = 45$

$90 + 45 = 135$

$\angle DEF$ measures about 135°.

Which is the angle measure?

1.

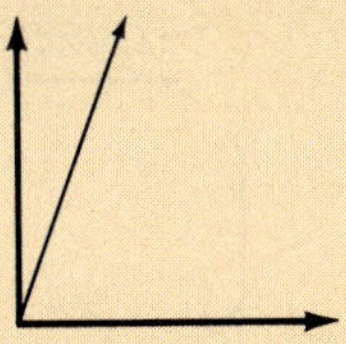

a. 30° **b.** 70° **c.** 90°

2.

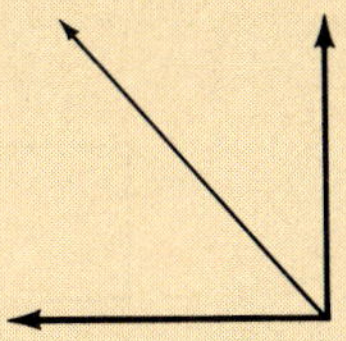

a. 20° **b.** 50° **c.** 80°

3.

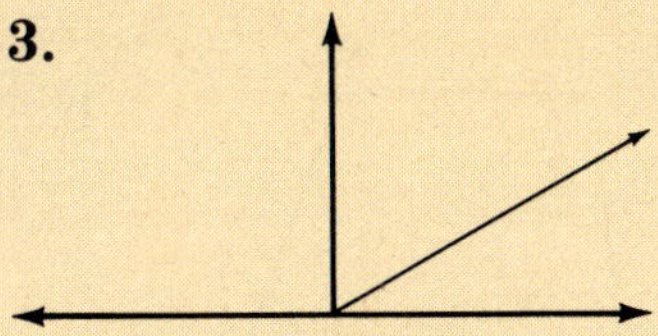

a. 105° **b.** 125° **c.** 150°

4.

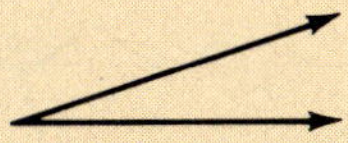

a. 18° **b.** 41° **c.** 66°

5.

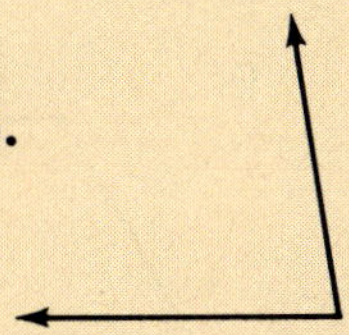

a. 44° **b.** 59° **c.** 82°

6.

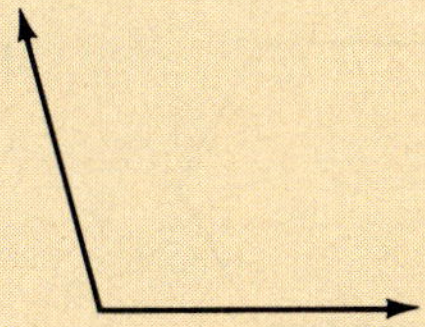

a. 103° **b.** 127° **c.** 162°

Estimate the measures of the angles. Use a protractor to check.

7.

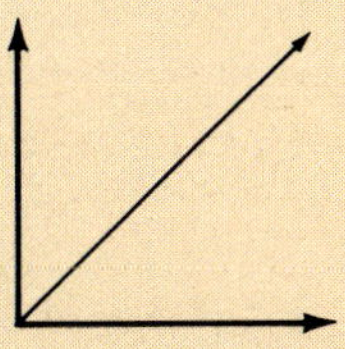

8.

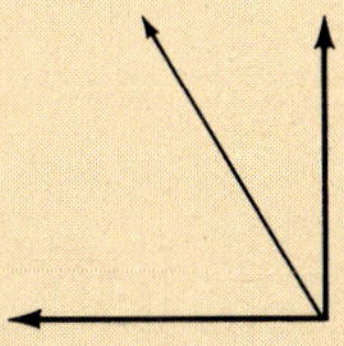

9.

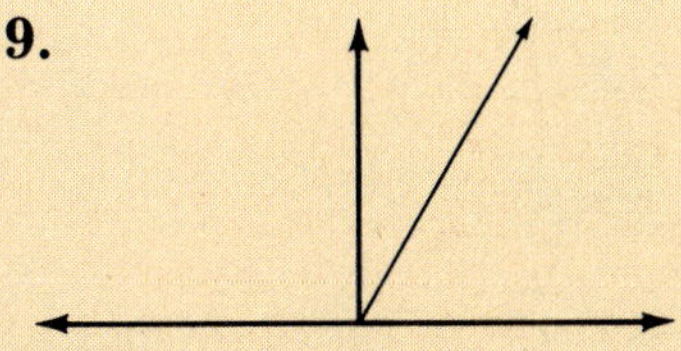

PreSkills Test

Add. [30, 62, 158]

1. $15 + 22 + 15 + 22$

2. $208 + 315 + 476 + 512 + 826$

3. $16.2 + 9.3 + 16.2 + 9.3$

4. $5.23 + 7.1 + 3.45 + 7.7 + 3.59$

5. $1\frac{3}{4} + 2\frac{1}{2} + 1\frac{3}{4} + 2\frac{1}{2}$

6. $5\frac{2}{3} + 3\frac{5}{9} + 3\frac{2}{3}$

Multiply. [36, 74, 146, 148]

7. 4×80

8. 4×4.03

9. 3×17.5

10. $6 \times 3.14 \times 19.2$

11. $4 \times \frac{1}{2}$

12. $14 \times 3\frac{1}{7}$

13. $6 \times \frac{22}{7}$

14. $2 \times \frac{22}{7} \times \frac{1}{4}$

Write as a product. [18]

15. 5^2

16. 8^2

17. 30^2

18. 600^2

19. 7^3

20. 10^3

21. 12^3

22. 20^3

Match each geometric figure with its name. [304, 318]

23. Square

24. Rectangle

25. Triangle

26. Regular hexagon

27. Trapezoid

28. Parallelogram

29. Pentagon

30. Regular octagon

a.
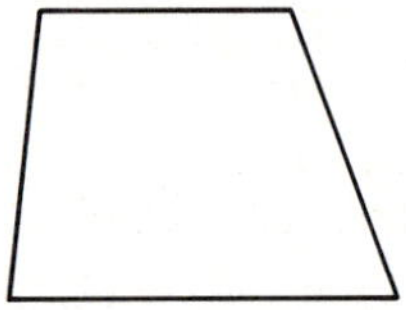

b.
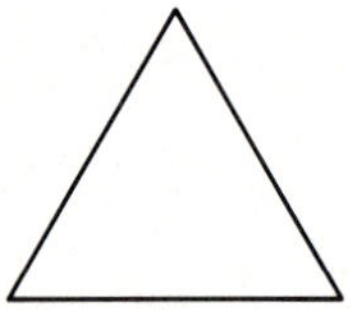

c.
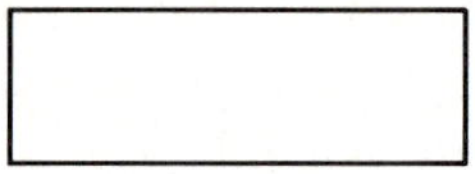

d.
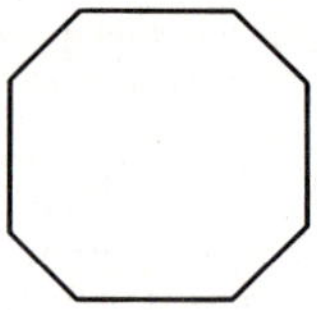

e.
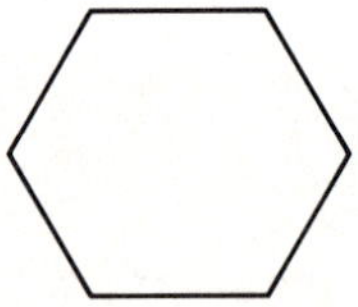

f.

g.
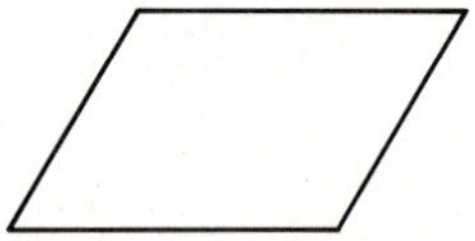

h.
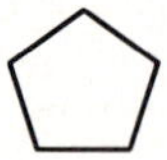

Chapter 12

PERIMETER, AREA, AND VOLUME

Road maps are very useful when traveling in unfamiliar areas. How are other kinds of maps used, and why are they important?

12.1 PERIMETER

The Culver Construction Company has just finished paving a new downtown parking lot. A diagram of the lot is shown at the right. The manager will use steel fencing to enclose the lot. How much fencing should he order?

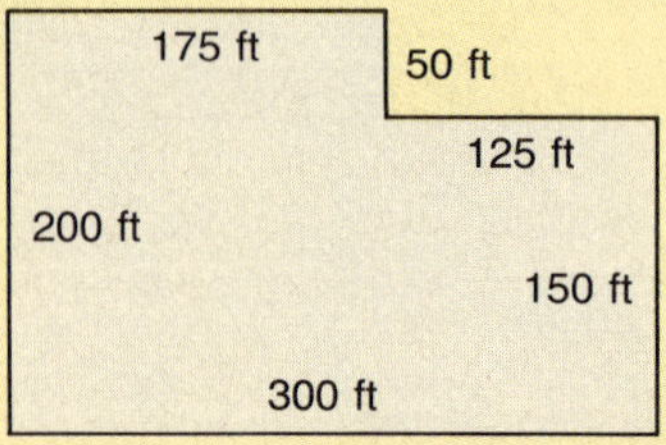

To find the correct amount of fencing, the manager needs to know the **perimeter,** or the distance around the parking lot.

EXAMPLE 1 Find the perimeter of the parking lot.

THINK: Add the lengths of the sides.

$P = 300 + 150 + 125 + 50 + 175 + 200$

$P = 1{,}000$

So, the manager should order 1,000 ft of fencing.

The opposite sides of a rectangle are congruent.

To find the perimeter of a rectangle:

1. Multiply the length by 2. $2 \times l$, or $2l$
2. Multiply the width by 2. $2 \times w$, or $2w$
3. Add the products. $P = 2l + 2w$

EXAMPLE 2 Find the perimeter.

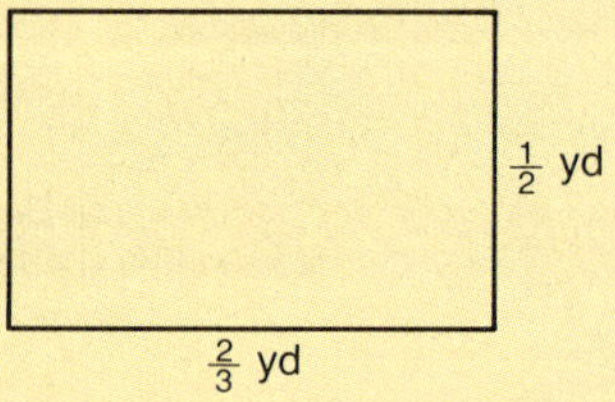

THINK: $P = 2l + 2w$

$P = 2\left(\frac{2}{3}\right) + 2\left(\frac{1}{2}\right)$

$P = 1\frac{1}{3} + 1 = 2\frac{1}{3}$

The perimeter of the rectangle is $2\frac{1}{3}$ yd.

In a regular polygon, all the sides are congruent.

To find the perimeter of a regular polygon, multiply the length of one side by the number of sides.

EXAMPLE 3 Find the perimeter.

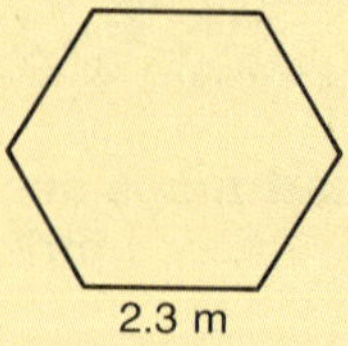

THINK: Each of the 6 sides is 2.3 m long.

$P = 6 \times 2.3 = 13.8$

So, the perimeter of the hexagon is 13.8 m.

PRACTICE EXERCISES

Find the perimeter.

1.

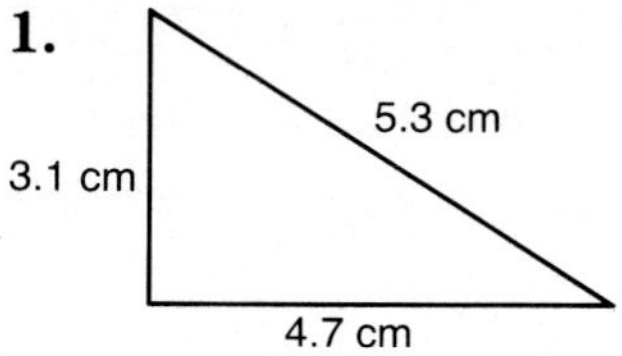

2.

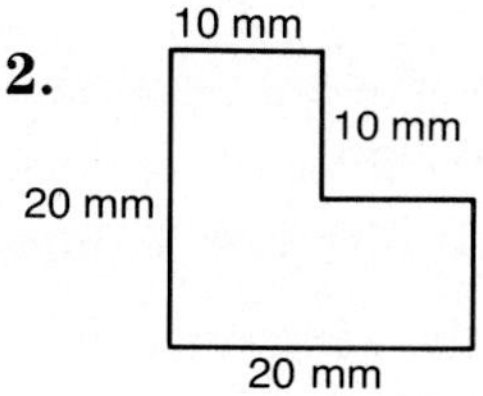

3.

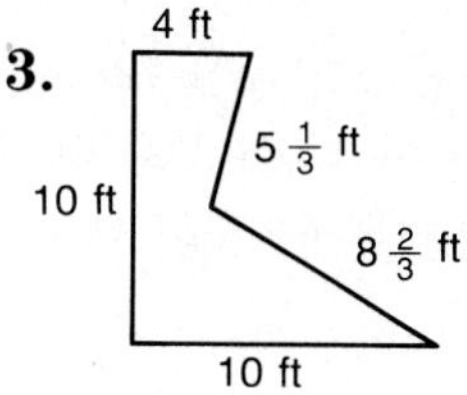

4.

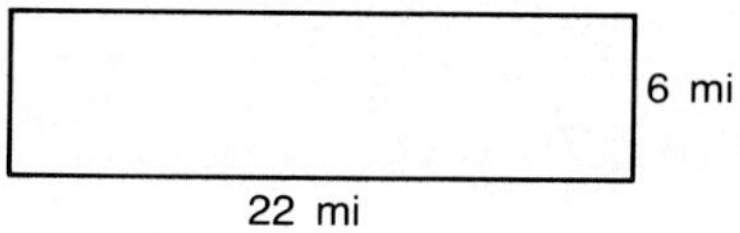

5.

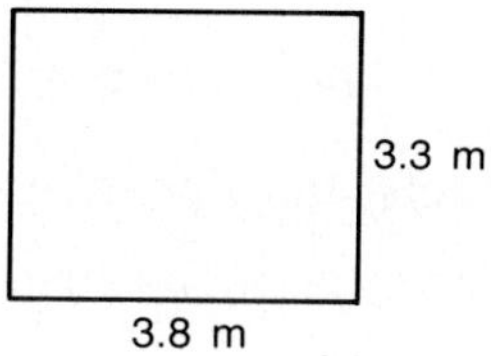

6.

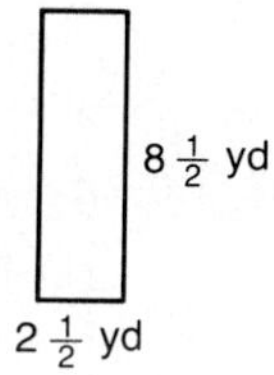

7.

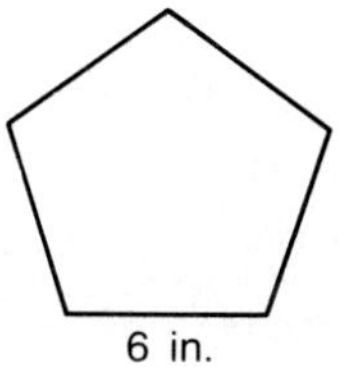

8.

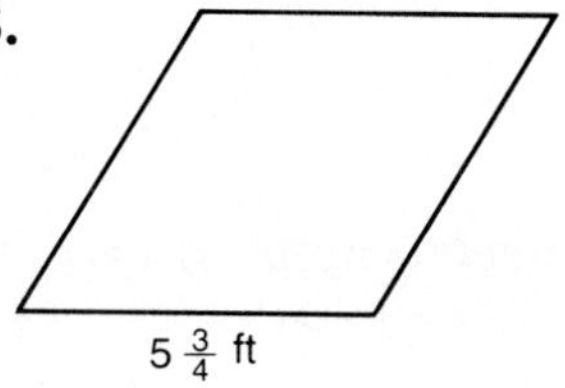

9.

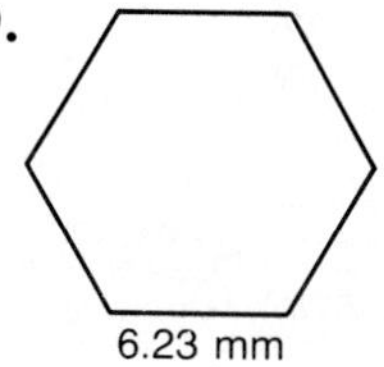

10. Rectangle: $l = 3\frac{3}{4}$ ft, $w = 2\frac{1}{8}$ ft

11. Rectangle: $l = 7$ in., $w = 1\frac{1}{4}$ in.

12. Rectangle: $l = 3.7$ mm, $w = 12$ mm

13. Rectangle: $l = 4.65$ m, $w = 1.34$ m

14. Square: $s = 23$ ft

15. Regular pentagon: $s = 123$ yd

16. Regular octagon: $s = 4.5$ m

17. Regular hexagon: $s = 16.3$ mm

Find the length of a side.

18. Square: $P = 36$ m

19. Regular octagon: $P = 2\frac{1}{2}$ yd

Solve.

20. How much fencing would it take to enclose a tennis court with a width of 16.7 m and a length of 35.3 m?

21. Culver Construction is building a patio in the shape of a regular octagon. Each side is $4\frac{3}{4}$ yd long. What is the perimeter of the patio?

22. Culver charges $1.25 per foot to install molding. How much would it cost to have Culver install molding in a room that is 72 ft around?

23. Last year, Culver had gross earnings of $3.2 million. Culver's net profit was $800,000. What percent of gross earnings was Culver's net profit?

12.2 CIRCUMFERENCE

The diameter of a bicycle tire is 65 cm. The tire makes one complete revolution. How far has the bicycle traveled?

To find the answer, you need to find the distance around the tire.

The distance around a circle is called the **circumference.** For all circles, the ratio of the circumference to the diameter is the same. The Greek letter π (pi) shows this ratio.

$\pi = \frac{\text{circumference of a circle}}{\text{length of the diameter}}$ $\quad\quad$ π is approximately 3.14 or $\frac{22}{7}$.

To find the circumference of a circle, multiply the diameter by π.

EXAMPLE 1

Find the circumference of the tire.
Use 3.14 for π.

THINK: $C = \pi \times d$

$C \approx 3.14 \times 65$

$C \approx 204.1$

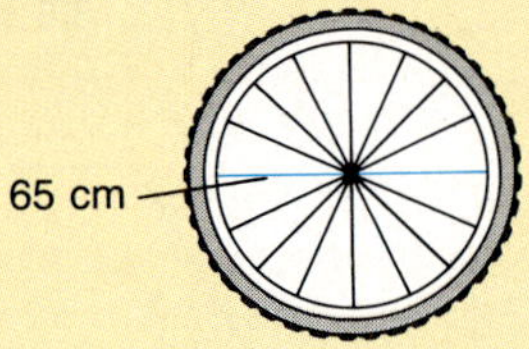

The bicycle has traveled approximately 204.1 cm.

A radius is half the length of a diameter. To find the circumference of a circle when you know only its radius, multiply the radius by π and then double the answer.

With fractions and mixed numbers, use $\frac{22}{7}$ for π.

EXAMPLE 2

Find the circumference of the circle shown below.

THINK: $C = 2 \times \pi \times r$

$C \approx 2 \times \frac{22}{7} \times \frac{1}{4}$

$C \approx \frac{11}{7} = 1\frac{4}{7}$

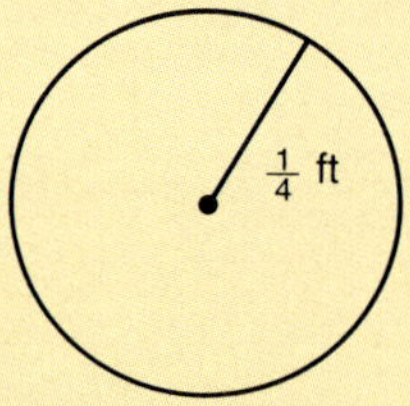

The circumference is approximately $1\frac{4}{7}$ ft.

CHECKPOINT

Write the letter of the correct answer.

Find the circumference.

1. $d = 8$ mm **a.** 11.14 mm **b.** 25.12 mm **c.** 50.24 mm **d.** 64 mm

2. $r = 1\frac{3}{4}$ ft **a.** $3\frac{1}{2}$ ft **b.** $5\frac{1}{2}$ ft **c.** 11 ft **d.** 22 ft

PRACTICE EXERCISES

Find the circumference. Use 3.14 for π with whole numbers and decimals. Use $\frac{22}{7}$ for π with fractions and mixed numbers.

1. 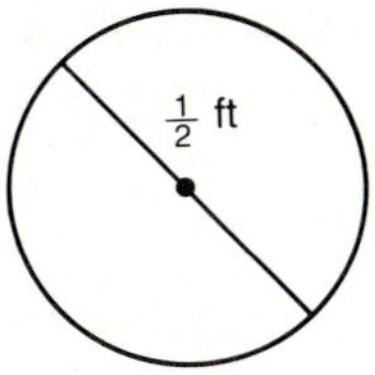

2. 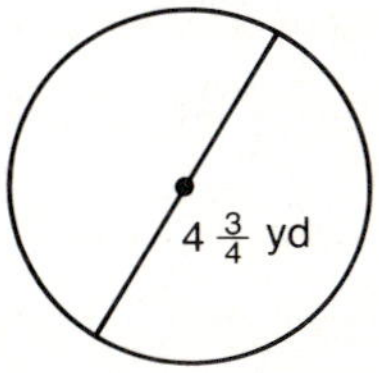

3. 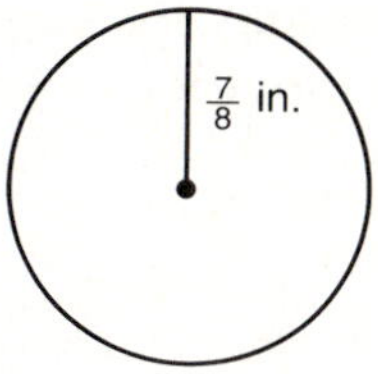

4. 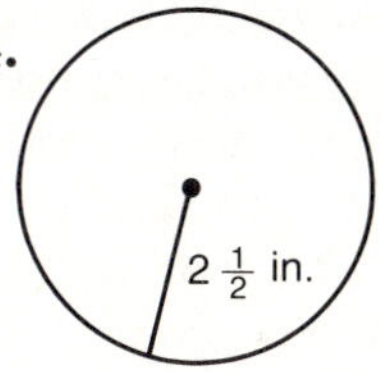

5. 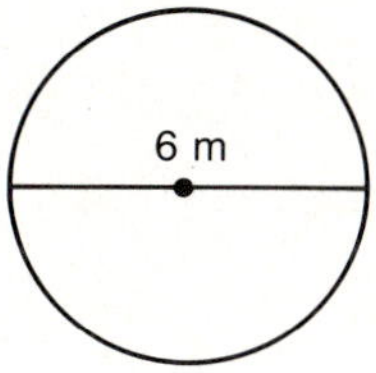

6. 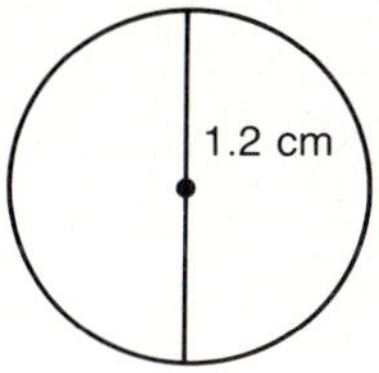

7. 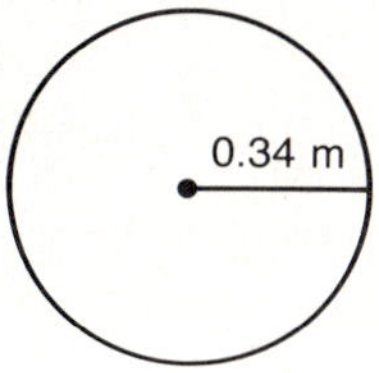

8. 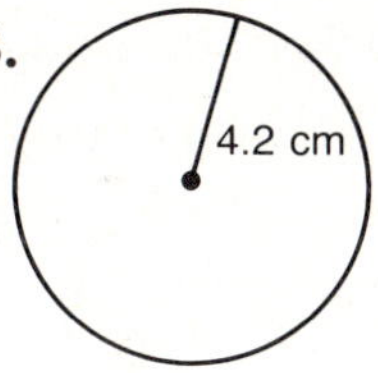

9. $d = \frac{7}{8}$ mi
10. $d = \frac{2}{3}$ yd
11. $d = 10\frac{1}{2}$ ft
12. $d = 6$ km
13. $d = 14.7$ m
14. $d = 15.65$ cm
15. $r = 2\frac{1}{3}$ yd
16. $r = \frac{5}{8}$ mi
17. $r = 3\frac{1}{2}$ ft
18. $r = 12.5$ cm
19. $r = 7$ m
20. $r = 0.5$ km

Solve.

21. Campbell Green is a rectangular park 2,300 yd long and 3,250 yd wide. How far is it to ride around Campbell Green?

22. Bicycle races take place in arenas called velodromes. The Campbell Green Velodrome has a circular track. Its diameter is 100 m. How many meters are there in one lap?

23. The Campbell Rally is a 36-mi road race for bicycles. This year's winner averaged 30 mph. What was the winning time?

24. Inside Campbell Green is a circular bicycle path 2,200 m around. What is its radius to the nearest meter?

12.3 AREAS OF RECTANGLES, SQUARES, AND PARALLELOGRAMS

Peter Rios works for a company that remodels kitchens. In one kitchen, he builds the rectangular island shown at the right. How much butcher block does he need to cover its countertop?

To find the amount of butcher block, you need to find the area of the rectangle. **Area** is the number of square units needed to cover the surface of a plane figure.

To find the area of a rectangle, multiply the length by the width.

EXAMPLE 1 Find the area of the countertop.

THINK: $A = l \times w$

$A = 5 \times 3 = 15$

So, Peter needs 15 ft^2 to cover its countertop.

A square is a rectangle whose length and width both have the same measure. To find its area, square the length of one side.

EXAMPLE 2 Find the area of the square.

THINK: $A = s^2$

$A = \left(4\frac{1}{2}\right)^2 = 4\frac{1}{2} \times 4\frac{1}{2} = 20\frac{1}{4}$

The area of the square is $20\frac{1}{4}$ ft^2.

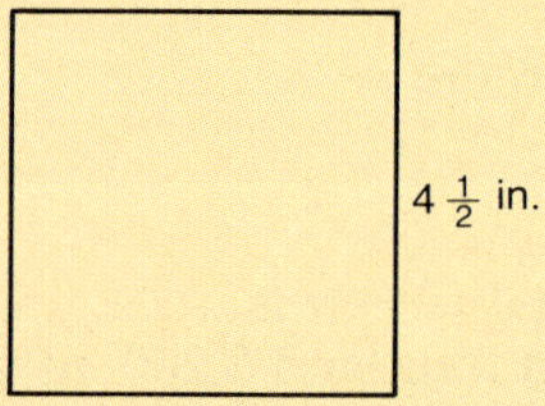

If a parallelogram and a rectangle have equal bases and equal heights, they are equal in area.

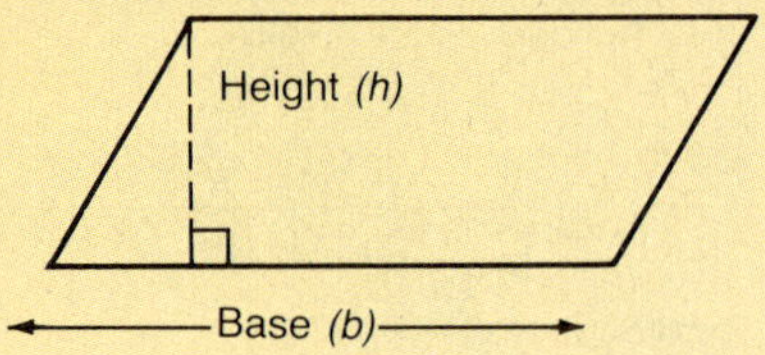

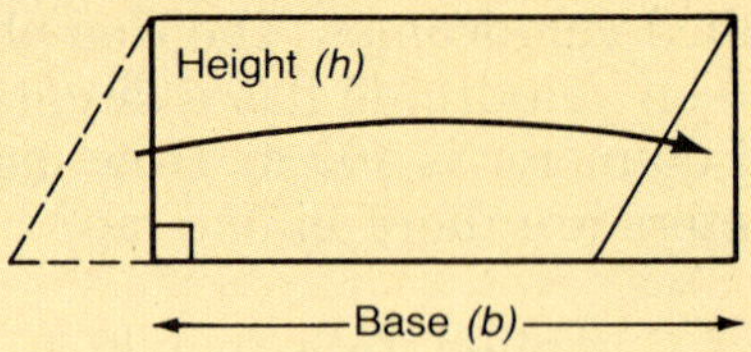

EXAMPLE 3 Find the area of the parallelogram.

THINK: Area (A) = base (b) × height (h)

$A = 8.3 \times 5.2 = 43.16$

The area of the parallelogram is 43.16 cm^2.

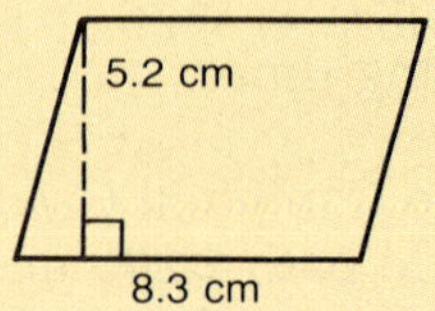

PRACTICE EXERCISES

Find the area of the rectangle.

1. 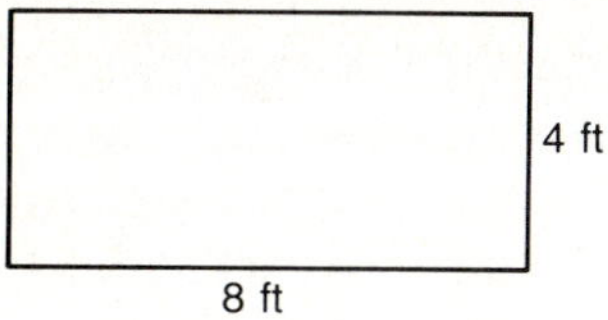

2. 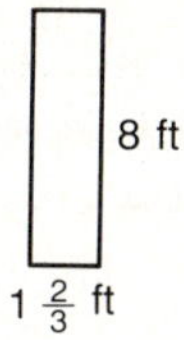

3. 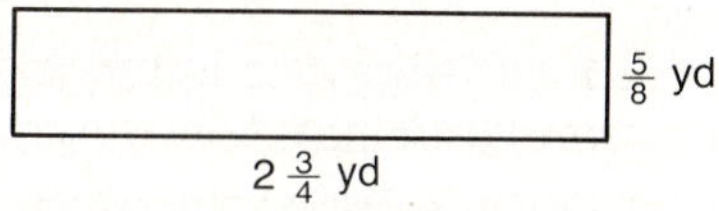

4. $l = 4.3$ mm, $w = 2.7$ mm **5.** $l = 3.4$ m, $w = 2.1$ m **6.** $l = 0.4$ m, $w = 2.2$ m

Find the area of the square.

7. 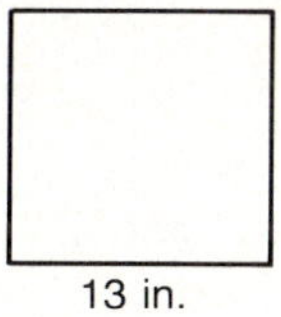

8. 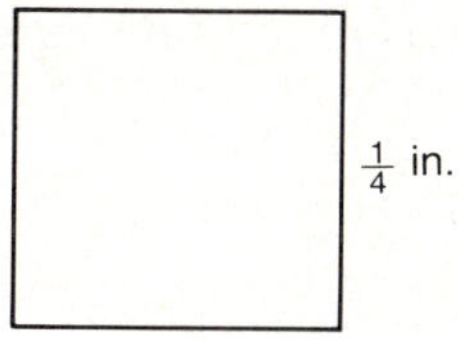

9. 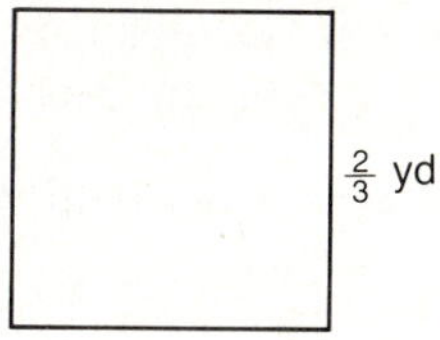

10. $s = 12$ m **11.** $s = 0.9$ cm **12.** $s = 2.3$ km

Find the area of the parallelogram.

13. 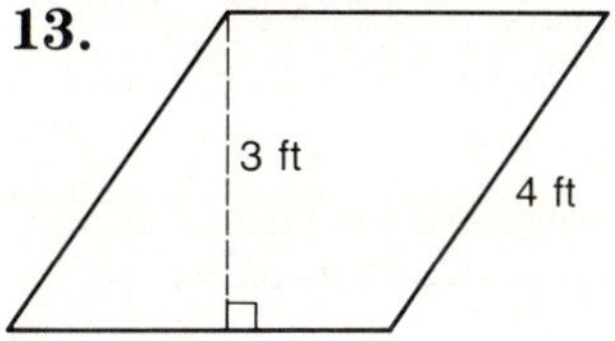

14. 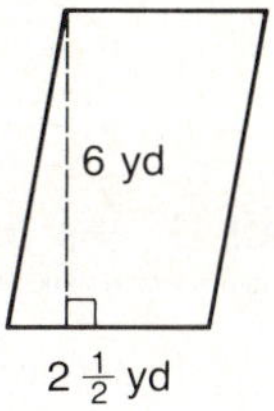

15. 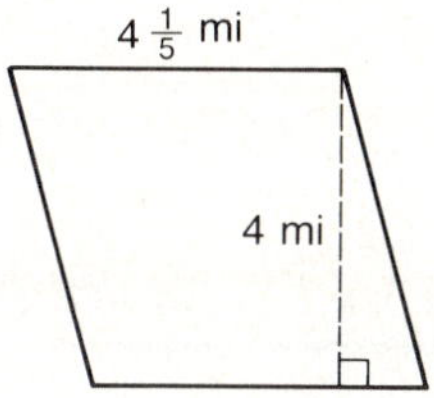

16. $b = 49$ m, $h = 3.2$ m **17.** $b = 0.8$ cm, $h = 4.1$ cm **18.** $b = 4.3$ m, $h = 5.1$ m

Solve.

19. Peter remodels a rectangular kitchen 12 ft by 8 ft. How many square yards of tile are needed to cover its floor? (*Hint:* Change feet to yards.)

20. The blueprint for remodeling this kitchen uses a scale of 1:8. The actual height of the kitchen ceiling is 7 ft. What is this distance on the blueprint?

21. One quart of paint covers 80 ft^2. How many gallons of paint are needed to cover 1,200 ft^2?

22. Peter is installing an unusual square sink. Its area is 576 $in.^2$ What is the sink's perimeter?

TIME OUT

The first floor of a town house has 929 ft^2 of floor space. Using graph paper, design a floor plan that includes the following rooms: living room (360 ft^2), den (225 ft^2), dining room (114 ft^2), kitchen (120 ft^2), bathroom (80 ft^2).

12.4 AREAS OF TRIANGLES AND TRAPEZOIDS

Laura Bergman is buying grass seed to repair the triangular patch of lawn shown at the right. How many square yards of lawn does she have to repair?

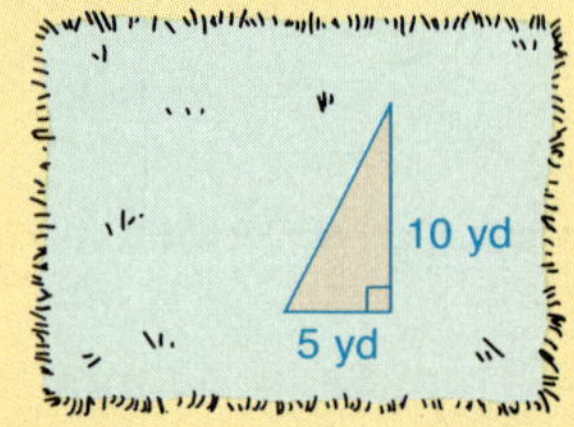

To find the amount of lawn to be repaired, you need to find the area of the triangle.

The figure at the right shows a parallelogram divided into two congruent triangles. Each triangle has an area equal to one-half the area of the parallelogram, or one-half the base times the height.

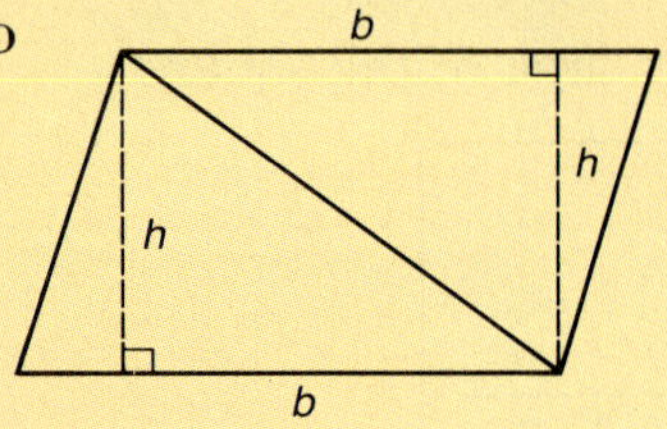

So, the area of a triangle is $\frac{1}{2} \times b \times h$.

EXAMPLE 1

Find the area of the triangular patch.

THINK: $A = \frac{1}{2} \times b \times h$

$A = \frac{1}{2} \times 5 \times 10$

$A = 25$

Laura Bergman has to repair 25 yd^2 of lawn.

In the figure below, a trapezoid is divided into two triangles. The area of the trapezoid is equal to the sum of the areas in the triangles.

$\text{Area } (\triangle_1) = \frac{1}{2} \times b_1 \times h$

$\text{Area } (\triangle_2) = \frac{1}{2} \times b_2 \times h$

$\text{Area (trapezoid)} = \frac{1}{2} \times (b_1 + b_2) \times h$

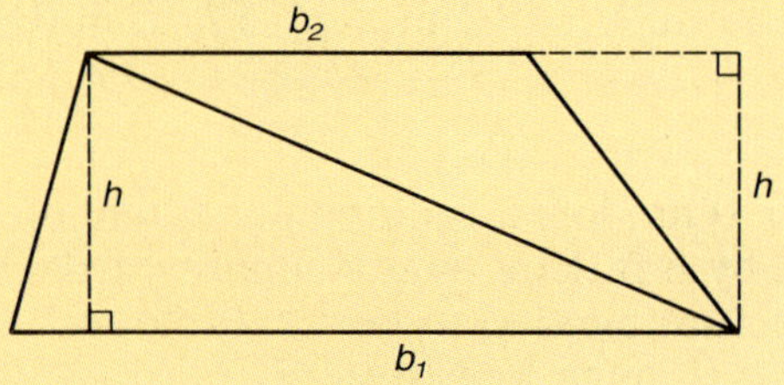

EXAMPLE 2

Find the area of the trapezoid.

THINK: $A = \frac{1}{2} \times (b_1 + b_2) \times h$

$A = \frac{1}{2} \times (7 + 3) \times 4$

$A = \frac{1}{2} \times 10 \times 4 = 20$

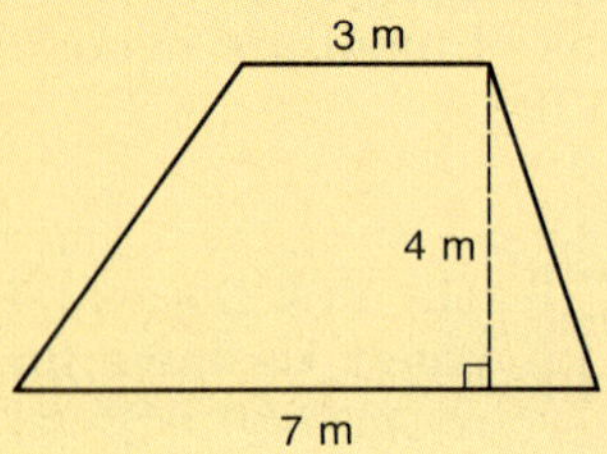

The area of the trapezoid is 20 m^2.

PRACTICE EXERCISES

Find the area of the triangle.

1.

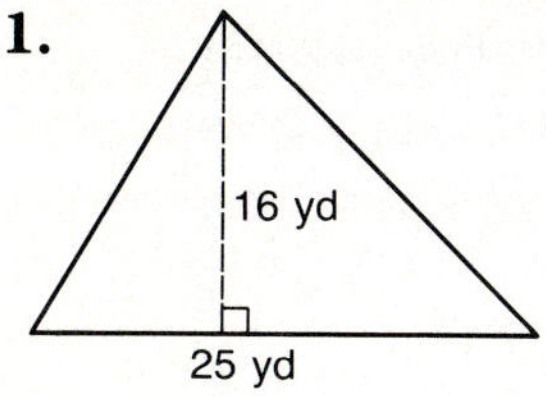

2.

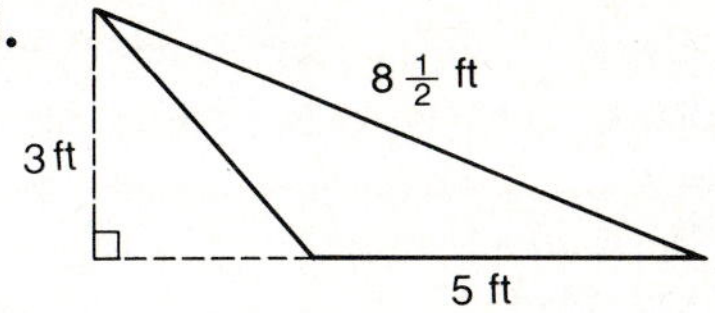

3.

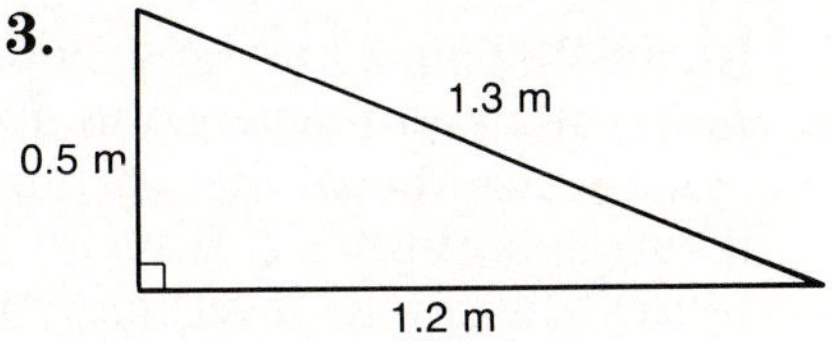

4. $b = 14$ in., $h = 12$ in. **5.** $b = 15.2$ cm, $h = 4$ cm **6.** $b = 2\frac{3}{8}$ yd, $h = 4$ yd

Find the area of the trapezoid.

7.

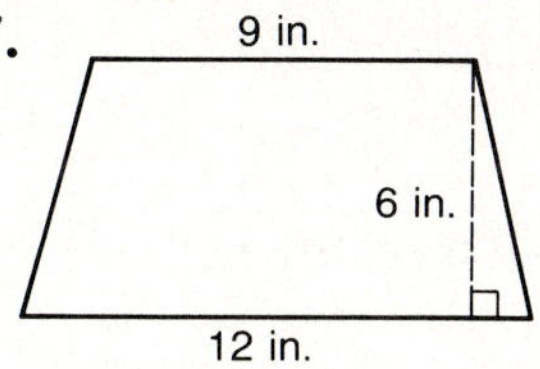

8.

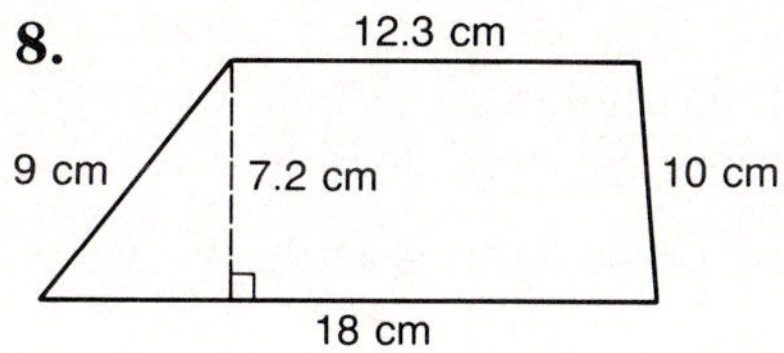

9.

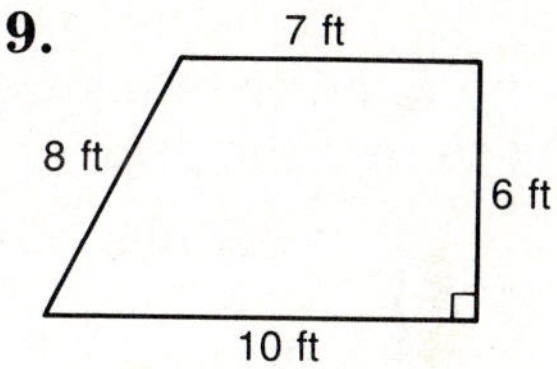

10. $b_1 = 3$ ft, $b_2 = 4$ ft, $h = 2$ ft

11. $b_1 = 5$ yd, $b_2 = 13$ yd, $h = 12$ yd

12. $b_1 = 1.8$ m, $b_2 = 6.2$ m, $h = 4.2$ m

13. $b_1 = 3\frac{1}{4}$ ft, $b_2 = 6\frac{3}{4}$ ft, $h = 8\frac{2}{3}$ ft

MIXED REVIEW

Find the number.

14. What number is 30% of 68?

15. 10% of what number is 8?

16. What percent of 12 is 8?

17. 24 is 75% of what number?

18. 12 is 4% of what number?

19. 25 is what percent of 100?

20. What number is 25% of 36?

21. What percent of 35 is 28?

Solve.

22. Laura keeps a $3\frac{1}{2}$ ft-high hedge clipped in the shape of a trapezoid 8 ft along the bottom and 6 ft on the top. When she sprays one side, what area does she have to cover?

23. When Laura fertilizes her zinnias, she mixes 1 oz of concentrate with enough water to make 1 gal of spray. In this spray, what is the ratio of concentrate to water?

24. Laura's flower bed is a regular octagon 4 ft on a side. She borders it with solid pieces cut from railroad ties 6 ft long. How many feet of scrap are left over?

25. Laura walks off a patch of garden for tomatoes. She walks 12 ft north, 5 ft west, and then 13 ft straight back to where she started. What is the area of Laura's tomato patch?

12.5 AREAS OF CIRCLES

Kevin Woollam is a landscaper. He is designing an underground sprinkler system for one of his clients. Each rotating sprinkler can water 40 ft of lawn in any direction. How many square feet of lawn can each sprinkler cover?

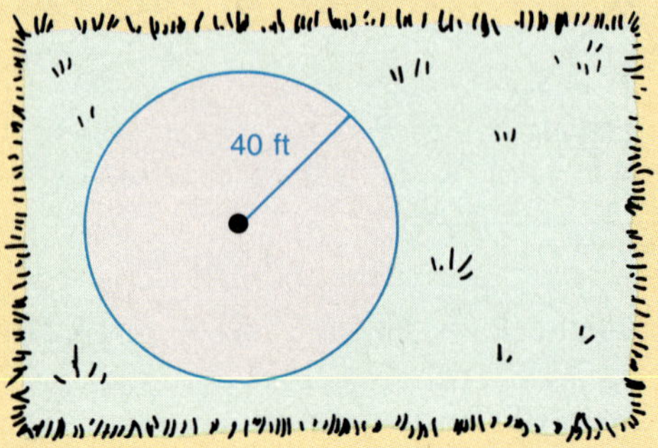

To find the number of square feet each sprinkler can cover, you need to find the area of the circle, which is $\pi \times r^2$.

EXAMPLE 1 Find the area of lawn each sprinkler can cover. Use 3.14 for π.

THINK: $A = \pi \times r^2$

$A \approx 3.14 \times 40^2$

$A \approx 5{,}024$

Each sprinkler can cover approximately 5,024 ft^2 of lawn.

With fractions and mixed numbers, use $\frac{22}{7}$ for π.

EXAMPLE 2 Find the area of a circle whose radius is $5\frac{1}{4}$ ft.

THINK: $A = \pi \times r^2$

$A \approx \frac{22}{7} \times \left(5\frac{1}{4}\right)^2$

$A \approx \frac{\overset{11}{\cancel{22}}}{\underset{1}{\cancel{7}}} \times \frac{\overset{3}{\cancel{21}}}{\underset{2}{\cancel{4}}} \times \frac{21}{4}$

$A \approx 86\frac{5}{8}$

The area of the circle is approximately $86\frac{5}{8}$ ft^2.

CHECKPOINT

Write the letter of the correct answer.

Find the area.

1. $r = 4$ km **a.** 50.24 km^2 **b.** 25.12 km^2 **c.** 12.56 km^2 **d.** 6.28 km^2

2. $r = 3\frac{1}{2}$ ft **a.** $38\frac{1}{2}$ ft^2 **b.** 25 ft^2 **c.** 11 ft^2 **d.** $5\frac{1}{2}$ ft^2

PRACTICE EXERCISES

Find the area. Use 3.14 for π with whole numbers and decimals. Use $\frac{22}{7}$ for π with fractions and mixed numbers.

1. 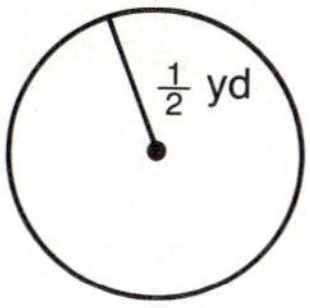

2. 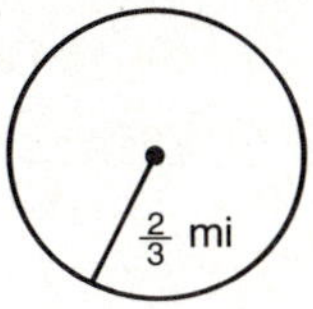

3. 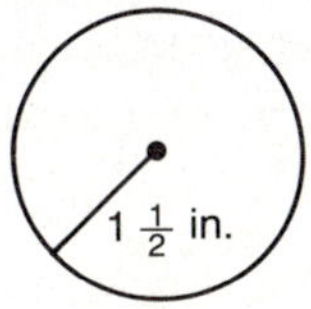

4. 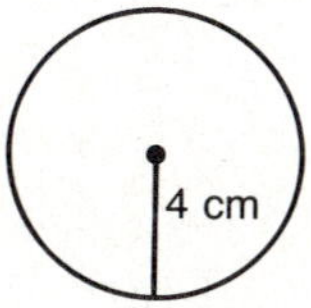

5. 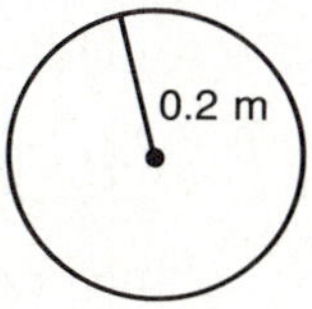

6. 

7. $r = 1$ in. **8.** $r = 4$ mi **9.** $r = 10$ ft **10.** $r = 7$ yd

11. $r = \frac{2}{3}$ ft **12.** $r = 2\frac{3}{4}$ yd **13.** $r = \frac{1}{8}$ in. **14.** $r = 3\frac{1}{4}$ ft

15. $r = 3$ cm **16.** $r = 0.1$ m **17.** $r = 0.5$ km **18.** $r = 13$ mm

19. $d = 10$ yd **20.** $d = 0.1$ m **21.** $d = 12$ in. **22.** $d = 5$ m

MIXED REVIEW

Complete.

23. 2 ft = ■ in. **24.** 12 qt = ■ gal **25.** 6 T = ■ lb

26. 2,500 m = ■ km **27.** 8 g = ■ mg **28.** 3.5 cm = ■ mm

Kevin Woollam's next project will be a fish pond for the Bergers. Use the diagram for Exercises 29–31.

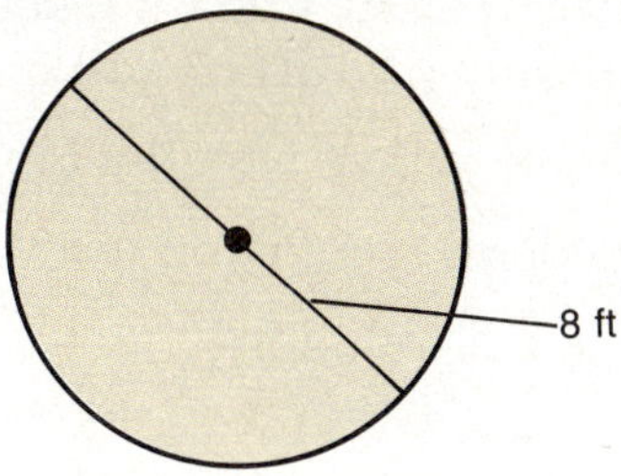

29. The Bergers have bought 60 ft^2 of tile for the pond. Do they have enough tile to cover the entire bottom of the pond?

30. Kevin plans to plant flowers along the border of the pond. What is the distance around the pond?

31. The Bergers want to stock the pond with goldfish so that there are 2 goldfish for every 5 ft^2 of pond. How many fish do they need?

PROBLEM Solving STRATEGY

12.6 SOLVING A SIMPLER PROBLEM

Situation:

Delray Condominiums hires Green Valley Contractors to plant 3,990 ivy cuttings along a stone boundary wall. In one day, 4 Green Valley workers plant 570 ivy cuttings each. How many cuttings remain to be planted?

Strategy:

Sometimes it helps to solve a simpler problem. Write subproblems using rounded numbers.

Applying the Strategy:

Subproblem 1: How many ivy cuttings were planted in all?

THINK: NUMBER OF WORKERS × NUMBER OF CUTTINGS

$$4 \times 600 = 2{,}400$$

Subproblem 2: How many ivy cuttings remain to be planted?

THINK: TOTAL NUMBER OF CUTTINGS − CUTTINGS ALREADY PLANTED

$$4{,}000 - 2{,}400 = 1{,}600$$

Now use the actual numbers.

Subproblem 1: $4 \times 570 = 2{,}280$ *Subproblem 2:* $3{,}990 - 2{,}280 = 1{,}710$

There are 1,710 ivy cuttings that remain to be planted.

FOR DISCUSSION

Green Valley needs 115 yd of molding for the Delray lobby. The molding costs $3.85 per yard. Green Valley's installation fee is $1.25 per yard. How much will Delray pay Green Valley for the lobby molding?

1. What rounded numbers would you use to solve the problem?
2. What subproblems would you use?
3. What is the actual solution?

PRACTICE EXERCISES

Write the letter of the better plan for simplifying the problem.

1. A parking lot at Delray needs a new fence. The lot is 52.5 m long and 36.75 m wide. Will the lot need more than 250 m of fencing?

   a. *Step 1:* $50 \times 40 = 2{,}000 \text{ m}^2$
   Step 2: $2{,}000 \text{ m}^2 > 250 \text{ m}$

   b. *Step 1:* $2 \times 50 + 2 \times 40 = 180 \text{ m}$
   Step 2: $180 \text{ m} < 250 \text{ m}$

2. It took 5 gardeners $2\frac{3}{4}$ hours to prepare a lawn area of $1{,}650 \text{ m}^2$ for seeding. How long would it take one gardener working at the same speed at prepare 330 m^2 of lawn for seeding?

   a. *Step 1:* $1{,}500 \div 3 = 500$
   Step 2: $500 \div 5 = 100$
   Step 3: $300 \div 100 = 3$ hours

   b. *Step 1:* $300 \times 3 = 900$
   Step 2: $900 \div 5 = 180$
   Step 3: $180 \div 3 = 60$ hours

Solve by writing a simpler problem.

3. Green Valley bought 50 pine trees for \$89.75 each. They sold the trees to Delray Condominiums for \$118.95 each. How much profit did Green Valley make on the pine trees?

4. Green Valley is planting 192 juniper bushes along a slope. One morning, 3 workers planted 18 bushes each. How many juniper bushes remain to be planted? 138 juniper bushes

5. It takes $2\frac{1}{4}$ cans of preservative to treat 3 dozen railroad ties. Green Valley plans to treat 512 ties. How much preservative will be needed?

6. Green Valley has workers paint the fence around the pool. The painters need $8\frac{1}{4}$ gal of paint for each 210 ft of fence. How much paint is needed to cover 840 ft of fence? 33 gal

7. Delray plans to add a new lawn of 775 m^2. The price to seed is \$1.95 per square meter. The price of sod is \$4.75 per square meter. How much money will be saved if seed is used instead of sod?

8. Green Valley uses a special soil mix for its planters. For each 120 lb of topsoil, Green Valley uses $1\frac{3}{4}$ bales of peat moss. A bale weighs 40 lb. How many pounds of peat moss are needed for 420 lb of topsoil?

MIDCHAPTER REVIEW

Find the perimeter and the area. [330, 334, 336]

1. $P = ■$ **2.** $A = ■$ **3.** $P = ■$ **4.** $A = ■$ **5.** $P = ■$ **6.** $A = ■$

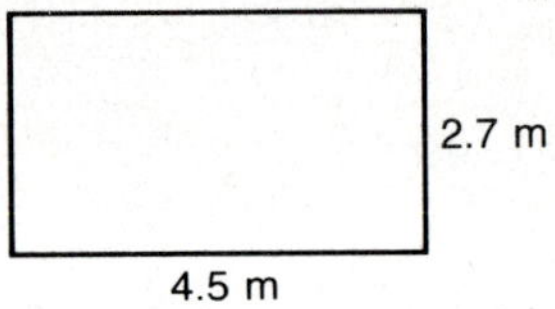

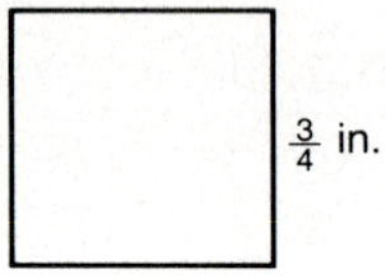

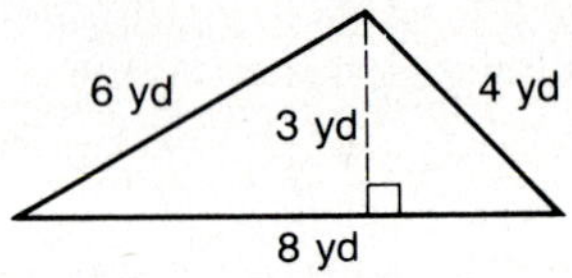

7. $P = ■$ **8.** $A = ■$ **9.** $P = ■$ **10.** $A = ■$ **11.** $P = ■$ **12.** $A = ■$

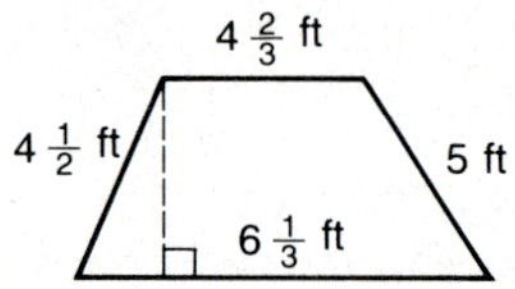

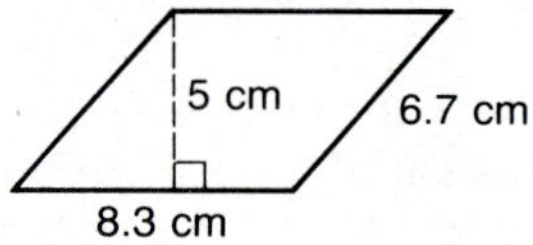

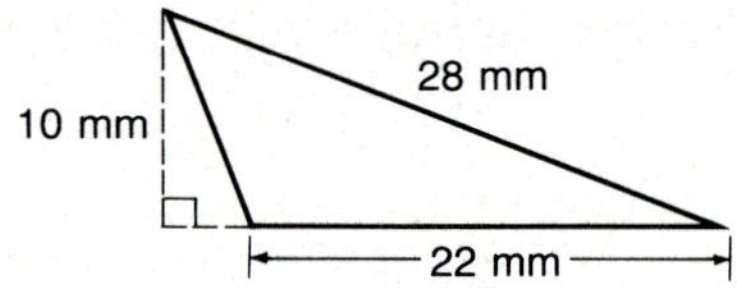

Find the circumference and the area. Use 3.14 for π with whole numbers and decimals. Use $\frac{22}{7}$ for π with fractions and mixed numbers. [332, 338]

13. $C = ■$ **14.** $A = ■$ **15.** $C = ■$ **16.** $A = ■$ **17.** $C = ■$ **18.** $A = ■$

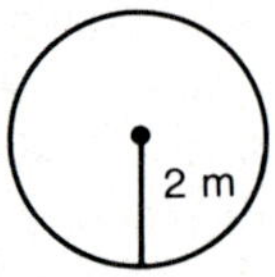

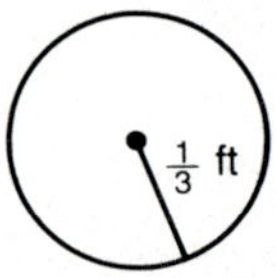

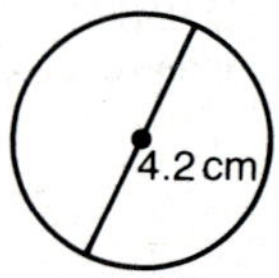

Solve. [340]

19. The Golden Hind is a deep-sea fishing boat out of Provincetown. It takes $2\frac{3}{4}$ hours for the Golden Hind to travel the 33 mi to the fishing grounds. How far can the Golden Hind travel in 2 hours?

20. The mate who runs the snack bar uses $13\frac{3}{4}$ oz of cheese to make 5 ham and cheese sandwiches. Assuming he has enough ham and bread, how many sandwiches can he make from $92\frac{1}{2}$ oz of cheese?

21. It takes the hook 42 seconds to reach the ocean floor 315 ft below. How long would it take the hook to reach the ocean floor at a depth of $247\frac{1}{2}$ ft?

22. A 6-lb cod produces two $2\frac{1}{4}$-lb fillets. The Mull family leaves the Golden Hind with $31\frac{1}{2}$ lb of cod. This is the equivalent of how many 6-lb fish?

PROBLEM Solving APPLICATION

12.7 CAREER: CARPET INSTALLER

Charles Monet is a carpet installer. The ability to measure accurately is important in wall-to-wall carpeting. Often, a room may consist of more than one rectangular section. Sometimes several rooms and a hallway are carpeted with the same type of carpet.

Find the number of square yards of carpeting needed.

1.

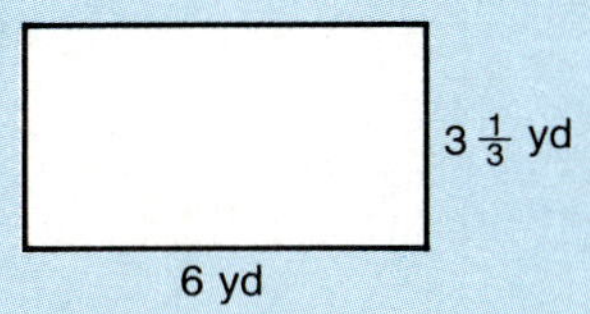

2.

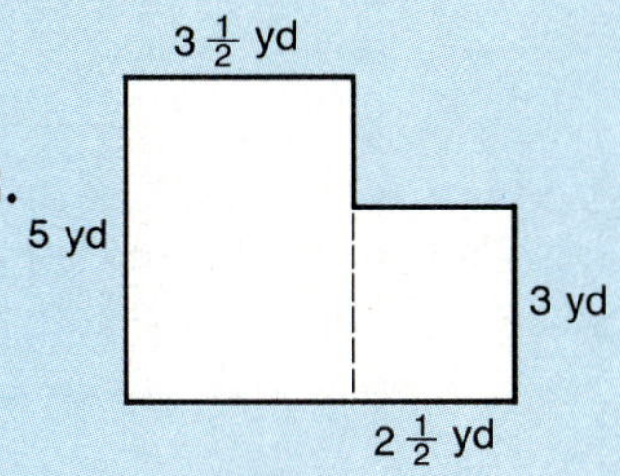

3.

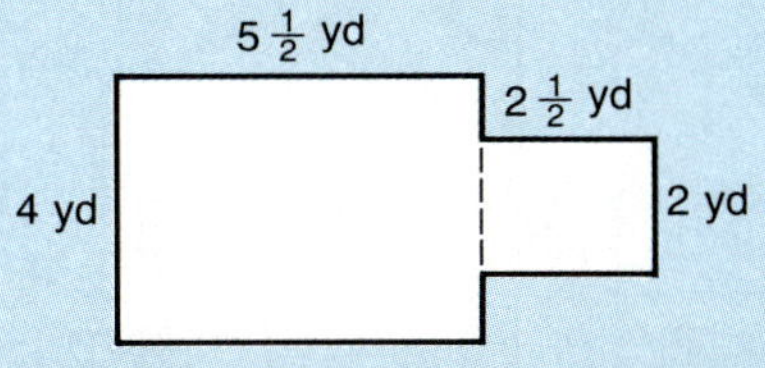

4.

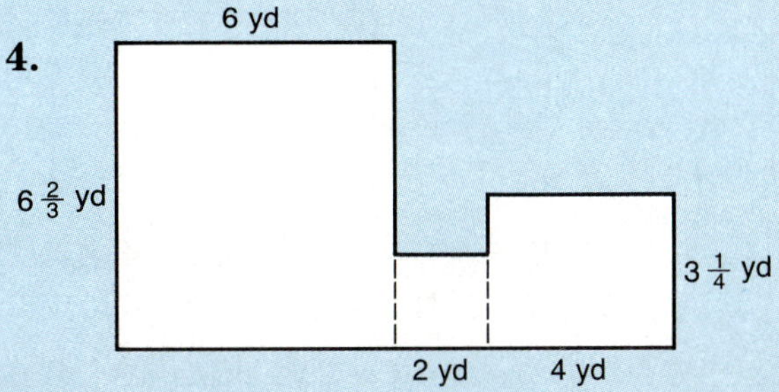

Solve.

5. Charles installs wall-to-wall carpeting in a rectangular room that is 6.5 yd long and 4.0 yd wide. What is the area covered by the carpet?

6. In the same house, Charles carpets a rectangular room 5.8 yd long and 4.2 yd wide. How much less is the area covered by this carpet than the area covered in Problem 5?

7. During a period of one week, Charles worked a total of 35 hours. Of this time, 24.5 hours were spent installing carpets. 10.5 hours were spent traveling to and from the clients' homes. What percent of Charles' time was spent traveling?

Charles' cousin, Ann, decides to buy wall-to-wall carpeting for a rectangular room 21 ft long and 18 ft wide.

8. What is the area of the floor in square yards?

9. When Ann orders the carpeting, should she order an amount that is less than, equal to, or greater than the area of the floor? Why?

12.8 SOLID FIGURES

Objects in the real world are called three-dimensional if they have length, width, and height. In mathematics, three-dimensional objects are known as **solid figures.**

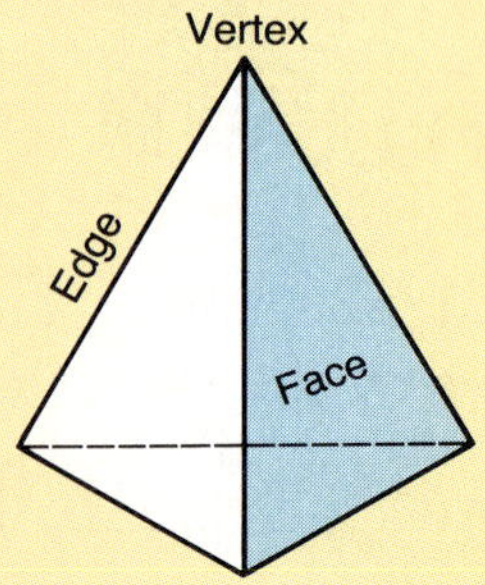

A **polyhedron** is a solid figure whose sides are all polygons. Each side is called a **face.** Two faces intersect in a line segment called an **edge.** Two edges intersect in a point called a **vertex.**

A **prism** is a polyhedron whose bases are congruent polygons and whose other faces are parallelograms. A prism is named according to the shape of its bases. A cube is a prism whose faces are all squares.

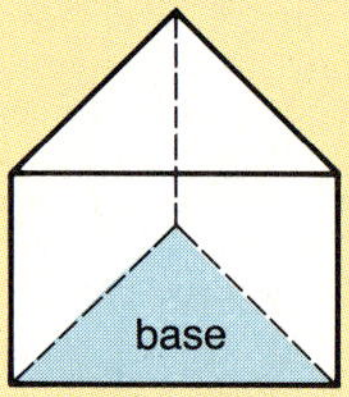

Triangular prism

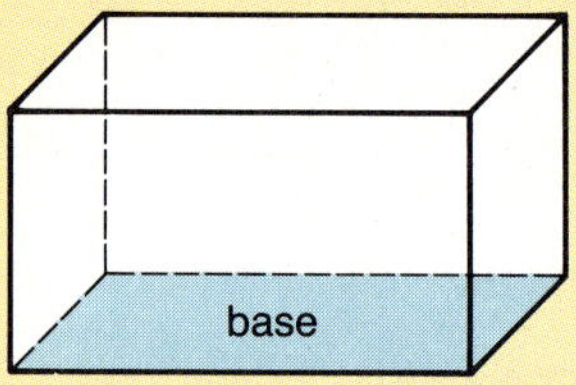

Rectangular prism

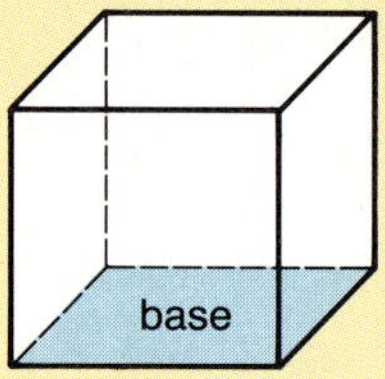

Cube

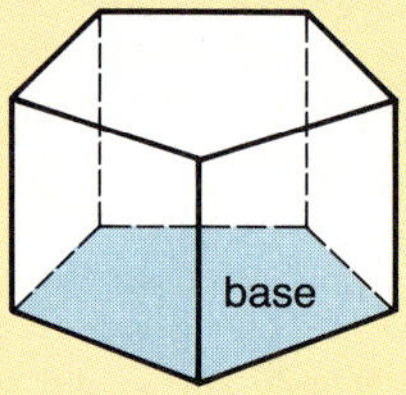

Pentagonal prism

EXAMPLE 1 Identify the figure. Count the numbers of faces, edges, and vertices.

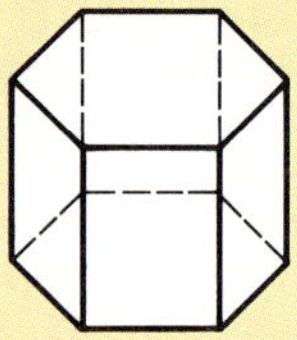

THINK: The sides are all polygons.
The figure is a prism.
Its bases are hexagons.

The figure is a hexagonal prism.
It has 8 faces, 18 edges, and 12 vertices.

A **pyramid** is a polyhedron whose base is any polygon and whose other faces are triangles that all meet at a common vertex. A pyramid is also named according to the shape of its base. The figure shown at the right is a square pyramid.

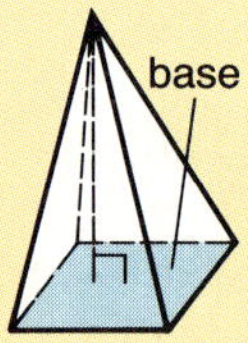

EXAMPLE 2 Identify the figure. Count the numbers of faces, edges, and vertices.

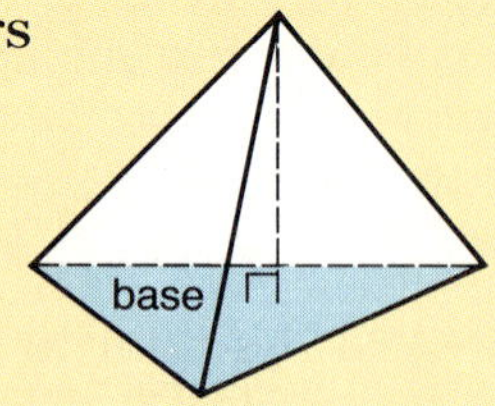

THINK: The sides are all triangles.
The figure is a pyramid.
Its base is a triangle.

The figure is a triangular pyramid.
It has 4 faces, 6 edges, and 4 vertices.

Other solid figures have curved surfaces.

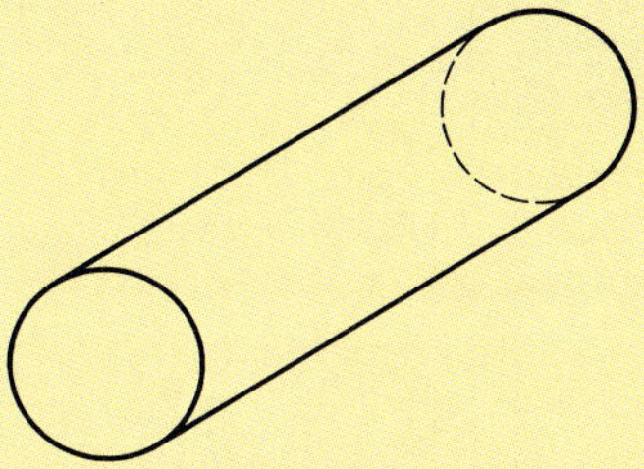

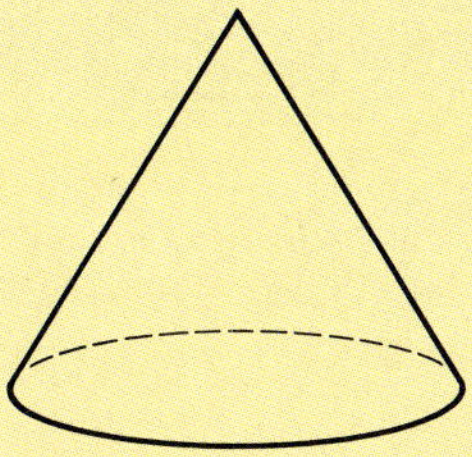

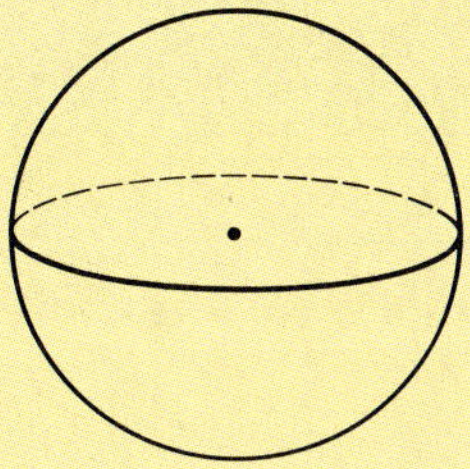

A **cylinder** has two circular bases that are congruent and parallel.

A **cone** has one circular base.

A **sphere** is a curved surface of points that are all the same distance from the center.

EXAMPLE 3 Write whether the following statement is *true* or *false*.

The circular bases of a cylinder can have different sizes.

THINK: According to the definition of a cylinder, the circular bases must be congruent.

The statement is false.

FOR DISCUSSION

Name objects from everyday life that have the following shapes: a rectangular prism, a cube, a square pyramid, a cylinder, a cone, a sphere.

PRACTICE EXERCISES

Identify the figure.

1.

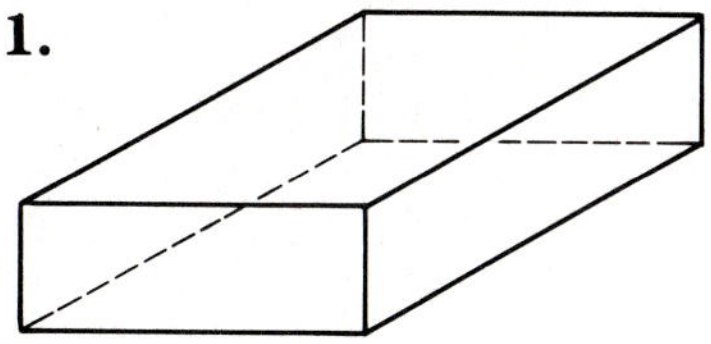

2.

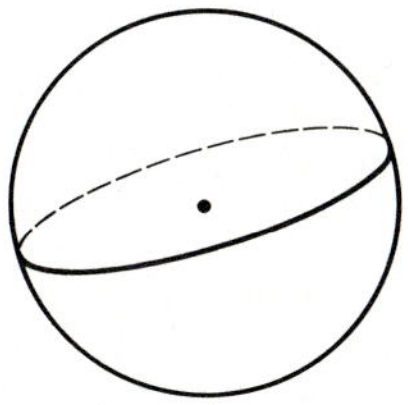

3.

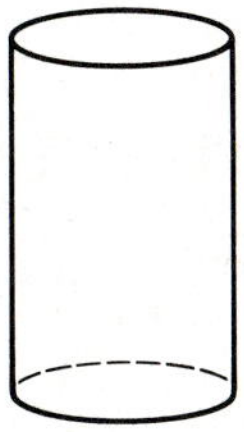

4.

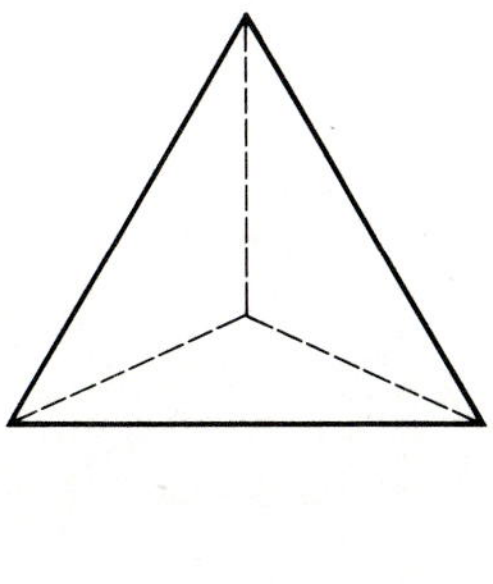

5.

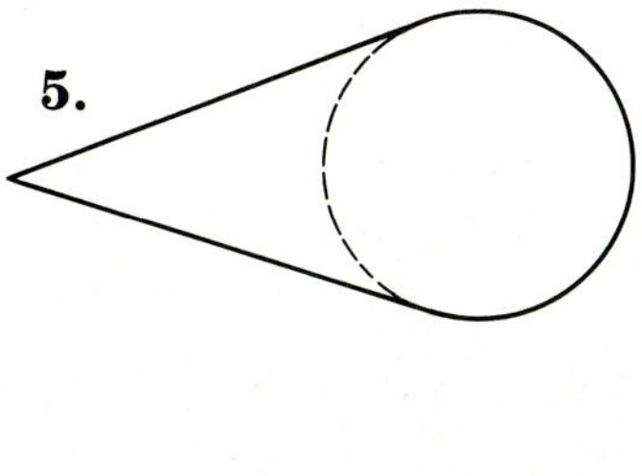

6.

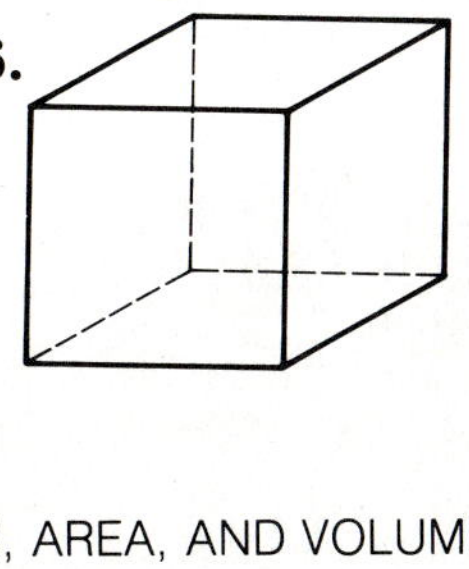

Copy and complete the table.

Polyhedron	Number of vertices	Number of faces	Number of edges
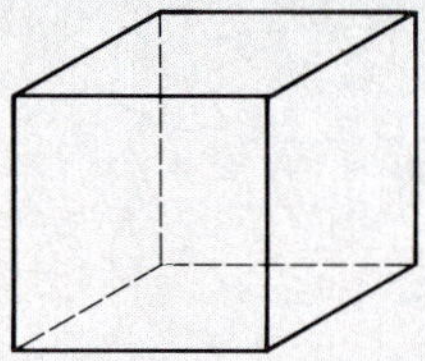	**7.** ■	**8.** ■	**9.** ■
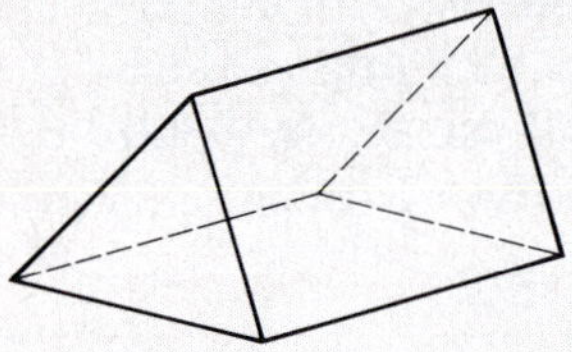	**10.** ■	**11.** ■	**12.** ■
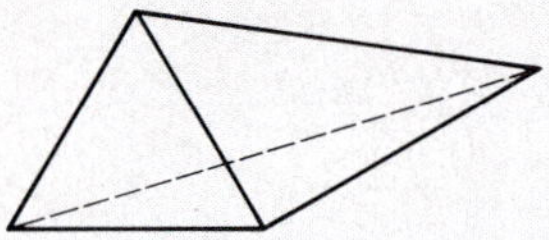	**13.** ■	**14.** ■	**15.** ■
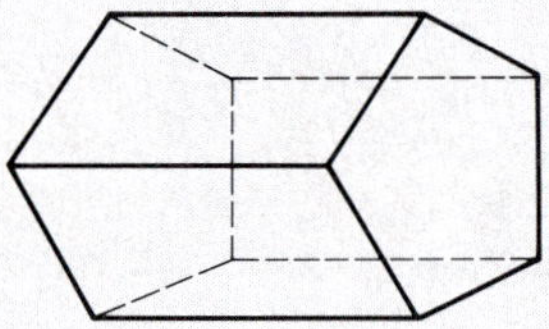	**16.** ■	**17.** ■	**18.** ■

Write whether the statement is *true* or *false.*

19. A sphere is a type of prism.

20. A cone is a circular pyramid.

21. All prisms have 6 faces and 8 vertices.

22. An octagonal prism has 8 edges.

23. A prism is a type of polyhedron.

24. A cylinder is a type of polyhedron.

25. A pyramid is named by the shape of its base.

26. A cube is a prism with 6 congruent square faces.

27. The faces of a polyhedron must all be rectangles.

TIME OUT

Using 12 unbroken matchsticks touching end to end, it is possible to form 6 squares. How?

12.9 SURFACE AREA

Deborah Kline owns an antique shop. She is refinishing the trunk shown at the right. The can of varnish she is using has enough liquid left in it to cover 30 ft^2. Is there enough varnish left in the can to refinish the trunk?

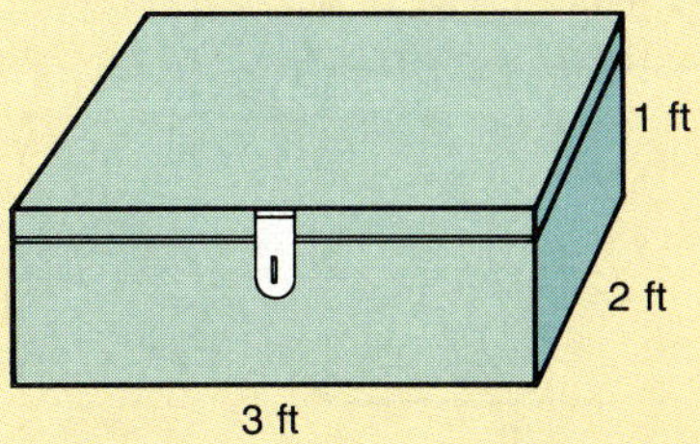

To find whether there is enough varnish left, you need to know the surface area of the trunk.

The **surface area** of a polyhedron is equal to the sum of the areas of its faces.

EXAMPLE 1 Find the surface area of the trunk.

1. Find the area of each face.

THINK: The faces are rectangles.

Use $A = l \times w$.

Top/bottom: $3 \times 2 = 6$ ft^2 each

Front/back: $3 \times 1 = 3$ ft^2 each

Sides: $2 \times 1 = 2$ ft^2 each

2. Add the areas of the faces.

$(2 \times 6) + (2 \times 3) + (2 \times 2) = 22$ ft^2

top/bottom **front/back** **sides**

The surface area of the trunk is 22 ft^2.
There is enough varnish left to refinish the trunk.

Because the 6 faces of a cube are all congruent squares, the surface area of a cube is 6 times the area of one face, or $6 \times s^2$.

EXAMPLE 2 Find the surface area of the cube.

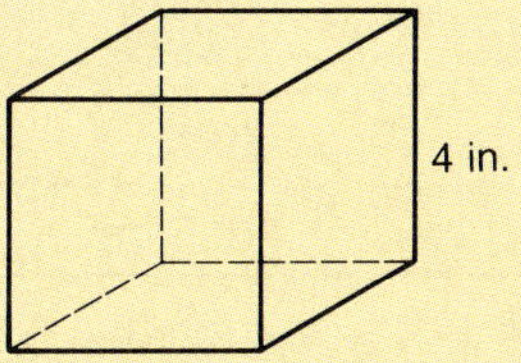

THINK: $A = 6 \times s^2$

$A = 6 \times 4^2 = 96$

The surface area of the cube is 96 $in.^2$.

You can use rounding to estimate the surface area of a prism.

EXAMPLE 3 Estimate the surface area of the cube.

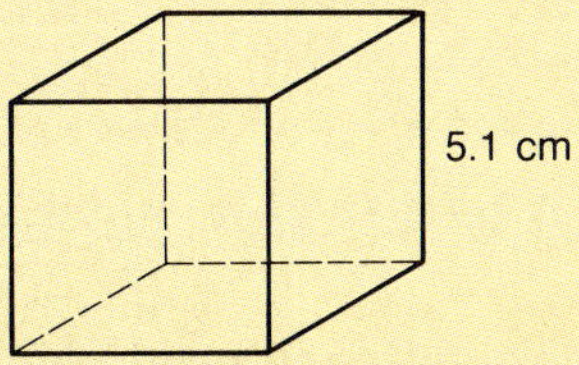

THINK: 5.1 rounds to 5.

Estimate: $A = 6 \times s^2$

$A = 6 \times 5^2 = 150$

The surface area of the cube is about 150 cm^2.

PRACTICE EXERCISES

Find the surface area.

1.

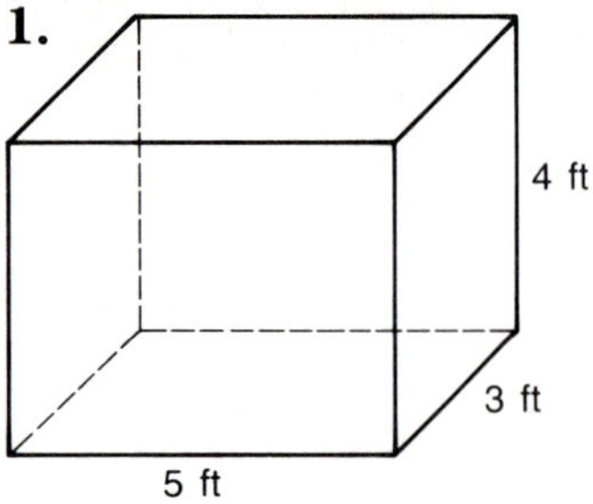

2.

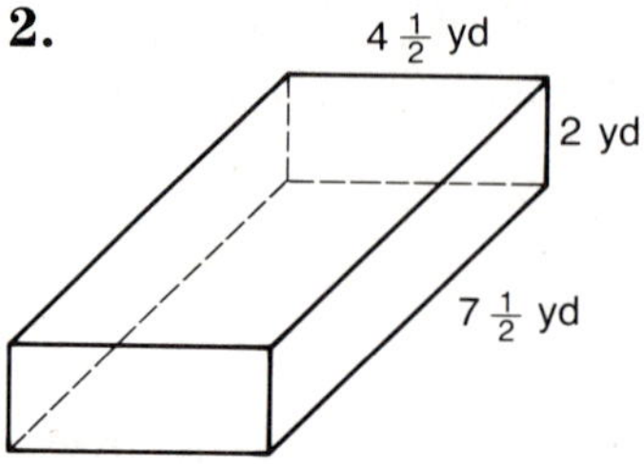

3.

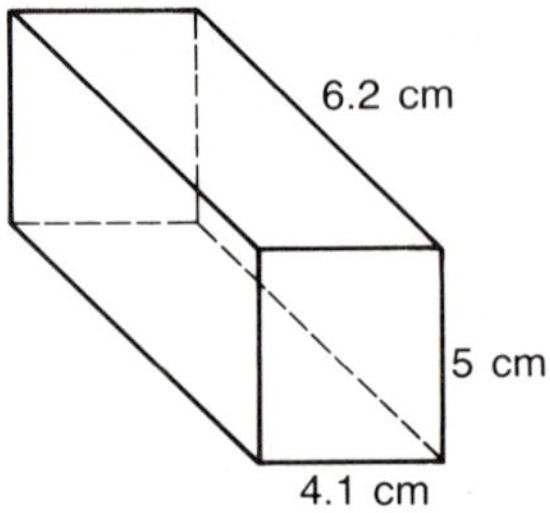

4.

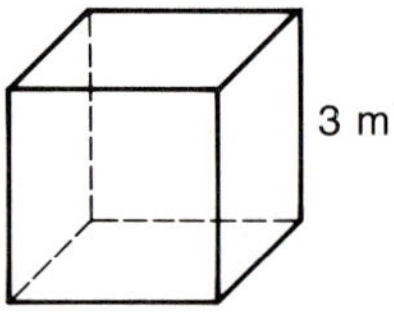

5.

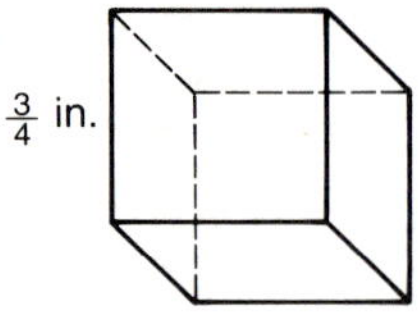

6.

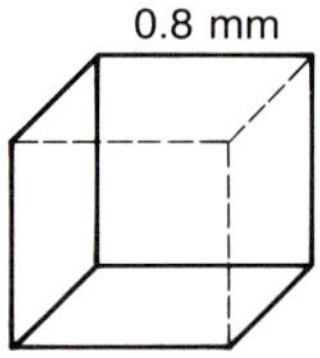

7.

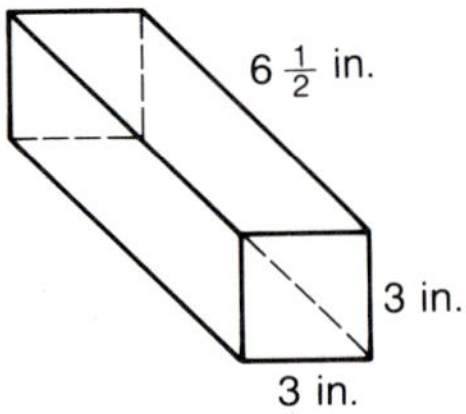

8.

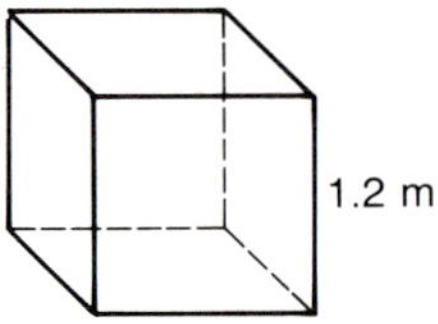

9.

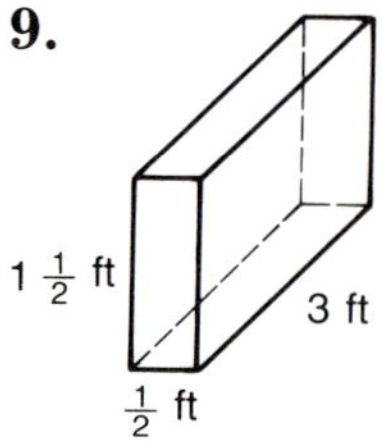

Complete the table.
Each set of measurements describes a rectangular prism.

Length	Width	Height	Estimated surface area
10. 62 ft	19 ft	101 ft	■
11. 10.2 cm	19.6 cm	72 cm	■
12. 3.06 m	8.97 m	6.18 m	■
13. $\frac{7}{16}$ in.	$\frac{7}{16}$ in.	$\frac{7}{16}$ in.	■

Solve.

14. Kline's Antiques ships sofas in plywood crates measuring 3 ft by $10\frac{1}{2}$ ft by 4 ft. How much plywood does it take to make each crate?

15. Kline's sells restored spice mills. Each mill is a cube 8 in. on a side. A pint of stain will cover 64 ft^2. How many mills will that same pint cover? (*Hint:* 1 ft^2 = 144 $in.^2$)

PROBLEM Solving APPLICATION

12.10 CONSUMER: GIFT WRAPPING A PACKAGE

A package has the shape of a rectangular prism, with the dimensions shown. You want to gift wrap the package. When wrapping a package, you need more than the minimum surface area. You need to use an additional 20% to allow paper for folding and overlapping. How many square inches of wrapping paper do you need to wrap the package?

1. Find the total surface area.

Top and bottom:	$2 \times (5 \times 3) =$	30
Front and back:	$2 \times (6 \times 5) =$	60
Sides:	$2 \times (6 \times 3) =$	36
Total:		126

2. Find 20% of the total surface area to allow for extra paper needed.

   THINK: $20\% = 0.20$ $\quad 0.20 \times 126 = 25.2$

3. Add to find the total amount of paper needed. $\quad 126 + 25.2 = 151.2 \approx 151$

So, you need about 151 in.^2 of gift wrapping paper.

Find the total amount, in square inches, of gift wrapping paper needed. Remember to allow for extra paper. Round the answer to the nearest in.^2

1.

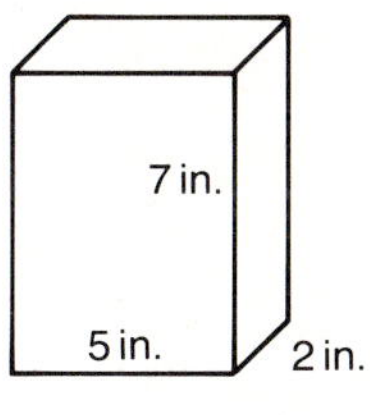

2.

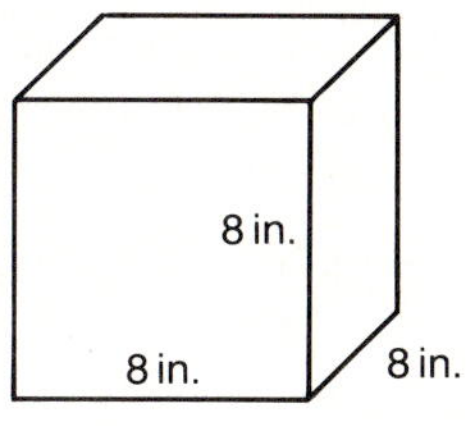

3.

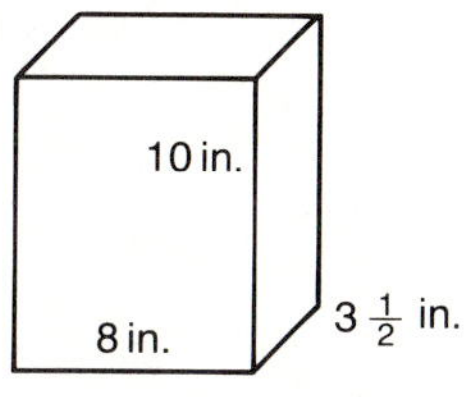

4. A box with a length of 2 ft, a width of 1 ft, and a height of 1 ft.

5. A box with a length of 1 ft 6 in., a width of 1 ft 6 in., and a height of 8 in.

You want to wrap a package with $l = 10$ in., $w = 6$ in., and $h = 4$ in.

6. Will 300 in.^2 of wrapping paper cover the package? Will you have any paper left over? If so, how much?

7. Will a sheet of wrapping paper measuring 12 in. by 15 in. cover the package? How many inches short or over will the sheet be?

12.11 VOLUMES OF PRISMS AND PYRAMIDS

Denise Grant keeps tropical fish. A model of her aquarium is shown at the right. Each fish needs at least 0.5 ft^3 of water. What is the maximum number of fish she can keep in the aquarium?

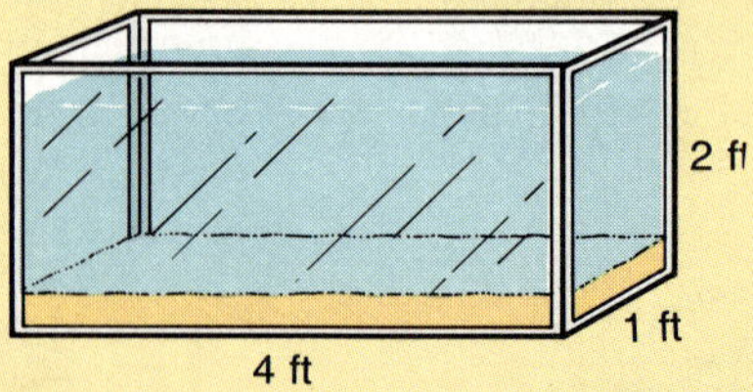

To find the answer, you need to find the volume of the aquarium.

Volume is the amount of space inside a solid figure. It is measured in cubic units such as cubic meters (m^3), cubic centimeters (cm^3), cubic yards (yd^3), cubic feet (ft^3), and cubic inches ($in.^3$).

To find the volume of a prism, multiply the area of its base by its height.

EXAMPLE 1

Find the volume of the aquarium.

THINK: The aquarium is a rectangular prism.

V = area of base (B) $\times$ height (h)

$V = (l \times w) \times h$

$V = (4 \times 1) \times 2$

$V = 8$

The aquarium holds 8 ft^3 of water. Since each fish needs at least 0.5 ft^3 of water, the aquarium can hold a maximum of 16 fish ($8 \div 0.5 = 16$).

The volume of a pyramid is $\frac{1}{3}$ the volume of the prism that has the same base and height.

EXAMPLE 2

Find the volume of the pyramid.

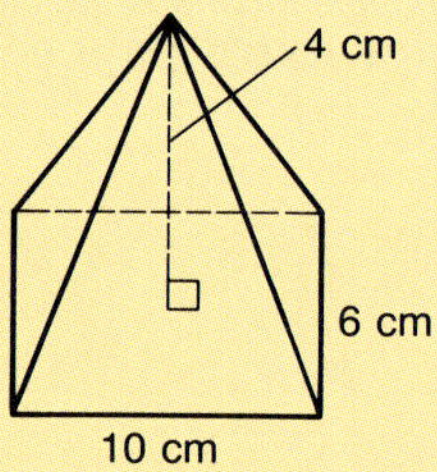

THINK: $V = \frac{1}{3} \times B \times h$

$V = \frac{1}{3} \times (l \times w) \times h$

$V = \frac{1}{3} \times (10 \times 6) \times 4$

$V = 80$

So, the volume of the pyramid is 80 cm^3.

FOR DISCUSSION

A number raised to the third power is called the cube of that number. Why is it called the cube?

PRACTICE EXERCISES

Find the volume of the rectangular prism.

1.

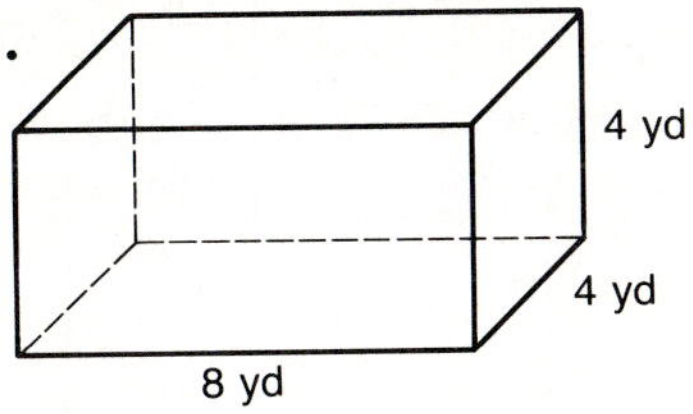

2.

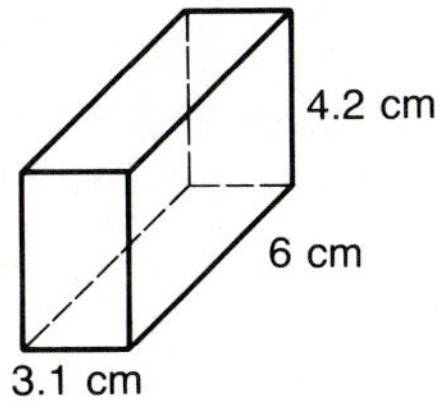

3.

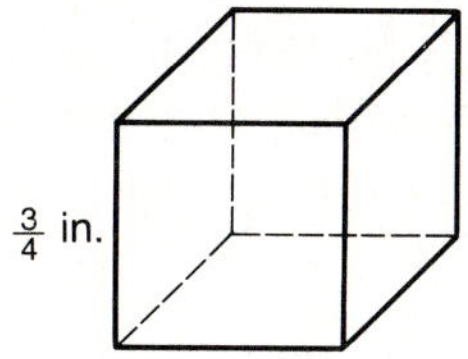

	Length	Width	Height		Length	Width	Height
4.	5 in.	4 in.	6 in.	**5.**	9 ft	9 ft	9 ft
6.	$1\frac{2}{3}$ yd	6 yd	$2\frac{1}{2}$ yd	**7.**	$\frac{1}{3}$ ft	$\frac{1}{3}$ ft	$\frac{1}{3}$ ft
8.	6 cm	6 cm	6 cm	**9.**	9 m	2 m	5 m
10.	3.4 cm	2 cm	5.2 cm	**11.**	12 m	4.5 m	0.5 m

Find the volume of the rectangular pyramid.

12.

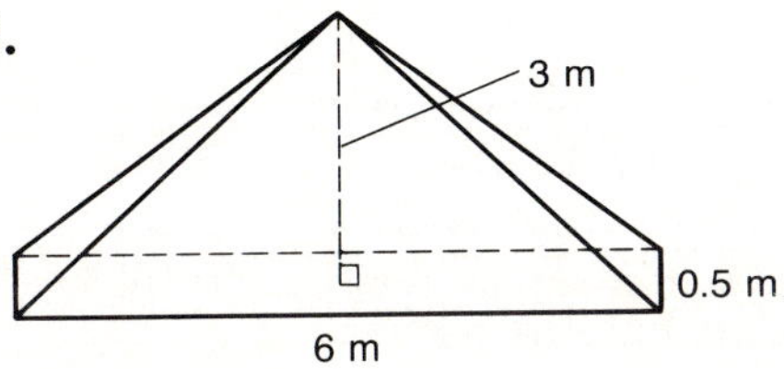

13.

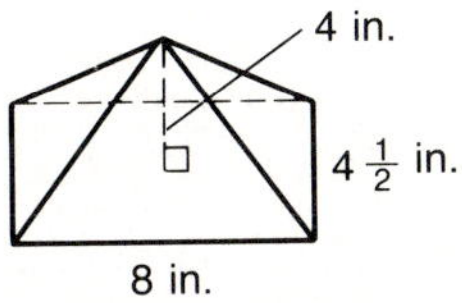

14.

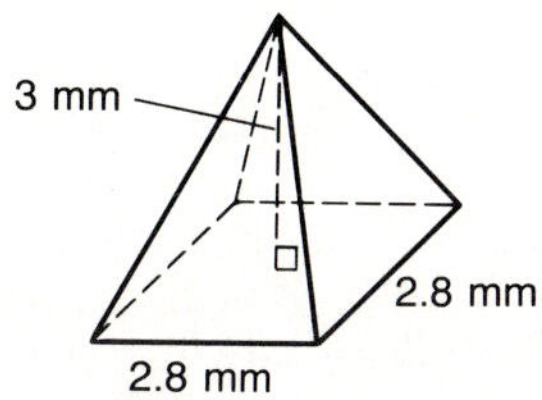

	Length	Width	Height		Length	Width	Height
15.	9 ft	5 ft	13 ft	**16.**	12 yd	3 yd	7 yd
17.	$2\frac{1}{2}$ in.	4 in.	3 in.	**18.**	$\frac{3}{4}$ ft	$\frac{3}{4}$ ft	16 ft
19.	8 m	3 m	4 m	**20.**	6 cm	6 cm	6 cm
21.	5.7 cm	3 cm	3.2 cm	**22.**	0.5 m	1.6 m	6 m

Solve.

23. The filter Denise Grant uses with her aquarium cleans 15 gal of water an hour. What part of a gallon does it clean in a minute?

24. The largest aquarium on sale at Fish U.S.A. is the 132-ft^3 Seascape model. The Seascape is 6 ft long and 4 ft high. How wide is the Seascape?

25. A display aquarium at Fish U.S.A. measures 36 in. by 24 in. by 18 in. To the nearest gallon, how many gallons does the tank hold? (*Hint:* 231 $in.^3$ hold 1 gal.)

26. Denise buys a saltwater tank 4 ft long, 2 ft wide, and 30 in. high. What is the area of the lamp hood that fits on top?

12.12 VOLUMES OF CYLINDERS AND CONES

Erik Henson spends his summers working for the Homer Pool Company. The pool shown at the right is the Odyssey model. What is its volume? What is its capacity in gallons?
(*Hint:* 1 ft^3 holds $7\frac{1}{2}$ gal.)

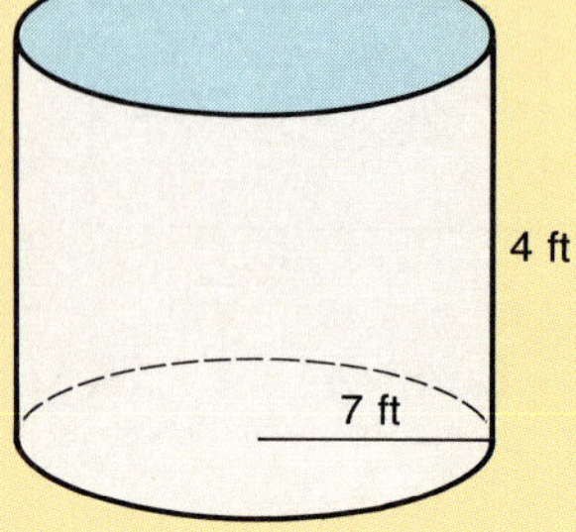

The pool is a cylinder. To find the capacity of the pool, you need to find the volume of the cylinder.

To find the volume of a cylinder, multiply the area of its base by its height.

EXAMPLE 1

Find the volume of the pool.

THINK: V = area of base (B) $\times$ height (h)

$$V = (\pi \times r^2) \times h$$

$$V \approx \left(\frac{22}{7} \times 7^2\right) \times 4$$

$$V \approx 616$$

The volume of the swimming pool is approximately 616 ft^3. Its capacity is approximately $7\frac{1}{2} \times 616$, or 4,620, gal.

The volume of a cone is $\frac{1}{3}$ the volume of the cylinder that has the same base and height.

EXAMPLE 2

Find the volume of the cone.

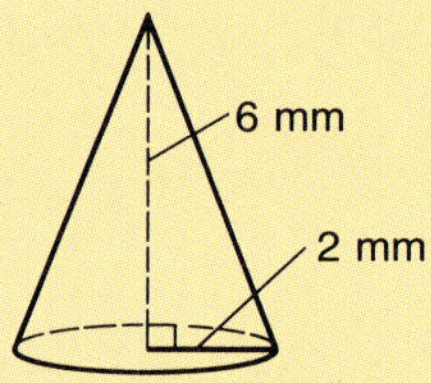

THINK: $V = \frac{1}{3} \times B \times h$

$$V = \frac{1}{3} \times (\pi \times r^2) \times h$$

$$V \approx \frac{1}{3} \times (3.14 \times 2^2) \times 6$$

$$V \approx 25.12$$

The volume of the cone is approximately 25.12 mm^3.

CHECKPOINT

Write the letter of the correct answer.

Find the volume of the figure.

1. Cylinder: $r = 3$ mi, $h = 7$ mi

a. 462 mi^3 **b.** 198 mi^3
c. 154 mi^3 **d.** 66 mi^3

2. Cone: $r = 10$ m, $h = 3$ m

a. 94.2 m^3 **b.** 282.6 m^3
c. 314 m^3 **d.** 942 m^3

PRACTICE EXERCISES

Find the volume of the cylinder. Use 3.14 for π with whole numbers and decimals. Use $\frac{22}{7}$ for π with fractions and mixed numbers.

1.

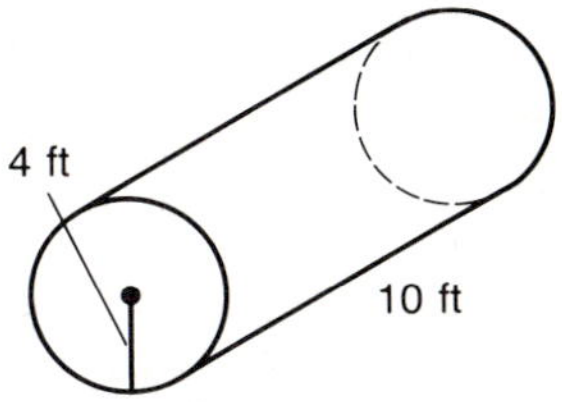

2.

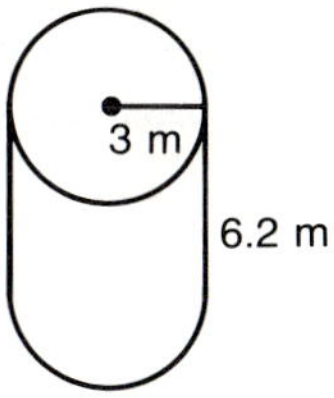

3.

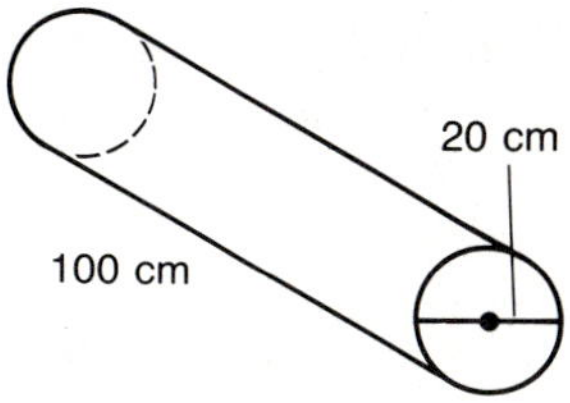

	Radius	Height
4.	2 ft	3 ft
7.	10 km	4.3 km
5.	$\frac{3}{4}$ yd	2 yd
8.	9 cm	3.2 cm
6.	1 yd	$3\frac{1}{2}$ yd
9.	0.5 m	5 m

Find the volume of the cone. Use 3.14 for π with whole numbers and decimals. Use $\frac{22}{7}$ for π with fractions and mixed numbers.

10.

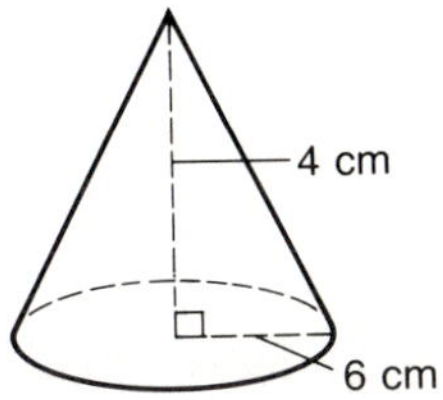

11.

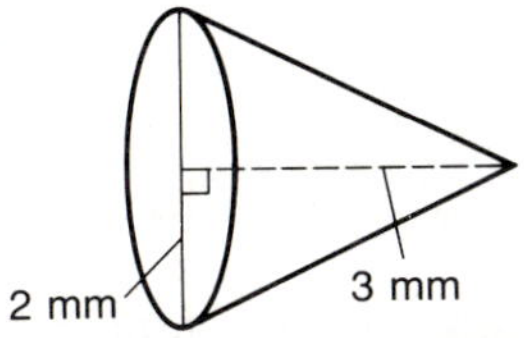

12.

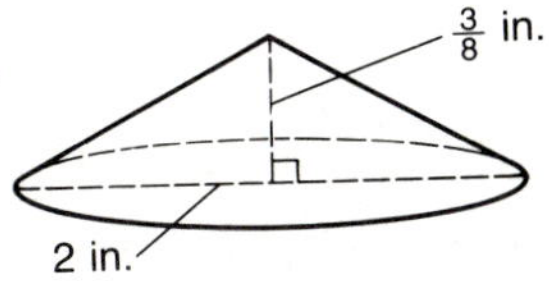

	Radius	Height
13.	3 ft	3 ft
16.	3 m	6.3 m
14.	$\frac{3}{4}$ yd	3 yd
17.	6 m	3.3 m
15.	24 ft	$1\frac{1}{2}$ ft
18.	0.3 cm	6 cm

Solve.

19. The Iliad is also a cylindrical swimming pool. Its capacity is 168 m^3. If the area of the bottom is 112 m^2, what is the depth of the pool?

20. The in-ground Hector is a rectangular prism 60 yd long, 30 yd wide, and 11 yd deep. How many square yards of tile are needed to cover the bottom?

21. The Homer Pool Company fills pools at the rate of \$0.015 per gallon. What would the charge be to fill a pool that holds 314 ft^3 of water? (*Hint:* 1 ft^3 holds $7\frac{1}{2}$ gal.)

22. The Penelope is a rectangular in-ground pool. It is 50 ft long and 17 ft wide. What is the perimeter of the Penelope?

PROBLEM Solving APPLICATION

12.13 CONSUMER: DECORATING AN APARTMENT

You want to paint your room, which is pictured below. It has the shape of a rectangular prism. Usually, a gallon of paint will cover about 450 ft^2. About how many gallons of paint will you need for your room?

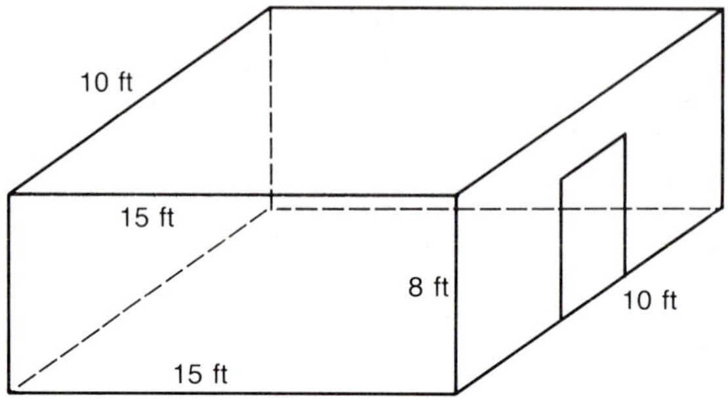

1. Find the total surface area.

Ceiling:	15×10	$= 150$
Longer walls:	$2 \times (8 \times 15)$	$= 240$
Shorter walls:	$2 \times (8 \times 10)$	$= 160$
Total:		550

2. Divide to find the approximate number of gallons needed. $550 \div 450 = 1.2$
3. Round up to next higher number. $1.2 \rightarrow 2$

So, you should buy 2 gal of paint for your room.

FOR DISCUSSION

A book about "do-it-yourself" home improvement says that preparation *before* painting should take 50% of your time. What do you think this means? What kinds of preparation are needed?

The area to be painted is given. Approximately how many gallons of paint are needed for one coat of paint? Give your answer to the nearest tenth.

1. 1,280 ft^2 **2.** 3,200 ft^2 **3.** 860 ft^2 **4.** 5,450 ft^2

Approximately how many gallons of paint are needed? Give your answer to the nearest tenth.

5. 1 coat of paint

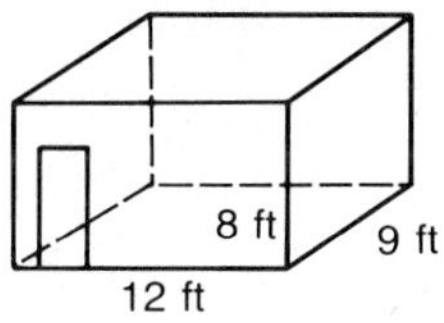

6. 2 coats of paint

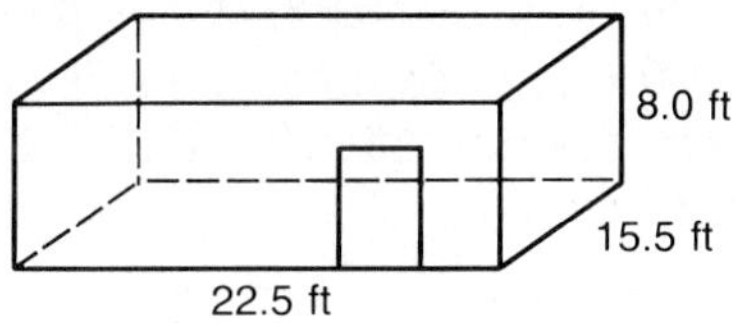

7. 3 coats of paint

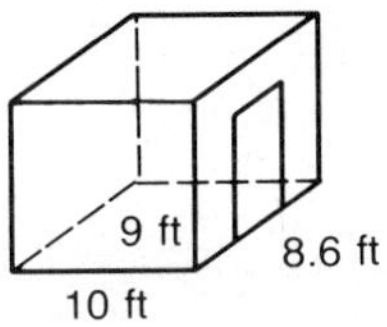

8. 2 coats of paint

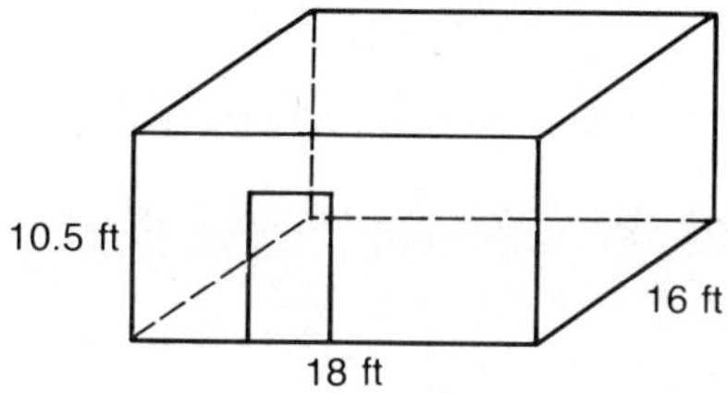

Tell how many 1-gal cans of paint you would buy for the room in:

9. Exercise 1. **10.** Exercise 2. **11.** Exercise 3. **12.** Exercise 4.

13. Exercise 5. **14.** Exercise 6. **15.** Exercise 7. **16.** Exercise 8.

Suppose you want to wallpaper the walls of your room. Wallpaper comes in different lengths and widths. However, a roll of American wallpaper usually contains about 20 ft^2 of usable wallpaper.

How many rolls of wallpaper are needed to wallpaper the room in:

17. Exercise 5? **18.** Exercise 7?

Find the total cost.

19. Paint the room in Exercise 6. Paint is $9.87 a gallon.

20. Wallpaper the walls in Exercise 6. Wallpaper is $8.25 a roll.

21. Paint the room in Exercise 1. Paint is $5.40 a gallon.

22. Wallpaper the walls in Exercise 8. Wallpaper is $10.75 a roll.

DECISION MAKING

12.14 DEVELOPING A BUDGET

Developing a budget for a club or organization requires group decisions, an understanding of the situation, and often compromise.

PROBLEM

The newly elected student officers of the Science Club are meeting to develop a budget for next year's activities. The budget must then be presented to the school's Funding Committee for approval. Since money is limited, the budget must be carefully done and include complete explanations of how the funds will be used. The students think that they can probably get more than the $500 the club got last year. Several of the 1,250 students in the school have submitted suggestions for what they would like to see done next year. The officers are studying each and estimating costs.

- Bus trip to the Science Museum for the Science Club members.

  Charter bus: $175 Admission tickets: $2 each for 37 members

- Help students pay for supplies for their Science Fair exhibits.

  About 40 students: $10 per student

- Sponsor a Health Fair at the school using volunteers from various health professions.

  Publicity: $50 Decorations, displays: $50 Refreshments: $150

- Build a simple greenhouse for students interested in horticulture.

  Materials: $200

- Plant trees in front of the school.

  Cost: 10 trees at about $25 each.

DECISION-MAKING FACTORS

Expense

Benefits to student body

Educational value

Benefits to the school

DECISION-MAKING COMPARISONS

Complete the table.

	Expense	Educational value	Benefits to student body	Benefits to school
Bus trip	**1.** ■	Excellent	**2.** ■	**3.** ■
Science Fair Aid	**4.** ■	**5.** ■	Limited	**6.** ■
Health Fair	**7.** ■	**8.** ■	**9.** ■	Yes
Greenhouse	$200	**10.** ■	**11.** ■	**12.** ■
Trees	**13.** ■	**14.** ■	**15.** ■	**16.** ■

MAKING THE DECISIONS

Which items should be in the budget:

17. if the plan was to benefit the school?

18. if the plan was to benefit the most students?

19. if the plan was to provide the most educational value?

20. The officers assumed that the Science Fair participants will compete even if they don't get money for supplies. What if half of them can't compete without financial aid. How does this affect the factor of "benefit to the school"?

21. What benefits could planting trees provide?

22. What benefits could building a greenhouse provide?

Select items and make a circle graph.

23. A budget that will most benefit all the students in the school.

24. A budget that will most benefit the school.

25. A budget that will most benefit the Science Club members.

CHAPTER REVIEW

Vocabulary Choose the letter of the word(s) that completes each statement.

1. Polygon is to perimeter as circle is to ■. [330, 332]

a. area **b.** circumference **c.** volume **d.** diameter

2. A solid figure whose faces are all polygons is a ■. [344]

a. cylinder **b.** sphere **c.** polyhedron **d.** polygon

3. The amount of space inside a solid figure is its ■. [350]

a. perimeter **b.** surface area **c.** height **d.** volume

Skills Find the area. [334, 336, 338]

Square: $s = 2.5$ km	*6.25 km²*

4. Rectangle: $l = 5$ in., $w = 7$ in.

5. Parallelogram: $b = 13$ ft, $h = 4$ ft

6. Triangle: $b = 63$ cm, $h = 18$ cm

7. Circle: $r = 7$ ft

8. Circle: $r = 3$ m

Find the surface area. [347]

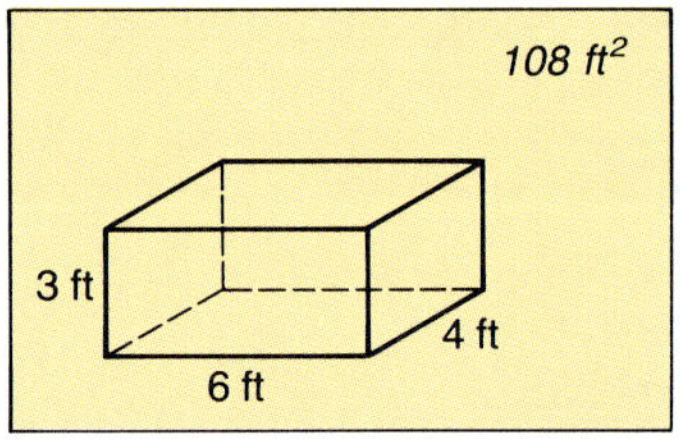

9.

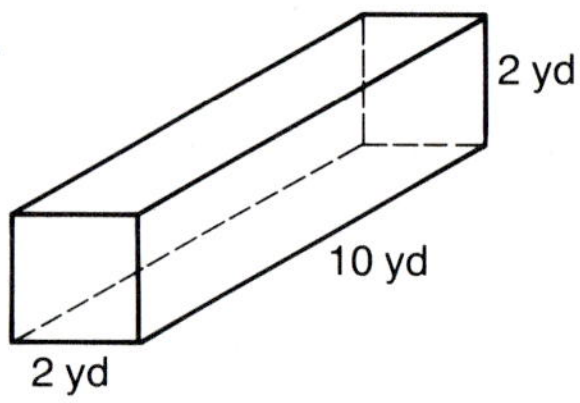

10.

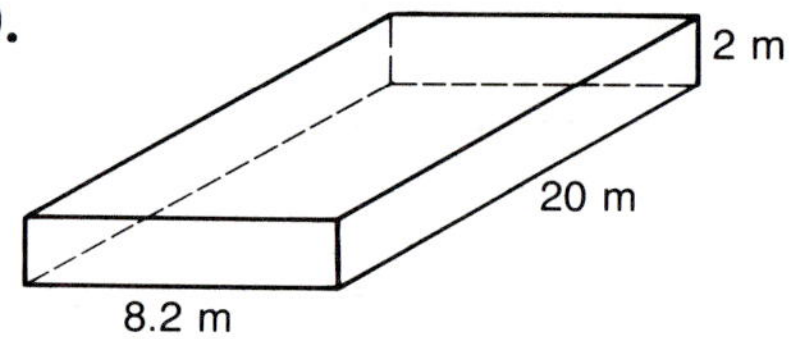

Find the volume. [350, 352]

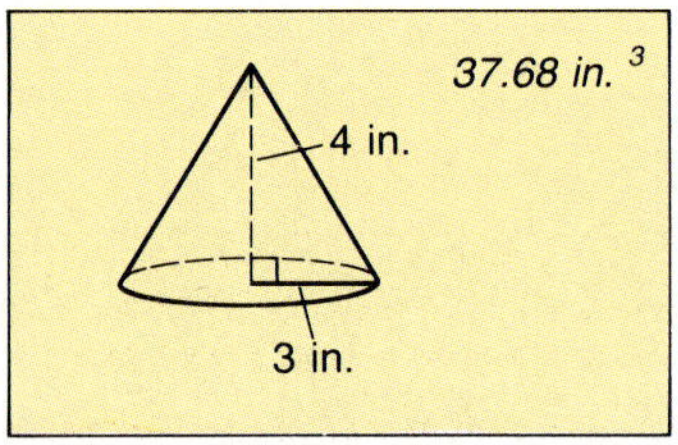

11.

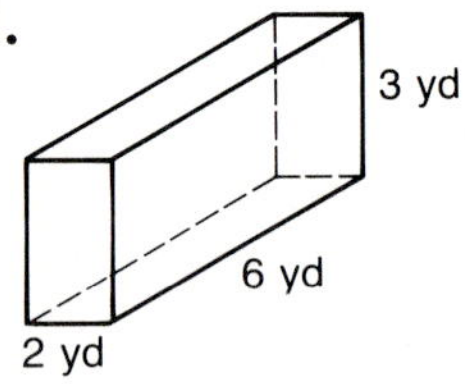

12.

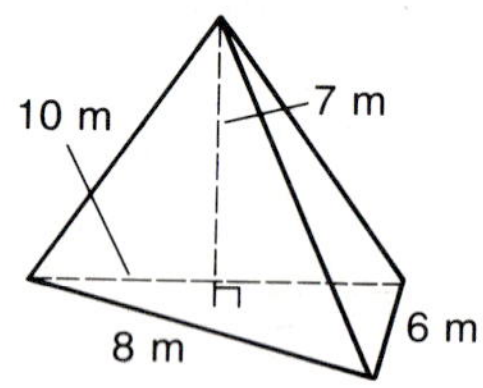

13. Cone: $r = 6$ ft, $h = 4$ ft

14. Square pyramid: $s = 6$ ft, $h = 3$ ft 36 ft³

15. Cylinder: $r = 1$ m, $h = 4.2$ m

16. Cube: $s = 0.1$ m

CHAPTER TEST

Use Figure A above for Exercises 1–6.

1. Name the figure.

2. Count the faces, edges, and vertices.

3. Find the perimeter of face ABCD.

4. Find the area of base DCGH.

5. Find the surface area.

6. Find the volume.

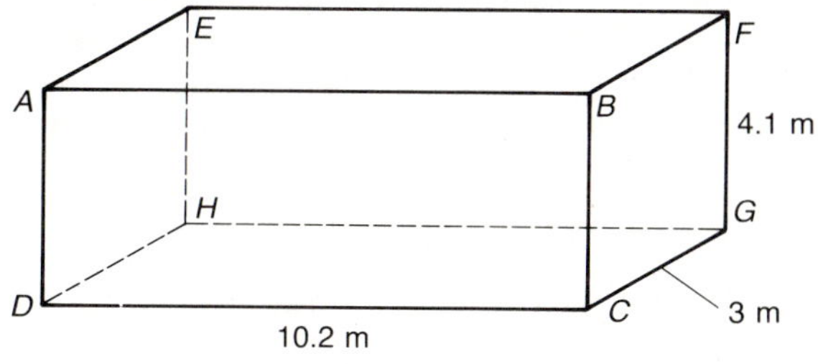

Figure A

Use Figure B above for Exercises 7–10.

7. Name the figure.

8. Find the perimeter of the base.

9. Find the area of face ABC.

10. Find the volume.

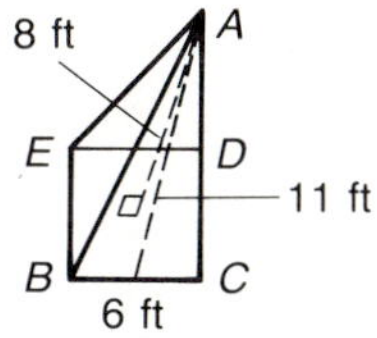

Figure B

Use Figure C above for Exercises 11–14.

11. Name the figure.

12. Find the circumference of the base.

13. Find the area of the base.

14. Find the volume.

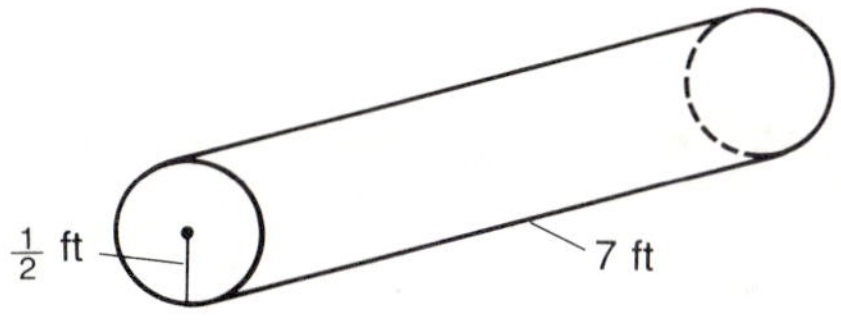

Figure C

Use Figure D above for Exercises 15–16.

15. Name the figure.

16. Find the volume.

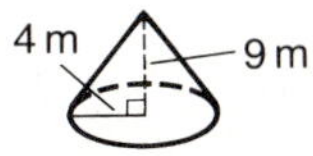

Figure D

Solve.

17. John Hutchins' library is 12 ft by 16 ft with an 8-ft ceiling. How many gallons of paint would he need to repaint it? (*Remember:* 1 gal covers about 450 ft^2.)

18. John uses 7 nails to put up each 4-ft length of ceiling molding. How many nails would he need to put up molding all around the library?

CUMULATIVE REVIEW

Choose the letter of the correct answer.

1. The sum of the measures of ■ angles is equal to 90°.

   a. acute **b.** complementary **c.** supplementary **d.** none of these

2. A pair of ■ lines do not intersect.

   a. perpendicular **b.** bisecting **c.** parallel **d.** none of these

3. A(n) ■ triangle has three angles each of which is less than 90°.

   a. obtuse **b.** right **c.** acute **d.** none of these

4. The relationship of a scale drawing to an object's actual size can be expressed as a ■.

   a. proportion **b.** range **c.** mean **d.** none of these

5. The ■ is the distance around a circle.

   a. radius **b.** chord **c.** diameter **d.** none of these

Select the best estimated answer.

6. $\frac{7}{8} + \frac{5}{7}$	**a.** 2	**b.** $\frac{7}{8}$	**c.** 4	**d.** $\frac{1}{4}$
7. $18\frac{2}{3} - 7\frac{4}{5}$	**a.** 25	**b.** 27	**c.** 1	**d.** 12
8. $2\frac{7}{8} \times 2\frac{1}{8}$	**a.** 6	**b.** 1	**c.** $\frac{3}{16}$	**d.** $\frac{1}{2}$
9. 4,875 × 53	**a.** 25,000	**b.** 250,000	**c.** 20,000	**d.** 2,500,000
10. 18% of 200	**a.** 4,000	**b.** 400	**c.** 40	**d.** 0.40

Find the answer.

11. $\frac{11}{8} \div 2\frac{1}{8}$	**a.** $\frac{11}{17}$	**b.** $\frac{8}{17}$	**c.** $\frac{22}{17}$	**d.** none of these
12. 5 m = ■ mm	**a.** 0.005	**b.** 500	**c.** 5,000	**d.** none of these
13. $7\frac{2}{3} - \frac{1}{2}$	**a.** $\frac{1}{6}$	**b.** $7\frac{1}{3}$	**c.** $7\frac{1}{6}$	**d.** none of these
14. $12\frac{1}{2}$% of 200	**a.** 2.5	**b.** 25	**c.** 0.125	**d.** none of these
15. $12\frac{1}{8} + 15\frac{1}{5}$	**a.** $27\frac{1}{40}$	**b.** $27\frac{3}{8}$	**c.** $27\frac{11}{40}$	**d.** none of these

Solve.

16. Rhonda buys a dishwasher at Eddie's Appliance Center for $410. She decides to pay in 12 monthly installments. The payments are $37.50 per month. What is the finance charge?

a. $450 **b.** $410 **c.** $37.50 **d.** none of these

17. Dan wants to carpet a hall that is $4\frac{1}{4}$ ft wide and 12 ft long. He also wants to carpet a room 17 ft wide and 19 ft long. How many square feet of carpet does he need?

a. 374 ft^2 **b.** $107\frac{2}{3}$ ft^2 **c.** $124\frac{2}{3}$ ft^2 **d.** none of these

18. Bella is 1.4 m tall. She is standing beside a telephone pole and casts a shadow 4.2 m long. If the telephone pole casts a shadow that is 31.5 m long, how tall is the pole?

a. 10.5 m **b.** 18.6 m **c.** 9.8 m **d.** none of these

19. A quality control technician for Call-Mate checks a random sample of 1,700 telephone answering machines. Of these, 34 are defective. What percent of the sample is defective?

a. 5% **b.** 2% **c.** 1.25% **d.** none of these

20. Vito buys a CD player that is listed in audio magazines at $265. He receives a 25% discount on the player at Claude's Audio Barn. How much money did Vito save?

a. $56.25 **b.** $198.75 **c.** $66.25 **d.** none of these

THINKING ABOUT MATH

1. Explain how you would quickly calculate the two remaining angle measures of a parallelogram if you knew the measure of two adjacent angles.

2. You need to create a cataloging system for your compact discs. You have decided on a 3-digit code. Describe a method for determining the number of different codes that are possible with your system.

3. Would you prefer to receive simple or compound interest on your savings account? Why?

4. Can two obtuse angles ever be supplementary? Explain.

5. Explain why store owners might offer discounts on their marked-up prices.

PreSkills TEST

Add. [30, 62, 158]

1. $24 + 76$ **2.** $20 + 45$ **3.** $365 + 294$

4. $\frac{2}{3} + \frac{1}{5}$ **5.** $\frac{3}{4} + \frac{2}{3}$ **6.** $\frac{5}{8} + \frac{1}{4}$

7. $0.34 + 1.2$ **8.** $4.98 + 7.5$ **9.** $0.2 + 7$

10. $823 + 560$ **11.** $\frac{3}{5} + \frac{7}{9}$ **12.** $6.3 + 0.306$

Subtract. [32, 64, 162]

13. $34 - 19$ **14.** $86 - 7$ **15.** $812 - 391$ **16.** $639 - 472$

17. $\frac{3}{4} - \frac{1}{8}$ **18.** $\frac{7}{8} - \frac{2}{5}$ **19.** $\frac{2}{3} - \frac{5}{9}$ **20.** $\frac{4}{5} - \frac{1}{3}$

21. $1.3 - 0.98$ **22.** $34.7 - 5.67$ **23.** $41.9 - 12.83$ **24.** $53 - 18.21$

Multiply. [36, 40, 42, 74, 146]

25. 23×93 **26.** 15×100 **27.** 59×26 **28.** 965×30

29. $\frac{7}{8} \times \frac{4}{5}$ **30.** $\frac{4}{7} \times \frac{3}{4}$ **31.** $\frac{1}{2} \times \frac{6}{8}$ **32.** $\frac{4}{5} \times \frac{2}{3}$

33. 12.3×0.8 **34.** 4.2×10.1 **35.** 5.23×7.7 **36.** 0.35×6.72

Divide. [36, 48, 80, 150]

37. $612 \div 17$ **38.** $444 \div 37$ **39.** $840 \div 24$ **40.** $920 \div 10$

41. $\frac{3}{7} \div \frac{4}{7}$ **42.** $\frac{7}{8} \div \frac{3}{4}$ **43.** $\frac{4}{5} \div \frac{1}{3}$ **44.** $\frac{5}{6} \div \frac{1}{4}$

45. $27.9 \div 0.9$ **46.** $0.12 \div 0.06$ **47.** $18.5 \div 0.5$ **48.** $6.3 \div 0.3$

Use the number line below for Exercises 49–52. [122]

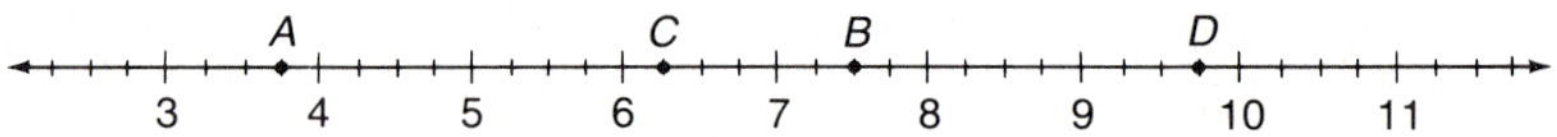

Write the number named by the point.

49. *A* **50.** *B* **51.** *C* **52.** *D*

Chapter 13 RATIONAL NUMBERS

Some businesses are owned by many people who hold stock in the company. What are the advantages and disadvantages of this kind of shared ownership?

13.1 INTEGERS

Whole numbers can be used to describe "how much" or "how many." When used with a positive sign (+) or a negative sign (−), whole numbers can describe opposite situations such as:

225 ft above sea level	+225	10 yd gain	+10
225 ft below sea level	−225	10 yd loss	−10

EXAMPLE 1 Write a positive or negative number to describe the situation.

a. Deposit $75 in your checking account.

THINK: You are increasing the amount of money in your account. Use the positive sign. +$75

b. Withdraw $100 from your checking account.

THINK: You are decreasing the amount of money in your account. Use the negative sign. −$100

Every positive whole number has an opposite, or negative whole number.

All positive and negative whole numbers make up a set of numbers called **integers.** Zero is also an integer, but it is neither positive nor negative.

Integers can be shown on a number line.

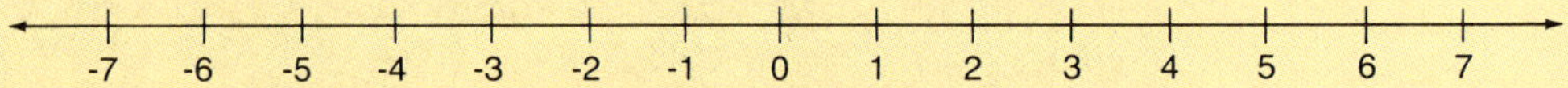

EXAMPLE 2 Name the opposite of the integer.

a. +4

THINK: +4 is 4 units to the right of zero. Find the integer that is 4 units to the left of zero.

The opposite of +4 is −4.

b. −6

THINK: −6 is 6 units to the left of zero. Find the integer that is 6 units to the right of zero.

The opposite of −6 is +6.

As you move right on the number line, integers increase in value. As you move left on the number line, integers decrease in value.

EXAMPLE 3 Compare the integers. Write $>$ or $<$.

a. +2 ● −6

THINK: +2 is to the right of −6.

$+2 > -6$

b. −5 ● −1

THINK: −5 is to the left of −1.

$-5 < -1$

The **absolute value** of an integer is the distance the integer is from zero. The symbol for absolute value is $|\;|$.

EXAMPLE 4 Find the absolute value.

a. $|+7|$

THINK: +7 is 7 units from 0.

$|+7| = 7$

b. $|-3|$

THINK: −3 is 3 units from 0.

$|-3| = 3$

PRACTICE EXERCISES

Write the integer that best describes the situation.

1. 40 degrees below zero
2. a loss of $15
3. the opposite of −100
4. 6 hours from now
5. 75 units to the left of zero
6. no gain on the play
7. 2 days ago
8. the opposite of +34
9. 2 weeks late
10. a loss of 2 points

Write the integer named by the point.

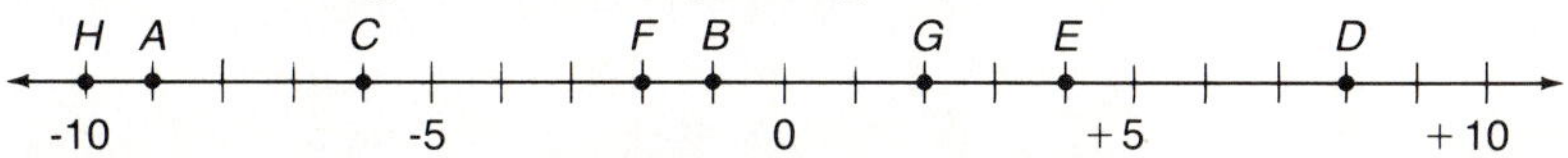

11. *C* **12.** *A* **13.** *E* **14.** *B* **15.** *D* **16.** *G* **17.** *H* **18.** *F*

Write the opposite of the number.

19. +16 **20.** −44 **21.** −16 **22.** +29 **23.** +9 **24.** −27 **25.** +89 **26.** −33

Compare the integers. Write > or <.

27. + 3 ● +5 **28.** −5 ● −6 **29.** 0 ● −5 **30.** −2 ● +3 **31.** −7 ● 6
32. +23 ● −17 **33.** −96 ● +100 **34.** −33 ● −98 **35.** 0 ● −45 **36.** −19 ● −23

Write the absolute value.

37. $|-1|$ **38.** $|-2|$ **39.** $|0|$ **40.** $|-5|$ **41.** $|-7|$
42. $|-9|$ **43.** $|-6|$ **44.** $|+89|$ **45.** $|-24|$ **46.** $|-17|$
47. $|+27|$ **48.** $|-8|$ **49.** $|+76|$ **50.** $|-91|$ **51.** $|-101|$

Use >, <, or = to make a true statement.

52. $|-3|$ ● $|+3|$ **53.** $|+23|$ ● $|-32|$ **54.** $|-5|$ ● $|0|$ **55.** $|+12|$ ● $|-13|$
56. $|-17|$ ● $|+17|$ **57.** −29 ● $|-29|$ **58.** $|-71|$ ● $|+71|$ **59.** $|0|$ ● $|-1|$

MIXED REVIEW

Identify the pair of angles as *complementary*, *supplementary*, or *neither*.

60. 56°, 34° **61.** 112°, 78° **62.** 81°, 19° **63.** 120°, 60°

The measures of two angles of a triangle are given. Find the measure of the third angle.

64.

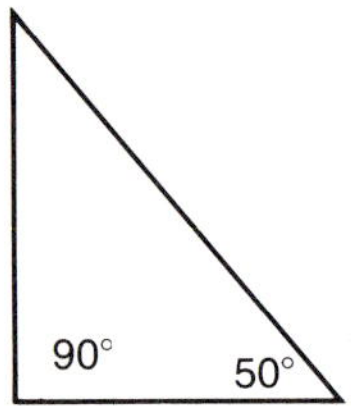

65.

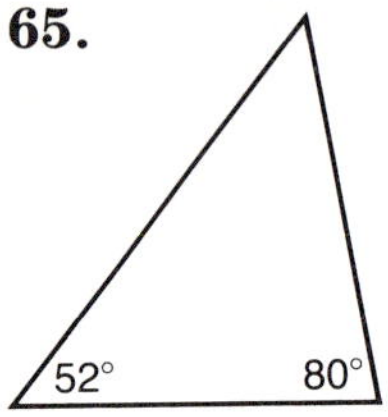

66.

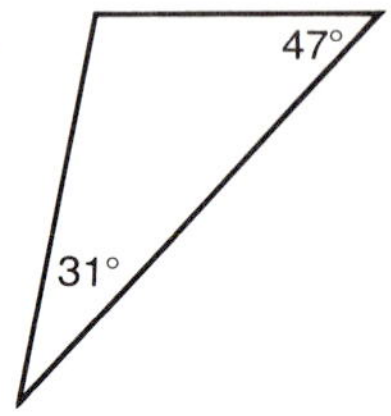

67.

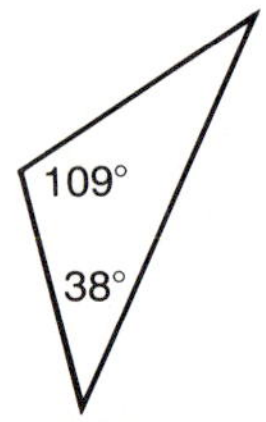

68. 46°, 69° **69.** 124°, 41° **70.** 90°, 80° **71.** 100°, 36°

Solve.

72. The highest point in the United States is Mount McKinley in Alaska. It is 20,320 ft above sea level. Write an integer to describe this altitude.

73. The lowest point in the United States is Death Valley, California. It is 282 ft below sea level. Write an integer to describe this altitude.

74. Borah Park in Idaho is 12,662 ft above sea level. Kings Peak in Utah is 13,528 ft above sea level. How much higher is Kings Peak than Borah Peak?

75. The highest point in the State of Texas is Guadalupe Peak at 8,749 ft above sea level. The lowest point is at the Gulf of Mexico which is sea level. What is the range of altitude in the state?

TIME OUT

Some months have 31 days.
Others have 30 days.
How many have 28 days?

13.2 ADDING INTEGERS

During the first two plays of the game, the Miami Sharks gained 7 yd and lost 5 yd. What was the result of the two plays?

To find the answer, you can add $+7$ and -5.
Usually, the $+$ sign is dropped when computing with positive integers.

EXAMPLE 1 Add $7 + (-5)$. Use the number line.

1. Locate 7.
2. Move 5 units to the left.

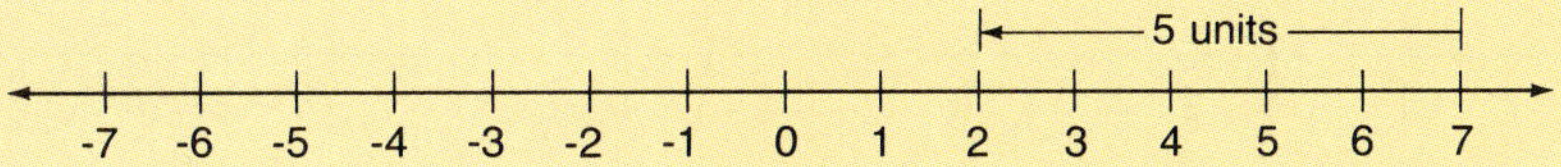

$7 + (-5) = 2$

So, the Sharks gained 2 yd.

You can add two integers without using a number line by following these rules.

- If the integers have the same signs, add their absolute values. The sum has the same sign as the addends.
- If the integers have different signs, subtract their absolute values. The sum has the sign of the addend with the greater absolute value.

EXAMPLE 2 Add the integers.

a. $-6 + (-4)$

1. Find the absolute values. $|-6| = 6 \quad |-4| = 4$
2. Add. $6 + 4 = 10$
3. Write the correct sign in the sum.

THINK: The addends have the same signs. So, the sum has the same sign. $-6 + (-4) = -10$

b. $3 + (-5)$

1. Find the absolute values. $|+3| = 3 \quad |-5| = 5$
2. Subtract.

   THINK: $5 > 3$ $5 - 3 = 2$
3. Write the correct sign in the sum.

THINK: The addends have different signs. Since -5 has the greater absolute value, the sum is negative. $3 + (-5) = -2$

CHECKPOINT

Write the letter of the correct answer.

Add.

1. $-8 + (-9)$ **a.** -1 **b.** 17 **c.** -17 **d.** 1

2. $4 + (-7)$ **a.** -3 **b.** -11 **c.** 11 **d.** 3

3. $-5 + 9$ **a.** -4 **b.** 4 **c.** -14 **d.** 14

PRACTICE EXERCISES

Add. Use the number line if necessary.

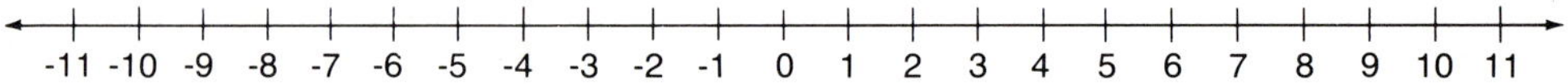

1. $0 + 3$ **2.** $-3 + (-7)$ **3.** $3 + (-3)$ **4.** $(-9) + 1$

5. $-3 + (-6)$ **6.** $8 + 2$ **7.** $-2 + (-4)$ **8.** $-7 + 9$

9. $-5 + 5$ **10.** $9 + (-9)$ **11.** $-8 + 11$ **12.** $10 + (-9)$

Add.

13. $13 + 15$ **14.** $25 + 36$ **15.** $-23 + (-24)$ **16.** $-35 + (-76)$

17. $6 + (-42)$ **18.** $17 + (-6)$ **19.** $-15 + 8$ **20.** $-9 + 33$

21. $-34 + 45$ **22.** $-52 + 52$ **23.** $72 + (-8)$ **24.** $-60 + 46$

25. $-88 + 76$ **26.** $-100 + (-23)$ **27.** $120 + (-75)$ **28.** $99 + (-99)$

Add. (*Hint:* Add the numbers with the same signs first.)

29. $-3 + (-5) + 6$ **30.** $-5 + 9 + 5$

31. $-6 + 7 + (-3)$ **32.** $-2 + 8 + (-6) + 7$

33. $-17 + 9 + (-16)$ **34.** $19 + (-3) + (-1) + (-9)$

35. $-45 + 13 + (-72)$ **36.** $9 + (-8) + (-5) + 1$

37. $-22 + (-34) + 22 + 34$ **38.** $-14 + (-12) + (-26)$

During the first six plays of the game, the Miami Sharks had the following results: 7-yd gain; 5-yd gain; 3-yd loss, 25-yd gain; 10-yd loss; 6-yd loss.

39. Express the gains using positive integers.

40. Express the losses using negative integers.

41. Find the result of the first three plays.

42. Find the result of the six plays.

Solve.

43. When the football game began, the temperature was 52°F. During the game, the temperature rose 5°. What was the temperature at the end of the game?

44. Sue Keller spent $12 on her ticket for the football game. She also bought a ticket for a friend, and spent $8.50 on food. How much money did Sue spend altogether?

45. During the first three plays, the Detroit Dolphins gained 4 yd, lost 6 yd, and gained 2 yd. What was the result of the three plays?

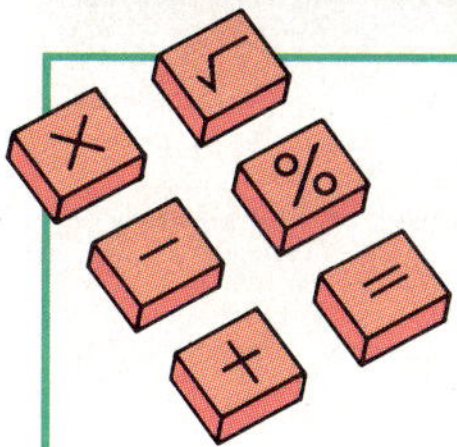

CALCULATOR

You can add integers on a calculator. Some calculators have a +/− key that changes a positive integer to a negative integer. If your calculator has this key, follow these steps to add −24 + 52:

Enter: [2] [4] [+/−] [+] [5] [2] [=] [28.]

If your calculator does not have a +/− key, follow these steps to add −24 + 52:

THINK: Press the − key before and after entering negative integers.

Enter: [−] [2] [4] [−] [+] [5] [2] [=] [28.]

Add. Use a calculator.

1. 7 + 12

2. 4 + (−6)

3. 5 + (−14)

4. −4 + 8

5. −5 + (−7) + (−12)

6. 8 + (−9) + (−3)

7. 17 + (−47) + (−98)

8. −43 + 87 + (−19)

9. 125 + (−50) + (−12)

10. 30 + 158 + (−101)

13.3 SUBTRACTING INTEGERS

At 3:00 A.M., the temperature was $-2°C$.
At 9:00 A.M., the temperature was $3°C$.
What was the change in temperature from 3:00 A.M. to 9:00 A.M.?

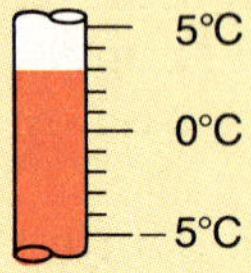

To find the temperature change, subtract $-2°$ from $3°$.
You can subtract integers using a number line.

EXAMPLE 1 Subtract: $3 - (-2)$

1. Locate the integer being subtracted.
2. Count the number of units moved to reach the first integer.
3. A move to the right is positive; a move to the left is negative.

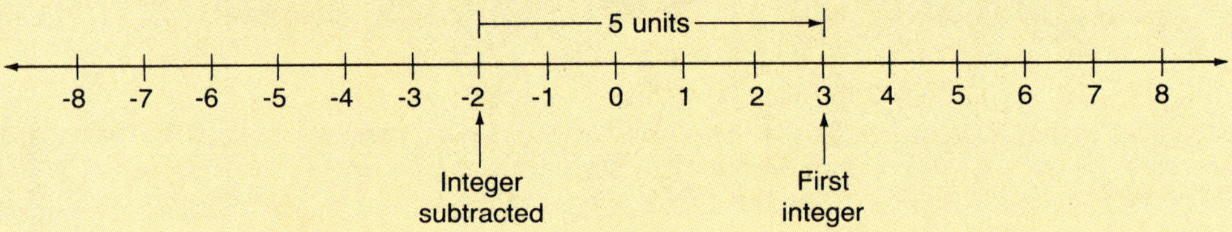

$$3 - (-2) = 5$$

So, the temperature rose 5°C.

Every subtraction example can be written as a related addition example. To write a related addition example, add the opposite of the number you are subtracting.

EXAMPLE 2 Write a related addition example.

a. $2 - 8$

THINK: The opposite of 8 is -8.

So, $2 - 8 = 2 + (-8)$.

b. $3 - (-4)$

THINK: The opposite of -4 is 4.

So, $3 - (-4) = 3 + 4$.

You can use opposites to subtract integers.

EXAMPLE 3 Subtract the integers.

a. $-10 - 5$

THINK: Add the opposite of the number being subtracted.

$-10 - 5$

$-10 + (-5) = -15$

So, $-10 - 5 = -15$.

b. $-6 - (-9)$

THINK: Add the opposite of the number being subtracted.

$-6 - (-9)$

$-6 + 9 = 3$

So, $-6 - (-9) = 3$.

CHECKPOINT

Choose the letter of the correct answer.

Subtract.

1. − 3 − 7	**a.** 10	**b.** −10	**c.** −4	**d.** 4
2. 6 − 9	**a.** −3	**b.** 3	**c.** −15	**d.** 15
3. −1 − (−8)	**a.** −9	**b.** 9	**c.** −7	**d.** 7

PRACTICE EXERCISES

Subtract. Use the number line if necessary.

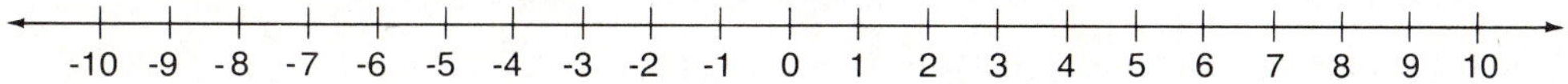

1. 10 − 3 **2.** 5 − 8 **3.** 2 − 9 **4.** 9 − 8

5. −3 − 5 **6.** −6 − 3 **7.** −2 − 9 **8.** −9 − 9

9. 4 − (−5) **10.** 6 − (−7) **11.** 9 − (−2) **12.** 0 − (−6)

13. −3 − (−3) **14.** −7 − (−9) **15.** −1 − (−5) **16.** −10 − (−4)

Complete.

17. −6 − 2 = −6 + ■ **18.** 12 − 4 = 12 + ■

19. 13 − (−28) = 13 + ■ **20.** −27 − (−100) = −27 + ■

21. 34 − 46 = 34 + ■ **22.** −89 − 45 = −89 + ■

Subtract.

23. 13 − 15 **24.** 14 − 25 **25.** 23 − 13 **26.** 19 − 29

27. −14 − 17 **28.** −16 − 13 **29.** 0 − 17 **30.** −19 −18

31. −17 − (−14) **32.** −6 − (−19) **33.** −22 − (−35) **34.** −45 − (−100)

35. 17 − (−9) **36.** −18 − 67 **37.** 98 − 112 **38.** −100 − (−134)

Use the table at the right to solve Exercises 39–41.

39. What is the difference between the lowest temperature in Caribou and that of Kansas City?

40. Which city has the lowest recorded temperature?

41. What is the difference between the lowest temperature in Chicago and that of Duluth?

RECORDED LOW TEMPERATURES

City	Temperature, °F
Caribou, Maine	−27
Chicago, Illinois	−22
Denver, Colorado	−19
Duluth, Minnesota	−30
Kansas City, Missouri	−7

13.4 MULTIPLYING INTEGERS

Karen DeSoto is in charge of all the statistics for the North Ridge High School football team. She needs to find the total loss of yards due to penalties in last week's game.

GAME 5

Number of Penalties	Yards Lost Per Penalty	Total Loss
7	−10	

To find the total number of yards lost, multiply 7 and -10.

To multiply two integers:

- Multiply the absolute values.
- Determine whether the product is positive or negative.
  The product of two integers having the same sign is positive.
  The product of two integers having different signs is negative.

EXAMPLE 1

Multiply: $7 \times (-10)$

1. Multiply the absolute values. $|7| \times |-10| = 7 \times 10 = 70$
2. Determine whether the product is positive or negative.

THINK: The signs are different. The product is negative.

$7 \times (-10) = -70$

So, North Ridge High School lost 70 yd due to penalties in Game 5.

When there are more than two integers, count the number of negative integers.

- The product is positive if there is an even number of negative factors.
- The product is negative if there is an odd number of negative factors.

EXAMPLE 2

Multiply:

a. $-3 \times 5 \times (-2)$

$|-3| \times |5| \times |-2| = 30$

THINK: There is an even number of negative factors. The product is positive.

$-3 \times 5 \times (-2) = 30$

b. $3 \times (-2) \times (-4) \times (-5)$

$|3| \times |-2| \times |-4| \times |-5| = 120$

THINK: There is an odd number of negative factors. The product is negative.

$3 \times (-2) \times (-4) \times (-5) = -120$

FOR DISCUSSION

Can you name this product at a glance? Explain.

$$-45 \times 36 \times (-78) \times (-86) \times 0 \times 14 \times (-9)$$

PRACTICE EXERCISES

Multiply.

1. 5×8
2. 8×11
3. 7×9
4. 23×42
5. $-7 \times (-2)$
6. $-9 \times (-2)$
7. $-6 \times (-14)$
8. $-26 \times (-45)$
9. $4 \times (-3)$
10. $8 \times (-9)$
11. -32×23
12. -41×52
13. $75 \times (-34)$
14. -83×61
15. $-80 \times (-90)$
16. 76×89
17. 123×0
18. $-126 \times (-97)$
19. 250×10
20. $310 \times (-25)$
21. $-1 \times (-2) \times 4$
22. $2 \times 3 \times (-3)$
23. $-3 \times (-1) \times (-4)$
24. $-2 \times (-4) \times 4$
25. $3 \times 2 \times 5$
26. $7 \times 0 \times (-9)$
27. $3 \times (-1) \times 5$
28. $-3 \times (-3) \times (-3)$
29. $7 \times (-4) \times (-2)$
30. $-2 \times (-4) \times (-7)$
31. $3 \times 8 \times 7$
32. $-1 \times 9 \times 78$
33. $12 \times 0 \times (-19)$
34. $-11 \times (-2) \times 1$
35. $-10 \times (-10) \times (-10)$
36. $-1 \times (-3) \times 4 \times (-2)$
37. $1 \times (-1) \times 1 \times (-1)$
38. $3 \times (-2) \times 0 \times (-1)$
39. $2 \times (-4) \times (-1) \times (-3)$
40. $4 \times (-2) \times 4 \times 2$
41. $-2 \times (-3) \times 3 \times (-4)$
42. $(-5) \times 5 \times (-1) \times (-2)$
43. $2 \times 4 \times 1 \times (-3)$
44. $3 \times 3 \times 3 \times 3$
45. $1 \times (-1) \times 2 \times (-1) \times 2$
46. $(-2) \times (-2) \times (-2) \times (-2) \times (-2)$
47. $3 \times (-2) \times 1 \times 0 \times (-7)$
48. $(-4) \times (-1) \times 2 \times (-1) \times (-2)$
49. $3 \times (-1) \times (-2) \times (-3) \times 1$
50. $3 \times (-2) \times 0 \times (-5) \times (-5)$

Solve.

51. In a recent football game the North Ridge team gained an average of 6 yd every time that the ball was run instead of passed. The team ran with the ball 35 times during the game. Write an integer for the total number of yards gained.

52. Football coach Mueller came to North Ridge High School 12 years ago. On the average his teams have lost only 4 games out of 24 per season. Write a negative integer for the number of games lost since Coach Mueller was hired.

53. The place kicker on the football team averaged 36.5 yd per kick. If he kicked the ball a total of 292 yd, how many times did he kick?

54. The halfback gained an average of 7.8 yd per carry. If he carried the ball 15 times, how many yards did he gain for the team?

13.5 DIVIDING INTEGERS

You already know that multiplication and division of whole numbers are related.

Multiplication	**Division**
$6 \times 8 = 48$	$48 \div 8 = 6$
	$48 \div 6 = 8$

The same is true for multiplication and division of integers.

Multiplication	**Division**
$(-7) \times (-5) = 35$	$35 \div (-7) = -5$
	$35 \div (-5) = -7$
$6 \times (-9) = -54$	$-54 \div (-9) = 6$
	$-54 \div 6 = -9$

The sign rules for dividing integers are identical to those for multiplying integers.

EXAMPLE 1

Divide: $-4{,}500 \div (-150)$

$|-4{,}500| \div |-150| = 30$

THINK: The signs are the same.
The quotient is positive.

$-4{,}500 \div (-150) = 30$

EXAMPLE 2

Divide: $-72 \div 9$

$|-72| \div |9| = 8$

THINK: The signs are different.
The quotient is negative.

$-72 \div 9 = -8$

CHECKPOINT

Write the letter of the correct answer.

Divide.

1. $16 \div 4$	**a.** 20	**b.** -4	**c.** 4	**d.** 12
2. $-15 \div (-5)$	**a.** -3	**b.** 3	**c.** -10	**d.** -20
3. $-12 \div 3$	**a.** 4	**b.** -4	**c.** -9	**d.** -36
4. $0 \div (-9)$	**a.** 9	**b.** -9	**c.** -90	**d.** 0

PRACTICE EXERCISES

Divide.

1. $6 \div 2$	**2.** $4 \div 4$	**3.** $6 \div 2$	**4.** $10 \div 5$
5. $-8 \div (-4)$	**6.** $8 \div (-4)$	**7.** $0 \div 9$	**8.** $-6 \div (-3)$
9. $12 \div (-3)$	**10.** $-15 \div 5$	**11.** $21 \div (-7)$	**12.** $-24 \div 6$
13. $36 \div (-6)$	**14.** $-28 \div 4$	**15.** $24 \div (-8)$	**16.** $-27 \div (-9)$
17. $-42 \div 7$	**18.** $42 \div (-6)$	**19.** $-39 \div 3$	**20.** $38 \div (-2)$
21. $44 \div 4$	**22.** $-48 \div 8$	**23.** $48 \div 6$	**24.** $-49 \div (-7)$
25. $56 \div (-8)$	**26.** $-64 \div (-8)$	**27.** $-72 \div 9$	**28.** $85 \div 5$
29. $-24 \div 12$	**30.** $36 \div (-18)$	**31.** $-48 \div (-16)$	**32.** $50 \div 10$
33. $-69 \div 23$	**34.** $60 \div (-15)$	**35.** $34 \div 17$	**36.** $-91 \div (-13)$
37. $98 \div (-14)$	**38.** $99 \div 33$	**39.** $-96 \div 16$	**40.** $78 \div (-13)$
41. $-75 \div (-15)$	**42.** $-88 \div (-11)$	**43.** $65 \div (-13)$	**44.** $0 \div (-56)$
45. $153 \div 9$	**46.** $-112 \div (-8)$	**47.** $224 \div (-7)$	**48.** $-207 \div 9$
49. $-156 \div (-13)$	**50.** $210 \div 14$	**51.** $-408 \div 17$	**52.** $304 \div (-19)$
53. $-357 \div 21$	**54.** $448 \div 32$	**55.** $-656 \div (-41)$	**56.** $627 \div (-3)$

MIXED REVIEW

Write the product.

57. 10^2	**58.** 10^4	**59.** 10^7	**60.** 10^3
61. 10^5	**62.** 10^8	**63.** 10^9	**64.** 10^6

Write in exponent form.

65. 10,000 **66.** 100,000 **67.** 1,000,000

TIME OUT

You have an important decision to make. Should you join the new Flying Fins Swim Club? A membership costs $20. Each swim is $2 for members and $5 for nonmembers. How many times would you have to swim to make joining the club a better buy than not joining?

PROBLEM Solving APPLICATION

13.6 CAREER: METEOROLOGIST

Ann Kurasawa is a meteorologist. As part of her work, she forecasts weather. Computers provide information about temperature, wind speeds, and barometric pressures.

Sometimes, the air outside feels colder than the actual temperature. As the wind speed increases, body heat is lost more rapidly and the air feels colder. This wind chill table shows how cold the outside air feels for various actual temperatures and wind speeds.

So, if the wind speed is 15 mph and the actual temperature is 10°F, then the wind chill is −18°F.

WIND CHILL TABLE

Wind speed (mph)	Temperatures (°F) 30	20	10	0	−10	−20	−30
5	27	16	7	−5	−15	−26	−36
15	9	−5	−18	−31	−45	−58	−72
25	1	−15	−29	−44	−59	−74	−88
35	−4	−20	−35	−52	−67	−82	−97
45	−6	−22	−38	−54	−70	−85	−102

Use the table. Find the wind chill.

Wind Speed	Temperature	Wind Chill	Wind Speed	Temperature	Wind Chill
25 mph	30°F	1. ■	35 mph	10°F	2. ■
45 mph	20°F	3. ■	5 mph	−10°F	4. ■
45 mph	0°F	5. ■	15 mph	−30°F	6. ■

Solve.

7. The thermometer outside Ann's window reads 20°F. She hears on the radio that the wind speed is 35 mph. What is the wind chill?

8. The actual temperature outside Ann's window is −20°F. A TV weather report says that the wind chill is −74°F. What is the wind speed?

9. Wind speed is 25 mph. Actual temperature is 10°F. How many degrees less than actual temperature does the air feel?

10. Which of the following conditions has a lower wind chill reading:
Wind speed: 5 mph; temperature: −20°F
Wind speed: 15 mph; temperature: −10°F
How much lower is it?

MIDCHAPTER REVIEW

Write the integer named by the point. [364]

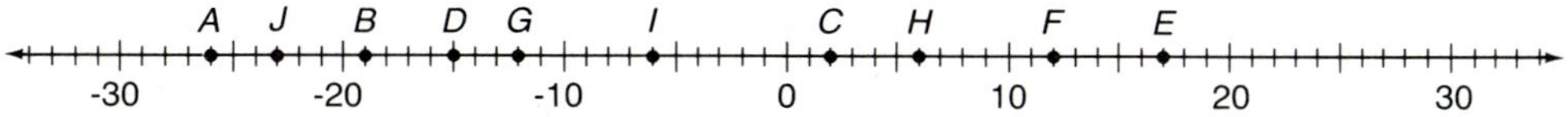

1. *J* **2.** *H* **3.** *F* **4.** *E* **5.** *C*

6. *B* **7.** *D* **8.** *G* **9.** *A* **10.** *I*

Compare. Write > or <. [364]

11. -3 ● 5 **12.** 6 ● -9 **13.** -3 ● -4 **14.** -12 ● -13

15. 35 ● 32 **16.** -54 ● -48 **17.** 23 ● -39 **18.** -42 ● -49

Add. [367]

19. $-3 + 5$ **20.** $6 + (-4)$ **21.** $0 + 2$ **22.** $-12 + 4$

23. $25 + (-26)$ **24.** $56 + (-62)$ **25.** $-32 + (-9)$ **26.** $-44 + (-12)$

27. $3 + (-2) + (-8)$ **28.** $-4 + (-12) + 21$ **29.** $2 + (-3) + 14$

Subtract. [370]

30. $8 - 5$ **31.** $7 - 9$ **32.** $3 - 5$ **33.** $5 - 12$

34. $5 - (-6)$ **35.** $7 - (-3)$ **36.** $-12 - (-14)$ **37.** $-24 - (-39)$

Multiply. [372]

38. $4 \times (-3)$ **39.** -6×12 **40.** $-4 \times (-4)$ **41.** $24 \times (-3)$

42. $-4 \times (-5) \times 20$ **43.** $-5 \times 9 \times (-7)$ **44.** $-12 \times (-4) \times (-10)$

Divide. [374]

45. $21 \div 7$ **46.** $-35 \div 5$ **47.** $-45 \div (-9)$ **48.** $-84 \div (-7)$

49. $-250 \div 25$ **50.** $400 \div 80$ **51.** $-330 \div (-11)$ **52.** $-110 \div (-10)$

Use the table for Exercises 53–54. [376]

53. The thermometer outside of Jerry's window reads 30°F. The weatherman on the television reports that the wind speed is 15 mph. What is the wind chill?

54. The temperature outside City Hall is 0°F. The wind speed is 25 mph. What is the wind chill?

WIND CHILL TABLE

Wind speed	Temperatures (°F)						
(mph)	30	20	10	0	−10	−20	−30
5	27	16	7	−5	−15	−26	−36
15	9	−5	−18	−31	−45	−58	−72
25	1	−15	−29	−44	−59	−74	−88

13.7 SCIENTIFIC NOTATION

Scientists frequently work with very large and very small numbers.

Distance from Earth to Alpha Centauri: About 30,000,000,000,000 mi

Mass of a molecule of water: About 0.00000000000000000000003 g

To simplify writing numerals for large and small numbers, scientists have developed a system called **scientific notation.** Numbers written in scientific notation have two factors. The first factor is a number between 1 and 10. The second factor is a power of 10.

EXAMPLE 1

Write in scientific notation.

a. 345,000

1. Write the first factor.

THINK: Move the decimal point to form a number between 1 and 10.

3.4 5, 0 0 0 → 3.45

5 places

2. Write the power of 10.

THINK: The decimal point was moved left 5 places. The exponent is positive.

$345{,}000 = 3.45 \times 10^5$

b. 0.00041

1. Write the first factor.

THINK: Move the decimal point to form a number between 1 and 10.

0 0 0 0 4.1 → 4.1

4 places

2. Write the power of 10.

THINK: The decimal point was moved right 4 places. The exponent is negative.

$0.00041 = 4.1 \times 10^{-4}$

EXAMPLE 2

Write in standard form.

a. 2.16×10^7

THINK: The exponent is positive. Move the decimal point 7 places to the right. Write additional zeros as needed.

2 1 6 0 0 0 0 0. → 21,600,000

$2.16 \times 10^7 = 21{,}600{,}000$

b. 6.1×10^{-3}

THINK: The exponent is negative. Move the decimal point 3 places to the left. Write additional zeros as needed.

0.0 0 6 1

$6.1 \times 10^{-3} = 0.0061$

FOR DISCUSSION

Is the number expressed in scientific notation? Explain.

1. 3×10^2 **2.** 8.2×10^{-5} **3.** 14×10^2 **4.** 0.21×10^5

PRACTICE EXERCISES

Write the missing exponent.

1. $34 = 3.4 \times 10^{\blacksquare}$

2. $542 = 5.42 \times 10^{\blacksquare}$

3. $6{,}226 = 6.226 \times 10^{\blacksquare}$

4. $0.03 = 3 \times 10^{\blacksquare}$

5. $0.0091 = 9.1 \times 10^{\blacksquare}$

6. $0.00005 = 5 \times 10^{\blacksquare}$

7. $21{,}100 = 2.1 \times 10^{\blacksquare}$

8. $200{,}000 = 2 \times 10^{\blacksquare}$

9. $0.000043 = 4.3 \times 10^{\blacksquare}$

10. $0.00000000712 = 7.12 \times 10^{\blacksquare}$

Write in scientific notation.

11. 25 **12.** 300 **13.** 456 **14.** 3,400

15. 0.3 **16.** 0.41 **17.** 0.05 **18.** 0.065

19. 30,000 **20.** 100,000 **21.** 125,000 **22.** 0.000000002

Write as a standard numeral.

23. 2.4×10^1 **24.** 3.1×10^2 **25.** 1.43×10^3 **26.** 9.43×10^4

27. 3.1×10^{-1} **28.** 4.7×10^{-2} **29.** 6.71×10^{-3} **30.** 8.76×10^{-4}

31. 2.3×10^8 **32.** 1.9×10^{-7} **33.** 6.54×10^{10} **34.** 2.43×10^{-9}

Solve.

35. Scientists estimate that the Sun is about 93,000,000 mi from Earth. Write this number in scientific notation.

36. By some estimates, the diameter of Mars is 4.2×10^3 mi. Write this number in standard form.

37. At the closest point of its rotation, the Moon is 221,463 mi away from Earth. At its farthest point, the Moon is 252,710 mi away. What is the average distance that the Moon revolves around Earth?

38. The surface gravity of Jupiter is 2.6 times that of Earth. If a man weighs 186 lb on Earth, what would he weigh on Jupiter?

13.8 GRAPHING POINTS IN A PLANE

The President of the United States is visiting Lakeville. To monitor the progress of the President's motorcade from the airport to city hall, the police have divided the city into quadrants. There is a stakeout in each quadrant.

Every point on a coordinate plane can be named by using an **ordered pair** of numbers. In an ordered pair, the first number tells how far the point is to the right or left of the **origin.** The second number tells how far the point is up or down. The numbers in an ordered pair are called **coordinates.**

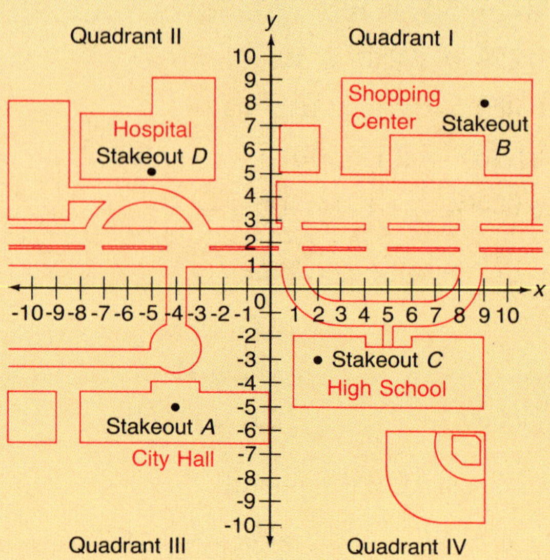

EXAMPLE 1 Use an ordered pair to name the stakeout in Quadrant III.

1. Find the x-coordinate.

   THINK: Move left 4 units from the origin. -4 is the x-coordinate.

2. Find the y-coordinate.

   THINK: Move down 5 units. -5 is the y-coordinate.

3. Write the ordered pair (x, y).

   THINK: The x-coordinate is -4. The y-coordinate is -5.

   The ordered pair is $(-4, -5)$.

$(-4, -5)$ names the location of the stakeout in Quadrant III.

You can use an ordered pair to locate a point on a coordinate plane.

EXAMPLE 2 What stakeout is located at $(2, -3)$?

1. Locate the x-coordinate.

   THINK: The x-coordinate is positive.
   Start at the origin. Move right two units.

2. Locate the y-coordinate.

   THINK: The y-coordinate is negative.
   Move down three units.

$(2, -3)$ names the location of Stakeout C.

FOR DISCUSSION

What can you say about all the numbers in each ordered pair in:

1. Quadrant I?

2. Quadrant II?

3. Quadrant III?

4. Quadrant IV?

PRACTICE EXERCISES

Use the coordinate grid at the right for Exercises 1–24.
Write the ordered pair.

1. Point A

2. Point B

3. Point C

4. Point D

5. Point E

6. Point F

7. Point G

8. Point H

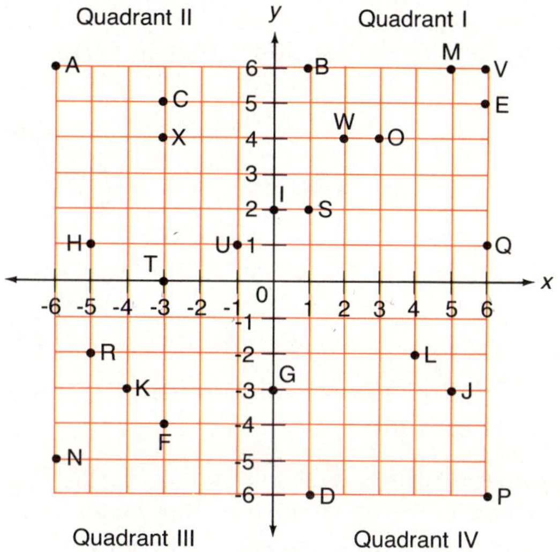

Write the point named by the ordered pair.

9. (1, 2) **10.** (3, 4) **11.** (5, 6) **12.** (6, 1)

13. (2, 4) **14.** (6, 6) **15.** (0, 2) **16.** $(-3, 0)$

17. $(-1, 1)$ **18.** $(-3, 4)$ **19.** $(6, -6)$ **20.** $(5, -3)$

21. $(4, -2)$ **22.** $(-6, -5)$ **23.** $(-4, -3)$ **24.** $(-5, -2)$

Complete.

25. If a point is located on the y-axis, its first coordinate must be ■.

26. If a point is located on the x-axis, its second coordinate must be ■.

27. The coordinates of the origin are ■.

28. If the first coordinate of a point is negative and the second coordinate is positive, then the point lies in Quadrant ■.

29. If the first coordinate of a point is positive and the second coordinate is negative, then the point lies in Quadrant ■.

PROBLEM Solving APPLICATION

13.9 LATITUDE AND LONGITUDE

Suppose you find this scrap of paper in a drawer filled with old papers:

What does 10°N, 105°W mean? In what ocean is the sunken treasure ship?

The location of points on the earth can be described in terms of latitude and longitude.

Latitude

Longitude

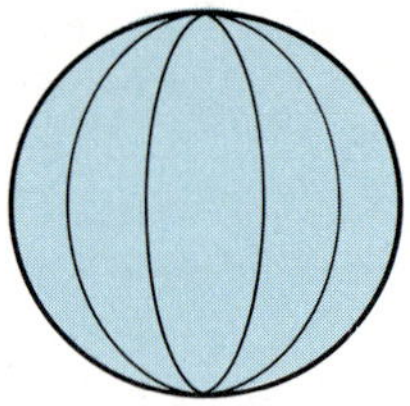

Latitude is used to measure degrees north and south of the equator. If a point is on the equator, its latitude is 0°.

Longitude is used to measure degrees east and west of the prime meridian. The **prime meridian** is a line or circle that passes through Greenwich, England, and extends in a north–south direction. If a point is on the prime meridian, its longitude is 0°.

In using pairs of numbers to give latitude and longitude, the latitude is given first.

So, 10°N, 105°W means that the location is 10° north of the equator and 105° west of the prime meridian.

Look at the map on page 383. Find the line for 10° north latitude. Then move to the intersection of this line with the line for 105° west longitude.

So, 10°N, 105°W means that the sunken treasure is located in the Pacific Ocean.

Use the map above to identify the country or body of water in which each point is located.

1. 15°S, 120°W

2. 45°S, 105°E

3. 60°N, 105°W

4. 30°N, 90°W

5. 30°N, 120°E

6. 20°N, 105°W

7. 30°S, 120°E

8. 60°N, 170°E

9. 60°N, 90°E

Use the map above to identify which point is closer to the equator.

10. 10°N, 130°W or 30°N, 130°W

11. 35°S, 120°E or 25°S, 120°E

12. 20°N, 101°W or 1°N, 120°W

13. 40°S, 120°E or 15°S, 130°E

Use the map to find the country in which each point is located.

14. 40°N, 100°W

15. 45°S, 175°E

16. 15°N, 120°E

17. 75°N, 45°W

18. 15°N, 105°E

19. 0°, 75°W

Solve. Use library resources if necessary.

20. The north–south distance around the earth is about 24,860 mi. What is the distance from the North Pole to the South Pole?

21. West longitudes can be any number between 0° and 180°. Why is it not possible for the longitude of a place to be 200°W?

13.10 RATIONAL NUMBERS

A **rational number** is a number that can be written as a ratio of two integers. The denominator cannot be zero. Rational numbers include:

Integers:	$15, -7$
Fractions and mixed numbers:	$\frac{5}{8}, -10\frac{3}{5}$
Terminating decimals:	$0.75, -2.84$
Repeating decimals:	$4.\overline{3}, -0.\overline{21}$

Every rational number can be shown on a number line.

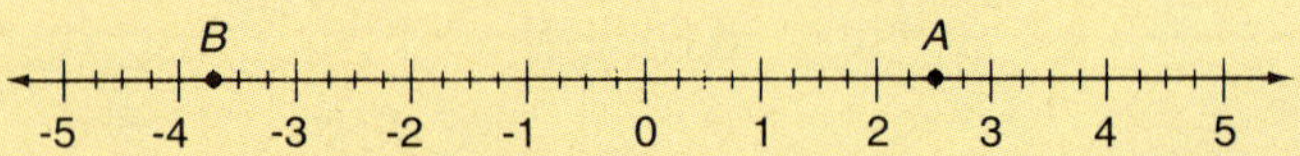

EXAMPLE 1 Write the rational number named by the point.

a. Point A

THINK: A is $\frac{1}{2}$ of the way between 2 and 3.

Point A names $2\frac{1}{2}$.

b. Point B

THINK: B is $\frac{3}{4}$ of the way between -3 and -4.

Point B names $-3\frac{3}{4}$.

You compare rational numbers the same way that you compare integers. Imagine a number line. As you move right, the numbers become greater. As you move left, the numbers become smaller.

EXAMPLE 2 Compare. Write $<$, $>$, or $=$.

a. $2 \bullet 0.75$

THINK: 2 is to the right of 0.75.

$2 > 0.75$

b. $-1\frac{2}{3} \bullet -0.86$

THINK: $-1\frac{2}{3}$ is to the left of -0.86.

$-1\frac{2}{3} < -0.86$

As with integers, the absolute value of a rational number is the distance the number is from zero.

EXAMPLE 3 Write the absolute value.

a. $|-6\frac{5}{8}|$

THINK: $-6\frac{5}{8}$ is $6\frac{5}{8}$ units from 0.

So, $|-6\frac{5}{8}|$ is $6\frac{5}{8}$.

b. $|12.7|$

THINK: 12.7 is 12.7 units from 0.

So, $|12.7|$ is 12.7.

PRACTICE EXERCISES

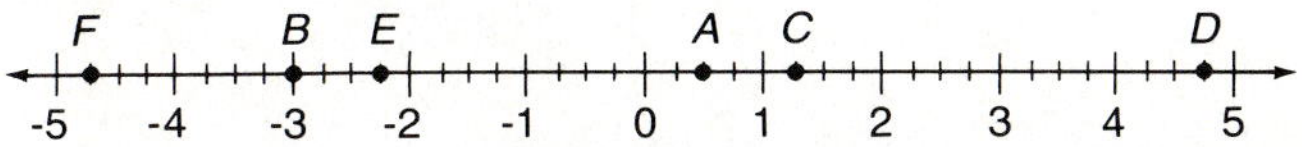

Use the number line above for Exercises 1–6.
Write the rational number named by the point.

1. *A* **2.** *B* **3.** *C* **4.** *D* **5.** *E* **6.** *F*

Compare. Write >, <, or =.

7. $\frac{1}{3}$ ● $\frac{2}{3}$ **8.** $\frac{3}{4}$ ● $\frac{1}{4}$ **9.** $-\frac{6}{10}$ ● $-\frac{3}{5}$ **10.** $-\frac{1}{5}$ ● $-\frac{4}{5}$

11. $\frac{3}{7}$ ● $\frac{5}{7}$ **12.** $-2\frac{3}{8}$ ● $-2\frac{7}{8}$ **13.** $-5\frac{4}{8}$ ● $-5\frac{1}{2}$ **14.** $-3\frac{6}{12}$ ● $-3\frac{1}{2}$

15. -0.6 ● -0.600 **16.** 0.73 ● -0.87 **17.** -0.15 ● 0.15 **18.** -746 ● -0.699

19. 2.34 ● -2.43 **20.** -4.07 ● -4.09 **21.** 7.002 ● 6.092 **22.** -100.3 ● -100.29

23. $-\frac{3}{4}$ ● -0.75 **24.** $2\frac{1}{4}$ ● -2.75 **25.** -5.25 ● $-5\frac{3}{8}$ **26.** 4.75 ● $-4\frac{7}{8}$

27. -4 ● $\frac{2}{3}$ **28.** $5\frac{1}{2}$ ● -6.9 **29.** -8 ● $-\frac{3}{5}$ **30.** $-5\frac{8}{9}$ ● -5.99

31. $-\frac{3}{8}$ ● $-\frac{6}{16}$ **32.** $6\frac{5}{10}$ ● -6.50 **33.** -34.098 ● -34.1 **34.** $7\frac{5}{8}$ ● 7.625

Write the absolute value.

35. $\left|\frac{3}{4}\right|$ **36.** $\left|-\frac{2}{3}\right|$ **37.** $\left|-\frac{6}{7}\right|$ **38.** $\left|\frac{7}{9}\right|$ **39.** $\left|-\frac{9}{10}\right|$

40. $|0.9|$ **41.** $|-0.3|$ **42.** $\left|-13\frac{2}{3}\right|$ **43.** $\left|4\frac{1}{2}\right|$ **44.** $|0|$

Draw a number line. Locate the following points.

45. Point *S* is an integer. It is neither positive nor negative.

46. Point *R* is located at -0.75.

47. Point *T* is located at the opposite of point *R*.

48. Point *V* is to the right of point *S*. The absolute value of point *V* is $2\frac{1}{2}$.

MIXED REVIEW

Find the number.

49. 58% of 200 is what number?

50. 14 is what percent of 40?

51. What is 40% of 200?

52. 102 is 17% of what number?

53. 147 is what percent of 245?

54. What is 63% of 1,100?

55. 86 is what percent of 43?

56. 187 is 85% of what number?

13.11 COMPUTING WITH RATIONAL NUMBERS

You can apply the rules for adding and subtracting integers to add and subtract rational numbers.

EXAMPLE 1

a. Add: $-34.56 + 125.8$

1. Subtract the absolute values.

THINK: $|-34.56| = 34.56$
$|125.8| = 125.8$ $\quad 125.8 - 34.56 = 91.24$

2. Write the correct sign.

THINK: 125.8 has the greater absolute value. The sum is positive.

$-34.56 + 125.8 = 91.24$

b. Subtract: $-\frac{1}{5} - \left(-\frac{2}{3}\right)$

THINK: Add the opposite of the number being subtracted.

$$-\frac{1}{5} - \left(-\frac{2}{3}\right) = -\frac{1}{5} + \frac{2}{3} = -\frac{3}{15} + \frac{10}{15} = \frac{7}{15}$$

$-\frac{1}{5} - \left(-\frac{2}{3}\right) = \frac{7}{15}$

You can also apply the rules for multiplying and dividing integers to multiply and divide rational numbers.

EXAMPLE 2

a. Multiply: $-5 \times \left(-\frac{4}{5}\right)$

THINK: The signs are the same. The product is positive.

$-\overset{1}{\cancel{5}} \times \left(-\frac{4}{\underset{1}{\cancel{5}}}\right) = 4$

b. Divide: $-0.34 \div 1.7$

THINK: The signs are different. The quotient is negative.

$-0.34 \div 1.7 = -0.2$

CHECKPOINT

Write the letter of the correct answer.

Compute.

1. $-\frac{3}{5} + \left(-2\frac{1}{10}\right)$ **a.** $-2\frac{4}{5}$ **b.** $2\frac{4}{5}$ **c.** $2\frac{7}{10}$ **d.** $-2\frac{7}{10}$

2. $-7.65 - 5.36$ **a.** 2.29 **b.** -2.29 **c.** -13.01 **d.** 13.01

3. $-31.2 \times (-0.07)$ **a.** 21.84 **b.** -21.84 **c.** 2.184 **d.** -2.184

4. $-\frac{3}{4} \div \left(-\frac{2}{3}\right)$ **a.** $-\frac{1}{2}$ **b.** $\frac{1}{2}$ **c.** $-1\frac{1}{8}$ **d.** $1\frac{1}{8}$

PRACTICE EXERCISES

Add or subtract.

1. $-65 + 42.87$ **2.** $\frac{1}{2} + \left(-\frac{7}{8}\right)$ **3.** $-4 + \frac{9}{16}$ **4.** $-3\frac{1}{5} + \frac{5}{8}$ **5.** $-12 + (-1.035)$

6. $-75 - 23.08$ **7.** $2\frac{9}{12} - \frac{7}{8}$ **8.** $178.09 - (-45)$ **9.** $-1\frac{1}{9} - \left(-3\frac{1}{3}\right)$ **10.** $\frac{5}{6} - 2\frac{1}{3}$

11. $-\frac{2}{9} + \left(-3\frac{5}{6}\right)$ **12.** $-15 + 75$ **13.** $-14 - 32$ **14.** $-0.75 + \frac{1}{2}$ **15.** $125 + (-178)$

16. $5 + \left(-\frac{2}{5}\right)$ **17.** $\frac{3}{7} - \frac{2}{14}$ **18.** $9.9 + (-0.07)$ **19.** $\frac{8}{12} - \left(-\frac{3}{4}\right)$ **20.** $4.67 + (-8.02)$

Multiply or divide.

21. 23×78 **22.** -45×-30 **23.** $\frac{1}{2} \times \frac{3}{4}$ **24.** $-40 \times \frac{1}{8}$ **25.** $2\frac{3}{4} \times 4$

26. $100 \div 40$ **27.** $-0.24 \div \frac{1}{6}$ **28.** $-0.75 \div 0.25$ **29.** $\frac{2}{3} \div \frac{4}{5}$ **30.** $-5\frac{5}{9} \div \frac{1}{3}$

31. $-18 \div (-3)$ **32.** $-\frac{5}{6} \times \left(-\frac{6}{5}\right)$ **33.** $-3.75 \div (-1.5)$ **34.** $45 \times \left(-\frac{1}{3}\right)$ **35.** 25×7

36. $-2\frac{1}{2} \div \left(-\frac{1}{2}\right)$ **37.** $-14 \div \left(-\frac{1}{7}\right)$ **38.** $-0.99 \div 0.33$ **39.** $-1.25 \times \frac{1}{2}$ **40.** $-12 \div (-0.04)$

TIME OUT

Scientists and weather forecasters often have to change temperatures between the Celsius and Fahrenheit systems. The procedure is easy now that you can compute with rational numbers.

14°F = ■°C

1. Subtract 32 from the temperature. $14 - 32 = -18$
2. Multiply the difference by $\frac{5}{9}$. $-18 \times \frac{5}{9} = -10$

So, 14°F = −10°C.

−25°C = ■°F

1. Multiply the temperature by $\frac{9}{5}$. $-25 \times \frac{9}{5} = -45$
2. Add 32 to the product. $-45 + 32 = -13$

So, −25°C = −13°F.

Write the corresponding Celsius or Fahrenheit temperature.

1. 77°F **2.** 50°F **3.** −13°F **4.** 40°C **5.** −15°C **6.** −50°C

PROBLEM *Solving* STRATEGY

13.12 CHOOSING A STRATEGY

Situation:

Suppose you walk into a party and see 4 of your friends. Each person shakes hands with the other 4 people. How many handshakes will there be among the 5 of you?

Strategy:

You can use more than one strategy to solve some problems.

Applying the Strategy:

A. You can solve the problem by drawing a diagram.

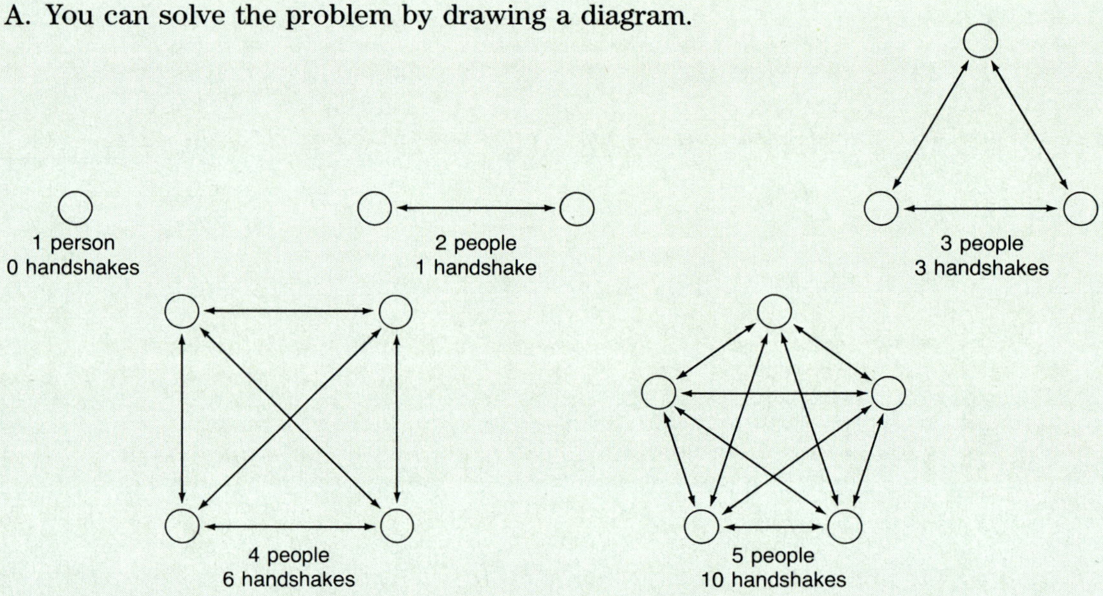

There will be 10 handshakes among the 5 of you.

B. You can also solve this problem by making a table and looking for a pattern.

Number of people	Number of handshakes	Pattern
1	0	0
2	1	1 + 0
3	3	2 + 1 + 0
4	6	3 + 2 + 1 + 0
5	10	4 + 3 + 2 + 1 + 0

There will be 10 handshakes among the 5 of you.

PRACTICE EXERCISES

Choose the most reasonable strategy to solve the problem.

1. Ms. Powell is planning a class trip to the museum for 56 students. The buses will cost $500. About how much money should Ms. Powell collect from each student to cover the cost of the buses?
a. Draw a Diagram
b. Guess-and-Check
c. Use Estimation

2. The museum employs 50 people. 10% of them are cashiers; 60% of them are guards; 20% of them are maintenance workers; and 10% of them are managers. How many employees are there in each category?
a. Interpret Remainders
b. Make a Table
c. Work Backward

3. In addition to 7 holidays, the museum is closed on Tuesdays. The average daily attendance is 1,543 people. How many days per year is the museum open?
a. Identify Needed Information
b. Look for a Pattern
c. Write Subproblems

4. The museum offers guided tours of its exhibits for groups of 30 people or more. If there are 186 people who want a guided tour, how many guides are needed?
a. Solve a Simpler Problem
b. Supply Missing Information
c. Make an Organized List

Solve the problem by using one or more problem-solving strategies.

5. A laboratory activity uses 5 g of salt for every 10 mL of water. How many milliliters of water do you need for 10 g of salt?

6. The area of the hamster cage is 600 cm^2. What are two possibilities for both the length and the width of the cage?

7. The admission to the science museum was $2 per student and $2.75 for an adult. Dorothy collected $150 for admissions. Can 65 students and 10 adults be admitted to the museum?

8. On Monday, the souvenir stand at the museum collected $3,500 and the cafeteria collected $5,775. Approximately the same amount of money was collected each day that week. About how much money was collected during the 5-day period?

PROBLEM Solving APPLICATION

13.13 CONSUMER: TRADING STOCKS

If you buy shares of stock in a corporation, you own part of the corporation. The prices of many different stocks are listed each day in most newspapers.

The table below shows data about six corporations. Notice that the price (in dollars) of a share of stock is written as a mixed number or a whole number.

52-Week						
High	**Low**	**Stock**	**High**	**Low**	**Last**	**Change**
$9\frac{1}{8}$	$6\frac{7}{8}$	Nasco	$8\frac{1}{2}$	$7\frac{3}{4}$	$8\frac{1}{4}$	$+\frac{1}{4}$
$62\frac{3}{8}$	55	NES	60	$56\frac{7}{8}$	$57\frac{1}{8}$	$-1\frac{1}{4}$
$115\frac{1}{8}$	$100\frac{1}{2}$	Nostro	$110\frac{5}{8}$	$105\frac{3}{8}$	107	$+1\frac{5}{8}$
$92\frac{1}{4}$	$83\frac{1}{2}$	Noxco	$87\frac{1}{2}$	$85\frac{1}{2}$	$86\frac{1}{2}$	-1
$94\frac{1}{2}$	$80\frac{1}{4}$	NR	92	$86\frac{7}{8}$	87	$-2\frac{1}{2}$
$45\frac{1}{2}$	$34\frac{1}{4}$	NYA	$42\frac{1}{4}$	$40\frac{1}{4}$	42	$+1\frac{1}{2}$
The highest and lowest prices during the last 52 weeks		Names of corporations often abbreviated	The highest and lowest prices of today		Last closing price of day	The change from yesterday's price to today's price +increase/−decrease

What was the closing price yesterday of a share of NYA stock?

THINK: To find yesterday's closing price, subtract the change from today's closing price.

Today's closing price of 1 share of NYA stock	−	Change from yesterday's price to today's price	=	Yesterday's closing price of 1 share of NYA stock
42	−	$1\frac{1}{2}$	=	$40\frac{1}{2}$

So, the closing price yesterday of a share of NYA stock was $40\frac{1}{2}$.

Find the closing price of the stock.

1. Nasco **2.** Noxco **3.** NES **4.** NR

Is today's price of a share of stock an increase or decrease from yesterday's?

5. NYA **6.** Nostro **7.** Noxco **8.** NES

Solve.

9. Which of the 6 stocks had the highest closing price today?

10. Which of the 6 stocks had the lowest closing price today?

11. How many of the 6 stocks decreased in price from yesterday to today?

12. Which stock increased in price by \$1.625 (1 dollar and $62\frac{1}{4}$ cents) from yesterday to today?

13. What were the 52-week high and low prices for Nasco?

14. What was the difference between the 52-week high and low prices for Nasco?

15. What is the price of 10 shares of NYA stock at the end of the day?

16. Was yesterday's closing price for Noxco higher than the closing price for NR?

17. If you bought 100 shares of Nostro stock yesterday and sold it at the end of today, how much would your profit or loss be?

18. The price of NES changes by $-1\frac{3}{4}$ on each of the next 3 days. On each of the following 2 days, it changes by $+\frac{1}{8}$. What will the price of the stock be?

CHAPTER REVIEW

Vocabulary Choose the letter of the word(s) that completes each statement.

1. A number in scientific notation has a first factor between 1 and 10 and a second factor that is a(n) ■. [378]

a. decimal **b.** power of 10 **c.** absolute **d.** square root

2. Every point on a coordinate plane can be named by using a(n) ■. [380]

a. axis **b.** absolute value **c.** ordered pair **d.** quadrant pair

3. A number that can be written as a ratio of two integers and does not have a denominator of zero is a ■ number. [384]

a. repeating **b.** rational **c.** rotating **d.** round

Skills Add. [367, 386]

$5 + 42$ *47*

4. $7 + (-13)$ **5.** $-12 + (-28)$ **6.** $-435 + 557$ **7.** $89 + (-34)$

8. $-8.7 + 4.58$ **9.** $12.5 + (-5.4)$ **10.** $\frac{3}{4} + \left(-\frac{4}{5}\right)$ **11.** $-14 + (-0.2)$ **12.** $\frac{1}{2} + (-1.25)$

Subtract. [370, 386]

$5 - 6$ *−1*

13. $-12 - 16$ **14.** $118 - (-101)$ **15.** $332 - (-33)$ **16.** $36 - (-34)$

17. $2.3 - 1.2$ **18.** $-43.5 - 22.8$ **19.** $\frac{5}{6} - \frac{7}{9}$ **20.** $-\frac{3}{4} - (-2.75)$ **21.** $-12.2 - 25.9$

Multiply. [372, 386]

$25 \times (-3)$ *−75*

22. -75×2 **23.** $0 \times (-25)$ **24.** $-12 \times (-12)$

25. $-20 \times (-2.5)$ **26.** -10.25×3 **27.** $-1 \times (-1.101)$ **28.** 0×1.3

Divide. [374, 386]

$-20 \div (-5)$ *4*

29. $-40 \div (-40)$ **30.** $63 \div (-9)$ **31.** $250 \div (-50)$

32. $-4\frac{3}{8} \div \frac{1}{2}$ **33.** $100.25 \div (-5)$ **34.** $\frac{5}{8} \div 3\frac{2}{5}$ **35.** $-20 \div \left(-1\frac{1}{5}\right)$

Write in scientific notation. [378]

350 *3.5×10^2*

36. 5,000 **37.** 250,000 **38.** 0.000005

CHAPTER TEST

Complete. Use $>$, $<$, or $=$ to make a true statement.

1. $|4| ● |-4|$ **2.** $\left|-\frac{2}{3}\right| ● \left|\frac{3}{4}\right|$ **3.** $|-45| ● |-44|$ **4.** $\left|-2\frac{3}{8}\right| ● \left|\frac{1}{2}\right|$ **5.** $|58.5| ● |-56.9|$

Add or subtract.

6. $12 + (-14)$ **7.** $-35 + (-27)$ **8.** $94 - (-63)$ **9.** $-123 + 0 + (-47)$

10. $-11 + (-15)$ **11.** $-32 - (-54)$ **12.** $-119 - 120$ **13.** $-87 - (-78)$

14. $\frac{6}{5} + \frac{8}{15}$ **15.** $-1\frac{1}{4} + \frac{2}{3} + \frac{4}{5}$ **16.** $0.45 + (-2.4)$ **17.** $4\frac{3}{5} - \left(-\frac{2}{5}\right)$

18. $1.25 - \left(-\frac{7}{10}\right)$ **19.** $-4.6 + 0.09$ **20.** $12.45 - (-4.2)$ **21.** $\frac{3}{7} - \left(-\frac{1}{2}\right)$

Multiply or divide.

22. 12×-40 **23.** $6 \div (-30)$ **24.** $-9 \times 12 \times (-5)$ **25.** $25 \times 0 \times -16$

26. $-125 \div (-25)$ **27.** -10×50 **28.** $200 \div (-25)$ **29.** $-150 \div -5$

30. $\frac{3}{10} + -\frac{5}{8}$ **31.** $1\frac{1}{3} \times (-3)$ **32.** 20×3.58 **33.** $15.5 \div -5$

34. $\frac{7}{8} \div -1\frac{1}{3}$ **35.** $0.03 \times -\frac{1}{2}$ **36.** -45×-10.05 **37.** $-350 \div 0.07$

Write in scientific notation.

38. 10,000 **39.** 25,000 **40.** 0.0000007 **41.** 0.00026 **42.** 30,000,000

Write as a standard numeral.

43. 5×10^3 **44.** 4.2×10^{-5} **45.** 8.3×10^8 **46.** 9.5×10^{-4}

Write the point named by the ordered pair.

47. (1, 3) **48.** (2, −3)

49. (−2, −4) **50.** (4, −3)

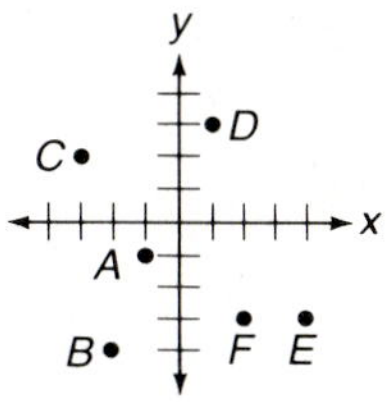

Solve.

51. ANEX stock sells for $60\frac{1}{2}$ per share today, up by $\frac{1}{2}$ from yesterday. If you bought 1,000 shares yesterday and sold it at the end of today, how much would your profit be?

52. You have won the Lucky Bucks Jackpot. You will receive \$1 on the first day, \$2 on the second, \$3 on the third day, and so on for 30-days. How much money will you have won in 30 days?

ENRICHMENT THE LAWS OF EXPONENTS

In the number 2^3, recall that 2 is the **base** and 3 is the **exponent.** The exponent tells you how many times the base occurs as a factor.

$$2^3 = 2 \times 2 \times 2$$

You can use the laws of exponents to help you multiply with exponents. When the bases are the same, add the exponents.

Example Multiply $8^3 \times 8^2$.

THINK: $8^3 \times 8^2 = (8 \times 8 \times 8) \times (8 \times 8) = 8^5$

So, $8^3 \times 8^2 = 8^{3+2}$ or 8^5.

To divide exponents with the same base, subtract the exponents.

Example Divide $7^5 \div 7^3$

THINK: $7^5 \div 7^3 = \frac{7 \times 7 \times 7 \times 7 \times 7}{7 \times 7 \times 7} = 7^2$

So, $7^5 \div 7^3 = 7^{5-3}$ or 7^2.

Multiply.

1. $10^3 \times 10^1$ **2.** $10^6 \times 10^2$ **3.** $10^5 \times 10^4$ **4.** $10^{17} \times 10^3$

5. $2^3 \times 2^3$ **6.** $3^2 \times 3^3$ **7.** $4^5 \times 4^3$ **8.** $5^7 \times 5^2$

9. $17^2 \times 17^{-3}$ **10.** $21^{-3} \times 21^{-4}$ **11.** $32^6 \times 32^{-3}$ **12.** $100^{-5} \times 100^8$

Divide.

13. $10^6 \div 10^3$ **14.** $10^5 \div 10^4$ **15.** $10^7 \div 10^5$ **16.** $10^{19} \div 10^{11}$

17. $4^3 \div 4^1$ **18.** $5^7 \div 5^2$ **19.** $6^9 \div 6^3$ **20.** $9^8 \div 9^4$

21. $15^4 \div 15^{-3}$ **22.** $23^{-6} \div 23^4$ **23.** $36^{-9} \div 36^5$ **24.** $100^8 \div 100^{-4}$

Multiply or divide.

25. $7^6 \div 7^3$ **26.** $11^4 \times 11^9$ **27.** $5^6 \times 5^{-6}$ **28.** $4^7 \div 4^3$

29. $16^{-10} \div 16^8$ **30.** $14^8 \times 14^{-2}$ **31.** $9^{12} \div 9^{-5}$ **32.** $3^{-7} \times 3^6$

Write the missing exponent.

33. $2^2 \times 2^{\blacksquare} = 2^6$ **34.** $10^{\blacksquare} \times 10^5 = 10^8$

35. $8^6 \div 8^{\blacksquare} = 8^3$ **36.** $6^{\blacksquare} \div 6^4 = 6^5$

37. $7^{\blacksquare} \times 7^{-3} = 7^8$ **38.** $12^{10} \div 12^{\blacksquare} = 12^{-2}$

MENTAL MATH USING THE PROPERTIES OF RATIONAL NUMBERS

These properties apply to the addition and multiplication of rational numbers.

COMMUTATIVE PROPERTY

The order of addends does not change the sum. $-2 + 8 = 8 + (-2)$

The order of factors does not change the product. $3 \times (-6) = -6 \times 3$

ASSOCIATIVE PROPERTY

The grouping of addends does not change the sum.

$-4 + [(-5) + 3] = [-4 + (-5)] + 3$

The grouping of factors does not change the product.

$[8 \times (-6)] \times 7 = 8 \times [(-6) \times 7]$

PROPERTY OF ZERO

The sum of a number and zero is the number. $-4.6 + 0 = -4.6$

DISTRIBUTIVE PROPERTY

To multiply a sum by a number, multiply each addend by the number, then add the products.

$3 \times [(-2) + 6] = [3 \times (-2)] + (3 \times 6)$

PROPERTY OF OPPOSITES

The sum of any number and its opposite is zero. $2.5 + (-2.5) = 0$

PROPERTY OF RECIPROCALS

The product of any number and its opposite is 1. $4 \times \frac{1}{4} = 1$

PROPERTY OF ONE

The product of any number and 1 is the number. $1 \times (-9) = -9$

You can use these properties to help you compute mentally.

Compute: $-67 + (-89) + 89$

THINK: -89 and 89 are opposites; their sum is 0.

$$-67 + \underbrace{(-89) + 89}$$
$$-67 + 0 = -67$$

The sum of -67 and 0 is -67.

Compute: $-4 \times \left(\frac{1}{2} \times 3.5\right)$

THINK: Use the Associative Property.

$$\begin{aligned} -4 \times \left(\tfrac{1}{2} \times 3.5\right) &= \left(-4 \times \tfrac{1}{2}\right) \times 3.5 \\ &= -2 \times 3.5 \\ &= -7 \end{aligned}$$

Compute mentally.

1. $47 + 12 + (-12)$

2. $6.5 + 8.6 + (-6.5)$

3. $-23 \times [5 + (-5)]$

4. $\left(\frac{1}{5} \times 5\right) \times 24$

5. $\frac{1}{4} \times [(-43) \times 4]$

6. $\left[\left(-\frac{2}{3}\right) + \frac{2}{3}\right] \times 36$

7. $8 \times \left(6 + \frac{1}{2}\right)$

8. $-18 \times \left[\frac{1}{3} + (-1)\right]$

9. $15 \times \left[\left(-\frac{1}{5}\right) + 2\right]$

10. $24 \times \left(\frac{1}{2} + \frac{1}{3}\right)$

11. $-12 \times \left(\frac{1}{6} + \frac{1}{4}\right)$

12. $60 \times \left[\left(-\frac{1}{10}\right) + \frac{1}{3}\right]$

13. $4 \times \left[\left(-\frac{3}{4}\right) + (-6)\right]$

14. $-20 \times \left(\frac{3}{10} + 1\right)$

15. $6 \times \left(\frac{2}{3} + \frac{1}{2}\right)$

PreSkills TEST

Add. [367]

1. $7 + (-12)$ **2.** $-8 + 4$ **3.** $-6 + (-5)$ **4.** $11 + (-19)$

5. $-4 + (-15)$ **6.** $-13 + 8$ **7.** $-4 + 10$ **8.** $-2 + (-13)$

Subtract. [370]

9. $3 - 9$ **10.** $-6 - 7$ **11.** $8 - (-5)$ **12.** $-4 - (-2)$

13. $-21 - 5$ **14.** $0 - 15$ **15.** $-3 - (-18)$ **16.** $2 - (-34)$

Multiply. [372]

17. $5 \times (-4)$ **18.** -7×9 **19.** $-3 \times (-6)$ **20.** $8 \times (-6)$

21. -12×0 **22.** $-2 \times (-13)$ **23.** -9×4 **24.** $-11 \times (-7)$

Divide. [374]

25. $16 \div (-4)$ **26.** $-35 \div 5$ **27.** $-56 \div (-8)$ **28.** $-63 \div 7$

29. $-36 \div (-3)$ **30.** $0 \div (-8)$ **31.** $54 \div (-9)$ **32.** $-48 \div (-8)$

Find the missing number.

33. $n + 8 = 11$ **34.** $n - 6 = 12$ **35.** $n + 7 = 15$ **36.** $n - 10 = 23$

37. $5 \times n = 40$ **38.** $-4 \times n = 28$ **39.** $n \div 6 = 3$ **40.** $-9 \div n = 45$

Write the ordered pair that names the point. [380]

41. *A* **42.** *B*

43. *C* **44.** *D*

45. *E* **46.** *F*

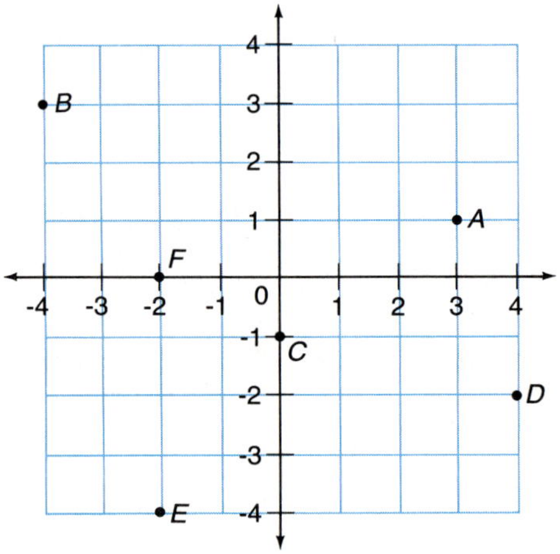

Chapter 14
EQUATIONS

People use a language, like English, to communicate. What other types of languages are in use today?

14.1 ORDER OF OPERATIONS

During an inventory, Carlos Moreno found 20 single-sided disks and 25 double-sided disks. How many sides could be used for word processing?

To find how many sides, Carlos wrote:

$20 + 25 \cdot 2$ **The raised dot means multiply.**

To find the answer, apply the rules for the Order of Operations.

- Raise to a power.
- Multiply or divide in order from left to right.
- Add or subtract in order from left to right.

EXAMPLE 1

Simplify:	$20 + 25 \cdot 2$
1. Multiply.	$20 + 50$
2. Add.	70

There are 70 usable sides for word processing.

To simplify expressions with parentheses, work within the parentheses first.

EXAMPLE 2

Simplify:	$8 + 3(4^2 - 6)$
1. Work within parentheses.	
THINK: Raise to a power, then subtract.	$8 + 3(16 - 6)$ $8 + 3(10)$
2. Multiply.	
THINK: 3(10) means 3 · 10	$8 + 30$
3. Add.	38

So, $8 + 3(4^2 - 6) = 38$.

To simplify expressions with a division bar, first do the work above the bar, then below the bar.

EXAMPLE 3

Simplify:	$\frac{40 - 30}{16 \div 8 + 3}$
1. Work above the bar.	
THINK: Subtract.	$\frac{10}{16 \div 8 + 3}$
2. Work below the bar.	
THINK: Divide, then add.	$\frac{10}{2 + 3} = \frac{10}{5}$
3. Divide.	2

So, $\frac{40 - 30}{16 \div 8 + 3} = 2$.

PRACTICE EXERCISES

Write the operation and the number(s) you would compute first to simplify.

1. $3 \cdot 5 + 6$ **•2.** $3(5 + 6)$ **3.** $3 + 5 \cdot 6$ **4.** $(3 + 5)6$

5. $12 - 4 - 6$ **6.** $(12 - 4)6$ **7.** $12 \cdot 4 \div 6$ **8.** $(12 - 4) - 6$

9. $2^2 - 3 \cdot 5$ **10.** $7 + 3^4 - 9$ **11.** $\frac{6 - 4 + 3}{12 + 4^2}$ **12.** $\frac{8(5^2 + 1)}{9 - 2}$

Simplify.

13. $11 - 4 + 7$ **14.** $5 + 7 \cdot 2$ **15.** $3^2 - 2^2 \div 4$ **16.** $(-2)^3 \div 2^2$

17. $2^3 + 3 \cdot 5$ **18.** $4^2 + 16 \div 2$ **19.** $8 + 3 \cdot 4^2$ **20.** $5 - 4^3 \div 8$

21. $5(8 - 3)$ **22.** $(16 - 4) \div 3$ **23.** $12 \div (8 - 5)$ **24.** $(2 - 10) \div (3 + 5)$

25. $3(5^2 - 20)$ **26.** $12(2^3 - 4)$ **27.** $(27 - 9) \div 3^2$ **28.** $(6 - 2)^2 - (5 - 9)$

29. $\frac{4 - 3}{6 + 2}$ **30.** $\frac{6(9 - 5)}{5 + 5^2}$ **31.** $\frac{7 - 2 \cdot 2}{4^2 - 1}$ **32.** $\frac{18 \div (3^2 - 18)}{3(6 - 5 + 2)}$

33. $\frac{7^2 - 9}{30 + 2 \cdot 5}$ **34.** $\frac{8 + 6 \cdot 2}{4^2 + 3^2}$ **35.** $\frac{16 - 12 \div 2 \cdot 3}{15 + 3(9 - 7)}$ **36.** $\frac{9 \div (6 - 15)}{6^2 - 5(4 + 2)}$

Write parentheses to make the sentence true.

37. $12 - 8 + 5 = -1$ **38.** $3 \cdot 2 - 6 \div 3 = 0$

39. $20 + 6 \cdot 2 \div 4 = 8$ **40.** $8 - 2 + 8 \div 2 = 3$

MIXED REVIEW

Measure the angle. Classify it as *acute, right, obtuse,* or *straight.*

41.

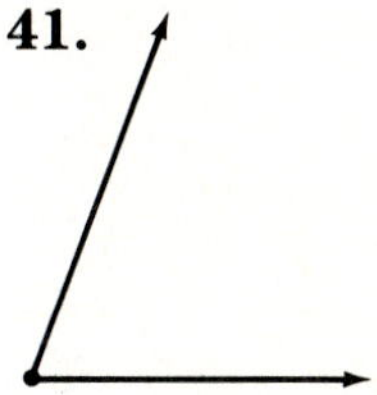

42.

43.

44.

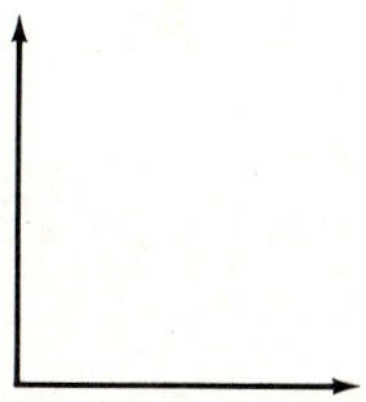

TIME OUT

Everyone knows that 5 twos are 10, $5 \cdot 2 = 10$. But did you know that 5 twos can equal zero? Check students' answers. Sample answers below.

$$2 - 2 \div 2 - 2 \div 2$$

$$2 - 1 - 1 = 0$$

Use the symbols $+$, $-$, $\cdot$, and $\div$ and the rules for the Order of Operations to write expressions that show how 5 twos can equal 1, 2, then 3.

14.2 ALGEBRAIC EXPRESSIONS

An **algebraic expression** uses a letter, or **variable,** to stand for one or more unknown numbers. Mathematical symbols are often used to translate word phrases into algebraic expressions.

Certain word phrases are associated with each operation.

$+$	$-$	$\times$	$\div$
Sum of	**Difference of**	**Product of**	**Quotient of**
Increased by	**Decreased by**	**Multiplied by**	**Divided by**
More than	**Less than**	**Twice**	**Ratio**

EXAMPLE 1 In the state basketball tournament, Dondero High School scored 15 fewer points than Ferndale High School. Write an algebraic expression for the number of points scored by Dondero High School.

THINK: Let n represent Ferndale's points. Write: n
Subtract to get Dondero's points. $n - 15$

EXAMPLE 2 Write an algebraic expression for the word phrase. Let n represent the number.

a. 6 more than twice a number
$2 \cdot n + 6$ or
$2n + 6$

b. 18 less than $\frac{1}{2}$ a number
$\frac{1}{2} \cdot n - 18$ or
$\frac{1}{2}n - 18$

You can also translate an algebraic expression into a word phrase.

EXAMPLE 3 Write a word phrase for the algebraic expression.

a. $3(n - 5)$

3 times the difference of a number minus 5

b. $\frac{r + 8}{3}$

The sum of a number and 8 divided by 3

CHECKPOINT

Write the letter of the correct answer.

Write an algebraic expression for the word phrase.

1. 11 less than one-fourth of a number **a.** $11 - 4n$ **b.** $\frac{1}{4}n - 11$ **c.** $11 - \frac{1}{4}n$
2. 19 times the difference of a number minus 2 **a.** $19(n - 2)$ **b.** $19n - 2$ **c.** $19(2 - n)$
3. A number divided by the sum of 6 and 8 **a.** $\frac{n}{6} + 8$ **b.** $\frac{n}{6 + 8}$ **c.** $\frac{6 + 8}{n}$

PRACTICE EXERCISES

Write an algebraic expression for the word phrase. Let n represent the number.

1. 6 added to a number

2. A number decreased by 14

3. Twice a number that is cubed

4. 7 times a number, divided by 3

5. 8 more than a number

6. 29 less than a number

7. The ratio of a number to 65

8. 34 divided by a number

9. The sum of 10 and a number

10. The product of a number and 41

11. 58 minus a number

12. A number divided by 17

13. 42 more than 8 times a number

14. 79 decreased by twice a number

15. The quotient of a number divided by 20

16. A number doubled and decreased by 72

17. The quotient of a number plus 90, divided by 47

18. The product of 56 and the sum of a number and 84

19. 3 times the sum of a number and 19, divided by 7

20. The fifth power of 2 increased by the product of a number and 4

Write a word phrase for the algebraic expression.

21. $r + 3$ **22.** $9 - a$ **23.** $64m$ **24.** $q - 14$

25. $17 + b$ **26.** $n - 6$ **27.** $5(3 + x)$ **28.** $4y - 6$

29. $11c - 2$ **30.** $13(9 - g)$ **31.** $28 - 7h$ **32.** $t^2 - 15$

33. $\frac{s - 7}{5}$ **34.** $\frac{3(x - 2)}{8}$ **35.** $\frac{60}{12 - z}$ **36.** $\frac{35}{2(r + 4)}$

Solve.

37. The Hillside High School debate team scored 3 more than twice as many points as their opponents. Write an algebraic expression for the number of points they scored.

38. The coach of the Tyler High School tennis team is 6 years older than twice the age of the youngest team member. Write the coach's age as an algebraic expression.

39. At the Glenwood High School cheerleading tryouts, $\frac{1}{2}$ of the contestants were eliminated in the first round. In the second round, $\frac{1}{2}$ of the remaining contestants were eliminated. Another $\frac{1}{2}$ were eliminated in the third round. Six cheerleaders were finally chosen. How many students originally tried out for the squad?

14.3 EVALUATING EXPRESSIONS AND FORMULAS

To **evaluate an expression,** first replace the variable by a given value. Then simplify the resulting numerical expression.

EXAMPLE 1 Evaluate: $4x + 7$ for $x = 5$

1. Replace x with 5. $4x + 7$

   $4(5) + 7$

2. Simplify. $20 + 7$

   27

Sometimes there is more than one variable in an expression.

EXAMPLE 2 Evaluate: $\frac{2x + y}{z}$ for $x = 2$, $y = 5$, and $z = 3$

1. Replace x with 2, y with 5, and z with 3. $\frac{2 \cdot 2 + 5}{3}$

2. Simplify. $\frac{4 + 5}{3} = 3$

A **formula** is an algebraic sentence. You can evaluate a formula by replacing the variables with a given value, then simplifying.

EXAMPLE 3 The pressure at a certain depth in the ocean can be approximated by the formula $P = 15 + \frac{1}{2}D$ in which P stands for the pressure in pounds per square inch and D is the depth in feet.

Find the amount of pressure at a depth of 40 ft.

1. Replace the variables with the given values. $P = 15 + \frac{1}{2}D$

   THINK: The depth D is 40 ft. $P = 15 + \frac{1}{2}(40)$

2. Simplify. $P = 15 + 20$

   $P = 35$

So, the water pressure at a depth of 40 ft is 35 lb per sq in.

CHECKPOINT Write the letter of the correct answer.

Evaluate.

1. $-3a + 4$ for $a = -2$ **a.** -2 **b.** -1 **c.** 3 **d.** 10
2. $7 - (x - y)$ for $x = 5, y = -6$ **a.** 4 **b.** -4 **c.** 6 **d.** 8

PRACTICE EXERCISES

Evaluate for $x = 4$.

1. $x + 7$ **2.** $x - 3$ **3.** $5x$ **4.** $-6x$

5. $-x + 9$ **6.** $x - 15$ **7.** $-7x$ **8.** $10x$

9. $13 - x$ **10.** $-22 + x$ **11.** $-14 - x$ **12.** $-3 + x - 4$

13. $\frac{x}{2}$ **14.** $\frac{16}{x}$ **15.** $\frac{x}{-8}$ **16.** $\frac{-x}{-2}$

Evaluate for $a = -2$ and $b = 5$.

17. $a + b$ **18.** $a - b$ **19.** $b - a$ **20.** ab

21. $6 - a + b$ **22.** $3 + a - b$ **23.** $2a - b$ **24.** $3b + a$

25. $\frac{ab}{10}$ **26.** $\frac{a}{b}$ **27.** $\frac{b}{a}$ **28.** $\frac{a + b}{3}$

29. $4a - b$ **30.** $b - 2a$ **31.** $5 + (b - a)$ **32.** $3(a - b)$

Evaluate for $x = 3$, $y = -1$, and $z = 2$.

33. $x + y + z$ **34.** $x - y - z$ **35.** $(x - y)z$ **36.** $x - (y + z)$

37. $\frac{x + z}{y}$ **38.** $\frac{x + y}{z}$ **39.** $\frac{y + z}{x}$ **40.** $\frac{xz}{y}$

41. $5(x + y)$ **42.** xyz **43.** $x(y + z)$ **44.** $z(x - y)$

45. $4x + y + 2z$ **46.** $-5xz$ **47.** $8(z - y)$ **48.** $7(2 + y)$

49. $\frac{3(y - z)}{x}$ **50.** $\frac{9(x - z)}{y + 4}$ **51.** $\frac{-4y + x}{1 + 3z}$ **52.** $\frac{3x + y}{z}$

The formula for the volume of a rectangular prism is $V = lwh$, where l is the length, w is the width, and h is the height. Find the volume.

53. $l = 4$ m; $w = 2$ m; $h = 5$ m

54. $l = 0.5$ m; $w = 0.2$ m; $h = 1.6$ m

55. $l = 1.4$ cm; $w = 7$ cm; $h = 2.3$ cm

56. $l = \frac{1}{2}$ ft; $w = \frac{3}{4}$ ft; $h = \frac{5}{6}$ ft

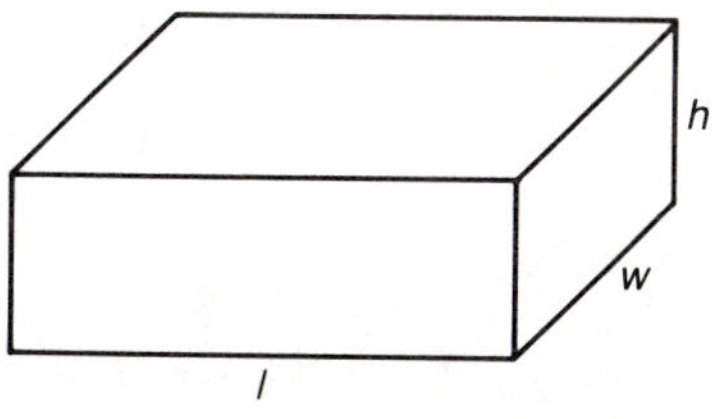

The formula for the area of a triangle is $A = \frac{1}{2}bh$ where b is the length of the base and h is the height. Find the area.

57. $b = 14$ m; $h = 3$ m

58. $b = 5.7$ m; $h = 6.2$ m

59. $b = 4.9$ cm; $h = 8.2$ cm

60. $b = \frac{20}{3}$ in.; $h = \frac{9}{2}$ in.

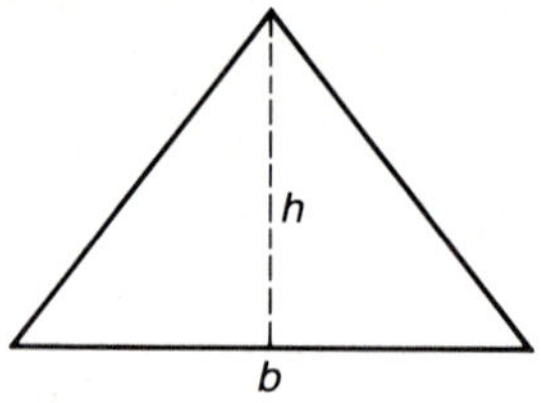

The formula for the area of a trapezoid is $A = \frac{1}{2}h(b_1 + b_2)$ where h is the height and b_1 and b_2 are the bases. Find the area.

61. $h = 16$ m; $b_1 = 7$ m; $b_2 = 13$ m

62. $h = 9.2$ mm; $b_1 = 3.8$ mm; $b_2 = 6.9$ mm

63. $h = 7.4$ cm; $b_1 = 4.9$ cm; $b_2 = 2.7$ cm

64. $h = \frac{26}{3}$ in.; $b_1 = \frac{43}{8}$ in.; $b_2 = \frac{58}{16}$ in.

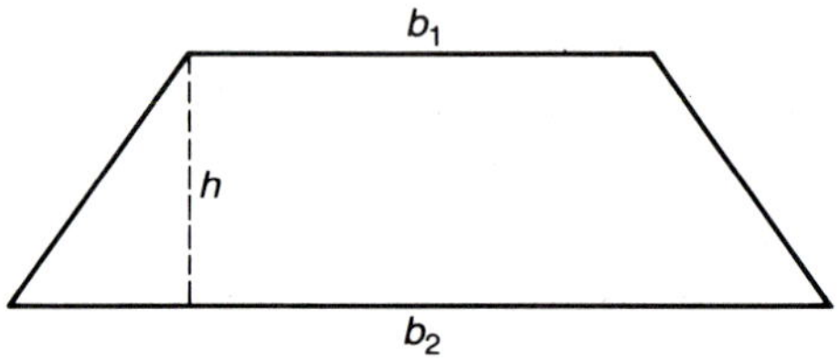

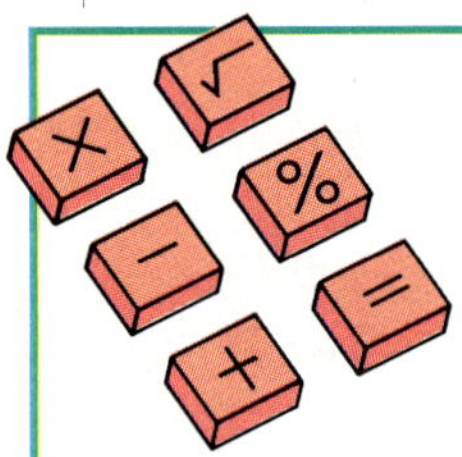

CALCULATOR

Calculators make it easy to discover number patterns. Try this experiment on your calculator.

1. Pick any number from 1 to 8. Divide it by 9, by 99, and by 999.

$8 \div 9 =$ ■ $8 \div 99 =$ ■ $8 \div 999 =$ ■

Describe the pattern.

2. Try two other numbers between 1 and 8. Does this pattern hold?

■ $\div\ 9 =$ ■	■ $\div\ 9 =$ ■
■ $\div\ 99 =$ ■	■ $\div\ 99 =$ ■
■ $\div\ 999 =$ ■	■ $\div\ 999 =$ ■

3. Does the pattern hold for 2-digit numbers?

Try this	**Your turn**
$47 \div 9 =$ ■	■ $\div\ 9 =$ ■
$47 \div 99 =$ ■	■ $\div\ 99 =$ ■
$47 \div 999 =$ ■	■ $\div\ 999 =$ ■

4. Does the pattern hold if you divide a 2-digit number by 9,999?

PROBLEM Solving APPLICATION

14.4 USING THE DISTANCE FORMULA

A Metroliner train is traveling at the rate of 84 mph. What distance will the train travel in 2.5 hours?

You can use the distance formula to solve problems involving time, rate (average speed), and distance.

$$\text{DISTANCE} = \text{RATE} \times \text{TIME}$$

$$d = rt$$

THINK: The rate and time are given. $r = 84$ mph, $t = 2.5$ h

1. Substitute the given numbers for each variable. $d = 84 \cdot 2.5$
2. Multiply. $d = 210$

So, the train travels 210 mi in 2.5 hours.

Find the missing amount.

1. $r = 76$ mph, $t = 5$ hours, $d =$ ■

2. $r = 52$ mph, $t = 6.5$ hours, $d =$ ■

3. $r =$ ■, $t = 3.2$ hours, $d = 384$ mi

4. $r =$ ■, $t = 2\frac{1}{4}$ hours, $d = 126$ mi

5. $r = 372$ mph, $t =$ ■, $d = 2{,}604$ mi

6. $r = 3\frac{1}{2}$ mph, $t =$ ■, $d = 6\frac{3}{10}$ mi

Solve. Remember to estimate whenever you use your calculator.

7. A jet plane travels 2,210 mi in 3.4 hours. What is its rate in miles per hour?

8. An Antelope ZR travels 459 mi at a rate of 54 mph. How long will the trip take?

9. The Concorde flew from Paris to New York in $3\frac{1}{2}$ h. Its average speed was 1,037.5 mph. What is the air distance between Paris and New York?

10. A train travels 403.62 mi between Moscow and Leningrad in 4.65 hours. A train travels between Wilmington and Baltimore in 46 minutes at a rate of 89.2 mph. Which train is faster? By how much?

11. A spacecraft orbiting the earth travels at a rate of 16,500 mph. How much more than 4 hours does it take to travel 80,850 mi?

14.5 EQUATIONS AND INEQUALITIES

Jeff Chan is shopping for summer clothes. He wants to buy a jacket that costs \$80. The cost of the jacket is ten dollars more than the cost of two shirts he also wants to buy.

An **equation** is a mathematical sentence that states that two expressions are equal.

EXAMPLE 1 Write an equation to show the relationship between the price of the shirts and the cost of the jacket.

THINK: Cost of the jacket is \$10 more than the cost of 2 shirts.

WRITE: $80 = 10 + 2s$

An **inequality** is a mathematical sentence that uses <, >, ≤, or ≥ to show the relationship between two expressions.

EXAMPLE 2 Write an inequality for the expression.

a. Five more than one-half of a number is greater than 13.

$5 + \frac{1}{2}n > 13$

b. 8 more than 3 times a number is less than or equal to 75.

$8 + 3n \leq 75$

FOR DISCUSSION

A rack of clothes had a sign that read, "No item more than \$5." How could the sign be written as an inequality? Which of these price tags could appear on the clothes: \$4, \$7.50, \$5, \$3.95, \$8.98?

PRACTICE EXERCISES

Write a mathematical sentence for the statement.

1. 18 times a number is 36.

2. 7 more than a number is 15.

3. 6 less than a number is 3.

4. The quotient of a number divided by 9 is 5.

5. 8 subtracted from 3 times a number is 16.

6. The sum of two-thirds of a number and 6 is 18.

7. 9 is 4 more than 3 times a number.

8. 17 is 42 minus one-half a number.

9. 8 times the sum of a number and 3 is 35.

10. 20 divided by the difference of 15 and a number is 3.

11. 7 times a number is less than 21.

12. 14 more than a number is greater than 56.

13. The quotient of a number divided by 14 is less than 93.

14. The difference between 87 and a number is greater than 34.

15. 15 times the sum of a number and 96 is less than 28.

16. 13 and twice a number is less than or equal to 61.

17. 4 more than a number is less than 38.

18. The quotient of 72 divided by a number is greater than or equal to 46.

19. 4 more than 5 times a number is less than or equal to 39.

20. The product of 12 and a number, divided by 6, is greater than 58.

MIXED REVIEW

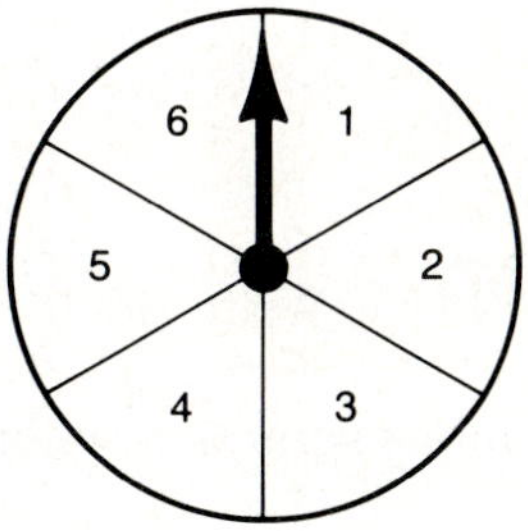

Find the probability of spinning:

21. a 4.

22. an even number.

The spinner is spun 600 times.
Predict how many times the spinner will land on:

23. an odd number.

24. a number greater than 1.

TIME OUT

In 1863 President Abraham Lincoln gave his famous Gettysburg Address. It starts "Fourscore and seven years ago. . . ." Fourscore and seven is equal to 87. How many years are in a score? What year was President Lincoln referring to?

14.6 GRAPHING POINTS ON A LINE

The solution to an equation or an inequality can be shown by a point or a set of points on a number line.

EXAMPLE 1 Graph the solution of $x + 6 = 8$.

1. Solve: $x + 6 = 8$

   THINK: $2 + 6 = 8 \qquad x = 2$

2. Graph: 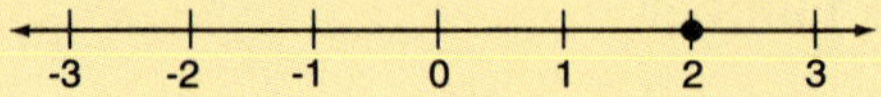

The solution of an inequality with the symbol $>$ or $<$ is shown by an open circle and a shaded line. The open circle indicates that the end point is not part of the solution.

EXAMPLE 2 Graph the solution of $x < 5$.

1. Solve: $x < 5$

   THINK: x can be any number less than 5.

2. Graph: 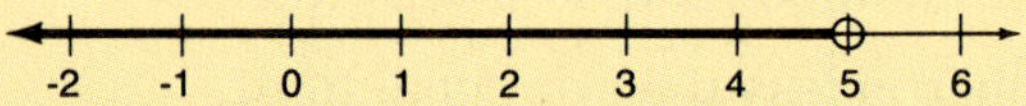

The solution to an inequality with the symbol $\geq$ or $\leq$ is shown by a closed circle and a shaded line.

EXAMPLE 3 Graph the solution of $x \geq -3$.

1. Solve: $x \geq -3$

   THINK: x is -3 or any number greater than -3.

2. Graph: 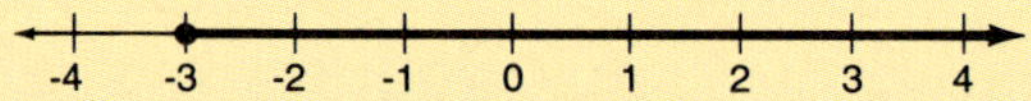

CHECKPOINT

Write the letter of the correct answer.

Choose the number sentence whose solution is graphed.

1. 

a. $x + 2 \geq 0$ **b.** $x < -2$

c. $x > -2$ **d.** $x \geq -2$

2. 

a. $x + 8 \leq 9$ **b.** $x \leq 1$

c. $x > 1$ **d.** $x < 1$

PRACTICE EXERCISES

Match the graph to the correct number sentence.

1. 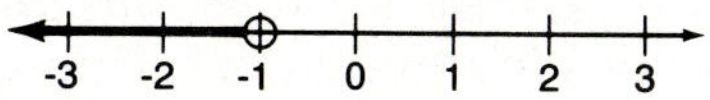

2. 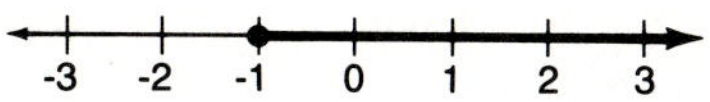

3. 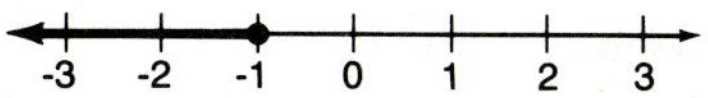

4. 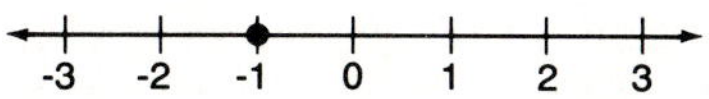

5. 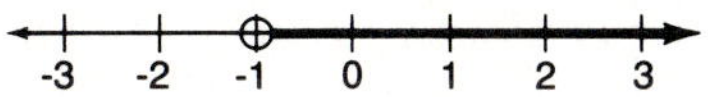

a. $x \leq -1$

b. $x = -1$

c. $x > -1$

d. $x < -1$

e. $x \geq -1$

Graph the solution.

6. $x + 5 = 7$ **7.** $x + 1 = 9$ **8.** $x + 4 = 10$ **9.** $x + 8 = 12$

10. $x - 3 = 6$ **11.** $x - 7 = 5$ **12.** $x - 2 = 9$ **13.** $x - 10 = 2$

14. $x < 6$ **15.** $x > 4$ **16.** $x > -1$ **17.** $x < -5$

18. $x \geq -2$ **19.** $x \geq 0$ **20.** $x \leq 3$ **21.** $x \leq -2$

22. $x \leq 3$ **23.** $x \geq -3$ **24.** $x \geq 2$ **25.** $x \leq -4$

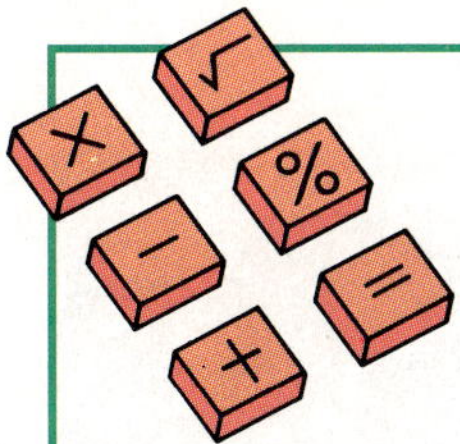

CALCULATOR

The sum of the digits of a 2-digit number is 11. If the digits are reversed, the number formed is 27 less than the original number. Find the original number.

Use a calculator and the guess-and-check method to find the number.

2-digit number	Reverse the digits. Subtract.	Difference of 27?
29, $2 + 9 = 11$	$92 - 29 = 63$	No
38, $3 + 8 = 11$	$83 - 38 = 45$	No

What is the number?

PROBLEM *Solving* APPLICATION

14.7 CAREER: PHARMACIST

Don Keefer is a pharmacist. He uses his knowledge of measurements and equations when preparing various solutions. For example, when working with grams and milliliters, Don uses the fact that 1 mL of water has a mass of 1 g.

How many grams of dextrose must Don use to prepare 3,000 mL of a solution that is 5% dextrose?

THINK: 5% of the 3,000-mL solution is dextrose. The number of grams of dextrose is equivalent to 5% of 3,000 g.

1. Write an equation.
   Let n represent the number of grams of dextrose. $\quad n = 5\%$ of 3,000
2. Multiply to find n. $\quad n = 0.05 \cdot 3{,}000$
   $n = 150$

So, Don must use 150 g of dextrose to make the 5% solution.

Find the number of grams of the chemical needed to make the solution.

	Chemical	Total volume of solution	% of chemical in solution	Number of grams needed
1.	A	4,000 mL	3%	■
2.	B	2,000 mL	4%	■
3.	C	1,500 mL	2.5%	■
4.	D	3,500 mL	3.5%	■

Solve. Remember to estimate whenever you use your calculator.

5. A 3,500-mL solution is 6% dextrose. How many grams of dextrose are used in the solution?

6. A 4,000-mL solution contains 2,800 L of alcohol. What percent of the solution is alcohol?

7. Don has 120 mL of a solution that is 30% iodine. He adds 80 mL of water to the solution. What percent of the new solution is iodine?

8. Don has 200 mL of a solution that is 40% iodine. He adds 20 g of iodine to the solution. What percent of the new solution is iodine?

MIDCHAPTER REVIEW

Simplify. [398]

1. $(5 + 7) \div 2$ **2.** $8 - (3 + 6)$ **3.** $4 - (8 + 2) \div 2$ **4.** $(5 - 4)3$

5. $12 - 6 - 5$ **6.** $9 + 4 - 2$ **7.** $12 \div 2 - 1$ **8.** $6 + 3 \cdot 5$

9. $14 \div 7 + 3$ **10.** $81 - 9 \div 3$ **11.** $54 \div 3^2 \cdot 2$ **12.** $72 - 6^2 \cdot 2$

Write as an algebraic expression. [400]

13. 2 more than a number

14. 8 times a number

15. 19 less than a number

16. a number divided by 6

17. 5 added to twice a number

18. half a number minus 11

Write as a word phrase. [400]

19. $3 + n$ **20.** $n - 12$ **21.** $2n - 4$ **22.** $73 - n$

23. $\frac{n-7}{5}$ **24.** $18 + n^3$ **25.** $16n + 9$ **26.** $12(n - 5)$

Evaluate for $x = 5, y = -3$. [402]

27. $x + y$ **28.** $x - y$ **29.** $y - x$ **30.** xy

31. $2x - y$ **32.** $2y - x$ **33.** $3(x + y)$ **34.** $(3 - y) - x$

Write a number sentence. [406]

35. 6 more than 4 times a number is 28.

36. 4 less than Kim's age is greater than 12.

37. The points in a game plus 10 are less than or equal to 64.

38. Fred's age 5 years ago was greater than or equal to 17.

Graph the solution on a number line. [408]

39. $x + 3 = 7$ **40.** $x - 2 = 8$ **41.** $x < 7$ **42.** $x \geq -5$

Solve. [405, 410]

43. Use the formula $d = rt$ to find the distance a jetliner would travel in 4.5 hours at an average speed of 560 mph.

44. A 5,000-mL solution is 15% dextrose. How many grams of dextrose are used in the solution? Remember, 1 g = 1 mL.

14.8 SOLVING EQUATIONS BY ADDITION OR SUBTRACTION

Addition and subtraction are inverse operations. One undoes the other. To solve an equation, use an inverse operation to get the variable by itself.

When a number is attached to the variable by addition, you can unattach it by subtracting that number from both sides of the equation.

EXAMPLE 1 Solve: $x + 9 = 14$. Check the solution.

1. Solve. $x + 9 = 14$

   THINK: 9 is attached by addition. Subtract 9 from both sides.

   $x + 9 - 9 = 14 - 9$
   $x = 5$

2. Check. $x + 9 = 14$

   THINK: Substitute the value of *x* into the equation.

   $5 + 9 \stackrel{?}{=} 14$
   $14 = 14$ It checks.

When a number is attached to the variable by subtraction, you can unattach it by adding that number to both sides of the equation.

EXAMPLE 2 Solve: $x - 7 = 8$. Check the solution.

1. Solve. $x - 7 = 8$

   THINK: 7 is attached by subtraction. Add 7 to both sides.

   $x - 7 + 7 = 8 + 7$
   $x = 15$

2. Check. $x - 7 = 8$

   THINK: Substitute the value of *x* into the equation.

   $15 - 7 \stackrel{?}{=} 8$
   $8 = 8$ It checks.

You can write and solve an equation to solve a problem.

EXAMPLE 3 Carrie Collins and Tom Kowalski own Magic Carpet Cleaners. They had 12 customers this week, 8 fewer than last week. How many customers did they have last week?

1. Write an equation.

   THINK: Eight fewer than the number last week is 12.

   WRITE: $n - 8 = 12$

2. Solve.

   $n - 8 = 12$
   $n - 8 + 8 = 12 + 8$
   $n = 20$

3. Check.

   $n - 8 = 12$
   $20 - 8 \stackrel{?}{=} 12$
   $12 = 12$. It checks.

Last week Magic Carpet Cleaners had 20 customers.

CHECKPOINT

Write the letter of the correct number.

Solve.

1. $x + 6 = 9$	**a.** -3	**b.** 3	**c.** -15	**d.** 15
2. $x - 4 = 12$	**a.** 8	**b.** -8	**c.** 16	**d.** -16
3. $-5 + x = 8$	**a.** 13	**b.** -13	**c.** -3	**d.** 3
4. $3 + x = 7$	**a.** 10	**b.** -10	**c.** 4	**d.** -4

PRACTICE EXERCISES

Solve and check.

1. $x + 7 = 12$ **2.** $x + 9 = 16$ **3.** $x + 4 = 23$ **4.** $x + 5 = 13$

5. $x + 2 = 50$ **6.** $8 + x = 16$ **7.** $3 + x = 27$ **8.** $6 + x = 21$

9. $2 + x = 19$ **10.** $4 + x = 35$ **11.** $x + 35 = 52$ **12.** $x + 19 = 83$

13. $x + 64 = 76$ **14.** $x + 42 = 95$ **15.** $x + 57 = 82$ **16.** $56 + x = 84$

17. $71 + x = 100$ **18.** $27 + x = 65$ **19.** $97 + x = 132$ **20.** $87 + x = 113$

21. $-5 + x = 7$ **22.** $-9 + x = 8$ **23.** $-6 + x = 13$ **24.** $-7 + x = 15$

25. $x - 6 = 11$ **26.** $x - 3 = 8$ **27.** $x - 4 = 12$ **28.** $x - 8 = 11$

29. $x - 2 = 18$ **30.** $x - 6 = 16$ **31.** $x - 32 = 18$ **32.** $x - 71 = 35$

33. $x - 58 = 29$ **34.** $x - 64 = 57$ **35.** $x - 23 = 46$ **36.** $x - 83 = 105$

37. $x + (-5) = 3$ **38.** $x + \frac{1}{2} = \frac{3}{4}$ **39.** $x - (-6) = 10$ **40.** $17.4 = x - 9.6$

41. $x - \frac{3}{8} = \frac{1}{5}$ **42.** $-18 + x = -40$ **43.** $x - (-26.8) = -18.2$ **44.** $x + \frac{3}{4} = -\frac{9}{10}$

45. $3 + x + 4 = 11$ **46.** $6 + x + 7 = 20$ **47.** $2 + x + 9 = 18$

48. $-4 + x - 18 = 37$ **49.** $-4 + x - 8 = 34$ **50.** $-3 + x - 2 = 41$

Solve. Write an equation as needed.

51. Magic Carpet Cleaners had $1,546.17 in its checking account. After Carrie made a deposit the balance was $1,823.39. How much money did Carrie deposit?

52. The odometer in one of the Magic Carpet Cleaners's vans read 794 on Monday. On Friday, it read 1,031. How many miles was the van driven that week?

53. Kim was cleaning a carpet that was 85.7 ft^2. After she had cleaned 80% of the carpet, the machine broke. How many square feet of carpet were left to clean?

54. A square rug and a rectangular rug are each 36 ft^2 in area. The length of the rectangular rug is twice that of the square rug. Find the dimensions of the rectangular rug.

14.9 SOLVING EQUATIONS BY MULTIPLICATION OR DIVISION

Multiplication and division are also inverse operations. When a number is attached to the variable by division, you can unattach it by multiplying both sides of the equation by that number.

EXAMPLE 1 Solve: $\frac{x}{6} = 7$. Check the solution.

1. Solve.

   THINK: 6 is attached by division. Multiply both sides by 6.

   $6 \cdot \frac{x}{6} = 6 \cdot 7$

   $x = 42$

2. Check.

   THINK: Substitute the value of *x* into the original equation.

   $\frac{42}{6} \stackrel{?}{=} 7$

   $7 = 7$ It checks.

When a number is attached to the variable by multiplication, you can unattach it by dividing both sides of the equation by that number.

EXAMPLE 2 Solve: $-4x = 20$. Check the solution.

1. Solve.

   THINK: -4 is attached by multiplication. Divide both sides by -4.

   $\frac{-4x}{-4} = \frac{20}{-4}$

   $x = -5$

2. Check.

   THINK: Substitute the value of *x* into the original equation.

   $-4(-5) \stackrel{?}{=} 20$

   $20 = 20$ It checks.

You can write and solve an equation to solve a problem.

EXAMPLE 3 At the end of his typing course, Steve typed 87 words per minute (wpm). This was 3 times faster than his rate at the beginning of the course. What was his rate at the beginning of the course?

1. Write an equation.

   THINK: Three times a number equals 87 words per minute (wpm). $3n = 87$

2. Solve.

   $3n = 87$

   $\frac{3n}{3} = \frac{87}{3}$

   $n = 29$

3. Check.

   $3n = 87$

   $3(29) \stackrel{?}{=} 87$

   $87 = 87$ It checks.

Steve typed 29 words per minute at the beginning of the course.

CHECKPOINT Write the letter of the correct answer.

Solve and check.

1. $\frac{x}{3} = 9$ a. 3 b. -3 c. 27 d. -27

2. $\frac{x}{0.34} = 7$ a. -2.38 b. 2.38 c. 7 d. 20.58

3. $-2x = 8$ a. 4 b. -4 c. 16 d. -16

4. $0.8x = 7.2$ a. -9 b. 0.9 c. 9 d. 90

PRACTICE EXERCISES

Solve and check.

1. $3x = 15$
2. $5x = 35$
3. $2x = 12$
4. $7x = 21$
5. $6x = 30$
6. $-8x = 56$
7. $4x = -16$
8. $-9x = 72$
9. $5x = 85$
10. $-3x = 96$
11. $\frac{x}{6} = 3$
12. $\frac{x}{4} = 8$
13. $\frac{x}{7} = 11$
14. $\frac{x}{2} = 9$
15. $\frac{x}{9} = 5$
16. $\frac{x}{5} = -12$
17. $\frac{x}{-3} = 15$
18. $\frac{x}{2} = -29$
19. $\frac{x}{-7} = 32$
20. $\frac{x}{8} = -18$
21. $0.75x = 3$
22. $0.5x = 7$
23. $0.2x = 8$
24. $0.8x = 20$
25. $0.4x = 12$
26. $0.6x = 18$
27. $0.75x = 36$
28. $0.3x = 27$
29. $0.7x = 14$
30. $0.04x = 3$
31. $\frac{x}{0.14} = 5$
32. $\frac{x}{0.08} = 12$
33. $\frac{x}{0.73} = 4$
34. $\frac{x}{0.29} = 6$
35. $\frac{x}{0.92} = 8$
36. $\frac{x}{0.31} = 0.4$
37. $\frac{x}{0.65} = 0.2$
38. $\frac{x}{0.97} = 1.6$
39. $\frac{x}{0.42} = 5.3$
40. $\frac{x}{0.86} = 3.9$

Solve. Write an equation as needed.

41. In typing class, the fastest student typed 60 wpm. That was three times as fast as the slowest student. How many words per minute did the slowest student type?

42. All the typing students were evenly divided among three classes. If each class had 23 students, how many students were taking typing?

43. In one week, Toni's typing rates were 63, 72, 59, 67, and 64 wpm. What was her average typing rate for the week?

14.10 SOLVING TWO-STEP EQUATIONS

Annette Coppola has a part-time job at the Frozen Fun Yogurt Stand. Last week she worked 4 days. She earned the same amount of money on each of the first 3 days and $32 on the fourth day. Her paycheck was $80. How much did Annette earn on each of the first 3 days?

You can write and solve an equation to find the answer.

EXAMPLE 1

Write and solve an equation to find how much money Annette earned on each of the first 3 days.

1. Write an equation.

THINK: Three times an amount plus $32 is $80. $3x + 32 = 80$

2. Solve: $3x + 32 = 80$

THINK: Unattach first by subtraction, then by division.

$$3x + 32 - 32 = 80 - 32$$
$$3x = 48$$
$$\frac{3x}{3} = \frac{48}{3}$$
$$x = 16$$

3. Check.

$$3x + 32 = 80$$
$$3(16) + 32 = 80$$
$$48 + 32 \stackrel{?}{=} 80$$
$$80 = 80 \quad \text{It checks.}$$

So, Annette earned $16 on each of the first three days.

CHECKPOINT

Write the letter of the correct answer.

Solve.

1. $\frac{x}{4} - 7 = 1$ **a.** -24 **b.** -2 **c.** 11 **d.** 32

2. $3x + 9 = 21$ **a.** -2 **b.** 4 **c.** 10 **d.** 36

PRACTICE EXERCISES

Solve and check.

1. $2x + 1 = 7$
2. $3x + 7 = 19$
3. $5x + 4 = 34$
4. $7x + 2 = 16$
5. $8x + 3 = 11$
6. $6x + 2 = 38$
7. $7x + 1 = 15$
8. $9x + 6 = 15$
9. $4x - 8 = 12$
10. $9x - 5 = 13$
11. $6x - 3 = 15$
12. $3x - 7 = 14$
13. $5x - 13 = 12$
14. $8x - 4 = 4$
15. $10x - 80 = 100$
16. $9x - 8 = 19$
17. $-3x + 2 = 11$
18. $-5x - 8 = 2$
19. $-4x + 10 = 50$
20. $-2x - 15 = 3$
21. $-8x - 6 = 2$
22. $-9x + 3 = 3$
23. $-12x - 9 = 27$
24. $-6x + 14 = 26$
25. $\frac{x}{3} - 4 = 8$
26. $\frac{x}{7} + 5 = 12$
27. $\frac{x}{11} - 2 = 8$
28. $\frac{x}{13} + 10 = 11$
29. $\frac{x}{9} + 3 = -5$
30. $\frac{x}{4} - 7 = -8$
31. $\frac{x}{6} + 9 = -1$
32. $\frac{x}{5} - 14 = 6$
33. $\frac{x}{6} - 5 = 1$
34. $\frac{x}{8} + 7 = 19$
35. $\frac{x}{3} - 19 = 1$
36. $\frac{x}{5} + 7 = 16$

MIXED REVIEW

Solve.

37. 20% of what number is 17?
38. 18 is what percent of 72?
39. What is $37\frac{1}{2}$% of 56?
40. 45 is 50% of what number?
41. What percent of 36 is 12?
42. $33\frac{1}{3}$% of 129 is what number?
43. 42 is what percent of 210?
44. 2% of what number is 27?
45. 1,000% of 26.2 is what number?
46. 14% of what number is 28?

Solve. Write an equation as needed.

47. Jim makes sodas and sundaes at the Frozen Fun Yogurt Stand. Last week the number of sodas he made was 4 fewer than 5 times the number of sundaes. If he made 96 sodas, how many sundaes did he make?
48. Annette ordered 263 sugar cones for the Frozen Fun Yogurt Stand. The number of sugar cones was 8 more than 3 times the number of waffle cones. How many waffle cones were ordered?
49. There are 8 flavors of yogurt at the Frozen Fun Yogurt Stand. If you order a triple scoop cone with no two scoops alike, how many combinations of flavors can you choose from?
50. The Frozen Fun Yogurt Stand had a special last Thursday. For each double scoop cone you bought, you got a single scoop cone free. Raoul served 33 scoops of yogurt on the special. How many free scoops did he serve?

14.11 GRAPHING EQUATIONS IN TWO VARIABLES

An equation with two variables has an ordered pair of numbers as a solution.

EXAMPLE 1 Find the solution of $x - 3y = 7$ when $x = -2$.

1. Substitute -2 for x.

$$x - 3y = 7$$
$$-2 - 3y = 7$$

2. Solve.

$$-2 + 2 - 3y = 7 + 2$$
$$\frac{3y}{-3} = \frac{9}{-3}$$
$$y = -3$$

3. Check.

THINK: $x = -2, y = -3$

$$x - 3y = 7$$
$$-2 - 3(-3) = 7$$
$$-2 + 9 = 7$$
$$7 = 7$$

So, $(-2, -3)$ is a solution of $x - 3y = 7$.

An equation with two variables usually has many solutions. You can make a **table of values** to show the solutions.

EXAMPLE 2 Make a table of values to find some solutions for the equation $y = 5 - x$.

1. Assign x different values.
2. Find the solutions.
3. Write the ordered pairs.

x	$5 - x$	y	(x, y)
-2	$5 - (-2)$	7	$(-2, 7)$
-1	$5 - (-1)$	6	$(-1, 6)$
0	$5 - (0)$	5	$(0, 5)$
1	$5 - (1)$	4	$(1, 4)$
2	$5 - (2)$	3	$(5, 3)$

You can draw the graph of an equation with two variables by graphing the ordered pairs and connecting them with a line.

EXAMPLE 3

Graph the equation $y = 2x - 1$.

1. Make a table to find the ordered pairs.

x	$2x - 1$	y	(x, y)
-1	$2(-1) - 1$	-3	$(-1, -3)$
0	$2(0) - 1$	-1	$(0, -1)$
1	$2(1) - 1$	1	$(1, 1)$
2	$2(2) - 1$	3	$(2, 3)$
3	$2(3) - 1$	5	$(3, 5)$

2. Graph each of the ordered pairs. Connect them with a line.

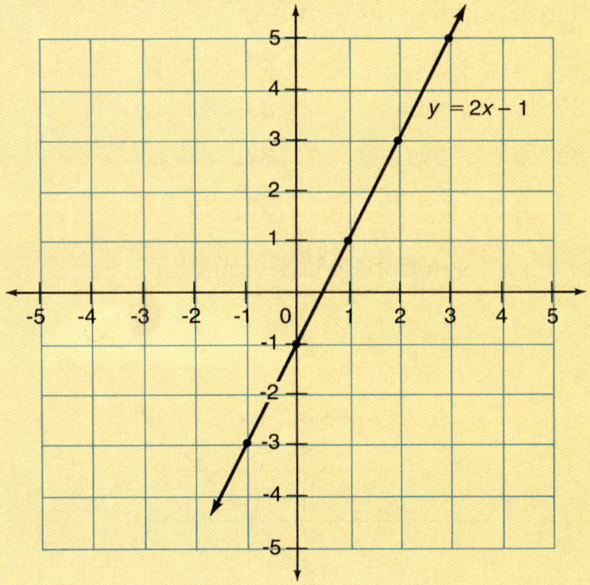

CHECKPOINT

Write the letter of the correct answer.

Find the value of the variable that is unknown.

1. $x + y = 8; x = 3$ **a.** -3 **b.** 4 **c.** 5 **d.** -5

2. $7 - x = y; x = -2$ **a.** 9 **b.** -9 **c.** 5 **d.** -5

3. $2x + y = 16; y = 10$ **a.** -3 **b.** 3 **c.** 5 **d.** 13

4. $y - 3x = 12; y = 6$ **a.** 6 **b.** -6 **c.** 2 **d.** -2

PRACTICE EXERCISES

Find the solution of the equation when $x = 0$, and then when $y = 0$.

1. $x + y = 13$ **2.** $3 - x = y$ **3.** $5 + y = x$ **4.** $y - x = 9$

5. $2x - 8 = y$ **6.** $5x + 3y = 15$ **7.** $3x - y = 12$ **8.** $-2y + 5x = 10$

9. $6 - 2x = y$ **10.** $4 + 2x = y$ **11.** $y - 3x = 6$ **12.** $-4y + x = 8$

Determine if the ordered pair is a solution of the equation.
Write *yes* or *no*.

13. $x - y = 11; (2, 9)$ **14.** $2 - x = y; (0, 2)$ **15.** $x + y = 9; (3, 6)$

16. $y = 2x - 4; (-2, -4)$ **17.** $x - 3y = 6; (6, 2)$ **18.** $2x + 5y = 20; (5, 2)$

19. $x - \frac{3y}{5} = 6; (6, -10)$ **20.** $\frac{3x}{4} + y = 12; (16, 10)$ **21.** $\frac{2x}{3} - 2y = 6; (9, 3)$

Graph the equation.
Make a table of values that has three values for x.

22. $y = x + 2$ **23.** $y = x - 3$ **24.** $x = y$ **25.** $y = 3x + 1$

26. $x + y = 7$ **27.** $x - y = 5$ **28.** $3x = y$ **29.** $-2x = y$

30. $y = 2x - 1$ **31.** $y = -3x$ **32.** $x - y = 3$ **33.** $x = 2y$

34. $3x - y = 8$ **35.** $2x + y = 10$ **36.** $3y = 4x$ **37.** $x - y = -2$

Use the graph to determine two solutions of the equation.

38.

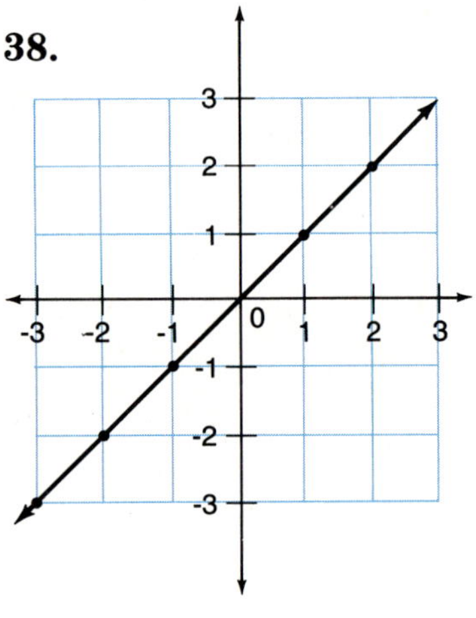

39.

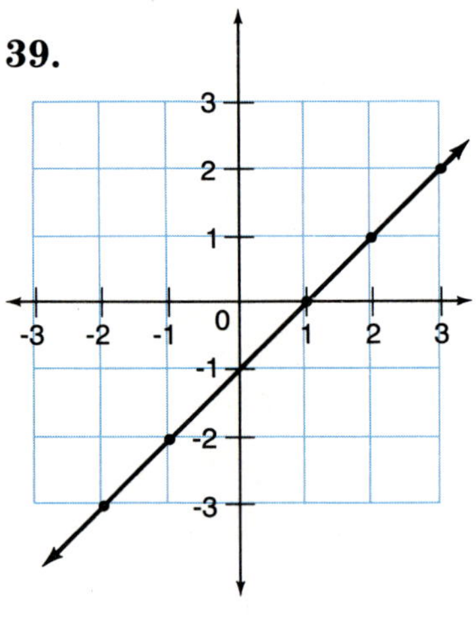

40.

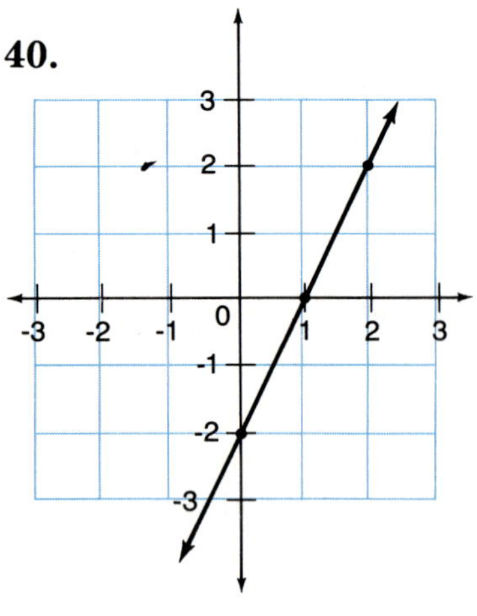

TIME OUT

If you add the square of Tom's age to Mary's age, the sum is 62.
If you add the square of Mary's age to Tom's age, the result is 176.
What are their ages?

PROBLEM Solving APPLICATION

14.12 CONSUMER: ENERGY CONSERVATION

The Johnson family owns a 3-bedroom home that is heated by oil. Their furnace uses an average of 90 gal per month.
A consumer magazine provides a report about a new thermostat that can cut the use of home heating oil by 15%.
The Johnsons are interested in purchasing this new thermostat.

How many gallons of oil will be saved using this new thermostat for 4 months?

1. Write an equation. Use the formula $s = 0.15 \cdot 90 \cdot t$.
   Let s represent the amount of oil saved and t the amount of time.

   THINK: Amount of oil saved is 4 times 15% of 90 gal

   WRITE: $s = 4\ (0.15 \cdot 90)$

2. Solve. $s = 4\ (13.5)$
   $s = 54$

So, using the new thermostat for 4 months will save 54 gal of oil.

FOR DISCUSSION

The Johnson family uses an *average* of 90 gal of oil per month. What does this statement mean?

Solve. Use the formula $s = 0.15 \cdot 90 \cdot t$.

1. How many gallons of oil would be saved in 36 months?

2. About how many months would it take to save 400 gal of oil?

3. About how many months would it take to save 650 gal of oil?

4. How many gallons of oil would be saved in $4\frac{1}{2}$ years?

Suppose oil costs $0.94 per gallon. Solve.

5. How much less money would you pay for oil in 12 months with the new thermostat?

6. How much money would be saved in 3.5 years with the new thermostat?

CHAPTER REVIEW

Vocabulary Choose the letter of the word that completes the statement.

1. An algebraic expression uses a ■ for an unknown number. [400]

   a. question mark **b.** blank **c.** variable **d.** dash

2. A mathematical sentence that uses $<$ to show the relationship between two expressions is an ■. [406]

   a. inequality **b.** equation **c.** equality **d.** graph

3. To solve the equation $5x - 3 = 7$, you would ■ first. [416]

   a. add **b.** subtract **c.** multiply **d.** divide

Skills Simplify. [398]

$(18 \div 2) - 4$ *5*

4. $(3 + 5)4$

5. $(12 - 2)^2 \div 5$

6. $48 - (7 - 1) + 2$

7. $(15 + 2 \cdot 6) - 3^2$

8. $\frac{12}{18 - 6}$

Write an algebraic expression for the word phrase. [400]

16 fewer than a number *$n - 16$*

9. 3 more than twice a number

10. 17 less than a number

11. A number increased by 5

12. Sue's age 7 years ago

13. Twice a number, decreased by 83

Evaluate for $x = -2$, $y = -5$. [402]

$3x - y$ *-1*

14. $x - y$

15. xy

16. $\frac{2x}{y - 1}$

17. $\frac{y + 5}{-8x}$

Write a mathematical sentence. [406]

2 more than $\frac{1}{4}$ of a number is 8. *$\frac{1}{4}n + 2 = 8$*

18. 6 more than a number is 17.

19. 3 times a number decreased by 4 is 8.

20. A number divided by 7 is less than 35.

Solve. [412, 414, 416]

$2x - 5 = 3$ *4*

21. $x - 13 = 25$

22. $x + 19 = 21$

23. $x - 5 = 93$

24. $-12x = -24$

25. $4x + 3 = 15$

26. $-3x - 4 = 14$

27. $\frac{x}{5} + 2 = -6$

CHAPTER TEST

Simplify.

1. $8 - 4 + 3$

2. $12 \div 6 \cdot 2$

3. $8 \cdot 3 \div 6 - 2^2$

4. $9 - 5 \cdot 2 + 1$

5. $(18 - 4)2 - 5$

6. $\frac{6(20 - 4)}{3}$

Write an algebraic expression.

7. The sum of a number and 17

8. 3 more than 6 times a number

Evaluate for $x = -1, y = 2, z = -3$.

9. $x + y + z$

10. $x - y - z$

11. $\frac{x + z}{-y}$

12. $\frac{2z - 3y}{4x}$

Write a mathematical sentence.

13. 8 fewer than a number is 41.

14. 3 times a number is less than 34.

Graph the solution.

15. $x < -2$

16. $x = 4$

17. $x \geq -1$

18. $x \leq 6$

Solve.

19. $x - 6 = 12$

20. $x + 13 = -4$

21. $9 + x = 8$

22. $-7 + x = -3$

23. $-7x = 35$

24. $\frac{x}{-2} = 16$

25. $9x = -81$

26. $\frac{x}{-5} = -8$

27. $7x + 4 = 32$

28. $-2x - 9 = 11$

29. $\frac{x}{4} - 13 = -15$

30. $\frac{x}{-11} + 8 = 7$

Find the solution of the equation when $x = -3$.

31. $x + y = 7$

32. $y = -2x + 4$

33. $3x - 4y = -9$

34. $-4x + 3y = 27$

Graph each equation. Make a table that has three values for x.

35. $y = x - 5$

36. $y = 2x - 1$

Solve. Use the formula $s = 0.20 \cdot 100 \cdot t$.
Let s stand for the amount of oil saved and t for a period of time.

37. The Garcia family uses an average of 100 gal of fuel oil per month. A new thermostat will cut the amount of oil used by 20%. How much oil will be saved by using the new thermostat for 3 months?

38. The Garcias hope that the new thermostat will allow them to save at least 200 gal of oil in a year. Is this possible? Use the formula above to find out.

ENRICHMENT COMBINING LIKE TERMS

In the expression $5x + 7a + 4x$, $5x$ and $4x$ are **like terms** because they have the same variable, x. Here are some other expressions with like terms.

$4a + 5b - 2a$	$8x^2 - 3x - 4x^2$	$6 + 3b - 2$	$7c + 5d^2 - 3c + 8d^2$
Like terms	**Like terms**	**Like terms**	**Like terms**

You can simplify some expressions by combining like terms.

Example Simplify: $-5b + 7a - 7b + 12a$

1. Identify the terms. $-5b + 7a - 7b + 12a$
2. Combine like terms. $-5b - 7b + 7a + 12a$

THINK: Group like terms together. $-12b + 19a$

You can solve some equations by combining like terms.

Example Solve: $7a - 4a = 24$. Check.

1. Combine like terms.

$$7a - 4a = 24$$
$$3a = 24$$

2. Solve for a.

$$\frac{3a}{3} = \frac{24}{3}$$
$$a = 8$$

3. Check.

$$7(8) - 4(8) \stackrel{?}{=} 24$$
$$56 - 32 = 24 \quad \text{It checks.}$$

Simplify by combining like terms.

1. $6x + 5x$ **2.** $9a - 6a$ **3.** $15c + 12c$ **4.** $18d - 9d$

5. $3a + a + 8a$ **6.** $9r - 2r - 3r$ **7.** $9n + 5n - 6n$ **8.** $15h + 6h - 8h$

9. $7c + 6 + 2c$ **10.** $8d - 2 + d$ **11.** $7y + 3y - 4z$ **12.** $12e - 7e + 4f$

13. $-y^2 + 7y^2 - 2y^2$ **14.** $-6ab + 6ab$ **15.** $17rs - (-6rs)$ **16.** $8x^2 - 6x^2 + 5x^2$

Solve by combining like terms.

17. $3x + 2x = 10$ **18.** $4b + 3b = 21$ **19.** $7k + 2k = -27$ **20.** $9m + m = 50$

21. $12d + 3d = 45$ **22.** $25a + 5a = -90$ **23.** $6c - 2c = 20$ **24.** $9y - 5y = 16$

25. $z - 8z = 28$ **26.** $-11t + 2t = 54$ **27.** $20r - 5r = 45$ **28.** $30x - 5x = 100$

CALCULATOR DIFFERENCE BETWEEN CALCULATORS

Do you have a scientific calculator or a nonscientific calculator? Enter the following example to find out.

Enter: [3] [+] [4] [×] [5] [=]

A scientific calculator uses the rules for the order of operations. It gives the correct answer of 23.
A nonscientific calculator performs the operations in the order in which they are entered. It gives an incorrect answer of 35.

Compute: $38 + 91 \div 13$

Scientific Calculator

Enter the numbers and the operation symbols in the order given.

Enter: [3] [8] [+] [9] [1] [÷] [1] [3] [=]

Display: [45.]

Nonscientific Calculator

Find the quotient $91 \div 13$ first. then add 38.

Enter: [9] [1] [÷] [1] [3] [=] [+] [3] [8] [=]

Display: [45.]

Evaluate the expression $7.8 + 1.5n$ when $n = 6$.

Scientific Calculator

Mentally substitute 6 for n. Enter the numbers and operation symbols in the order given.

Enter: [7] [.] [8] [+] [1] [.] [5] [×] [6] [=]

Display: [16.8]

Nonscientific Calculator

Mentally substitute 6 for n.
Multiply 1.5×6 first.
Then add 7.8.

Enter: [1] [.] [5] [×] [6] [=] [+] [7] [.] [8] [=]

Display: [16.8]

Compute. Use a calculator.

1. $29 + 6 \times 53$
2. $37 + 54 \div 6$
3. $360 + 19 \times 82$
4. $76 + 851 \div 37$
5. $4.7 + 3.6 \times 4.5$
6. $8.4 + 31.9 \div 29$
7. $0.28 + 0.32 \times 140$
8. $0.4 + 58.5 \div 6.5$
9. $6.29 + 1.9 \times 2.7$

Evaluate each expression for the given value of n. Use a calculator.

10. $14 + 8n; n = 12$
11. $49 + n \div 2; n = 68$
12. $16 + 25n; n = 11$
13. $17 + n \div 14; n = 98$
14. $3.6 + 39n; n = 2.1$
15. $0.6 + n \div 8; n = 5$
16. $17 + 12 \div n; n = 2$
17. $22 + \frac{n}{19}; n = 152$
18. $7 + \frac{396}{n}; n = 36$
19. $12.8 + 7n; n = 4.2$
20. $3.6 + n \div 1.5; n = 6$
21. $20 + \frac{196}{n}; n = 7$

EXTRA PRACTICE

CHAPTER 1, PAGES 4–5

Find the place and the value of the underlined digit.

1. $\underline{2}21$ **2.** $9\underline{3}5{,}456{,}820$ **3.** $8\underline{4}{,}618$ **4.** $\underline{5}08{,}912{,}631{,}008$

5. $7\underline{4}3{,}117$ **6.** $\underline{6}{,}284{,}122$ **7.** $286{,}1\underline{9}3$ **8.** $5\underline{9}2{,}816{,}134{,}726$

Write the number.

9. 491 thousand, 224

10. 93 billion, 641

11. 6 million, 803

12. 825 million, 187

13. Three hundred five thousand, nine hundred forty-three

14. Twenty-three million, seventy-five thousand, eight hundred twenty

15. Sixty-two billion, two hundred four million, forty-six

16. One hundred sixteen billion, eighty-six million, thirty-nine

CHAPTER 1, PAGES 6–7

Find the place and the value of the underlined digit.

1. $62.\underline{6}$ **2.** $25.061\underline{2}84$ **3.** $0.978\underline{4}5$ **4.** $628.3\underline{1}4$

5. $4.3875\underline{9}1$ **6.** $78.20\underline{3}62$ **7.** $29.04026\underline{1}3$ **8.** $55.69\underline{9}$

Write the number.

9. 8 and 42 hundredths

10. 2,624 ten thousandths

11. 6 hundred and 13 thousandths

12. 766 millionths

CHAPTER 1, PAGES 8–9

Compare. Write >, <, or =.

1. 18.0023 ● 18.023 **2.** 8,392 ● 8,382 **3.** 0.8003 ● 0.80030 **4.** 2.8734 ● 2.87343

5. 650,412,731.662 ● 650,412,731.652

6. 7.02386 ● 7.02836

7. 42,861.73 ● 42,861.23

8. 738.002 ● 737.111

Arrange in order from greatest to least.

9. 24.863; 73.52; 104; 17.6

10. 54.183; 541.8; 54,183; 60.18

11. 43.09; 409; 93.04; 30.49

12. 0.0382; 0.382; 0.8302; 0.2803

13. 92; 430; 4,329; 29,034; 409,320

14. 0.417; 0.41; 70.14; 0.4

CHAPTER 1, PAGES 14–15

Solve, if possible. If not, identify the needed information.

1. In John Adams High School there are 2,334 students. 1,620 students come to school by bus. The rest walk or ride bikes. How many students walk to school?

2. Cal's odometer reads 1,920.8 mi when he begins his road trip. After the first day, it reads 2,295.0 mi. After the second day, it reads 2,801.3 mi. How far did Cal drive the second day?

3. According to a survey, 861 people entered the park on Saturday. 222 picnicked and 149 used the 19 tennis courts. The rest played baseball. How many played baseball?

4. The camp cook at Lake Wakega knows he can make 50 pancakes out of 1 lb of pancake mix. He has 5 lb of pancake mix. Can he feed 204 campers?

CHAPTER 1, PAGES 16–17

Round to the nearest hundred and to the nearest thousand.

1. 5,642 **2.** 3,299 **3.** 48,201 **4.** 71,045

5. 756,892 **6.** 201,638 **7.** 231,575 **8.** 615,123

Round to the nearest tenth and to the nearest hundredth.

9. 18.103 **10.** 10.305 **11.** 33.4444 **12.** 19.555

13. 0.092 **14.** 0.0167 **15.** 1.567 **16.** 7.00346

CHAPTER 1, PAGES 18–19

Write in exponent form.

1. 8 squared **2.** 9 cubed **3.** 15 to the 4th power

4. 26 cubed **5.** 100 squared **6.** 10 to the 12th power

7. $7 \times 7 \times 7 \times 7 \times 7$ **8.** $11 \times 11 \times 11$ **9.** $2 \times 2 \times 2 \times 2 \times 2$

Write as a product.

10. 3^4 **11.** 4^2 **12.** 25^0 **13.** 7^2 **14.** 2^3 **15.** 9^0

Write in exponent form.

16. 1,000 **17.** 100,000 **18.** 1,000,000 **19.** 10,000,000,000

Write as a product.

20. 1^7 **21.** 10^9 **22.** 1^{12} **23.** 10^2 **24.** 10^6 **25.** 10^0

CHAPTER 2, PAGES 28–29

Estimate by rounding.

1. 482 + 27
2. 390 + 226
3. 5,970 + 4,001
4. 32,031 + 551 + 56,264
5. 1,609 + 6,312 + 2,049

6. 54 − 27
7. 795 − 362
8. 3,804 − 768
9. 4,809 − 1,245
10. 26,274 − 16,108

11. 6,483 + 612
12. 2,096 − 1,245
13. 72,546 − 18,609
14. 91,111 + 8,867
15. 76,452 + 2,367 + 24,786

Estimate by using front-end estimation.

16. 342 + 874
17. 2,590 + 28 + 1,078
18. 3,619 + 4,456
19. 343,332 + 141,608 + 9,009
20. 59,401 + 34,822

21. 73,108 − 18,945
22. 9,345 + 5,777
23. 13,367 + 91,862
24. 89,806 − 5,712
25. 319,450 + 74,433 + 822,101

26. 567,981 − 390,417
27. 39,560 − 11,053
28. 3,815 − 762
29. 426 − 193
30. 15,687 − 9,200

CHAPTER 2, PAGES 30–31

Add.

1. 166 + 555
2. 892 + 439
3. 458 + 675
4. 3,689 + 796
5. 7,043 + 8,907

6. 44,036 + 3,016
7. 9,933 + 2,448
8. 534,856 + 760,455
9. 745,912 + 312,276
10. 949,624 + 62,083

11. 469 + 10,823 + 7,118
12. 16,127 + 528 + 123,045 + 8,003
13. 412,065 + 7,783 + 26,655 + 607,027
14. 6,817 + 19,341 + 943 + 7,726
15. 23,747 + 178,019 + 6,913 + 666

16. 56,453 + 18,352 + 26,308 + 31,325
17. 21,912 + 66,909 + 34,865 + 37,809
18. 308,416 + 403,769 + 293,008 + 185,444
19. 125,645 + 912,863 + 377,715 + 714,543
20. 647,792 + 351,115 + 662,258 + 523,087

21. 708 + 93 + 215
22. 87 + 1,257 + 38,409
23. 834,906 + 2,176 + 45,573

CHAPTER 2, PAGES 32–34

Subtract. Check your answer.

1. $\begin{array}{r} 460 \\ -183 \\ \hline \end{array}$	**2.** $\begin{array}{r} 851 \\ -438 \\ \hline \end{array}$	**3.** $\begin{array}{r} 12{,}804 \\ -\ 8{,}691 \\ \hline \end{array}$	**4.** $\begin{array}{r} 4{,}283 \\ -2{,}624 \\ \hline \end{array}$	**5.** $\begin{array}{r} 2{,}820 \\ -\ 543 \\ \hline \end{array}$
6. $\begin{array}{r} 69{,}005 \\ -40{,}326 \\ \hline \end{array}$	**7.** $\begin{array}{r} 36{,}112 \\ -18{,}813 \\ \hline \end{array}$	**8.** $\begin{array}{r} 281{,}113 \\ -\ 79{,}465 \\ \hline \end{array}$	**9.** $\begin{array}{r} 110{,}311 \\ -106{,}413 \\ \hline \end{array}$	**10.** $\begin{array}{r} 683{,}117 \\ -229{,}032 \\ \hline \end{array}$
11. $\begin{array}{r} 45{,}082 \\ -\ 8{,}762 \\ \hline \end{array}$	**12.** $\begin{array}{r} 19{,}411 \\ -14{,}708 \\ \hline \end{array}$	**13.** $\begin{array}{r} 612{,}734 \\ -589{,}037 \\ \hline \end{array}$	**14.** $\begin{array}{r} 810{,}056 \\ -311{,}086 \\ \hline \end{array}$	**15.** $\begin{array}{r} 300{,}000 \\ -134{,}082 \\ \hline \end{array}$
16. $\begin{array}{r} 789{,}135 \\ -\ 9{,}456 \\ \hline \end{array}$	**17.** $\begin{array}{r} 70{,}003 \\ -\ 874 \\ \hline \end{array}$	**18.** $\begin{array}{r} 230{,}049 \\ -209{,}876 \\ \hline \end{array}$	**19.** $\begin{array}{r} 5{,}123{,}452 \\ -\ 358{,}118 \\ \hline \end{array}$	**20.** $\begin{array}{r} 6{,}019{,}012 \\ -5{,}923{,}714 \\ \hline \end{array}$

First estimate by rounding. Then find the exact answer.

21. 87 − 52 **22.** 683 − 179 **23.** 7,821 − 1,684

24. 18,891 − 12,963 **25.** 402,264 − 295,611 **26.** 892,126 − 602,820

CHAPTER 2, PAGES 36–37

Multiply.

1. 30 × 10 **2.** 4,000 × 80 **3.** 300 × 200 **4.** 268 × 1,000

5. 640 × 10 **6.** 3,260 × 1,000 **7.** 2,000 × 500 **8.** 6,000 × 8,000

Divide.

9. 90,000 ÷ 30 **10.** 35,000 ÷ 700 **11.** 8,000,000 ÷ 2,000

12. 30,000 ÷ 300 **13.** 429,000 ÷ 1,000 **14.** 1,612,000 ÷ 100

CHAPTER 2, PAGES 38–39

Estimate by rounding.

1. $\begin{array}{r} 72 \\ \times\ 8 \\ \hline \end{array}$	**2.** $\begin{array}{r} 631 \\ \times\ 5 \\ \hline \end{array}$	**3.** $\begin{array}{r} 382 \\ \times\ 9 \\ \hline \end{array}$	**4.** $\begin{array}{r} 6{,}220 \\ \times\ 4 \\ \hline \end{array}$	**5.** $\begin{array}{r} 1{,}892 \\ \times\ 7 \\ \hline \end{array}$
6. $\begin{array}{r} 56 \\ \times 46 \\ \hline \end{array}$	**7.** $\begin{array}{r} 29 \\ \times 73 \\ \hline \end{array}$	**8.** $\begin{array}{r} 118 \\ \times\ 93 \\ \hline \end{array}$	**9.** $\begin{array}{r} 745 \\ \times\ 81 \\ \hline \end{array}$	**10.** $\begin{array}{r} 3{,}085 \\ \times\ 12 \\ \hline \end{array}$
11. $\begin{array}{r} 4{,}780 \\ \times\ 57 \\ \hline \end{array}$	**12.** $\begin{array}{r} 3{,}089 \\ \times\ 33 \\ \hline \end{array}$	**13.** $\begin{array}{r} 22{,}135 \\ \times\ 68 \\ \hline \end{array}$	**14.** $\begin{array}{r} 58{,}036 \\ \times\ 88 \\ \hline \end{array}$	**15.** $\begin{array}{r} 73{,}586 \\ \times\ 52 \\ \hline \end{array}$
16. $4\overline{)389}$	**17.** $7\overline{)664}$	**18.** $5\overline{)4{,}377}$	**19.** $3\overline{)8{,}672}$	**20.** $8\overline{)4{,}469}$
21. $9\overline{)86{,}800}$	**22.** $6\overline{)57{,}060}$	**23.** $49\overline{)964}$	**24.** $33\overline{)286}$	**25.** $58\overline{)5{,}710}$

Estimate by rounding. Which answer is reasonable?

1. 3 × 787 **a.** 2,361 **b.** 231 **c.** 1,492

2. 24 × 96 **a.** 225 **b.** 9,820 **c.** 2,304

3. 78 × 3,120 **a.** 243,360 **b.** 120,000 **c.** 24,350

CHAPTER 2, PAGES 40–41

Multiply.

4. $\begin{array}{r} 132 \\ \times \; 5 \\ \hline \end{array}$ **5.** $\begin{array}{r} 612 \\ \times \; 7 \\ \hline \end{array}$ **6.** $\begin{array}{r} 2{,}391 \\ \times \; 6 \\ \hline \end{array}$ **7.** $\begin{array}{r} 4{,}748 \\ \times \; 3 \\ \hline \end{array}$ **8.** $\begin{array}{r} 1{,}955 \\ \times \; 9 \\ \hline \end{array}$

9. $\begin{array}{r} 62 \\ \times 51 \\ \hline \end{array}$ **10.** $\begin{array}{r} 392 \\ \times \; 12 \\ \hline \end{array}$ **11.** $\begin{array}{r} 826 \\ \times \; 23 \\ \hline \end{array}$ **12.** $\begin{array}{r} 5{,}431 \\ \times \; 89 \\ \hline \end{array}$ **13.** $\begin{array}{r} 7{,}622 \\ \times \; 43 \\ \hline \end{array}$

14. 9 × 18 **15.** 2 × 75 **16.** 7 × 63 **17.** 4 × 36 **18.** 8 × 781

First estimate using compatible numbers. Then find the exact answer.

19. $\begin{array}{r} 8{,}920 \\ \times \; 3 \\ \hline \end{array}$ **20.** $\begin{array}{r} 2{,}384 \\ \times \; 6 \\ \hline \end{array}$ **21.** $\begin{array}{r} 27 \\ \times 61 \\ \hline \end{array}$ **22.** $\begin{array}{r} 114 \\ \times \; 77 \\ \hline \end{array}$ **23.** $\begin{array}{r} 439 \\ \times \; 56 \\ \hline \end{array}$

24. $\begin{array}{r} 42 \\ \times 43 \\ \hline \end{array}$ **25.** $\begin{array}{r} 513 \\ \times \; 60 \\ \hline \end{array}$ **26.** $\begin{array}{r} 285 \\ \times \; 52 \\ \hline \end{array}$ **27.** $\begin{array}{r} 9{,}130 \\ \times \; 19 \\ \hline \end{array}$ **28.** $\begin{array}{r} 6{,}120 \\ \times \; 103 \\ \hline \end{array}$

CHAPTER 2, PAGES 42–43

Multiply.

1. 408 × 725 **2.** 826 × 506 **3.** 6,105 × 464 **4.** 516 × 742

5. 3,620 × 428 **6.** 6,094 × 532 **7.** 907 × 963 **8.** 2,635 × 380

9. 8,003 × 951 **10.** 218 × 585 **11.** 4,306 × 888 **12.** 6,720 × 634

13. 268 × 423 **14.** 7,962 × 111 **15.** 207 × 896 **16.** 10,082 × 627

CHAPTER 2, PAGES 46–47

Divide.

1. $5\overline{)495}$ **2.** $8\overline{)247}$ **3.** $7\overline{)840}$ **4.** $4\overline{)7{,}592}$ **5.** $2\overline{)6{,}441}$

6. $6\overline{)608}$ **7.** $3\overline{)4{,}067}$ **8.** $8\overline{)1{,}234}$ **9.** $7\overline{)4{,}002}$ **10.** $6\overline{)1{,}111}$

11. $5\overline{)12{,}310}$ **12.** $9\overline{)21{,}567}$ **13.** $6\overline{)20{,}003}$ **14.** $8\overline{)186{,}423}$ **15.** $9\overline{)201{,}005}$

16. 612 ÷ 4 **17.** 1,701 ÷ 7 **18.** 8,696 ÷ 9 **19.** 20,600 ÷ 5 **20.** 52,783 ÷ 6

21. 71,003 ÷ 5 **22.** 22,344 ÷ 3 **23.** 142,098 ÷ 2 **24.** 687,834 ÷ 4 **25.** 501,013 ÷ 7

CHAPTER 2, PAGES 48–49

Divide.

1. $26\overline{)364}$ **2.** $13\overline{)793}$ **3.** $52\overline{)936}$ **4.** $81\overline{)891}$ **5.** $34\overline{)820}$

6. $77\overline{)462}$ **7.** $48\overline{)727}$ **8.** $22\overline{)6{,}527}$ **9.** $63\overline{)2{,}961}$ **10.** $39\overline{)4{,}875}$

11. $57\overline{)7{,}296}$ **12.** $98\overline{)8{,}140}$ **13.** $40\overline{)33{,}800}$ **14.** $91\overline{)59{,}332}$ **15.** $42\overline{)84{,}378}$

First estimate using compatible numbers. Then find the exact answer.

16. $821 \div 44$ **17.** $3{,}429 \div 83$ **18.** $928 \div 25$ **19.** $47{,}924 \div 62$

20. $367 \div 51$ **21.** $6{,}272 \div 12$ **22.** $2{,}098 \div 41$ **23.** $78{,}429 \div 79$

CHAPTER 2, PAGES 50–51

Divide.

1. $1{,}320 \div 258$ **2.** $7{,}500 \div 314$ **3.** $61{,}679 \div 593$ **4.** $4{,}561 \div 291$

5. $41{,}535 \div 585$ **6.** $970{,}062 \div 642$ **7.** $154{,}921 \div 701$

8. $244{,}089 \div 312$ **9.** $608{,}145 \div 450$ **10.** $824{,}567 \div 809$

11. $378\overline{)759{,}636}$ **12.** $175\overline{)67{,}200}$ **13.** $147\overline{)802{,}428}$ **14.** $198\overline{)402}$

15. $174\overline{)29{,}600}$ **16.** $305\overline{)6{,}557}$ **17.** $183\overline{)11{,}520}$ **18.** $983\overline{)99{,}156}$

19. $211\overline{)5{,}087}$ **20.** $437\overline{)40{,}970}$ **21.** $256\overline{)302{,}145}$ **22.** $710\overline{)2{,}304{,}568}$

CHAPTER 2, PAGES 52–53

Solve.

1. Jennifer has saved $73.00. On lay-away, she can buy a pair of roller skates for $7.00 a week for 12 weeks. Does she have enough money to buy the skates? If not, how much more does she need?

2. It costs the Seabrook Inn $3.50, on the average, to serve a meal. How much will it cost to serve 75 people? If they charge an average of $6.25 a meal, how much profit will they make?

3. Derrick needs to buy 4 shock absorbers and 4 tires. The shock absorbers each cost $17.50. The tires each cost $79.00. How much money does Derrick need to buy the shock absorbers and the tires?

4. Marta wants summer clothes that cost $112.00. She earns $2.00 per hour baby-sitting. How many hours must she baby-sit to buy the clothes? If she baby-sits 7 hours per week, how many weeks must she work to earn enough money?

CHAPTER 3, PAGES 62–63

Add.

1. 18.302 + 1.9	**2.** 24.0005 + 66.1	**3.** 5.418 + 2.2334	**4.** 0.76 + 1.69	**5.** 543.00 + 46.12
6. 513.04 6.38 + 0.93	**7.** 2.6 119.73 + 88.1	**8.** 734.61 3.289 + 76.43	**9.** 17.561 6.092 + 0.836	**10.** $4.83 2.99 + 0.57

First estimate by rounding, then find the exact answer.

11. 26.2 + 17.38	**12.** 404.89 + 316.27	**13.** 0.8318 + 0.3634	**14.** 0.071 + 0.7329	**15.** 1.007 + 5.8261

Solve for n.

16. $6.12 + 21.4 + 8 = n$ **17.** $0.563 + 0.91 = n$ **18.** $48.2 + 7 = n$

CHAPTER 3, PAGES 64–65

Subtract.

1. 8.2 − 6.3	**2.** 24.7 − 18.3	**3.** 7.639 − 1.217	**4.** 12.008 − 6.51	**5.** 89 − 0.673
6. 281.063 − 41.026	**7.** 3 − 0.57	**8.** 7.005 − 1.217	**9.** 0.9382 − 0.41	**10.** 68 − 33.064

First estimate by rounding, then find the exact answer.

11. $601.00 − 2.89 **12.** 24.024 − 11.82 **13.** 0.38 − 0.19

Solve for n.

14. $66.814 - 53.107 = n$ **15.** $0.4028 - n = 0.3241$

16. $8.603 - n = 5.94$ **17.** $0.923 - 0.058 = n$ **18.** $421.9 - 2 = n$

CHAPTER 3, PAGES 68–69

Multiply.

1. 8.4×10 **2.** 0.89×10 **3.** 62.7×100 **4.** 5.3×100

5. $73.8 \times 1{,}000$ **6.** $5.36 \times 1{,}000$ **7.** $0.22 \times 1{,}000$ **8.** $25.3 \times 10{,}000$

Divide.

9. $86.4 \div 10$ **10.** $5.3 \div 10$ **11.** $320 \div 100$ **12.** $0.03 \div 100$

13. $90.3 \div 1{,}000$ **14.** $22 \div 1{,}000$ **15.** $56.9 \div 10{,}000$ **16.** $3.42 \div 10{,}000$

CHAPTER 3, PAGES 70–71

Solve the problem by using the guess-and-check method.

1. Tom spent \$2.72 on 14 stamps. He bought 22¢ stamps and 4¢ stamps. How many of each kind of stamp did Tom buy?

2. Mrs. Brown bought clothing for her children. She bought Sean 4 more items than Mike. She bought 18 items in all. How many things did she buy for Mike?

3. Marie bought milk and juice at the store. Her change was \$1.87 in quarters, dimes, and pennies. She received 10 coins. How many were dimes?

4. Sue and Jan ran a total of 21 km. Sue ran 5 more kilometers than Jan. How many kilometers did each girl run?

CHAPTER 3, PAGES 74–75

Multiply.

1. $\begin{array}{r} 38.5 \\ \times\ 16 \\ \hline \end{array}$ **2.** $\begin{array}{r} 79.32 \\ \times\ 2.08 \\ \hline \end{array}$ **3.** $\begin{array}{r} 1.09 \\ \times 0.86 \\ \hline \end{array}$ **4.** $\begin{array}{r} 6.044 \\ \times 0.432 \\ \hline \end{array}$ **5.** $\begin{array}{r} 95.19 \\ \times 0.002 \\ \hline \end{array}$

6. $\begin{array}{r} 10.003 \\ \times\ 7.2 \\ \hline \end{array}$ **7.** $\begin{array}{r} 594 \\ \times\ 0.3 \\ \hline \end{array}$ **8.** $\begin{array}{r} 0.471 \\ \times 0.002 \\ \hline \end{array}$ **9.** $\begin{array}{r} 17.4 \\ \times\ 8.8 \\ \hline \end{array}$ **10.** $\begin{array}{r} 6.863 \\ \times 0.027 \\ \hline \end{array}$

First estimate by rounding. Then find the exact answer.

11. $\begin{array}{r} 898 \\ \times\ 0.5 \\ \hline \end{array}$ **12.** $\begin{array}{r} 4.03 \\ \times 0.92 \\ \hline \end{array}$ **13.** $\begin{array}{r} 90.22 \\ \times\ 0.74 \\ \hline \end{array}$ **14.** $\begin{array}{r} 19.8 \\ \times\ 5.2 \\ \hline \end{array}$ **15.** $\begin{array}{r} 0.9203 \\ \times\ 0.471 \\ \hline \end{array}$

16. 6.3×0.074 **17.** 0.31×0.006 **18.** 2.9×0.48

CHAPTER 3, PAGES 78–79

Divide.

1. $5\overline{)26.55}$ **2.** $8\overline{)64.88}$ **3.** $7\overline{)18.2}$ **4.** $42\overline{)14.196}$

5. $6\overline{)0.0192}$ **6.** $3\overline{)0.171}$ **7.** $18\overline{)1.9836}$ **8.** $2\overline{)0.2014}$

First estimate using compatible numbers. Then find the exact quotient.

9. $154.32 \div 24$ **10.** $4.6718 \div 71$ **11.** $280.854 \div 42$

12. $27.339 \div 30$ **13.** $64.944 \div 82$ **14.** $78.032 \div 20$

Solve for n.

15. $51.66 \div 3 = n$ **16.** $3.114 \div 4 = n$ **17.** $0.0456 \div 8 = n$

CHAPTER 3, PAGES 80–82

Name the power of 10 needed to make the divisor a whole number. Write 10; 100; 1,000; or 10,000.

1. $2.78\overline{)0.434}$ **2.** $2.5\overline{)625}$ **3.** $0.0008\overline{)640.64}$ **4.** $0.73\overline{)9.83}$

5. $0.035\overline{)130.2}$ **6.** $9.5\overline{)76.2}$ **7.** $0.549\overline{)64.1}$ **8.** $0.084\overline{)726}$

Divide.

9. $0.3\overline{)21}$ **10.** $0.6\overline{)1.8}$ **11.** $0.25\overline{)5.75}$ **12.** $0.015\overline{)18.6}$

13. $29.1\overline{)6{,}896.7}$ **14.** $20.3\overline{)103.124}$ **15.** $7.2\overline{)17.28}$

Divide. Round to the nearest hundredth.

16. $1.032\overline{)6.40}$ **17.** $0.87\overline{)8.64}$ **18.** $61.2\overline{)7.15}$

First estimate the quotient by using compatible numbers. Then find the exact answer.

19. $0.18\overline{)9.945}$ **20.** $3.7\overline{)29.304}$ **21.** $5.1\overline{)6.273}$ **22.** $0.05\overline{)20.25}$

23. $0.32\overline{)58.4}$ **24.** $7.5\overline{)6.9525}$ **25.** $0.24\overline{)2.088}$ **26.** $8.9\overline{)81.88}$

CHAPTER 4, PAGES 96–97

Use the bar graph below to answer Exercises 1–3.

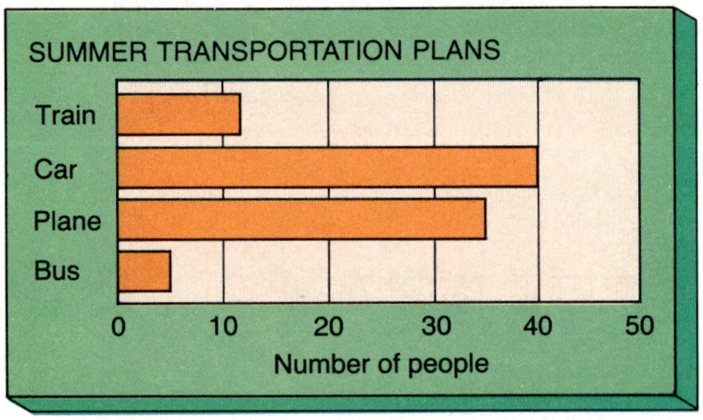

1. Which form of transportation are most people planning to use this summer?

2. About how many more people are planning to travel by plane than by bus?

3. About how many people are planning to use ground transportation this summer?

CHAPTER 4, PAGES 98–99

Find the mean. Round to the nearest tenth.

1. 2, 5, 10, 15

2. 28, 64, 37, 79, 88, 92

3. 208, 173, 180, 225, 158

4. 1,489; 2,045; 1,372; 3,004; 2,178

5. 3,567; 815; 2,003; 935; 267; 1,003

6. 4,004; 9,156; 10,487; 12,348

7. 24.6; 18.3; 7.9; 12.4; 9.04

8. 20.3, 19.5, 14, 26.2, 30.1

9. 326.8; 172.7; 419.5; 296.3; 509.7

10. 105.4, 267.3, 450.3, 336.4, 612, 708.3

CHAPTER 4, PAGES 100–101

Find the statistical measures for each set of data.
Round the answer to the nearest tenth.

Data	Range	Median	Mode	Mean
23; 19; 6; 19; 42; 3	**1.** ▒	**2.** ▒	**3.** ▒	**4.** ▒
7; 2.3; 9.2; 8.4; 9.2; 6.6; 9.2	**5.** ▒	**6.** ▒	**7.** ▒	**8.** ▒
64; 71.5; 58; 64; 62; 61; 70.8	**9.** ▒	**10.** ▒	**11.** ▒	**12.** ▒
600; 525; 743.6; 82.9; 310; 298.1	**13.** ▒	**14.** ▒	**15.** ▒	**16.** ▒
730; 725.2; 730; 735; 734.6; 738	**17.** ▒	**18.** ▒	**19.** ▒	**20.** ▒
88; 123; 117.7; 121.5; 115; 102.3; 97	**21.** ▒	**22.** ▒	**23.** ▒	**24.** ▒
315; 297.3; 304.5; 315; 299; 304.5	**25.** ▒	**26.** ▒	**27.** ▒	**28.** ▒

CHAPTER 5, PAGES 122–123

Write a fraction for the shaded part.

1. 

2. 

3. 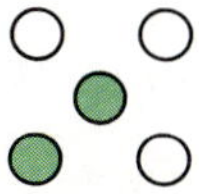

4. 

5. 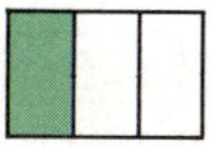

6. 

7. 

8. 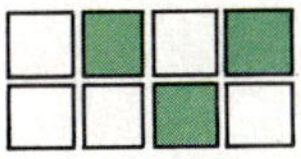

Write a fraction for the point.

9. 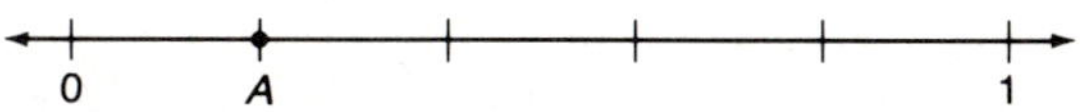

10. 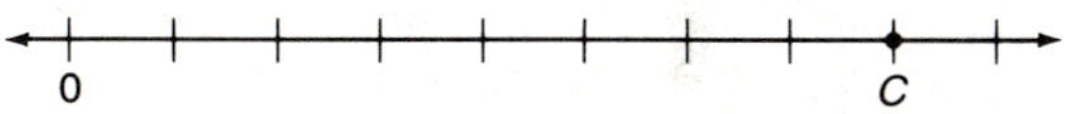

11. 

12. 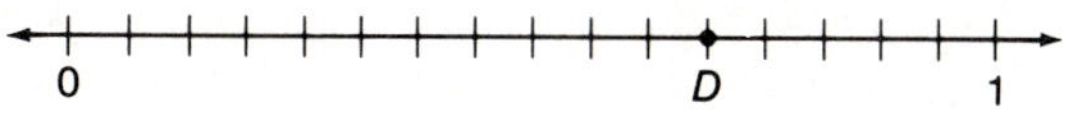

CHAPTER 5, PAGES 124–125

Find the greatest common factor (GCF).

1. 6, 25 **2.** 12, 20 **3.** 9, 36 **4.** 18, 27 **5.** 35, 49

6. 15, 40 **7.** 24, 81 **8.** 100, 95 **9.** 96, 64 **10.** 28, 12

11. 14, 21 **12.** 9, 33 **13.** 15, 22 **14.** 16, 22 **15.** 32, 48

Find the least common multiple (LCM).

16. 3, 8 **17.** 6, 10 **18.** 7, 9 **19.** 3, 12 **20.** 4, 5

21. 9, 21 **22.** 15, 4 **23.** 4, 16 **24.** 4, 10 **25.** 8, 20

26. 7, 8 **27.** 17, 4 **28.** 4, 12 **29.** 11, 12 **30.** 6, 14

CHAPTER 5, PAGES 126–127

Write the fraction in lowest terms.

1. $\frac{12}{20}$ **2.** $\frac{18}{24}$ **3.** $\frac{63}{81}$ **4.** $\frac{8}{28}$ **5.** $\frac{7}{21}$ **6.** $\frac{30}{54}$ **7.** $\frac{16}{66}$ **8.** $\frac{6}{9}$

9. $\frac{55}{95}$ **10.** $\frac{28}{38}$ **11.** $\frac{7}{49}$ **12.** $\frac{15}{90}$ **13.** $\frac{9}{21}$ **14.** $\frac{20}{90}$ **15.** $\frac{3}{27}$ **16.** $\frac{12}{42}$

Complete.

17. $\frac{1}{2} = \frac{\blacksquare}{14}$ **18.** $\frac{12}{27} = \frac{\blacksquare}{9}$ **19.** $\frac{2}{9} = \frac{\blacksquare}{45}$ **20.** $\frac{16}{24} = \frac{\blacksquare}{3}$ **21.** $\frac{5}{8} = \frac{25}{\blacksquare}$

22. $\frac{9}{60} = \frac{\blacksquare}{20}$ **23.** $\frac{20}{25} = \frac{4}{\blacksquare}$ **24.** $\frac{2}{7} = \frac{\blacksquare}{35}$ **25.** $\frac{3}{4} = \frac{\blacksquare}{20}$ **26.** $\frac{12}{48} = \frac{1}{\blacksquare}$

Using cross products, determine if the fractions are equivalent. Write $=$ or $\neq$.

27. $\frac{5}{6} \bullet \frac{8}{9}$ **28.** $\frac{1}{7} \bullet \frac{4}{28}$ **29.** $\frac{3}{9} \bullet \frac{1}{3}$ **30.** $\frac{11}{16} \bullet \frac{5}{8}$ **31.** $\frac{5}{12} \bullet \frac{125}{300}$

CHAPTER 5, PAGES 128–130

Compare using equivalent fractions. Write $>$, $<$, or $=$.

1. $\frac{2}{5} \bullet \frac{4}{5}$ **2.** $\frac{4}{7} \bullet \frac{2}{7}$ **3.** $\frac{9}{14} \bullet \frac{11}{14}$ **4.** $\frac{4}{13} \bullet \frac{5}{26}$ **5.** $\frac{1}{2} \bullet \frac{1}{3}$

6. $\frac{5}{8} \bullet \frac{3}{4}$ **7.** $\frac{9}{10} \bullet \frac{5}{6}$ **8.** $\frac{9}{27} \bullet \frac{1}{3}$ **9.** $\frac{1}{8} \bullet \frac{1}{7}$ **10.** $\frac{5}{10} \bullet \frac{1}{2}$

11. $\frac{5}{7} \bullet \frac{1}{2}$ **12.** $\frac{2}{30} \bullet \frac{1}{10}$ **13.** $\frac{26}{50} \bullet \frac{26}{100}$ **14.** $\frac{21}{40} \bullet \frac{4}{20}$ **15.** $\frac{5}{25} \bullet \frac{1}{5}$

Arrange in order from greatest to least.

16. $\frac{4}{7}, \frac{5}{7}, \frac{1}{7}, \frac{6}{7}, \frac{2}{7}$ **17.** $\frac{2}{8}, \frac{2}{4}, \frac{2}{10}, \frac{2}{6}, \frac{2}{12}$ **18.** $\frac{5}{12}, \frac{2}{7}, \frac{3}{8}, \frac{4}{5}, \frac{1}{10}$

19. $\frac{3}{4}, \frac{3}{9}, \frac{3}{8}, \frac{3}{6}, \frac{3}{10}$ **20.** $\frac{5}{8}, \frac{0}{8}, \frac{2}{8}, \frac{7}{8}, \frac{6}{8}$ **21.** $\frac{2}{5}, \frac{6}{7}, \frac{3}{4}, \frac{1}{3}, \frac{1}{6}$

CHAPTER 5, PAGES 132–133

Write as an improper fraction.

1. $4\frac{1}{9}$ **2.** $3\frac{1}{2}$ **3.** $7\frac{5}{6}$ **4.** $6\frac{2}{3}$ **5.** $3\frac{2}{9}$

6. $6\frac{5}{12}$ **7.** $2\frac{3}{4}$ **8.** $3\frac{7}{8}$ **9.** $1\frac{3}{4}$ **10.** $5\frac{2}{3}$

11. $3\frac{2}{3}$ **12.** $9\frac{1}{6}$ **13.** $5\frac{3}{16}$ **14.** $2\frac{1}{6}$ **15.** $3\frac{1}{12}$

16. $6\frac{3}{5}$ **17.** $9\frac{3}{7}$ **18.** $4\frac{5}{12}$ **19.** $10\frac{3}{10}$ **20.** $3\frac{9}{16}$

CHAPTER 5, PAGES 134–135

Solve.

1. Eric's lasagne recipe feeds 8 people and calls for 8 oz of cheese. He invited 22 friends for lasagne. How many pans of lasagne should he make?

2. The Tygart City Police Academy accepts 75 recruits per class. 348 recruits are ready to begin training. How many must wait for a new class to form?

Frank's Auto Repair ordered 400 fan belts. The company sent 36 fan belts per crate, and the remainder were in a package.

3. How many crates of fan belts arrived?

4. How many fan belts were in the package?

CHAPTER 5, PAGES 136–137

Write as a fraction or mixed number in simplest form.

1. 4.2	**2.** 0.05	**3.** 3.4	**4.** 0.015	**5.** 1.54	**6.** 6.125
7. 0.3	**8.** 0.875	**9.** 2.44	**10.** 5.25	**11.** 0.065	**12.** 9.625
13. 0.025	**14.** 10.8	**15.** 30.78	**16.** 7.03	**17.** 0.55	**18.** 12.375

Write the fraction or mixed number as a decimal.

19. $\frac{2}{5}$	**20.** $\frac{3}{50}$	**21.** $\frac{1}{10}$	**22.** $\frac{5}{16}$	**23.** $\frac{2}{25}$	**24.** $6\frac{7}{8}$
25. $4\frac{1}{2}$	**26.** $2\frac{1}{4}$	**27.** $7\frac{9}{10}$	**28.** $1\frac{3}{20}$	**29.** $5\frac{5}{8}$	**30.** $8\frac{3}{4}$
31. $5\frac{3}{20}$	**32.** $7\frac{1}{8}$	**33.** $16\frac{1}{2}$	**34.** $24\frac{5}{8}$	**35.** $12\frac{7}{25}$	**36.** $20\frac{17}{50}$

Write the fraction or mixed number as a decimal. Round to the nearest hundredth.

37. $\frac{1}{14}$	**38.** $\frac{4}{11}$	**39.** $\frac{7}{9}$	**40.** $\frac{6}{7}$	**41.** $\frac{7}{12}$	**42.** $\frac{1}{11}$
43. $\frac{5}{13}$	**44.** $\frac{7}{15}$	**45.** $\frac{8}{11}$	**46.** $\frac{5}{6}$	**47.** $5\frac{4}{13}$	**48.** $9\frac{9}{11}$
49. $2\frac{1}{3}$	**50.** $3\frac{4}{7}$	**51.** $1\frac{1}{12}$	**52.** $4\frac{2}{13}$	**53.** $7\frac{2}{3}$	**54.** $6\frac{3}{7}$

Write the fraction or mixed number as a decimal. Use the bar notation.

55. $\frac{4}{9}$	**56.** $\frac{7}{15}$	**57.** $\frac{49}{90}$	**58.** $\frac{2}{9}$	**59.** $\frac{5}{6}$	**60.** $\frac{10}{11}$
61. $2\frac{8}{9}$	**62.** $1\frac{2}{3}$	**63.** $1\frac{7}{11}$	**64.** $3\frac{5}{9}$	**65.** $40\frac{2}{3}$	**66.** $19\frac{3}{7}$

CHAPTER 6, PAGES 146–147

Multiply. Write the product in simplest form.

1. $\frac{1}{4} \times \frac{3}{5}$ **2.** $\frac{2}{3} \times \frac{1}{6}$ **3.** $\frac{2}{3} \times \frac{3}{4}$ **4.** $\frac{12}{25} \times \frac{5}{6}$ **5.** $\frac{1}{16} \times \frac{1}{2}$ **6.** $\frac{1}{6} \times \frac{9}{10}$

7. $\frac{1}{3} \times \frac{5}{8}$ **8.** $\frac{1}{6} \times \frac{2}{3}$ **9.** $\frac{2}{4} \times \frac{2}{9}$ **10.** $\frac{1}{7} \times \frac{4}{5}$ **11.** $\frac{2}{5} \times \frac{1}{4}$ **12.** $\frac{2}{7} \times \frac{8}{9}$

Multiply. Use the shortcut when possible.

13. $\frac{7}{15} \times \frac{10}{21} \times \frac{1}{2}$ **14.** $\frac{2}{5} \times \frac{3}{4} \times \frac{5}{7}$ **15.** $\frac{6}{7} \times \frac{7}{9} \times \frac{1}{8}$ **16.** $\frac{1}{6} \times \frac{2}{3} \times \frac{9}{11}$

17. $\frac{5}{8} \times \frac{2}{3} \times \frac{2}{15}$ **18.** $\frac{6}{5} \times \frac{3}{18} \times \frac{1}{2}$ **19.** $\frac{6}{7} \times \frac{14}{24} \times \frac{1}{3}$ **20.** $\frac{3}{10} \times \frac{5}{8} \times \frac{4}{7}$

CHAPTER 6, PAGES 148–149

Multiply. Write the product in simplest form.

1. $7 \times \frac{3}{5}$ **2.** $\frac{1}{8} \times 12$ **3.** $10\frac{1}{3} \times \frac{5}{6}$ **4.** $3\frac{1}{5} \times \frac{7}{8}$ **5.** $6\frac{1}{3} \times 10\frac{1}{4}$

6. $5\frac{5}{12} \times 3\frac{3}{4}$ **7.** $4\frac{2}{3} \times 3\frac{1}{2}$ **8.** $8\frac{2}{5} \times 6\frac{1}{3}$ **9.** $3\frac{1}{3} \times 1\frac{1}{8}$ **10.** $1\frac{4}{5} \times 5\frac{5}{8}$

11. $3\frac{3}{4} \times \frac{1}{3} \times \frac{3}{5}$ **12.** $2\frac{1}{4} \times 1\frac{2}{3} \times 4\frac{4}{5}$ **13.** $6\frac{1}{3} \times 2\frac{4}{10} \times 3\frac{3}{8}$

14. $\frac{1}{2} \times 2 \times \frac{1}{3}$ **15.** $\frac{7}{8} \times 8\frac{1}{4} \times \frac{2}{7}$ **16.** $3\frac{1}{7} \times 4\frac{2}{3} \times \frac{7}{8}$

17. $3\frac{1}{7} \times 3\frac{1}{4} \times \frac{1}{3}$ **18.** $\frac{5}{6} \times 2\frac{1}{6} \times \frac{3}{10}$ **19.** $\frac{4}{9} \times 12\frac{1}{4} \times 3\frac{2}{7}$

20. $\left(3\frac{4}{5} + \frac{1}{10}\right) \times \frac{2}{3}$ **21.** $\left(2\frac{3}{4} + 3\frac{1}{6}\right) \times \frac{4}{5}$ **22.** $\left(1\frac{8}{10} + 2\frac{2}{5}\right) \times \frac{3}{7}$

CHAPTER 6, PAGES 150–151

Find the reciprocal of the number.

1. 8 **2.** 14 **3.** 27 **4.** 100 **5.** $\frac{1}{9}$ **6.** $\frac{5}{7}$

7. $\frac{8}{3}$ **8.** $\frac{10}{9}$ **9.** $4\frac{9}{10}$ **10.** $6\frac{7}{8}$ **11.** $2\frac{1}{5}$ **12.** $12\frac{2}{3}$

Divide. Write the quotient in simplest form.

13. $\frac{1}{2} \div \frac{1}{8}$ **14.** $\frac{3}{4} \div \frac{3}{8}$ **15.** $\frac{2}{3} \div \frac{4}{9}$ **16.** $\frac{9}{10} \div \frac{5}{6}$ **17.** $\frac{8}{17} \div 4$

18. $\frac{4}{5} \div \frac{8}{15}$ **19.** $\frac{8}{3} \div \frac{4}{5}$ **20.** $\frac{6}{7} \div 3$ **21.** $\frac{1}{3} \div \frac{1}{9}$ **22.** $\frac{4}{9} \div \frac{2}{3}$

23. $7 \div \frac{1}{5}$ **24.** $9 \div \frac{3}{4}$ **25.** $\frac{1}{6} \div \frac{2}{3}$ **26.** $\frac{3}{5} \div \frac{1}{10}$ **27.** $\frac{5}{9} \div \frac{1}{4}$

CHAPTER 6, PAGES 152–153

Divide. Write the quotient in simplest form.

1. $4\frac{3}{4} \div 2$ **2.** $8\frac{9}{10} \div 4$ **3.** $3 \div 3\frac{1}{2}$ **4.** $1\frac{1}{3} \div 5$ **5.** $6\frac{3}{4} \div \frac{1}{8}$

6. $\frac{1}{10} \div 9\frac{4}{5}$ **7.** $3\frac{7}{8} \div \frac{1}{32}$ **8.** $2\frac{9}{25} \div \frac{1}{100}$ **9.** $\frac{2}{3} \div 2\frac{2}{3}$ **10.** $2\frac{1}{5} \div \frac{3}{5}$

11. $6 \div 2\frac{2}{3}$ **12.** $12 \div 1\frac{3}{4}$ **13.** $\frac{1}{9} \div 4\frac{1}{2}$ **14.** $6\frac{1}{5} \div \frac{1}{10}$ **15.** $8\frac{1}{3} \div 5$

CHAPTER 6, PAGES 154–155

Solve each problem by working backward.

1. On Tuesday Jan made one-half as many school safety posters as she made on Thursday. She made six on Tuesday. How many posters did she make on Thursday?

2. Keith walked his dog four times as far on Saturday as he did on Sunday. He walked the dog 1.4 km on Sunday. How far did he walk on Saturday?

3. Bart had $25.00 on Friday. By Sunday, he had $9.05 left. He went to a movie on Saturday and then out to eat. His meal cost $7.75. How much money did Bart spend at the movies?

4. Ira worked $1\frac{1}{3}$ times as many hours last week than this week. This week he worked 39 hours. How many hours did Ira work last week?

CHAPTER 6, PAGES 158–159

Add. Write the sum in simplest form.

1. $\frac{5}{8} + \frac{1}{8}$ **2.** $\frac{3}{8} + \frac{5}{32}$ **3.** $\frac{2}{5} + \frac{1}{10}$ **4.** $\frac{3}{4} + \frac{1}{6}$ **5.** $\frac{1}{3} + \frac{7}{12}$

6. $\frac{3}{5} + \frac{1}{3}$ **7.** $\frac{3}{5} + \frac{1}{2}$ **8.** $\frac{1}{7} + \frac{8}{9}$ **9.** $\frac{5}{8} + \frac{11}{12}$ **10.** $\frac{2}{5} + \frac{5}{6}$

CHAPTER 6, PAGES 160–161

Add. Write the sum in simplest form.

1. $2\frac{5}{6} + 1\frac{5}{8}$ **2.** $9\frac{3}{5} + 3\frac{2}{20}$ **3.** $10\frac{3}{10} + 8\frac{9}{10}$ **4.** $12\frac{7}{8} + 11\frac{9}{16}$

5. $7\frac{1}{4} + 8\frac{7}{16} + 6\frac{3}{16}$

6. $7\frac{1}{4} + 8\frac{1}{3} + 5\frac{1}{6}$

7. $3\frac{2}{3} + 6\frac{1}{2} + 8\frac{2}{5}$

8. $4\frac{1}{4} + 2\frac{5}{6} + 10\frac{1}{2}$

9. $6\frac{1}{2} + 4\frac{2}{5} + 9\frac{3}{10}$

10. $3\frac{5}{8} + 6\frac{1}{4} + 2\frac{1}{5}$

CHAPTER 6, PAGES 162–163

Subtract. Write the difference in simplest form.

1. $\frac{7}{8} - \frac{5}{8}$ **2.** $\frac{3}{4} - \frac{2}{5}$ **3.** $\frac{1}{4} - \frac{1}{9}$ **4.** $\frac{9}{10} - \frac{5}{6}$ **5.** $\frac{7}{10} - \frac{3}{8}$

6. $\frac{5}{6} - \frac{1}{6}$ **7.** $\frac{7}{9} - \frac{3}{5}$ **8.** $\frac{2}{3} - \frac{3}{10}$ **9.** $\frac{1}{2} - \frac{1}{4}$ **10.** $\frac{15}{36} - \frac{2}{9}$

CHAPTER 6, PAGES 164–166

Subtract. Write the difference in simplest form.

1. $3\frac{1}{2} - 1\frac{1}{4}$ **2.** $6\frac{1}{8} - 4\frac{3}{4}$ **3.** $23\frac{5}{8} - 12\frac{1}{2}$ **4.** $19\frac{5}{16} - 5\frac{1}{4}$

5. $9\frac{7}{12} - 6\frac{11}{12}$ **6.** $14\frac{2}{5} - 8\frac{1}{10}$ **7.** $8\frac{2}{3} - 5\frac{1}{2}$ **8.** $11 - 9\frac{1}{4}$ **9.** $13\frac{1}{3} - 10\frac{3}{4}$ **10.** $5\frac{3}{10} - 4\frac{3}{4}$

11. $5\frac{3}{4} - \frac{5}{8}$ **12.** $18\frac{5}{9} - 8\frac{5}{6}$ **13.** $14 - 6\frac{1}{2}$ **14.** $20 - 15\frac{3}{16}$ **15.** $6\frac{1}{12} - \frac{2}{3}$ **16.** $27\frac{3}{8} - 6\frac{2}{3}$

CHAPTER 6, PAGES 168–169

Round to the nearest whole number.

1. $56\frac{4}{5}$ **2.** $12\frac{1}{2}$ **3.** $23\frac{4}{9}$ **4.** $17\frac{1}{6}$ **5.** $45\frac{5}{8}$

6. $11\frac{1}{3}$ **7.** $25\frac{7}{9}$ **8.** $24\frac{9}{16}$ **9.** $49\frac{11}{12}$ **10.** $18\frac{3}{8}$

Estimate.

11. $8\frac{3}{8} + 5\frac{1}{4}$ **12.** $7\frac{3}{5} + 5\frac{1}{4}$ **13.** $15\frac{3}{8} + 14\frac{4}{5}$ **14.** $10\frac{1}{3} + 8\frac{3}{4}$

15. $9\frac{1}{10} - 5\frac{3}{4}$ **16.** $20\frac{1}{20} - 10\frac{1}{40}$ **17.** $17\frac{39}{50} - 8\frac{3}{4}$ **18.** $32\frac{1}{5} - 19\frac{2}{3}$

19. $6\frac{1}{6} \times 4\frac{4}{5}$ **20.** $2\frac{3}{4} \times 5\frac{2}{3}$ **21.** $3\frac{3}{4} \times 1\frac{1}{8}$ **22.** $7\frac{5}{6} \times 2\frac{1}{3}$

23. $10\frac{3}{10} \div 5\frac{1}{5}$ **24.** $4\frac{4}{9} \div 1\frac{2}{3}$ **25.** $5\frac{3}{5} \div 2\frac{4}{15}$ **26.** $2\frac{7}{8} \div \frac{15}{16}$

27. $19\frac{3}{5} - 4\frac{1}{3}$ **28.** $31\frac{2}{3} \div 7\frac{8}{9}$ **29.** $4\frac{1}{5} \times 2\frac{1}{10}$ **30.** $12\frac{3}{4} + 8\frac{7}{9}$

CHAPTER 7, PAGES 180–181

Complete.

1. 9 in. = ■ yd	**2.** 7,920 ft = ■ mi	**3.** $5\frac{1}{4}$ ft = ■ in.
4. 3 mi = ■ ft	**5.** 540 in. = ■ yd	**6.** 69 yd = ■ ft
7. 90 in. = ■ ft	**8.** $14\frac{1}{2}$ yd = ■ in.	**9.** 126 in. = ■ ft
10. 7,040 yd = ■ mi	**11.** 40 ft = ■ in.	**12.** 18 yd = ■ in.
13. 3 in. = ■ yd	**14.** 2,640 ft = ■ mi	**15.** 3 mi = ■ yd
16. 6 mi = ■ ft	**17.** 96 yd = ■ ft	**18.** 936 in. = ■ yd

CHAPTER 7, PAGES 182–183

Complete.

1. 7 pt = ■ c	**2.** 8,000 lb = ■ T	**3.** 6 gal = ■ qt
4. $\frac{1}{2}$ qt = ■ fl oz	**5.** $4\frac{3}{4}$ gal = ■ pt	**6.** $24\frac{1}{2}$ lb = ■ oz
7. 400 fl oz = ■ gal	**8.** 13 pt = ■ qt	**9.** 17 c = ■ pt
10. 10 c = ■ qt	**11.** 3 gal = ■ fl oz	**12.** 48 oz = ■ lb
13. 18 qt = ■ pt	**14.** 30 fl oz = ■ c	**15.** $2\frac{1}{4}$ T = ■ lb
16. 13 pt = ■ c	**17.** 1,000 lb = ■ T	**18.** 10 gal = ■ qt
19. 3 qt = ■ fl oz	**20.** 7 gal = ■ pt	**21.** 18 lb = ■ oz

CHAPTER 7, PAGES 184–185

Add or subtract.

1. 2 pt 1 c +3 pt 1 c	**2.** 6 yd 1 ft −2 yd 2 ft	**3.** 8 mi 1,680 ft +2 mi 5,000 ft	**4.** 12 lb 5 oz − 6 lb 9 oz
5. 9 ft 6 in. −1 ft 8 in.	**6.** 7 gal −3 gal 3 qt	**7.** 6 T 1,000 lb + 1,500 lb	**8.** 3 c 7 fl oz +2 c 6 fl oz
9. 3 mi 900 yd +6 mi 900 yd	**10.** 10 qt − 6 qt 1 pt	**11.** 17 lb 4 oz −10 lb 12 oz	**12.** 6 gal 2 qt +11 gal 3 qt
13. 6 pt 1 c +7 pt 1 c	**14.** 9 yd 2 ft −7 yd 1 ft	**15.** 4 mi 340 ft +10 mi 920 ft	**16.** 18 lb 3 oz −16 lb 14 oz
17. 15 ft 1 in. −12 ft 4 in.	**18.** 13 gal − 2 gal 1 qt	**19.** 5 T 500 lb + 1,700 lb	**20.** 1 c 6 fl oz +4 c 2 fl oz

CHAPTER 7, PAGES 186–187

Solve.

Your plane leaves New York at 12:48 P.M. How long does it take you to reach Atlanta if you arrive at:

1. 2:04 P.M.? **2.** 1:42 P.M.? **3.** 4:32 P.M.? **4.** 3:12 P.M.?

The baseball game began at 3:30 P.M. How long did the game last if it ended at:

5. 5:54 P.M.? **6.** 6:49 P.M.? **7.** 6:09 P.M.? **8.** 7:50 P.M.?

The Saturday double feature at the Plaza Theater begins at 1:50 P.M. How long will it last if it ends at:

9. 5:15 P.M.? **10.** 6:02 P.M.? **11.** 6:35 P.M.? **12.** 5:53 P.M.?

CHAPTER 7, PAGES 192–193

Choose the most reasonable estimate.

1. The height of a coffee mug	**a.** 4 km	**b.** 6 cm	**c.** 6 mm
2. The length of a football field	**a.** 91 m	**b.** 91 cm	**c.** 91 mm
3. The height of a grandfather's clock	**a.** 3 m	**b.** 3 cm	**c.** 3 mm
4. The length of a dog's tail	**a.** 15 m	**b.** 15 cm	**c.** 15 mm
5. The width of a piano	**a.** 120 m	**b.** 120 mm	**c.** 120 cm
6. The thickness of a sandwich	**a.** 20 m	**b.** 20 km	**c.** 20 mm
7. The thickness of a pencil	**a.** 5 m	**b.** 5 cm	**c.** 5 mm

CHAPTER 7, PAGES 194–196

Choose the most reasonable estimate.

1. The mass of a tube of lipstick	**a.** 100 kg	**b.** 100 g	**c.** 100 mg
2. The mass of a table lamp	**a.** 7 kg	**b.** 30 g	**c.** 10 g
3. The capacity of a thermos bottle	**a.** 5 kL	**b.** 1 L	**c.** 6 mL
4. The capacity of an eyedropper	**a.** 1 kL	**b.** 3 L	**c.** 5 mL
5. The amount of water in a bathtub	**a.** 10 kL	**b.** 100 L	**c.** 1,000 mL
6. The mass of a bird feather	**a.** 10 kg	**b.** 30 g	**c.** 20 mg
7. The mass of an apple	**a.** 25 g	**b.** 250 g	**c.** 25 kg

CHAPTER 7, PAGES 198–199

Complete.

1. 4 m = ■ mm	**2.** 2 cm = ■ mm	**3.** 800 cm = ■ m
4. 6 km = ■ m	**5.** 300 mm = ■ m	**6.** 70 m = ■ km
7. 3 kg = ■ g	**8.** 73,000 L = ■ kL	**9.** 5,000 mg = ■ g
10. 160 g = ■ kg	**11.** 7 L = ■ mL	**12.** 2 kL = ■ mL
13. 6 kg = ■ mg	**14.** 820 mL = ■ L	**15.** 82.3 mm = ■ cm
16. 60 mg = ■ g	**17.** 0.434 kg = ■ g	**18.** 6.888 km = ■ m

CHAPTER 7, PAGES 200–201

Choose the most reasonable temperature for each of the following.

1. A radiator	**a.** 32°F	**b.** 230°F	**c.** 55°C
2. Beef stew cooking	**a.** 350°F	**b.** 37°C	**c.** 60°C
3. A sick baby	**a.** 2°C	**b.** 40°F	**c.** 40°C
4. A day for sledding	**a.** 60°F	**b.** 28°F	**c.** 20°C
5. Steam from boiling water	**a.** 50°F	**b.** 95°C	**c.** 75°F
6. A freezer	**a.** −100°F	**b.** −5°F	**c.** 22°F

CHAPTER 7, PAGES 202–203

Solve each problem by drawing a diagram.

Jason's room measures 6 m × 5 m. Two windows take up a total of 2.5 m on the south wall which is 5 m in length. The door takes up 1.5 m on the north wall. His desk measures 2 m across, his bed is 2.2 m, and his dresser is 1.25 m.

1. Will all his furniture fit along the west wall?

2. How much space will be left along the wall with the door?

Lincoln Airport is built in the shape of a square. It measures 6 km on each side. A fuel truck enters the airport at the center of the north side and travels 2 km south. Then it travels 2.5 km west, then 0.5 km south.

3. How far is the truck from the western border of the airport?

4. How far is the truck from the southern border of the airport?

CHAPTER 8, PAGES 210–211

Complete to form an equivalent ratio.

1. $\frac{4}{5} = \frac{\blacksquare}{20}$ **2.** $\frac{20}{150} = \frac{2}{\blacksquare}$ **3.** $\frac{3}{13} = \frac{9}{\blacksquare}$ **4.** $\frac{1}{6} = \frac{\blacksquare}{72}$ **5.** $\frac{22}{14} = \frac{\blacksquare}{7}$

Use these denominators to write a ratio equivalent to $\frac{2}{3}$.

6. 18 **7.** 36 **8.** 48 **9.** 96 **10.** 24 **11.** 108 **12.** 72 **13.** 54

Use these numerators to write a ratio equivalent to $\frac{40}{160}$.

14. 25 **15.** 16 **16.** 80 **17.** 7 **18.** 22 **19.** 15 **20.** 6 **21.** 32

Write the ratio in simplest form.

22. $\frac{3}{21}$ **23.** $\frac{36}{63}$ **24.** $\frac{6}{84}$ **25.** $\frac{50}{75}$ **26.** $\frac{25}{45}$ **27.** $\frac{6}{66}$ **28.** $\frac{42}{54}$ **29.** $\frac{48}{144}$

CHAPTER 8, PAGES 212–213

Write as a rate in simplest form.

1. 14 balls in 3 innings
2. 95 words in 1 minute
3. 2 mi in 6 minutes
4. 16 apples for \$3.16

Compare the rates. Write >, <, or =.

5. $\frac{\text{12 pens}}{\$0.90}$ ● $\frac{\text{2 pens}}{\$0.15}$ **6.** $\frac{\text{8 players}}{\text{4 games}}$ ● $\frac{\text{5 players}}{\text{7 games}}$ **7.** $\frac{\text{6 eggs}}{\$0.59}$ ● $\frac{\text{12 eggs}}{\$0.89}$

8. $\frac{\text{21 cats}}{\text{14 dogs}}$ ● $\frac{\text{32 cats}}{\text{23 dogs}}$ **9.** $\frac{\text{60 mi}}{\text{2 hours}}$ ● $\frac{\text{30 mi}}{\text{1 hour}}$ **10.** $\frac{\text{5 shows}}{\text{2 weeks}}$ ● $\frac{\text{14 shows}}{\text{20 weeks}}$

CHAPTER 8, PAGES 216–217

Determine if the ratios form a proportion. Write *yes* or *no*.

1. $\frac{4}{2} \stackrel{?}{=} \frac{6}{3}$ **2.** $\frac{6.8}{3.4} \stackrel{?}{=} \frac{24.2}{8.6}$ **3.** $\frac{12}{3} \stackrel{?}{=} \frac{116}{29}$ **4.** $\frac{67}{4} \stackrel{?}{=} \frac{33}{2}$

5. $\frac{5}{9} \stackrel{?}{=} \frac{7}{13}$ **6.** $\frac{1.2}{2.5} \stackrel{?}{=} \frac{2.0}{4.5}$ **7.** $\frac{25}{10} \stackrel{?}{=} \frac{100}{40}$ **8.** $\frac{16}{56} \stackrel{?}{=} \frac{30}{105}$

Find n.

9. $\frac{n}{32} = \frac{5}{8}$ **10.** $\frac{34}{20} = \frac{17}{n}$ **11.** $\frac{19}{11} = \frac{n}{44}$ **12.** $\frac{2}{3} = \frac{200}{n}$

13. $\frac{6.3}{2} = \frac{n}{8.2}$ **14.** $\frac{90}{1.5} = \frac{n}{6}$ **15.** $\frac{0.5}{50} = \frac{4.5}{n}$ **16.** $\frac{6}{n} = \frac{9}{30.6}$

17. $\frac{16}{3.2} = \frac{n}{20}$ **18.** $\frac{3.9}{n} = \frac{4.5}{7.5}$ **19.** $\frac{n}{8.2} = \frac{20}{0.5}$ **20.** $\frac{20}{35} = \frac{6.2}{n}$

CHAPTER 8, PAGES 218–219

Which is the better estimate?

1. Suzanne wants to buy a skirt for \$14.99, shoes for \$22.84, socks for \$1.99, and a shirt for \$6.56. How much money should she have?

   a. \$14 + \$22 + \$1 + \$6 = \$43 **b.** \$15 + \$23 + \$2 + \$7 = \$47

Solve.

RAY'S PRODUCE	
Apples	6/\$1.75
Cucumbers	12/\$2.99
Avocados	\$0.89

2. Jim wants to buy 4 avocados and 6 apples. Estimate the cost.
3. Carol buys 12 cucumbers and 10 avocados. Estimate the cost.
4. Wanda's mother sends her to Ray's for 5 avocados, 6 cucumbers, and 6 apples. Estimate the cost.

CHAPTER 8, PAGES 224–225

A model airplane was made to the scale of 1 in. : 3 ft. Find the actual measures.

Airplane	Length	Wingspan	Length of one wing	Height of tail fin
Actual measure of model	1. ■	2. ■	3. ■	4. ■
Scale measure of model	8 in.	6 in.	3 in.	1 in.

CHAPTER 9, PAGES 234–235

Write as a decimal.

1. 5%	**2.** 8%	**3.** 66%	**4.** 23%	**5.** 535%
6. 16%	**7.** 162%	**8.** 325%	**9.** 2.3%	**10.** 1.5%
11. $\frac{1}{4}$%	**12.** $\frac{7}{8}$%	**13.** $\frac{3}{5}$%	**14.** $37\frac{1}{2}$%	**15.** $40\frac{3}{4}$%

Write as a percent.

16. 0.14	**17.** 0.61	**18.** 0.98	**19.** 0.8	**20.** 0.887
21. 0.125	**22.** 0.333	**23.** 0.0725	**24.** 0.125	**25.** 0.00346
26. 4.3	**27.** 3.9	**28.** 6.8	**29.** 7.25	**30.** 4.375

CHAPTER 9, PAGES 236–237

Write as a fraction, whole number, or mixed number in simplest form.

1. 81% **2.** 3% **3.** 19% **4.** 63% **5.** 29%

6. 15% **7.** 20% **8.** 90% **9.** 56% **10.** 78%

11. 200% **12.** 300% **13.** 600% **14.** 4,000% **15.** 7,000%

16. 850% **17.** 594% **18.** 330% **19.** 944% **20.** 120%

21. 222% **22.** 1,470% **23.** 108% **24.** 3,240% **25.** 125%

26. 675% **27.** 549% **28.** 6,680% **29.** 8,635% **30.** 9,766%

CHAPTER 9, PAGES 238–239

Write as a percent.

1. $\frac{1}{10}$ **2.** $\frac{4}{25}$ **3.** $\frac{1}{2}$ **4.** $\frac{3}{50}$ **5.** $\frac{1}{4}$ **6.** $\frac{7}{20}$

7. $\frac{3}{8}$ **8.** $\frac{5}{12}$ **9.** $\frac{4}{5}$ **10.** $\frac{1}{3}$ **11.** $\frac{2}{9}$ **12.** $\frac{5}{6}$

13. 5 **14.** 2 **15.** 14 **16.** 30 **17.** 3 **18.** 17

19. $1\frac{1}{5}$ **20.** $2\frac{4}{5}$ **21.** $4\frac{5}{6}$ **22.** $9\frac{3}{8}$ **23.** $3\frac{3}{5}$ **24.** $1\frac{7}{8}$

25. $5\frac{7}{20}$ **26.** $7\frac{3}{4}$ **27.** $6\frac{2}{3}$ **28.** $3\frac{14}{25}$ **29.** $2\frac{13}{20}$ **30.** $2\frac{1}{3}$

CHAPTER 9, PAGES 240–241

Find the percent of the number.

1. 3% of 90 **2.** 90% of 900 **3.** 100% of 50 **4.** 5.2% of 14

5. 4.4% of 600 **6.** $\frac{1}{4}$% of 36 **7.** $\frac{3}{8}$% of 48 **8.** $12\frac{1}{2}$% of 64

Compute.

9. What is 4% of 500?

10. What is 10% of 20?

11. What is $8\frac{1}{2}$% of 40?

12. What number is 0.17% of 92?

13. What is 6% of 450?

14. 8.3% of 60 is what number?

15. What is 200% of 45?

16. 50.8% of 520 is what number?

17. What is $7\frac{1}{4}$% of 500?

18. What number is 130% of 800?

CHAPTER 9, PAGES 246–247

Find the missing number.

1. What percent of 20 is 12?

2. What percent of 16 is 5.6?

3. 8 is what percent of 10?

4. 9 is what percent of 45?

5. What percent of 400 is 3.2?

6. What percent of 72 is 70.2?

7. 75 is what percent of 300?

8. 65 is what percent of 32.5?

Estimate.

9. What percent of 10 is 5.4?

10. What percent of 8 is 6.2?

11. 19.4 is what percent of 59.8?

12. What percent of 20 is 1.1?

13. 3.2 is what percent of 12.4?

14. 3.9 is what percent of 5?

CHAPTER 9, PAGES 248–249

Find the number. Change the percent to a decimal.

1. 7% of what number is 28?

2. 90 is 18% of what number?

3. 15% of what number is 30?

4. 3% of what number is 1.8?

Find the number. Change the percent to a fraction.

5. 12% of what number is 9?

6. 15 is 6% of what number?

7. 30 is $37\frac{1}{2}\%$ of what number?

8. 30 is $83\frac{1}{3}\%$ of what number?

Find the number.

9. 20% of what number is 20?

10. 160 is 400% of what number?

11. 125 is 125% of what number?

12. 2 is 0.4% of what number?

13. 2.5% of what number is 30?

14. 130 is 65% of what number?

CHAPTER 9, PAGES 250–251

Use a proportion to solve.

1. What number is 4% of 15?

2. 63% of 300 is what number?

3. 44 is what percent of 66?

4. 3 is 40% of what number?

5. 25 is what percent of 2.5?

6. What number is 8.5% of 60?

7. 180 is what percent of 120?

8. 45% of 280 is what number?

CHAPTER 9, PAGES 252–253

Solve by supplying the missing information.

1. Jenny needs 6 ft of material to make a skirt. How many yards of material should she buy?

2. Paul uses 3 oz of yeast for every two loaves of bread he bakes. How many pounds of yeast does he need for 32 loaves of bread?

3. It takes the Decorating Committee 45 minutes to decorate $\frac{1}{8}$ of the gym for the school dance. How many hours will it take to finish the job?

4. Brian goes to visit his grandfather every summer. This year he leaves on June 28 and stays through August 31. How many days will he be at his grandfather's?

5. Dan is buying refreshments for the school dance. He expects 300 students to come to the dance and for each of them to drink three 8-oz cups of punch. How many quarts of punch does he buy?

6. Leslie figures that she can earn $45 per week working part time in the bookstore. If she works for an entire year, how much money will Leslie earn? How much will she earn per month?

CHAPTER 9, PAGES 258–259

Complete.

	Original amount	New amount	Percent change
1.	$100	$150	▪
2.	8	5	▪
3.	15	20	▪
4.	25	15	▪
5.	6	24	▪
6.	5,600	4,200	▪
7.	25	75	▪
8.	2,000	1,200	▪
9.	420	▪	10% increase
10.	300	▪	5% decrease
11.	700	▪	25% increase
12.	850	▪	16% decrease

CHAPTER 10, PAGES 268–270

A clothes dryer contains 20 socks. Socks are picked at random. Refer to the chart at the right and find the probability.

PAIRS OF SOCKS

	Large	Small
Blue	3	2
White	2	3

1. *P*(blue)

2. *P*(white)

3. *P*(large blue)

4. *P*(large white)

5. *P*(small blue)

6. *P*(small white)

CHAPTER 10, PAGES 272–273

City Park Officials surveyed 500 park users on a Saturday in May. Here are the resuts. For the attendance figure, predict how many people will choose each activity listed at the right.

ACTIVITY

Jogging	60
Baseball	210
Picnic	120
Other	110

Attendance	Jogging	Baseball	Picnic	Other
1,000	**1.** ■	**2.** ■	**3.** ■	**4.** ■
1,200	**5.** ■	**6.** ■	**7.** ■	**8.** ■
1,500	**9.** ■	**10.** ■	**11.** ■	**12.** ■

A spinner is divided into eight equal sections. Two sections are green, three are red, and three are blue. It is spun 240 times.

13. Predict how many times a red section will be spun.

14. Predict how many times a blue section or a green section will be spun.

CHAPTER 10, PAGES 276–277

Solve.

1. Jim has 3 shirts in his closet and 5 pairs of pants. He is going to choose one shirt and one pair of pants. How many different outfits can he choose?

2. Ava has 4 skirts and 6 blouses in her closet. She is going to choose one skirt and one blouse. How many different outfits can she choose?

3. There are 3 routes from Milltown to Sawville and 3 routes from Sawville to Elmont. How many routes are there from Milltown to Elmont?

4. There are 4 routes from Ellis to Danton and 5 routes from Danton to Lincoln. How many routes are there from Ellis to Lincoln?

Write *independent* or *dependent* to describe the series of events.

1. Two players' moves in a chess game

2. Tossing a penny five times

3. Drawing numbers in a bingo game and replacing them

4. Dialing a telephone number correctly

5. Tossing a basketball through a basket 10 times

Linda put 10 marbles in a can. There are 4 white ones, 3 blue ones, 2 red ones, and 1 yellow one. Without looking, she will pick marbles and not replace them. Find the probability.

6. *P*(yellow)

7. *P*(white, then white)

8. *P*(blue, then yellow, then red)

9. *P*(red, then white or blue)

10. *P*(white, then blue, then red, then yellow)

CHAPTER 11, PAGES 292–293

Draw and label the figure.

1. $\overleftrightarrow{DC}$ **2.** $\angle ABC$ **3.** $\overline{CD}$ **4.** $\overrightarrow{XY}$ **5.** $\angle MNO$ **6.** plane *XYZ*

Identify the geometric figure.

7. $\angle L$ **8.** $\overline{EF}$ **9.** $\angle CDE$ **10.** $\overleftrightarrow{ST}$ **11.** *Q* **12.** $\overrightarrow{MR}$

Name the ray in three ways.

13.

14.

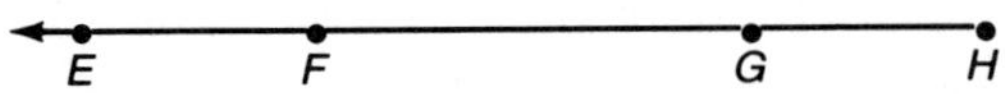

Name the six line segments.

15.

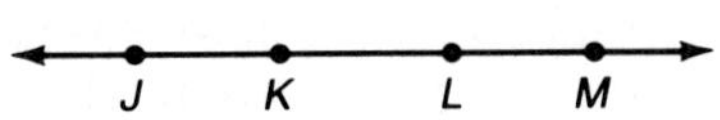

Name the angle in four ways.

16.

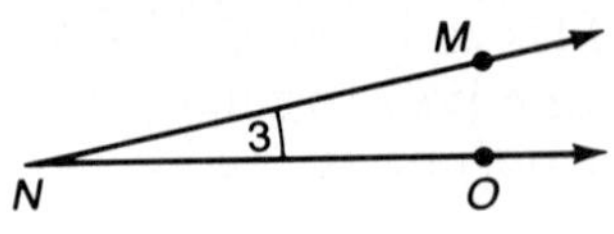

Write *true* or *false*.

17. The length of a ray can be measured.

18. A line segment can be named in at least two ways.

CHAPTER 11, PAGES 296–297

Classify the angle as *acute, right, obtuse,* or *straight.*

1. 30° **2.** 43° **3.** 98° **4.** 90° **5.** 9° **6.** 180°

7. 104° **8.** 25° **9.** 120° **10.** 145° **11.** 12° **12.** 16°

Identify the angle pair as *complementary, supplementary,* or *neither.*

13. 70°, 110° **14.** 45°, 30° **15.** 60°, 30° **16.** 15°, 165°

17. 40°, 60° **18.** 65°, 25° **19.** 72°, 18° **20.** 100°, 75°

Find the complement of the angle.

21. 49° **22.** 70° **23.** 5° **24.** 82° **25.** 45° **26.** 35°

27. 10° **28.** 33° **29.** 28° **30.** 75° **31.** 56° **32.** 15°

Find the supplement of the angle.

33. 160° **34.** 16° **35.** 39° **36.** 8° **37.** 168° **38.** 115°

39. 125° **40.** 6° **41.** 27° **42.** 140° **43.** 89° **44.** 175°

CHAPTER 11, PAGES 298–300

Solve.

1. Use a ruler to draw a 6-cm line segment. Label it $\overline{CD}$. Construct $\overline{EF}$ so that $\overline{EF} \cong \overline{CD}$.

2. Use a protractor to draw a 30° angle. Label it $\angle SRO$. Construct $\angle TUV$ so that $m\angle TUV \cong m\angle SRO$.

CHAPTER 11, PAGES 301–303

Solve.

1. Draw a 6-in. line segment. Label it $\overline{RS}$. Use a compass and straightedge to bisect $\overline{RS}$ at point T. Bisect $\overline{RT}$ and $\overline{TS}$ at points U and V. The length of $\overline{RV}$ is what fractional part of the length of $\overline{RS}$?

Trace the line segment and then construct the perpendicular bisector.

2. 

3. 

4. 

5. 

CHAPTER 11, PAGES 304–305

Identify and name the polygons.

1.
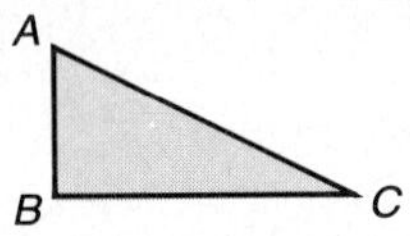

2.
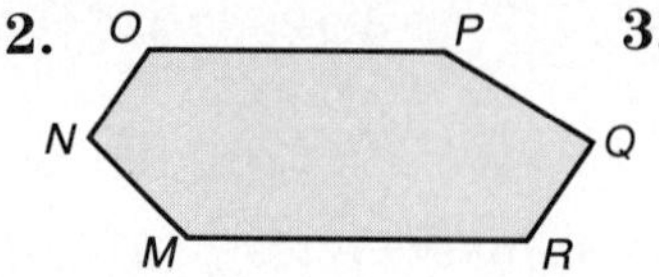

3.
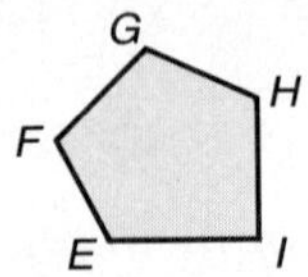

4.
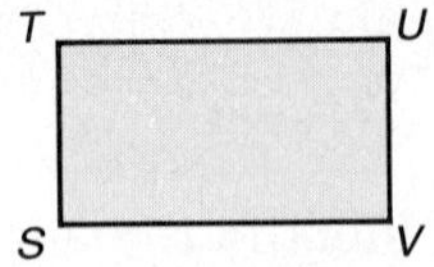

Find the missing length.

5. $r = 2.5$ cm, $d =$ ■ **6.** $r = 2\frac{3}{4}$ in., $d =$ ■ **7.** $r = 1\frac{1}{2}$ ft, $d =$ ■

8. $d = 9$ in., $r =$ ■ **9.** $d = 3.2$ m, $r =$ ■ **10.** $d = 6\frac{1}{4}$ ft, $r =$ ■

11. $d = 6.3$ cm, $r =$ ■ **12.** $d = 7\frac{1}{2}$ in., $r =$ ■ **13.** $d = 10.4$ ft, $r =$ ■

CHAPTER 11, PAGES 312–313

Find each square root.

1. $\sqrt{16}$ **2.** $\sqrt{25}$ **3.** $\sqrt{36}$ **4.** $\sqrt{49}$ **5.** $\sqrt{121}$ **6.** $\sqrt{144}$

7. $\sqrt{17}$ **8.** $\sqrt{20}$ **9.** $\sqrt{42}$ **10.** $\sqrt{53}$ **11.** $\sqrt{74}$ **12.** $\sqrt{96}$

Find each square root. Round to the nearest tenth.

13. $\sqrt{14}$ **14.** $\sqrt{19}$ **15.** $\sqrt{24}$ **16.** $\sqrt{31}$ **17.** $\sqrt{46}$ **18.** $\sqrt{52}$

19. $\sqrt{58}$ **20.** $\sqrt{67}$ **21.** $\sqrt{75}$ **22.** $\sqrt{88}$ **23.** $\sqrt{92}$ **24.** $\sqrt{98}$

Find each square root. Use the table on page 313.

25. $\sqrt{324}$ **26.** $\sqrt{729}$ **27.** $\sqrt{1{,}156}$ **28.** $\sqrt{1{,}764}$ **29.** $\sqrt{2{,}209}$

30. $\sqrt{3{,}136}$ **31.** $\sqrt{4{,}356}$ **32.** $\sqrt{5{,}329}$ **33.** $\sqrt{7{,}569}$ **34.** $\sqrt{8{,}281}$

35. $\sqrt{2{,}601}$ **36.** $\sqrt{5{,}041}$ **37.** $\sqrt{8{,}649}$ **38.** $\sqrt{4{,}624}$ **39.** $\sqrt{5{,}776}$

CHAPTER 11, PAGES 314–315

Determine whether each set of numbers can form a right triangle.

1. $a = 13$, $b = 12$, $c = 5$

2. $a = 6$, $b = 7$, $c = 10$

3. $a = 9$, $b = 12$, $c = 15$

4. $a = 30$, $b = 40$, $c = 50$

5. $a = 5$, $b = 5$, $c = 10$

Find the missing length in the right triangle.

6. $a = 12$, $b = 16$, $c =$ ■

7. $a = 10$, $b = 24$, $c =$ ■

8. $a = 6$, $b =$ ■, $c = 10$

9. $a = 3$, $b = 4$, $c =$ ■

10. $a =$ ■, $b = 21$, $c = 29$

CHAPTER 12, PAGES 330–331

Find the perimeter.

1. rectangle: $l = 4\frac{1}{2}$ ft, $w = 6\frac{3}{8}$ ft

2. square: $s = 41\frac{7}{10}$ ft

3. rectangle: $l = 10$ in., $w = 3\frac{1}{4}$ in.

4. square: $s = 7.85$ m

5. regular pentagon: $s = 223$ yd

6. regular octagon: $s = 14.2$ mm

7. regular hexagon: $s = 5$ m

8. regular octagon: $s = 8.92$ mm

CHAPTER 12, PAGES 332–333

Find the circumference of the circle. Use 3.14 for π with whole numbers and decimals. Use $\frac{22}{7}$ for π with fractions and mixed numbers.

1. $d = \frac{1}{4}$ yd

2. $d = \frac{1}{2}$ mi

3. $d = 12\frac{3}{4}$ ft

4. $d = 7$ km

5. $d = 11.2$ m

6. $d = 24.8$ cm

7. $r = 6\frac{5}{8}$ yd

8. $r = \frac{1}{6}$ mi

9. $r = \frac{14}{15}$ ft

10. $r = 8.5$ cm

11. $r = 6$ m

12. $r = 0.2$ km

CHAPTER 12, PAGES 334–335

Find the area of the rectangle.

1. $l = 11$ ft, $w = 3$ ft

2. $l = 3\frac{4}{5}$ ft, $w = 6$ ft

3. $l = 1\frac{2}{3}$ yd, $w = \frac{1}{3}$ yd

4. $l = 3.6$ mm, $w = 2$ mm

5. $l = 5.1$ m, $w = 1.8$ m

6. $l = 0.9$ m, $w = 4$ m

7. $l = 4.25$ m, $w = 6.4$ m

8. $l = 0.3$ km, $w = 1.1$ km

9. $l = 17.6$ m, $w = 6.3$ m

Find the area of the square.

10. $s = 23$ in.

11. $s = \frac{7}{8}$ in.

12. $s = \frac{2}{5}$ yd

13. $s = 14$ m

14. $s = 0.3$ cm

15. $s = 1.1$ km

16. $s = 0.45$ km

17. $s = 20.8$ m

18. $s = 3.12$ m

Find the area of the parallelogram.

19. $b = 8$ ft, $h = 4$ ft

20. $b = 7\frac{1}{4}$ yd, $h = 5$ yd

21. $b = 9\frac{1}{3}$ m, $h = 2$ m

22. $b = 52$ m, $h = 4.1$ m

23. $b = 0.5$ cm, $h = 3$ cm

24. $b = 6.4$ m, $h = 7$ m

25. $b = 7.25$ m, $h = 3.6$ m

26. $b = 5.83$ m, $h = 12.4$ m

27. $b = 41.2$ m, $h = 7.5$ m

CHAPTER 12, PAGES 336–337

Find the area of the triangle.

1. $b = 10$ in., $h = 7$ in.
2. $b = 12.3$ cm, $h = 5$ cm
3. $b = 1\frac{2}{7}$ m, $h = 4$ m
4. $b = 2\frac{1}{2}$ ft, $h = 4\frac{2}{3}$ ft
5. $b = 8.7$ mm, $h = 4.5$ mm
6. $b = \frac{11}{12}$ ft, $h = \frac{5}{6}$ ft

Find the area of the trapezoid.

7. $b_1 = 6$ ft, $b_2 = 8$ ft, $h = 4$ ft
8. $b_1 = 3$ yd, $b_2 = 15$ yd, $h = 11$ yd
9. $b_1 = 2.7$ m, $b_2 = 17.8$ m, $h = 4.5$ m
10. $b_1 = 2.6$ m, $b_2 = 9.5$ m, $h = 3.2$ m
11. $b_1 = 0.6$ km, $b_2 = 0.8$ km, $h = 1.2$ km
12. $b_1 = 2$ m, $b_2 = 0.4$ m, $h = 3.5$ m

CHAPTER 12, PAGES 338–339

Find the area of the circle. Use 3.14 for π with whole numbers and decimals. Use $\frac{22}{7}$ for π with fractions and mixed numbers.

1. $r = 16$ in.
2. $r = 6$ mi
3. $r = 12$ ft
4. $r = 9$ yd
5. $r = \frac{1}{5}$ ft
6. $r = 4\frac{1}{2}$ yd
7. $r = 0.6$ in.
8. $r = 0.4$ ft

Find the area of the circle. Remember that the measure of the diameter is twice the measure of the radius.

9. $d = 13$ yd
10. $d = 0.2$ m
11. $d = 48$ in.
12. $d = 32$ mm
13. $d = 14$ m
14. $d = 4\frac{1}{2}$ in.
15. $d = 6.8$ cm
16. $d = \frac{7}{12}$ ft

CHAPTER 12, PAGES 340–341

Solve by writing a simpler problem.

1. Barbra Hunter has 187 chickens. Each chicken eats $\frac{1}{4}$ lb of feed each day. Each pound costs $0.20. How much does Barbra spend on feed each day?

2. Chris Hunter has 6 corn fields. He uses 3,000 lb of fertilizer on each one. If fertilizer costs $59.10 per ton, what will his total cost be?

3. Christy raises evergreen trees. Each tree needs 6.4 m by 4.5 m to grow. She wants to plant 115 trees. How many square meters does she need?

4. Christy charges $29.95 for a Scotch pine tree, $19.95 for a Balsam tree, and $39.95 for a Douglas fir tree. She has an order for 24 balsams, 18 pines, and 48 firs. How much will she charge?

CHAPTER 12, PAGES 344–346

Identify the figure.

1. 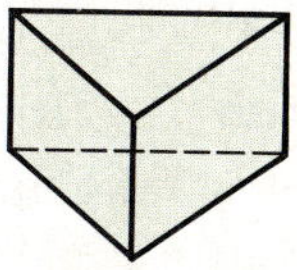

2. 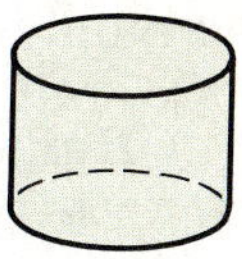

3. 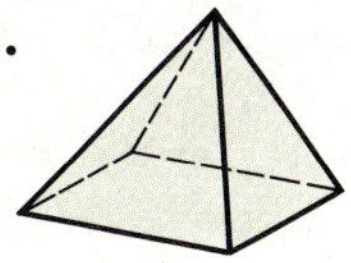

4. 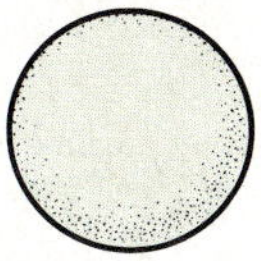

5. 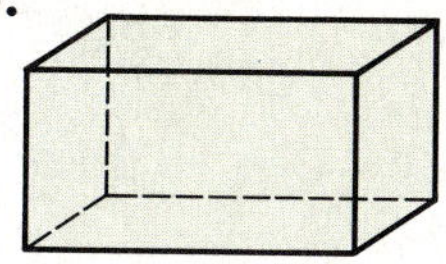

6. 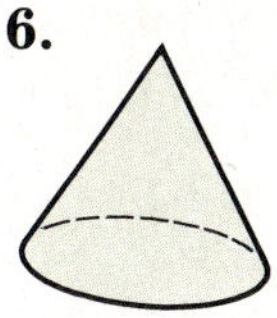

CHAPTER 12, PAGES 347–348

Copy and complete the chart. The set of measurements describes a rectangular prism.

Length	Width	Height	Surface area
1. 6 ft	8 ft	9 ft	■
2. $2\frac{1}{2}$ yd	$6\frac{1}{2}$ yd	3 yd	■
3. 3.2 cm	5.8 cm	7 cm	■
4. 0.9 m	0.9 m	0.9 m	■

CHAPTER 12, PAGES 350–351

Find the volume of the rectangular prism.

	Length	Width	Height		Length	Width	Height
1.	10 ft	4 ft	15 ft	**2.**	8 ft	8 ft	8 ft
3.	2.5 cm	7 cm	5.6 cm	**4.**	11 m	3 m	6 m
5.	0.1 m	0.1 m	0.1 m	**6.**	10 m	6.2 m	0.5 m
7.	$\frac{1}{2}$ ft	$3\frac{1}{2}$ ft	$\frac{3}{4}$ ft	**8.**	$2\frac{1}{3}$ in.	$\frac{1}{2}$ in.	$4\frac{2}{3}$ in.

Find the volume of the rectangular pyramid.

	Length	Width	Height		Length	Width	Height
9.	10 ft	4 ft	15 ft	**10.**	13 yd	4 yd	8 yd
11.	$3\frac{1}{2}$ in.	5 in.	4 in.	**12.**	$\frac{1}{2}$ ft	$\frac{1}{2}$ ft	14 ft
13.	9 m	2 m	5 m	**14.**	7 cm	7 cm	7 cm
15.	6.2 cm	4 cm	5.1 cm	**16.**	0.4 m	1.2 m	5 m

CHAPTER 12, PAGES 352–353

Find the volume of the cylinder. Use 3.14 for π with whole numbers and decimals. Use $\frac{22}{7}$ for π with fractions and mixed numbers.

	Radius	Height		Radius	Height		Radius	Height
1.	2 ft	8 ft	**2.**	1 ft	4 ft	**3.**	6 m	8 m
4.	$\frac{1}{2}$ yd	6 yd	**5.**	2 yd	$5\frac{1}{4}$ yd	**6.**	0.3 m	4 m
7.	3 m	9.1 m	**8.**	9 km	3.2 km	**9.**	8 cm	2.4 cm

Find the volume of the cone. Use 3.14 for π for whole numbers and decimals. Use $\frac{22}{7}$ for π with fractions and mixed numbers.

	Radius	Height		Radius	Height		Radius	Height
10.	$4\frac{3}{4}$ ft	4 ft	**11.**	$\frac{1}{4}$ yd	4 yd	**12.**	22 ft	$2\frac{1}{2}$ ft
13.	6 m	9.2 m	**14.**	8 m	5.4 m	**15.**	0.5 cm	3 cm

CHAPTER 13, PAGES 364–366

Write the opposite of the number.

1. $+18$ **2.** -55 **3.** $+37$ **4.** $+6$ **5.** -21 **6.** $+74$ **7.** -49

Compare the integers. Write $>$ or $<$.

8. $+2$ ● $+6$ **9.** -88 ● $+92$ **10.** -45 ● -76 **11.** -1 ● $+5$ **12.** -9 ● $+3$

13. -4 ● 0 **14.** -7 ● -8 **15.** $+19$ ● -22 **16.** -36 ● $+53$ **17.** -2 ● -3

Write the absolute value.

18. $|-2|$ **19.** $|-8|$ **20.** $|+75|$ **21.** $|-15|$ **22.** $|+134|$

Use $>$, $<$, or $=$ to make a true statement.

23. $|-4|$ ● $|+4|$ **24.** $|+2|$ ● $|-6|$ **25.** $|-7|$ ● $|0|$ **26.** $|+10|$ ● $|+9|$

27. $|-18|$ ● $|+18|$ **28.** $|-31|$ ● $|-31|$ **29.** $|+14|$ ● $|-11|$ **30.** $|0|$ ● $|1|$

CHAPTER 13, PAGES 367–369

Add.

1. $18 + 12$ **2.** $43 + (-7)$ **3.** $-33 + (-35)$ **4.** $-52 + (-82)$

5. $-64 + 64$ **6.** $-19 + 9$ **7.** $22 + (-6)$ **8.** $152 + (-50)$

9. $-70 + 44$ **10.** $-14 + 25$ **11.** $-98 + (-30)$ **12.** $86 + (-11)$

CHAPTER 13, PAGES 370–371

Complete each statement.

1. $-5 - 3 = -5 + \blacksquare$

2. $10 - 6 = 10 + \blacksquare$

3. $-32 - (-47) = -32 + \blacksquare$

4. $22 - (-73) = 22 + \blacksquare$

5. $-64 - (-50) = -64 + \blacksquare$

6. $-86 - (-21) = -86 + \blacksquare$

7. $73 - (-59) = 73 + \blacksquare$

8. $-116 - (-65) = -116 + \blacksquare$

9. $-253 - 304 = -253 + \blacksquare$

10. $-876 - 91 = -876 + \blacksquare$

Subtract.

11. $6 - 3$
12. $4 - 5$
13. $1 - 8$
14. $-7 - 5$
15. $-2 - 2$
16. $-8 - 3$
17. $0 - (-12)$
18. $1 - (-5)$
19. $9 - (-7)$
20. $-7 - (-2)$
21. $-10 - (-3)$
22. $-6 - (-4)$
23. $17 - 30$
24. $-21 - 7$
25. $0 - 29$
26. $-8 - (-31)$
27. $-19 - (-45)$
28. $-44 - 72$
29. $-83 - (-98)$
30. $4 - (-13)$
31. $-23 - 37$
32. $-56 - (-77)$
33. $92 - (-87)$
34. $-36 - 57$

CHAPTER 13, PAGES 372–373

Multiply.

1. 3×9
2. -3×11
3. $-8 \times (-5)$
4. $-46 \times (-61)$
5. $4 \times (-14)$
6. $-35 \times (-5)$
7. 82×6
8. -27×9
9. $15 \times (-100)$
10. $55 \times (-14)$
11. -176×23
12. -16×52
13. $2 \times (-3) \times 6$
14. $-4 \times (-12) \times 1$
15. $-15 \times (-3) \times (-2)$
16. $6 \times 13 \times 5$
17. $-9 \times (-3) \times 2$
18. $10 \times (-1) \times (-7)$
19. $-5 \times 3 \times (-4)$
20. $12 \times 2 \times (-7)$
21. $-8 \times 0 \times -17$

CHAPTER 13, PAGES 374–375

Divide.

1. $7 \div 3$
2. $8 \div 4$
3. $16 \div 4$
4. $20 \div 5$
5. $-9 \div (-3)$
6. $-64 \div (-16)$
7. $-100 \div (-25)$
8. $-46 \div (-2)$
9. $60 \div (-3)$
10. $36 \div (-6)$
11. $63 \div (-7)$
12. $51 \div (-17)$
13. $-90 \div 5$
14. $-44 \div 11$
15. $-81 \div 9$
16. $-75 \div 15$
17. $84 \div (-12)$
18. $-224 \div (-14)$
19. $176 \div 22$
20. $-420 \div (-28)$
21. $-392 \div 7$
22. $-168 \div (-21)$
23. $255 \div (-3)$
24. $-540 \div (-45)$

CHAPTER 13, PAGES 378–379

Find the missing exponent.

1. $62 = 6.2 \times 10^{\blacksquare}$ **2.** $3{,}121 = 3.121 \times 10^{\blacksquare}$

3. $0.005 = 5 \times 10^{\blacksquare}$ **4.** $0.00085 = 8.5 \times 10^{\blacksquare}$

5. $7{,}000{,}000 = 7 \times 10^{\blacksquare}$ **6.** $76{,}400 = 7.64 \times 10^{\blacksquare}$

7. $0.000055 = 5.5 \times 10^{\blacksquare}$ **8.** $0.000000932 = 9.32 \times 10^{\blacksquare}$

Write in scientific notation.

9. 36 **10.** 230 **11.** 8,500 **12.** 0.5

13. 0.62 **14.** 0.019 **15.** 1,500,000 **16.** 0.0000001

Write as a standard numeral.

17. 4.1×10^3 **18.** 5.85×10^8 **19.** 3.71×10^{-7}

CHAPTER 13, PAGES 384–385

Compare. Write >, <, or =.

1. $\frac{3}{8}$ ● $\frac{5}{8}$ **2.** $-\frac{2}{7}$ ● $-\frac{4}{7}$ **3.** $-\frac{5}{20}$ ● $-\frac{1}{4}$ **4.** $-5\frac{4}{8}$ ● $-5\frac{1}{2}$

5. -0.8 ● -0.80 **6.** -0.27 ● 0.27 **7.** -83 ● -0.349 **8.** 6.28 ● -6.8

Write the absolute value.

9. $\left|\frac{4}{5}\right|$ **10.** $\left|\frac{1}{15}\right|$ **11.** $\left|-2\frac{7}{8}\right|$ **12.** $|-0.6|$ **13.** $|1.6|$

CHAPTER 13, PAGES 386–387

Add or subtract.

1. $\frac{1}{6} + \left(-\frac{7}{12}\right)$ **2.** $-2\frac{1}{3} + \frac{8}{9}$ **3.** $-0.80 + \frac{4}{5}$ **4.** $-7\frac{1}{8} - \left(1\frac{1}{4}\right)$

5. $-48 + 23.14$ **6.** $-83 - 13.07$ **7.** $10.2 + (-0.08)$ **8.** $5.89 + (-14.37)$

9. $4 - \left(-1\frac{2}{3}\right)$ **10.** $-17.04 + (-0.258)$ **11.** $-60.05 - 23.47$ **12.** $-2\frac{1}{8} + \frac{3}{4}$

Multiply or divide.

13. $\frac{2}{3} \times \frac{5}{6}$ **14.** $6\frac{1}{8} \times 3$ **15.** $-1.2 \div \frac{3}{5}$ **16.** $60 \times \left(-\frac{1}{4}\right)$

17. $-70 \times (-35)$ **18.** $24 \div (-8)$ **19.** $1{,}000 \div 300$ **20.** $-16 \times (-0.02)$

21. $0.2 \times (-3.7)$ **22.** $-2\frac{1}{5} \div \left(-\frac{1}{3}\right)$ **23.** $1.96 \div (-0.04)$ **24.** $-\frac{3}{8} \times \frac{2}{5}$

CHAPTER 13, PAGES 388–389

Solve each problem by using one or more problem-solving strategies.

1. The Special Events Office advertises horseback rides through the park for 15 people at a time. They must have exactly 15 people in order to form a ride. If 72 people sign up, how many rides can they form? How many people will not ride?

2. The Parks Department rents out rowboats for $12 per hour. Ten boats are available between 10 A.M. and 5 P.M. Two people must ride in each boat. 140 people want to rent boats the first day. How much money does the Parks Department earn in boat rentals that day?

3. Fly-Hi Airlines sold 1,024 tickets for flights from Los Angeles to New York. Each plane holds 320 passengers. How many planes can Fly-Hi fill? How many passengers are on the unfilled flight?

4. Marion needs to make 2 sawhorses to hold up her table top. It takes 5 pieces of lumber each 95 cm long to make one sawhorse. How much lumber will be left over if she buys ten 1-m lengths?

5. Calvin works in a bottling plant. His boss has told him to increase his output to 1,500 bottles a day. Calvin bottles 525 bottles the first day, 600 bottles the second day, 675 bottles the third day, and so on. At this rate, how long will it take Calvin to reach his goal of 1,500 bottles per day?

6. An airport was built in the shape of a square. It measures 4 km on each side. A fuel truck entered the airport at the center of the north side and traveled 1.5 km south. It then traveled 1.5 km west, then 0.5 km south. At that point how far was the truck from the western border of the airport?

CHAPTER 14, PAGES 398–399

Write the operation and the number(s) you would compute first to simplify.

1. $2 \cdot 4 + 3$ **2.** $5(7 + 8)$ **3.** $2 + 1 \cdot 4$ **4.** $(9 + 6)7$

5. $11 - 2 - 8$ **6.** $(12 - 5) - 3$ **7.** $8 - 5 + 4$ **8.** $13 - 1 \cdot 7$

9. $11 - 7 + 3$ **10.** $15 \div 3 + 6$ **11.** $15 + 8^2$ **12.** $(10 - 2)^2$

Simplify.

13. $(5 - 2) \times 4$ **14.** $(7 - 3) - 1$ **15.** $10 \div (9 - 4)$ **16.** $8 - 2 + 6$

17. $22 - 4^2 + 1$ **18.** $40 + 9 - 3$ **19.** $4^2 - (15 - 5)$ **20.** $(2 + 6) \div 4$

21. $\frac{10 - 3 + 5}{6}$ **22.** $4 + (16 - 8)^2$ **23.** $\frac{(2^3 + 4)7}{4}$ **24.** $\frac{96 - (2 + 4)^2}{15}$

Write parentheses to make each sentence true.

25. $32 \div 8 - 5 - 4 = -5$ **26.** $15 + 5 \div 4 + 1 = 4$

27. $144 \div 12 - 3 \cdot 2 = 18$ **28.** $12 + 96 \div 6 - 4 = 54$

CHAPTER 14, PAGES 400–401

Write an algebraic expression for the word phrase. Let n represent the number.

1. The quotient of a number divided by 70

2. 5 added to a number

3. 83 decreased by twice a number

4. 44 minus a number

5. The product of a number and 15

6. A number decreased by 20

7. The ratio of a number to 34

8. A number divided by 18

9. The product of 35 and the difference of a number minus 12

10. The quotient of a number plus 8, divided by 21

CHAPTER 14, PAGES 402–404

Evaluate for $a = -3$ and $b = 4$.

1. $b - a$

2. $5 + a - b$

3. $4b + a$

4. $2(a - b)$

5. $6a - b$

6. $4 + (a - b)$

7. $a - 2b$

8. $3b \div a$

Evaluate for $x = 2$, $y = -2$, and $z = 3$.

9. $(x - y)z$

10. $\frac{8(z - y)}{4x}$

11. $\frac{y + z}{x}$

12. $\frac{xz}{y}$

13. $\frac{2(y - z)}{x}$

The formula for the area of a triangle is $A = \frac{1}{2}bh$, where b is the length of the base and h is the height. Find the area.

14. $b = 12$ m; $h = 2$ m

15. $b = 3.2$ ft; $h = 6.4$ ft

16. $b = 5$ m; $h = 8.7$ m

17. $b = 4.5$ cm; $h = 2.3$ cm

Some salespeople are paid a commission based on their total amount of sales. The formula for commission is $C = R \times A$, where C is the commission, R is the rate, and A is the amount of sales. Find the commission.

18. $R = 5\%$; $A = \$30,000$

19. $R = 7\%$; $A = \$50,000$

20. $R = 2\%$; $A = \$10,000$

CHAPTER 14, PAGES 408–409

Match the graph to the correct number sentence.

1.

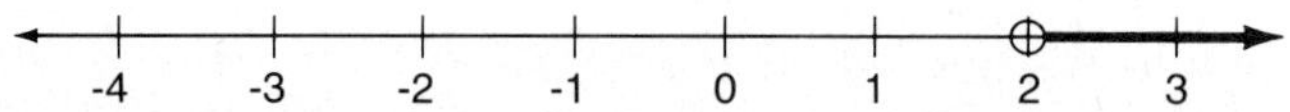

2.

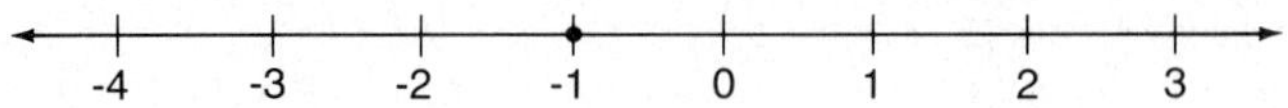

3.

4.

5.

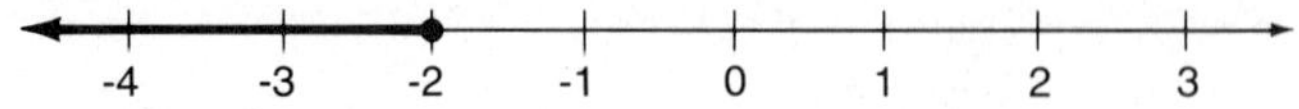

a. $x \leq -2$

b. $x = -1$

c. $x > 2$

d. $x < 0$

e. $x = 3$

CHAPTER 14, PAGES 412–413

Solve and check.

1. $x + 3 = 10$ **2.** $x + 8 = 14$ **3.** $x + 5 = 20$ **4.** $x + 4 = 17$

5. $x + 6 = 33$ **6.** $9 + x = 18$ **7.** $42 + x = 56$ **8.** $x + 37 = 62$

9. $x + 55 = 73$ **10.** $x + 47 = 98$ **11.** $x + 66 = 89$ **12.** $54 + x = 75$

13. $76 + x = 79$ **14.** $32 + x = 69$ **15.** $5 + x = 99$ **16.** $9 + x = 123$

17. $-4 + x = 8$ **18.** $-7 + x = 9$ **19.** $-5 + x = 13$ **20.** $-6 + x = 14$

21. $x - 3 = 16$ **22.** $x - 5 = 18$ **23.** $x - 33 = 13$ **24.** $x - 74 = 38$

25. $2 + x + 5 = 15$ **26.** $8 + x + 9 = 22$ **27.** $11 + x + 7 = 277$

CHAPTER 14, PAGES 414–415

Solve.

1. $4x = 16$ **2.** $3x = 30$ **3.** $5x = 10$ **4.** $8x = 32$ **5.** $2x = 18$

6. $-3x = 48$ **7.** $9x = -27$ **8.** $-7x = 56$ **9.** $4x = 64$ **10.** $-2x = 86$

11. $-\frac{y}{2} = -22$ **12.** $\frac{x}{6} = 16$ **13.** $\frac{x}{2} = -38$ **14.** $\frac{x}{5} = -13$ **15.** $-\frac{x}{4} = 20$

16. $0.01x = 2$ **17.** $0.5x = 7$ **18.** $0.02x = 16$ **19.** $0.25x = 11$ **20.** $0.9x = 81$

CHAPTER 14, PAGES 416–417

Solve and check.

1. $3x + 2 = 11$ **2.** $2x + 9 = 19$ **3.** $4x + 3 = 27$ **4.** $8x + 1 = 33$

5. $2x - 4 = 14$ **6.** $8x - 3 = 37$ **7.** $5x - 6 = 29$ **8.** $7x - 4 = 10$

9. $2x - 4 = 196$ **10.** $-3x - 13 = -64$ **11.** $-6x - 12 = -102$ **12.** $4x + 1 = 9$

13. $\frac{x}{12} - 5 = 2$ **14.** $\frac{x}{7} + 11 = 6$ **15.** $\frac{x}{4} - 17 = 3$ **16.** $\frac{x}{9} + 7 = 18$

CHAPTER 14, PAGES 418–420

Find the solution to the equation when $x = 3$, then when $y = -2$.

1. $x + y = 14$ **2.** $x - y = 1$ **3.** $xy = 18$ **4.** $4x = y$

5. $6 - 2x = y$ **6.** $4x + 1 = y$ **7.** $2x - y = 4$ **8.** $-3y + x = 6$

Determine if the ordered pair is a solution of the equation. Write *yes* or *no*.

9. $x + y = 6\ (7, -1)$ **10.** $8 - x = y\ (6, 2)$ **11.** $\frac{3x}{y} = 10\ (10, 6)$

12. $4x - 2y = 0\ (2, 4)$ **13.** $\frac{x}{2y} = 6\ (-36, -3)$ **14.** $-3x + 2y = 7\ (-1, -2)$

GLOSSARY

absolute value The distance a number is from zero on the number line. [365]

acute angle An angle that measures more than 0° but less than 90°. [296]

acute triangle A triangle that has three acute angles. [306]

addends The numbers being added. In $3 + 6 = 9$, 3 and 6 are the addends. [30]

algebraic expression A mathematical phrase that uses symbols and variables to stand for one or more unknown numbers. [400]

angle Two rays with the same endpoint. [292]

angle bisector A ray that divides an angle into two parts. [299]

area The number of square units needed to cover a surface of a plane figure. [334]

bar graph A graph used to compare quantities by using horizontal or vertical bars. [106]

bisect To divide into two equal parts. [299]

broken-line graph A graph using points connected by line segments to show change over time. [108]

central angle An angle whose vertex is at the center of the circle. [304]

chord A line segment that has endpoints on the circle. [304]

circle The set of all points in a plane that are the same distance from the center. [304]

circle graph A graph using a divided circle to represent the parts of the whole. [317]

circumference The distance around a circle. [332]

complementary angles Two angles where the sum of their measures is 90°. [296]

compound interest Interest computed on the principal and on the interest previously earned. [261]

cone A three-dimensional figure with one circular base and a curved lateral surface that comes to a point. [345]

congruent Geometric figures that have the same shape and size are congruent. [298]

coordinates The ordered pair that corresponds to a point located on a grid. [380]

cube A rectangular prism with six congruent square faces. [344]

cylinder A three-dimensional figure with two congruent circular bases in parallel planes and a curved lateral surface. [345]

decimal A base 10 numeral that uses place value and contains a decimal point. [6]

degree A standard unit of measure for angles. [294]

denominator The bottom number of a fraction that identifies the total number of equal parts. In $\frac{1}{2}$, 2 is the denominator. [122]

dependent events When the outcome of the first event affects the outcome of the second event. [278]

diameter A chord that passes through the center of a circle. [304]

difference The result when subtracting numbers. In $5 - 3 = 2$, the difference is 2. [28]

digits The 10 symbols used to form numbers. 0, 1, 2, 3, 4, 5, 6, 7, 8, and 9 are digits. [4]

discount The amount that is subtracted from the list price of an item. [256]

dividend The number to be divided. In $40 \div 8 = 5$, the dividend is 40. [46]

divisible When one number divides another without a remainder. 20 is divisible by 4. [58]

divisor The number that is used to divide another number. In $20 \div 4 = 5$, the divisor is 4. [46]

edge A line segment in a polyhedron formed where two faces intersect. [344]

equation A statement of equality between two quantities using the "=" sign. [406]

equiangular triangle A triangle whose angles are of equal measure. [306]

equilateral triangle A triangle whose sides are of equal length. [306]

equivalent Having the same value. $2\frac{1}{2}$ is equivalent to $\frac{5}{2}$. [126]

estimate To approximate a value. [28]

exponent A number that tells how many times the base is used as a factor. [18]

face A side of a polyhedron. [344]

factors Numbers to be multiplied. In $4 \times 3 = 12$, 4 and 3 are factors of 12. [40]

formula An equation that expresses a relationship between two or more quantities involving more than one variable. [402]

fraction A part of a whole, or the expressed division of two whole numbers. [122]

greatest common factor (GCF) The greatest number which is a common factor of two or more numbers. [124]

hexagon A six-sided polygon. [304]

histogram A bar graph in which the bars represent the frequency of occurrence. The bars are placed next to each other to show changes in data in different intervals (periods of time). [92]

hypotenuse In the right triangle, the longest side which is also opposite the right angle. [314]

improper fraction A fraction with a numerator equal to or greater than the denominator. $\frac{22}{5}$ is an improper fraction. [132]

independent events When the outcome of the first event does not affect the outcome of a second event. [278]

inequality When two quantities do not have the same value. The symbols $>$, $<$, $\geq$, and $\leq$ designate this condition. [406]

integer A positive or negative whole number or zero. [364]

interest A charge for the use of money. Usually a percent of the amount borrowed, invested, or loaned. [260]

inverse operation Operations that "undo" each other. Addition and subtraction are inverse operations since $5 + 2 = 7$ and $7 - 2 = 5$. Similarly, multiplication and division are inverse operations since $4 \times 2 = 8$ and $8 \div 2 = 4$. [412, 414]

isosceles triangle A triangle with two sides of equal length. [306]

key On a map or graph shows the symbol, its meaning, and the amount it represents. [112]

least common denominator (LCD) The least number divisible by a set of denominators. The LCD of $\frac{1}{2}$, $\frac{1}{4}$, and $\frac{1}{5}$ is 20. [158]

least common multiple (LCM) number evenly divisible by two or more factors. [124]

legs In a right triangle, they are the two sides which are not the hypotenuse. [314]

line A set of points that is a straight path extending in opposite directions without end. [292]

line segment Part of a line, consisting of two endpoints and all the points in between. [292]

mean Statistical measure of the average of a set of data, found by dividing the sum of the items by the number of items. [98]

median Statistical measure of the middle value when a set of numbers is listed in order. [100]

mixed number A whole number and a fraction. [132]

mode Statistical measure of the value that occurs most often in a set of data. [100]

multiple The product of two factors. 6 is a multiple of 3 and 2. [124]

negative integers Numbers to the left of zero on the number line; the opposites of the positive integers. [364]

numerator The top part of the fraction that tells how many parts are being counted. In $\frac{3}{5}$, the numerator is 3. [122]

obtuse angle An angle that measures more than 90° but less than 180°. [296]

obtuse triangle A triangle that contains one obtuse angle. [306]

octagon An eight-sided polygon. [304]

odds A ratio describing the likelihood of an event; the ratio of favorable outcomes to unfavorable outcomes. [269]

opposites Any pair of integers whose sum is zero. [364]

order of operations To simplify an expression using the proper sequence of operations. First raise to a power, then multiply or divide from left to right. Finally add or subtract from left to right. [398]

ordered pair Two numbers that name a specific point in a coordinate plane. [380]

origin The point assigned to zero on the number line or the point where the horizontal and the vertical axes meet. [380]

outcome A result of an event. [268]

parallel lines Two lines in a plane that do not intersect. [301]

parallelogram A quadrilateral with opposite sides parallel. [318]

pentagon A five-sided polygon. [304]

percent Per hundred. The ratio of a given number to 100; "hundredths" or "out of 100." [234]

percent of change The amount of change divided by the original amount and renamed as a percent. [258]

perimeter The distance around a polygon. [330]

perpendicular bisector A line perpendicular to a line segment that divides the segment into two congruent parts. [302]

perpendicular lines Lines that intersect and form a right angle. [301]

pi (π) The number that is the ratio of the circumference of a circle to its diameter. It is approximately equal to $\frac{22}{7}$ or 3.14. [332]

pictograph A graph in which a symbol is used to represent a definite number. [112]

plane A set of points forming a flat surface that extends in all directions without end. [292]

point A location in space with no size. [292]

polygon A simple closed curve consisting entirely of line segments. [304]

polyhedron A three-dimensional figure with faces as polygonal regions, edges as line segments, and vertices as points. [344]

positive integers Numbers to the right of zero on the number line; the opposites of the negative integers. [364]

power of 10 The number 10 or any other number that can be formed by using 10 as a factor two or more times. [36]

prism A three-dimensional figure that has two parallel bases of identical size and shape and sides that are parallelograms. [344]

probability The chance or likelihood that an event will occur. It is expressed as a fraction between 0 and 1, including 0 and 1; the ratio of favorable outcomes to the total number of possible outcomes. [268]

product The result of multiplying two numbers. In $2 \times 3 = 6$, the product is 6. [38]

proportion Two equivalent ratios. [216]

pyramid A three-dimensional figure whose base is any polygon, and which has triangular lateral faces that meet at a common vertex. [344]

Pythagorean theorem The relationship among the legs of a right triangle where the square of the length of the hypotenuse is equal to the sum of the squares of the lengths of the legs ($a^2 + b^2 = c^2$). [314]

quadrilateral A four-sided polygon. [304]

quotient The result obtained by dividing one number by another. In $42 \div 7 = 6$, the quotient is 6. [38]

radius A line segment that goes from the center of the circle to a point on it. [304]

range A measure of spread or dispersion, found by taking the difference between the highest and lowest numbers in a given group of numbers. [100]

rate A ratio that compares different units. [212]

ratio A comparison of two like quantities by division. The ratio of a to b is written $\frac{a}{b}$ or a:b. [210]

rational number Any number that can be expressed as the ratio of an integer to a nonzero integer. 9, 0, $3\frac{1}{4}$, $-4.\overline{7}$ are all rational numbers. [384]

ray A part of a line with one endpoint that extends indefinitely in one direction. [292]

reciprocals Two numbers with products of 1. $\frac{1}{2}$ and 2 are reciprocals since $\frac{1}{2} \times 2 = 1$. [150]

rectangle A parallelogram with all right angles. [318]

regular polygon A polygon with sides of equal length and with angles of equal measure. [304]

remainder The amount left over in a division problem. In $23 \div 5 = 4$ R3, the remainder is 3. [46]

repeating decimal A decimal that repeats a digit *or* a group of digits without end. $0.\overline{3}$ and $7.\overline{142857}$ are repeating decimals. [136]

rhombus A parallelogram with sides of equal length. [318]

right angle An angle which measures exactly 90°. [296]

right triangle A triangle with a right angle. [306]

sample space A listing of all the possible outcomes in a given situation. [276]

scale drawing A drawing of an object drawn in proportion to the actual object it represents. [224]

scalene triangle A triangle with no two sides of equal length. [306]

scientific notation A number between 1 and 10 and the second factor is a power of 10. For example, $50{,}000{,}000 = 5 \times 10^7$ and $0.0000045 = 4.5 \times 10^{-6}$. [378]

similar figures Figures that have the same shape but may not have the same size. Their corresponding angles are congruent, and the ratios of the lengths of the corresponding sides are equal. [320]

solid figure A three-dimensional figure having length, width, and height. [344]

sphere A three-dimensional figure with a curved surface of points that are the same distance from the center. [345]

square A rectangle with sides of equal length. [318]

square root A factor which multiplied by itself yields a product. In $8^2 = 64$, the square root of 64 is 8. [312]

straight angle An angle that measures exactly 180°. [296]

sum The result of adding 2 or more numbers. In $2 + 3 = 5$, the sum is 5. [28]

supplementary angles Two angles where the sum of their measures is 180°. [296]

surface area The total area of the surface of a solid figure. [347]

transversal In a plane, a line that intersects two or more lines at a different point. [326]

trapezoid A quadrilateral with only one pair of parallel sides. [318]

triangle A three-sided polygon. [304]

tree diagram A drawing that shows all possible outcomes using a "branch" to represent each choice or possible outcome. [276]

variable A letter or symbol used to represent a number. [400]

vertex of an angle The common endpoint of the two rays of an angle. [292]

vertices of a polygon The points formed by the intersection of the sides. [304]

vertices of a polyhedron The points where the edges intersect. [344]

volume The amount of space inside a three-dimensional figure, measured in cubic units. [350]

whole number A number such as 0, 1, 2, ... Each successive number is formed by adding 1 to the previous number. [4]

SELECTED ANSWERS

CHAPTER 1

Page 3 **1.** How much? **2.** measure **21.** code **22.** location

Page 5 **1.** tens; 40 **2.** hundreds; 600 **21.** one hundred seventy-five **22.** three hundred fifty-eight **37.** 336,154 **38.** 893,201

Page 7 **1.** tenths; 0.1 **2.** hundredths; 0.06 **21.** five-tenths **22.** one and eight-tenths **37.** 4.27 **38.** 13.63 **53.** two thousand one hundred five and seventy-five hundredths

Page 9 **1.** $<$ **9.** $<$ **29.** b **33.** 9,742; 7,492; 4,279; 2,972 **37.** 36,081; 36,180; 38,016; 306,810 **40.** Paris and Istanbul

Page 11 **1.** 5,927 **5.** 2,749 **9.** 23,416 **10.** 23,436

Page 12 **1.** \$0.56 **7.** \$2.74

Page 15 **1.** a **2.** b

Page 17 **1.** 2,800; 3,000 **2.** 4,000; 4,000 **17.** 0; 0.02 **18.** 0.1; 0.08 **33.** \$5.00 **34.** \$40.00 **41.** \$0.86 **42.** \$1.90 **49.** 4,000 **50.** 49.5 **57.** \$0.36

Page 19 **1.** 2^2 **2.** 7^3 **13.** 4 **14.** 1 **20.** 10^2 **24.** 1 **25.** 1,000

Page 21 **1.** \$11.50 **4.** 3 items **10.** \$11.00

CHAPTER 2

Page 29 **1.** 1,100 **21.** 200 **31.** 1,100 **51.** 200

Page 31 **1.** b **5.** 110 **6.** 51 **40.** 710,700; 744,670 **44.** R: 8,000 mi; FE: 5,000 mi

Page 33 **1.** d **5.** 50 **6.** 14 **30.** 200; 190 **31.** 5,000; 5,172

Page 34 **39.** 700,000 **40.** 8,000,000 **45.** $>$ **46.** $<$ **51.** 15,524 cars

Page 35 **1.** 2 dimes, 1 penny = 21¢ **2.** 3 quarters, 1 nickel, 3 pennies = 83¢

Page 37 **1.** 200 **2.** 4,200 **21.** 80 **22.** 20

Page 39 **1.** 350 **21.** 40 **41.** 40 **47.** 70 **53.** 8,000 cans

Page 41 **1.** b **5.** 78 **30.** 2,241 **45.** 285 **49.** 1,924 **57.** 10,000; 11,540 **59.** 2,400; 2,166

Page 43 **1.** 96,228 **21.** 347,985 **41.** 193,824 **65.** \$65,403

Page 45 **1.** 1,840 **3.** 108

Page 47 **1.** 137 R2 **11.** 105 R2 **56.** 62 **57.** 224

Page 49 **1.** 8 R12 **6.** 6 R55 **46.** 10; 9 **47.** 30; 31

Page 51 **1.** 38 **13.** 20 R8 **41.** 1,062 **42.** 1,112 **49.** 95 boxes

Page 53 **1.** 1,488 cal **2.** 248 cal

Page 55 **1.** interested **2.** somewhat interested **10.** Band **11.** Dance Club, Future Teachers' Club, Science Club

CHAPTER 3

Page 63 **1.** 76.8 **6.** 91.583 **27.** 880; 873.704 **28.** 810; 810.94 **37.** 107.45 **41.** Snow

Page 65 **1.** 2.3 **6.** 11.629 **24.** 4; 3.57 **25.** 0; 0.0241 **29.** 7.847 **30.** 1.751 **35.** 1.99 seconds

Page 67 **1.** $280.00 **5.** $144.69

Page 69 **1.** 504 **2.** 72 **21.** 0.55 **22.** 0.9 **41.** 600.2 **42.** 0.006002 **53.** 17,000 **59.** $23.00

Page 71 **1.** b **2.** c

Page 73 **1.** $425.85 **2.** $370.60

Page 75 **1.** 44.24 **2.** 33.28 **31.** 480; 475.2 **32.** 1.4; 1.251 **44.** 2.002 **47.** $302.95 billion

Page 76 **1.** $72.00 **5.** $109.20

Page 77 **1.** 1.192 **3.** 0.139 L

Page 79 **1.** 6.83 **5.** 1.6 **37.** 6; 6.2 **38.** 1; 0.91

Page 81 **1.** 10 **2.** 100 **9.** 2 **10.** 12.008

Page 82 **39.** 7,000; 7,420 **40.** 40; 40.95 **59.** 11,458 **60.** 15,022 **75.** 38.7 weeks

Page 83 **1.** 26 **2.** 17.5 **9.** 13.5 mpg

Page 85 **1. moderate 2. expensive** **13.** Discount **14.** Exclusive

CHAPTER 4

Page 93 **1.** 22 **2.** 31

Page 94 **6.** 2–3 **13.** 36.4 **14.** 12.22

Page 95 **1.** $12,009

Page 97 **1.** eggs **2.** milk **9.** 30 **10.** 35

Page 99 **1.** 7 **2.** 26 **13.** 118.88 **14.** 22.376

Page 101 **1.** 9 **2.** 6 **3.** 6 **4.** 6 **29.** 36 mph

Page 103 **1.** a **2.** b

Page 105 **1.** 26–50 **2.** over 50

Page 107 **1.**

Page 109 **1.**

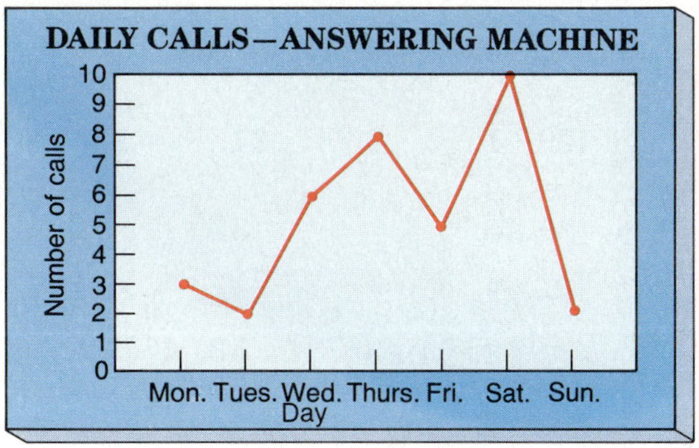

Page 111 **1.** $74.00 **2.** $41.00 **7.** $109.30 **9.** $100

Page 113 **1.** 10,000 **2.** green
7.

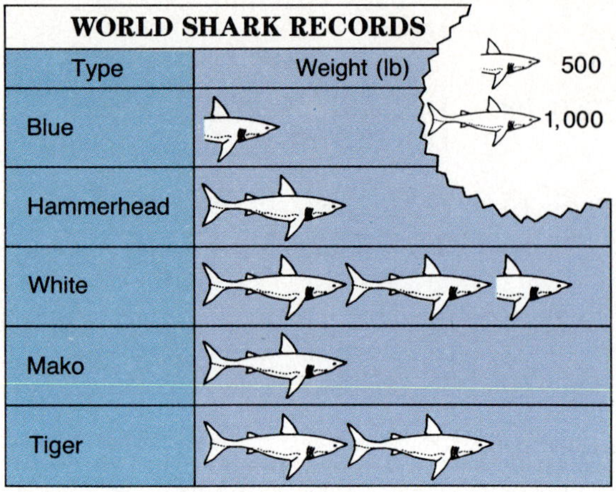

Page 115 **1.** \$325 **2.** \$374 **7.** \$3,718
11. \$2,992

CHAPTER 5

Page 123 **1.** $\frac{5}{6}$ **2.** $\frac{1}{4}$ **13.** $\frac{1}{4}$ **14.** $\frac{5}{12}$
19. $\frac{5}{6}$

Page 125 **1.** 2 **2.** 4 **11.** 2 **12.** 3
31. 6 **32.** 4 **41.** 12 **42.** 15

Page 127 **1.** $\frac{1}{3}$ **2.** $\frac{1}{2}$ **33.** 3 **34.** 6
53. $=$ **54.** $\neq$ **63.** $\frac{2}{8}, \frac{1}{4}$

Page 129 **1.** $<$ **2.** $<$ **21.** $<$ **22.** $<$

Page 130 **36.** $\frac{7}{8}, \frac{5}{8}, \frac{3}{8}$ **37.** $\frac{8}{10}, \frac{7}{10}, \frac{4}{10}$
44. $\frac{0}{6}, \frac{1}{6}, \frac{5}{6}$ **45.** $\frac{9}{24}, \frac{11}{24}, \frac{17}{24}$
52. 11,149 **53.** 15,039
68. blue paint **69.** No

Page 133 **1.** $\frac{7}{3}$ **2.** $\frac{5}{3}$ **33.** $1\frac{1}{2}$ **34.** $1\frac{1}{3}$
65. 14 **69.** 7

Page 134 **1.** b **2.** c

Page 137 **1.** $\frac{3}{10}$ **6.** $2\frac{7}{10}$ **25.** 0.5
33. 1.6 **49.** 0.33 **50.** 0.83
57. $0.\overline{3}$ **58.** $0.\overline{7}$

Page 138 **1.** $\frac{2}{5}$ **5.** transportation
10. housing

Page 139 **1.** 4 **2.** 5
7.

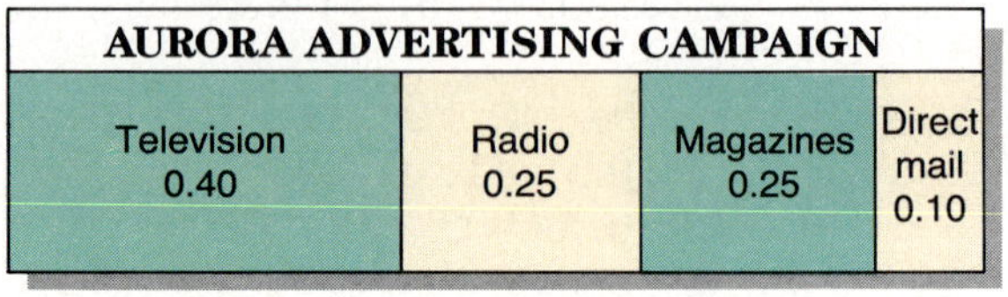

CHAPTER 6

Page 147 **1.** $\frac{1}{6}$ **13.** $\frac{3}{5}$ **37.** $\frac{1}{15}$ **38.** $\frac{1}{4}$
45. 415 **50.** 19

Page 149 **1.** 1 **11.** 3 **26.** 3
58. 27 mph

Page 151 **1.** $\frac{1}{3}$ **2.** $\frac{1}{7}$ **25.** $\frac{8}{9}$ **43.** $\frac{1}{4}$
55. 6 stakes

Page 153 **1.** 3 **6.** $\frac{1}{2}$ **11.** $\frac{4}{9}$ **21.** $\frac{3}{4}$

Page 155 **1.** a. 3; b. 27 **2.** a. \$55; b. \$85

Page 157 **1.** 12 **3.** $1\frac{1}{4}$ ft **7.** $\frac{3}{4}$ yd

Page 159 **1.** $\frac{3}{5}$ **11.** $\frac{9}{10}$ **16.** $1\frac{5}{12}$ **41.** $1\frac{1}{4}$
47. $\frac{11}{24}$

Page 161 **1.** $8\frac{1}{2}$ **6.** $7\frac{5}{8}$ **36.** $14\frac{3}{4}$
41. $5\frac{9}{10}$ mi

Page 163 **1.** $\frac{1}{3}$ **11.** $\frac{1}{4}$ **31.** $\frac{2}{3}$ **41.** 207
42. 7,480 **51.** 5 minutes

Page 165 **1.** $2\frac{1}{3}$ **6.** $3\frac{5}{8}$ **16.** $4\frac{11}{12}$ **26.** $3\frac{1}{4}$
40. 10.91 **41.** 48.374

Page 166 **49.** $2\frac{3}{8}$ lb **50.** $\frac{7}{8}$ lb

Page 167 **1.** \$2.55 **5.** \$2.25

Page 169 **1.** 5 **11.** 8 **16.** 1 **21.** 10 **26.** 3 **41.** 13 ft

Page 171 **1.** \$523.20; \$43.20 **9.** \$1,046.40; \$86.40

Page 173 **1.** about \$3,150 **2.** \$3,775 **9.** Ace; ourselves **10.** no

CHAPTER 7

Page 181 **1.** 24 **10.** 3 **28.** $\frac{3}{4}$ in. **29.** $2\frac{3}{16}$ in. **33.** Randolph

Page 183 **1.** 32 **10.** 5 **43.** $1\frac{1}{2}$ T

Page 185 **1.** 7 yd 2 ft **9.** 2 ft 1 in. **29.** 2 mi 1,110 yd

Page 187 **1.** 17 minutes **9.** 10:30 P.M. **15.** 3 hours 33 minutes **17.** 5:54 A.M. **21.** 12:43 P.M.

Page 188 **1.** 2 P.M.

Page 189 **7.** 8:30 A.M. **11.** no

Page 191 **1.** \$210.70 **2.** \$239.20 **9.** \$235.20

Page 193 **1.** b **2.** b **11.** 27 mm **12.** 66 mm **20.** centimeters

Page 195 **1.** b **7.** c

Page 196 **23.** $\frac{1}{6}$ **31.** $7\frac{7}{8}$ **39.** 45 kL is too much

Page 197 **1.** 9.5 cm **5.** 2 cm; 0.5 cm

Page 199 **1.** 2,000 **4.** 4 **37.** 32 **38.** 20 **39.** 17 **40.** 12 **53.** 74.4 kL

Page 201 **1.** 12°C **2.** 78°C **9.** c **10.** b **21.** no

Page 202 **1.** 67 cm **2.** 5

CHAPTER 8

Page 211 **1.** $\frac{6}{1}$ **2.** 6:1 **9.** 15 **10.** 16 **24.** 9 **25.** 18 **32.** 12 **33.** 4 **40.** $\frac{1}{3}$ **41.** $\frac{1}{6}$ **56.** $\frac{3}{1}$

Page 213 **1.** $\frac{\$10}{3}$ **2.** $\frac{7}{9}$ **19.** = **20.** > **28.** 8 pencils per pack **29.** 28 people per bus **38.** 3,280 **42.** 9.86 **46.** \$1.13

Page 215 **1.** \$0.11 per ounce **2.** \$0.065 per foot **7.** \$0.156 per ounce **8.** \$0.672 per can **12.** 3 lb for \$1.89 **13.** 6-oz jar for \$1.53

Page 217 **1.** no **2.** yes **9.** 15 **10.** 49 **25.** 0.25 **26.** 0.9 **31.** 14 cups

Page 219 **1.** b **2.** yes

Page 221 **1.** 290 mi **2.** 260 mi **7.** 5 **8.** 12.5 **13.** 378 mi

Page 223 **1.** 12.5 **7.** 9 in.

Page 225 **1.** 8 m **2.** 6 m **17.** 24 ft **18.** 18 ft **22.** 0.7 cm

Page 227 **1.** 5.0 cm **2.** 111 km **9.** 100 km **13.** 53 km **17.** 238 km

CHAPTER 9

Page 235 **1.** 0.28 **11.** 1.3 **16.** 0.0035 **26.** 0.0025 **31.** 0.035 **36.** 23% **46.** 370% **51.** $66\frac{2}{3}$% **56.** $8\frac{9}{11}$%

Page 237 **1.** $\frac{3}{100}$ **11.** $\frac{3}{25}$ **21.** 2 **31.** $1\frac{11}{100}$ **56.** 3.35 **57.** 2.16 **64.** 21 hours 48 minutes

Page 239 **1.** 50% **13.** $33\frac{1}{3}\%$ **25.** 300% **37.** 375% **55.** 2,776

Page 241 **1.** 1.2 **9.** 4 **17.** 4 **23.** 486 **31.** 20 **32.** 4 **39.** $\frac{5}{6}$

Page 243 **1.** \$450 **2.** \$1,050 **13.** \$986.25; \$130.25 **14.** \$9,747.50; \$3,897.50 **17.** \$6,765.50

Page 245 **1.** \$12.00; \$36.00 **2.** \$4.50; \$19.50

Page 247 **1.** 40% **5.** 75% **9.** 200% **29.** 20% **33.** 50% **37.** $28\frac{4}{7}\%$

Page 249 **1.** 50 **2.** 20 **13.** 230 **14.** 200 **25.** 4.5 **26.** 12 **39.** 8,000 bits

Page 251 **1.** 7.2 **7.** 75% **13.** 40 **31.** \$17,500

Page 253 **1.** b **3.** 108 ski poles

Page 255 **1.** \$2,400 **7.** 3% **11.** \$760

Page 257 **1.** \$40 **2.** \$9 **11.** \$13.50; \$76.50 **12.** \$26.00; \$104 **16.** \$44

Page 259 **1.** 50% inc. **2.** $33\frac{1}{3}\%$ inc. **7.** 40% dec. **8.** 30% dec. **33.** 120 **34.** 28 **39.** 14%

Page 261 **1.** \$60.00 **2.** \$112.00 **21.** \$522.84

CHAPTER 10

Page 269 **1.** $\frac{1}{4}$ **2.** $\frac{1}{12}$ **13.** 1:4 **14.** 3:2

Page 270 **25.** 18,000 **26.** 38 **37.** $\frac{1}{7}$

Page 271 **1.** 60.8 **5.** 76.6

Page 273 **1.** 25 **4.** 160 **5.** 60 **8.** 4,000 **9.** 3,000 **20.** 3,750 **21.** 8,750 **24.** $\frac{12}{25}$

Page 274 **1.** 250 **4.** 1,875 **7.** 0.25

Page 277 **1.** Aisle; Center; Window; Smoking, Center; Smoking, Window; Nonsmoking, Aisle; Nonsmoking, Center; Nonsmoking, Window **2.** 2 **3.** 3 **8.** 10,000 **9.** 625 **14.** 1,000 **15.** 17,576 **20.** $\frac{17}{20}$

Page 279 **1.** independent **4.** $\frac{1}{48}$ **5.** $\frac{1}{12}$ **13.** $\frac{2}{17}$ **14.** $\frac{1}{102}$ **23.** 27 **24.** 26.5

Page 281

1. Possible Combinations

1st digit	*2nd digit*
• 3	5
• 3	7
• 5	3
• 5	7
• 7	3
• 7	5

5. Tennis Matches

• Cathy	Ann
• Cathy	Wes
• Cathy	Glen
• Ann	Wes
• Ann	Glen
• Wes	Glen

Page 283 **1.** 60 **5.** 356 calculators **6.** 4,410 calculators

Page 285 **1.** \$1,238 **4.** 0 **6.** 0 **9.** \$1,238 **19.** car **20.** plane

CHAPTER 11

Page 293 **1.**

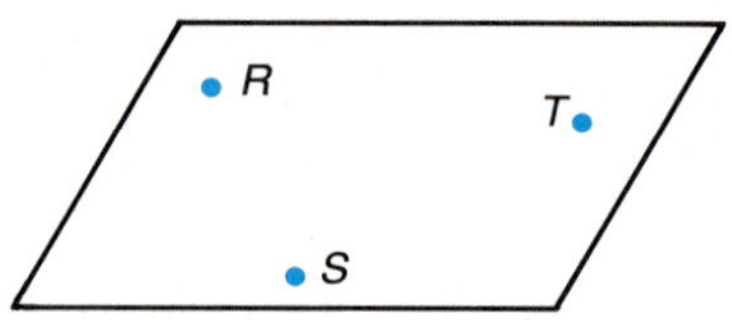

2.

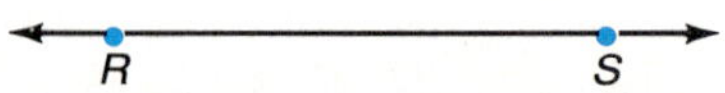

7. angle T **8.** line segment BC **13.** line P **14.** angle 4 **21.** $\overrightarrow{AD}$, $\overrightarrow{AC}$, $\overrightarrow{AB}$ **25.** false

Page 295 **1.** 30° **2.** 45° **10.** 45° **11.** 130°

18.

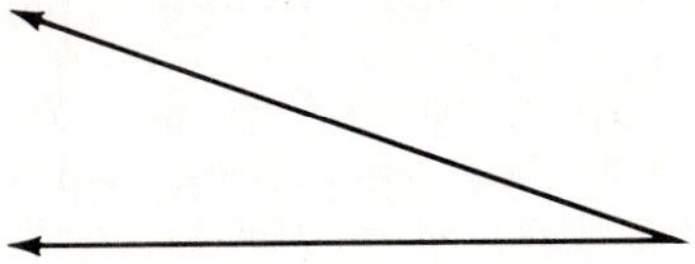

19.

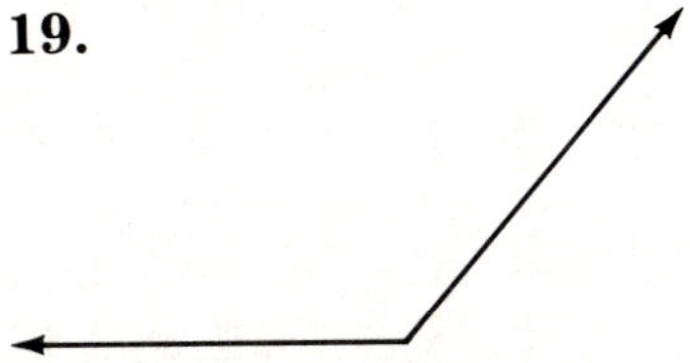

24. 110°

Page 297 **1.** obtuse **2.** acute **5.** acute **6.** acute **11.** neither **12.** complementary **16.** complementary **20.** 70° **21.** 35° **26.** 120° **27.** 40°

Page 299 **1.**

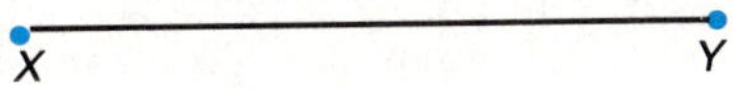

Page 300 **5.**

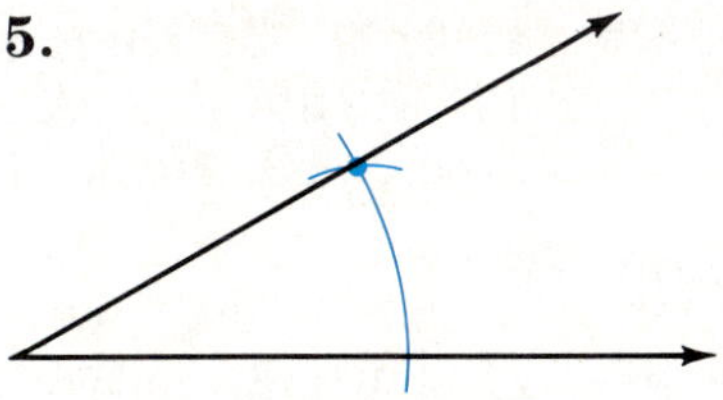

11.

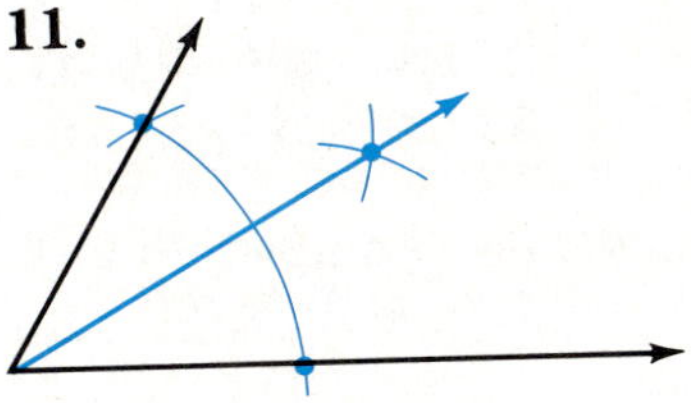

17. 36 **18.** 14

33.

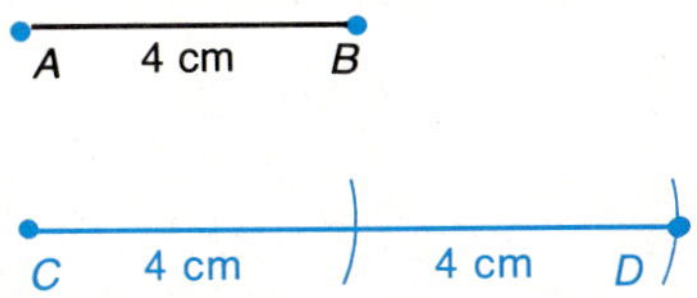

Page 303 **1.**

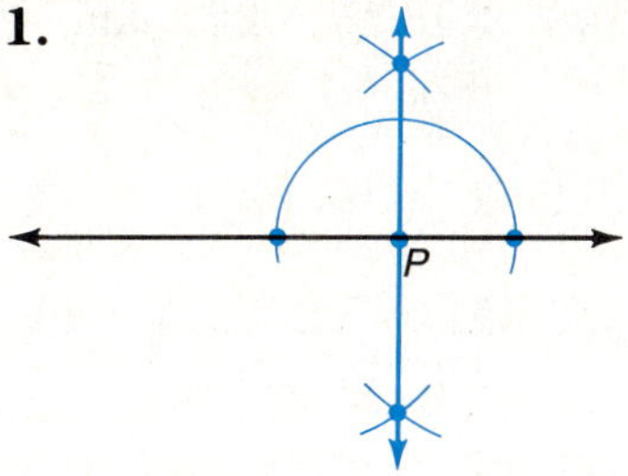

6.

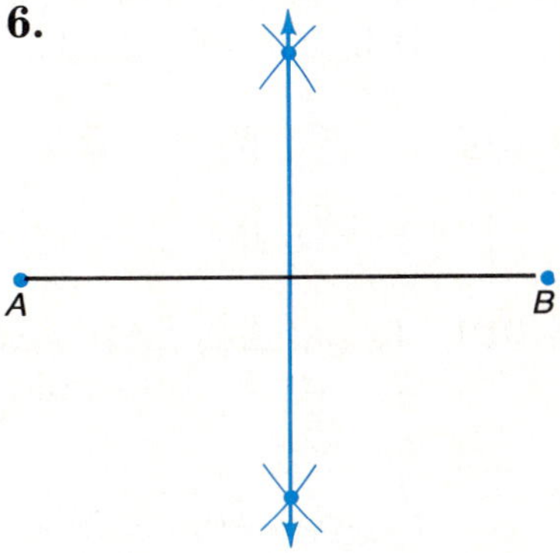

10.

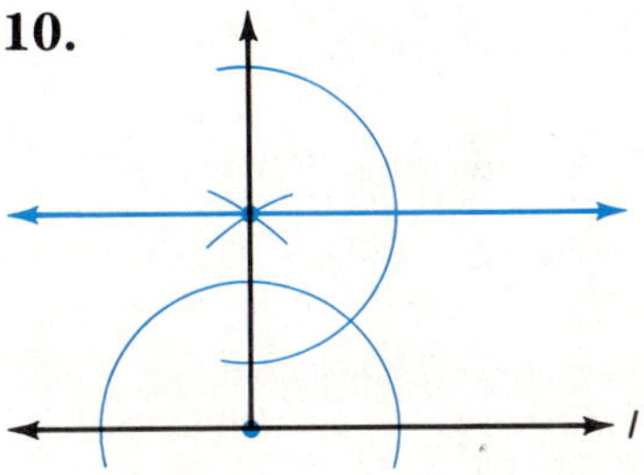

Page 305 **1.** triangle ABC **5.** diameter **9.** 10 cm **15.** 4 ft **21.** 4

Page 307 **1.** isosceles **2.** equilateral **9.** acute **10.** equiangular

17.

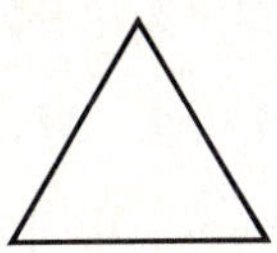

18.

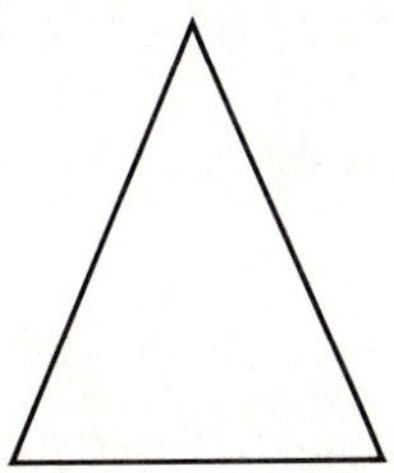

Page 308 **23.** 45° **27.** 65° **28.** 15° **35.** 4.2 **36.** 7.5 **43.** 35°; 110°

Page 311 **1.** multiply by 2 **2.** 128; 256

Page 313 **1.** 2 **2.** 3 **13.** 3.9 **14.** 4.2 **25.** 14 **26.** 24

Page 315 **1.** yes **2.** no **6.** 17 in. **10.** 25 **15.** 2 **16.** 4

Page 316 **1.** 27.5 ft **2.** 57 ft

Page 317 **1.** \$86.25 **2.** \$316.25

Page 319 **1.** parallelogram **2.** square **7.** all **8.** trapezoid

Page 321 **1.** $\angle X$ **2.** $\angle Y$ **7.** similar; congruent corresponding angles; ratio of sides equal **9.** 3 in.

Page 322 **1.** 9.1 m

Page 323 **5.** 12 m

CHAPTER 12

Page 331 **1.** 13.1 cm **4.** 56 mi **7.** 30 in. **10.** $11\frac{3}{4}$ ft **14.** 92 ft **18.** 9 m **20.** 104 m

Page 333 **1.** $1\frac{4}{7}$ ft **3.** $5\frac{1}{2}$ yd **9.** $2\frac{3}{4}$ mi **15.** $14\frac{2}{3}$ yd **21.** 11,100 yd

Page 335 **1.** 32 ft^2 **7.** 169 $in.^2$ **13.** 12 ft^2 **19.** $10\frac{2}{3}$ yd^2

Page 337 **1.** 200 yd^2 **7.** 63 $in.^2$ **14.** 20.4 **15.** 80 **22.** $24\frac{1}{2}$ ft^2

Page 339 **1.** $\frac{11}{14}$ yd^2 **2.** $1\frac{25}{63}$ mi^2 **7.** 3.14 $in.^2$ **8.** 50.24 mi^2 **19.** 78.5 yd^2 **23.** 24 **24.** 3

Page 341 **1.** b **2.** a

Page 343 **1.** 20 yd^2 **5.** 26 yd^2

Page 345 **1.** rectangular prism **2.** sphere

Page 346 **7.** 8 **8.** 6 **9.** 12 **19.** false **20.** false

Page 348 **1.** 94 ft^2 **4.** 54 m^2 **10.** 18,400 ft^2 **11.** 4,600 cm^2

Page 349 **1.** 142 $in.^2$ **4.** 1,728 $in.^2$

Page 351 **1.** 128 yd^3 **4.** 120 $in.^3$ **5.** 729 ft^3 **12.** 3 m^3 **15.** 195 ft^3 **16.** 84 yd^3 **23.** $\frac{1}{4}$ gal

Page 353 **1.** 502.4 ft^3 **4.** 37.68 ft^3 **5.** $3\frac{15}{28}$ yd^3 **10.** 150.72 cm^3 **13.** 28.26 ft^3 **14.** $1\frac{43}{56}$ yd^3 **19.** 1.5 m

Page 355 **1.** 2.8 gal **5.** 1 gal **9.** 3 **10.** 8 **19.** \$49.35

Page 357 **1.** \$249 **2.** Limited **17.** Health Fair, Greenhouse, Trees **18.** Health Fair

CHAPTER 13

Page 365 **1.** -40 **2.** $-\$15$ **11.** -6 **12.** -9 **19.** -16 **20.** $+44$ **27.** $<$ **28.** $>$ **37.** 1 **38.** 2 **52.** $=$ **53.** $<$

Page 366 **60.** complementary **64.** 40° **72.** $+20,320$

Page 368 **1.** 3 **2.** -10 **13.** 28 **14.** 61 **29.** -2 **30.** 9 **39.** 37

Page 369 **43.** 57°F

Page 371 **1.** 7 **2.** -3 **17.** -2 **18.** -4 **23.** -2 **24.** -11 **39.** 20°

Page 373 **1.** 40 **2.** 88 **21.** 8 **22.** −18 **36.** −24 **37.** 1 **45.** 4 **46.** −32 **51.** +210

Page 375 **1.** 3 **5.** 2 **9.** −4 **57.** 100 **65.** 10^4

Page 376 **1.** 1°F **2.** −35°F **7.** −20°F

Page 379 **1.** 1 **2.** 2 **11.** 2.5×10^1 **12.** 3×10^2 **23.** 24 **27.** 0.31 **35.** 9.3×10^7

Page 381 **1.** (−6, 6) **2.** (1, 6) **9.** *S* **10.** O **25.** 0

Page 383 **1.** Pacific Ocean **2.** Indian Ocean **10.** 10°N, 130°W **14.** United States **15.** New Zealand

Page 385 **1.** $\frac{1}{2}$ **2.** −3 **7.** < **15.** = **23.** = **35.** $\frac{3}{4}$ **36.** $\frac{2}{3}$

45. 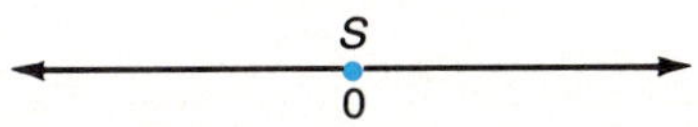

49. 116 **50.** 35

Page 387 **1.** −22.13 **2.** $-\frac{3}{8}$ **6.** −98.08 **11.** $-4\frac{1}{18}$ **21.** 1,794 **22.** 1,350 **30.** $-16\frac{2}{3}$ **31.** 6

Page 389 **1.** c **5.** 20 mL

Page 390 **1.** $8\frac{1}{4}$ **5.** increase

Page 391 **9.** Nostro

CHAPTER 14

Page 399 **1.** 3 · 5 **2.** 5 + 6 **13.** 14 **14.** 19 **37.** 12 − (8 + 5) = −1 **41.** 70°, acute

Page 401 **1.** $n + 6$ **2.** $n - 14$ **21.** 3 more than a number **22.** 9 minus a number **37.** $2n + 3$

Page 403 **1.** 11 **2.** 1 **17.** 3 **18.** −7 **33.** 4 **34.** 2 **53.** 40 m^3

Page 404 **57.** 21 m^2 **61.** 160 m^2

Page 405 **1.** 380 mi **2.** 338 mi **7.** 650 mph

Page 407 **1.** $18n = 36$ **2.** $n + 7 = 15$ **21.** $\frac{1}{6}$

Page 409 **1.** d **2.** e

6. 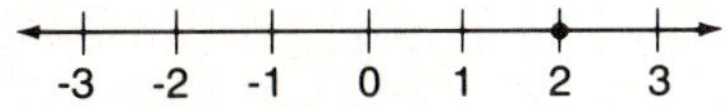

14.

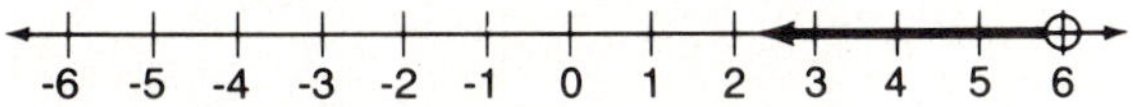

Page 410 **1.** 120 g **5.** 210 g

Page 413 **1.** 5 **2.** 7 **25.** 17 **26.** 11 **51.** $277.22

Page 415 **1.** 5 **2.** 7 **11.** 18 **12.** 32 **41.** 20 wpm

Page 417 **1.** 3 **2.** 4 **25.** 36 **26.** 49 **37.** 85 **38.** 25% **47.** 20 sundaes

Page 420 **1.** 13, 13 **5.** −8, 4 **13.** no **16.** no

22. 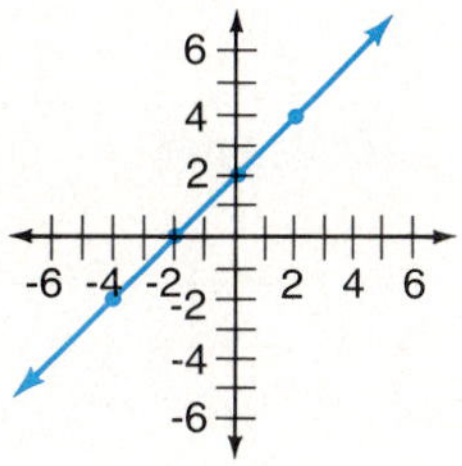

23. 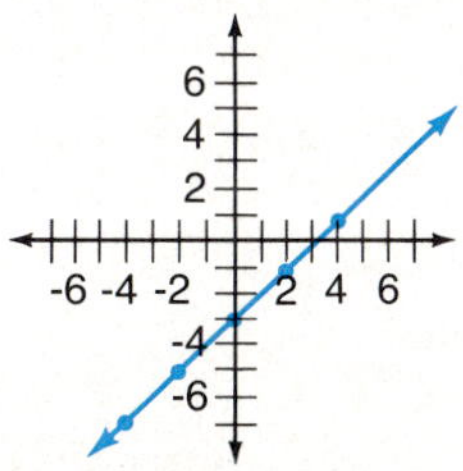

Page 421 **1.** 486 gal **5.** $152.28

TABLES

FORMULAS

Perimeter	Polygon	P = sum of the sides
	Rectangle	$P = 2l + 2w$
	Square	$P = 4s$
Circumference	Circle	$C = 2\pi r$ or $C = \pi d$
Area	Parallelogram	$A = bh$
	Rectangle	$A = lw$
	Square	$A = s^2$
	Triangle	$A = \frac{1}{2}bh$
	Trapezoid	$A = \frac{1}{2}(b_1 + b_2)h$
	Circle	$A = \pi r^2$
Volume	Rectangular prism	$V = lwh$
	Pyramid	$V = \frac{1}{3}Bh$
	Cube	$V = e^3$
	Cylinder	$V = Bh$, or $V = \pi r^2 h$
	Cone	$V = \frac{1}{3}Bh$, or $V = \frac{1}{3}\pi r^2 h$
Other	Diameter	$d = 2r$
	Pythagorean rule	$c^2 = a^2 + b^2$
Consumer	Distance	$d = rt$
	Interest (simple)	$i = prt$

METRIC SYSTEM OF MEASURES

Length		
10 millimeters (mm)	=	1 centimeter (cm)
10 centimeters	=	1 decimeter (dm)
10 decimeters	=	1 meter (m)
100 centimeters	=	1 meter
1,000 millimeters	=	1 meter
1,000 meters	=	1 kilometer (km)

Capacity		
1,000 milliliters (mL)	=	1 liter (L)
1,000 liters	=	1 kiloliter (kL)

Mass		
1,000 milligrams (mg)	=	1 gram (g)
1,000 grams	=	1 kilogram (kg)
1,000 kilograms	=	1 metric ton (t)

Temperature	
Freezing point of water:	0°C
Boiling point of water:	100°C
Normal body temperature:	37°C

U.S. CUSTOMARY SYSTEM OF MEASURES

Length
12 inches (in.) = 1 foot (ft)
3 feet = 1 yard (yd)
36 inches = 1 yard
1,760 yards = 1 mile (mi)
5,280 feet = 1 mile

Capacity
8 fluid ounces (fl oz) = 1 cup (c)
2 cups = 1 pint (pt)
2 pints = 1 quart (qt)
4 quarts = 1 gallon (gal)

Weight
16 ounces (oz) = 1 pound (lb)
2,000 pounds = 1 ton (T)

Temperature	
Freezing point of water:	32°F
Boiling point of water:	212°F
Normal body temperature:	98.6°F

SYMBOLS

$<$	is less than
$>$	is greater than
$\leq$	is less than or equal to
$\geq$	is greater than or equal to
$\neq$	is not equal to
$\approx$	is approximately equal to
$4 \cdot 8$	4 times 8
$12 \div 3$	12 divided by 3
4^5	4 to the fifth power
$0.\overline{16}$	0.16161616 . . .
$\sqrt{\ }$	square root
-8	negative 8
$\lvert -7 \rvert$	absolute value of negative 7
9^{-4}	9 to the negative fourth power
%	percent
7:2	the ratio of 7 to 2
$6/hour	the rate of $6 per hour
$^\circ$	degree
A	point A
$\overleftrightarrow{AB}$	line AB
$\overrightarrow{AB}$	ray AB
$\overline{AB}$	line segment AB
$\angle ABC$	angle ABC
$m\angle D$	measure of $\angle D$
$\triangle XYZ$	triangle XYZ
$\perp$	is perpendicular to
$\parallel$	is parallel to
$\cong$	is congruent to
$\sim$	is similar to
π	pi (about 3.14)
(6, 4)	the ordered pair 6, 4
$P(2)$	the probability of the outcome 2
5!	5 factorial or $5 \times 4 \times 3 \times 2 \times 1$
${}_8P_5$	the number of permutations of 8 objects taken 5 at a time $8 \times 7 \times 6 \times 5 \times 4$

TABLE OF SQUARES AND SQUARE ROOTS

n	n^2	$\sqrt{n}$	n	n^2	$\sqrt{n}$	n	n^2	$\sqrt{n}$
1	1	1	35	1,225	5.916	68	4,624	8.246
2	4	1.414	36	1,296	6	69	4,761	8.307
3	9	1.732	37	1,369	6.083	70	4,900	8.367
4	16	2	38	1,444	6.164	71	5,041	8.426
5	25	2.236	39	1,521	6.245	72	5,184	8.485
6	36	2.449	40	1,600	6.325	73	5,329	8.544
7	49	2.646	41	1,681	6.403	74	5,476	8.602
8	64	2.828	42	1,764	6.481	75	5,625	8.660
9	81	3	43	1,849	6.557	76	5,776	8.718
10	100	3.162	44	1,936	6.633	77	5,929	8.775
11	121	3.317	45	2,025	6.708	78	6,084	8.832
12	144	3.464	46	2,116	6.782	79	6,241	8.888
13	169	3.606	47	2,209	6.856	80	6,400	8.944
14	196	3.742	48	2,304	6.928	81	6,561	9
15	225	3.873	49	2,401	7	82	6,724	9.055
16	256	4	50	2,500	7.071	83	6,889	9.110
17	289	4.123	51	2,601	7.141	84	7,056	9.165
18	324	4.243	52	2,704	7.211	85	7,225	9.220
19	361	4.359	53	2,809	7.280	86	7,396	9.274
20	400	4.472	54	2,916	7.348	87	7,569	9.327
21	441	4.583	55	3,025	7.416	88	7,744	9.381
22	484	4.690	56	3,136	7.483	89	7,921	9.434
23	529	4.796	57	3,249	7.550	90	8,100	9.487
24	576	4.899	58	3,364	7.616	91	8,281	9.539
25	625	5	59	3,481	7.681	92	8,464	9.592
26	676	5.099	60	3,600	7.746	93	8,649	9.644
27	729	5.196	61	3,721	7.810	94	8,836	9.695
28	784	5.292	62	3,844	7.874	95	9,025	9.747
29	841	5.385	63	3,969	7.937	96	9,216	9.798
30	900	5.477	64	4,096	8	97	9,409	9.849
31	961	5.568	65	4,225	8.062	98	9,604	9.899
32	1,024	5.657	66	4,356	8.124	99	9,801	9.950
33	1,089	5.745	67	4,489	8.185	100	10,000	10
34	1,156	5.831						

FRACTIONS AND PERCENTS

$\frac{1}{2} = 50\%$					
$\frac{1}{3} = 33\frac{1}{3}\%$	$\frac{2}{3} = 66\frac{2}{3}\%$				
$\frac{1}{4} = 25\%$	$\frac{3}{4} = 75\%$				
$\frac{1}{5} = 20\%$	$\frac{2}{5} = 40\%$	$\frac{3}{5} = 60\%$	$\frac{4}{5} = 80\%$		
$\frac{1}{6} = 16\frac{2}{3}\%$	$\frac{5}{6} = 83\frac{1}{3}\%$				
$\frac{1}{7} = 14\frac{2}{7}\%$	$\frac{2}{7} = 28\frac{4}{7}\%$	$\frac{3}{7} = 42\frac{6}{7}\%$	$\frac{4}{7} = 57\frac{1}{7}\%$	$\frac{5}{7} = 71\frac{3}{7}\%$	$\frac{6}{7} = 85\frac{5}{7}\%$
$\frac{1}{8} = 12\frac{1}{2}\%$	$\frac{3}{8} = 37\frac{1}{2}\%$	$\frac{5}{8} = 62\frac{1}{2}\%$	$\frac{7}{8} = 87\frac{1}{2}\%$		
$\frac{1}{9} = 11\frac{1}{9}\%$	$\frac{2}{9} = 22\frac{2}{9}\%$	$\frac{4}{9} = 44\frac{4}{9}\%$	$\frac{5}{9} = 55\frac{5}{9}\%$	$\frac{7}{9} = 77\frac{7}{9}\%$	$\frac{8}{9} = 88\frac{8}{9}\%$
$\frac{1}{10} = 10\%$	$\frac{3}{10} = 30\%$	$\frac{7}{10} = 70\%$	$\frac{9}{10} = 90\%$		

INDEX

D

E

M

N

O

P

Q

R

S

T

Credits

Book Design and Production: Textart Inc.
Cover Design: Textart Inc.
Cover Photo: Robert Fishman
Technical Art: Network Graphics

Illustrations: *Mark Giglio* 98, 110, 230, 234; *Steve Marchesi* 78, 127, 196, 202, 212; *Samantha Smith* 115, 180, 192, 194, 241, 273, 375

Picture Credits:
Contents Pages: III, IV–V, VI–VII, VIII–IX HRW Photo Ken Karp

Chapter 1 Opener: Page 1 HRW Photo Richard Haynes; **10–11** HRW Photo Ken Karp; **12** HRW Photo Richard Haynes; **15** James Blank/Archive Pictures Inc.; **17** Robert McElroy/Woodfin Camp & Assoc.; **21** Robert Ferrick/Odyssey Prod.
Chapter 2 Opener: Page 27 HRW Photo Richard Haynes; **35** HRW Photo Ken Karp; **45** Richard Hirneisi/Stock Shop; **52–53** HRW Photo Ken Karp; **55** HRW Photo Ken Lax
Chapter 3 Opener: Page 61 HRW Photo Michal Heron; **70–71** HRW Photo Ken Karp; **77** Sepp Seitz/Rainbow; **83** Jose Fernandez/Woodfin Camp & Assoc.
Chapter 4 Opener: Page 91 Thomas Sobolik; **92** Guido Alberto/Image Bank; **95** Richard Vadnai/Stock Market; **101** Philip Jon Bailey/The Picture Cube; **103** HRW Photo Ken Karp; **111** Bill Eppridge/DOT; **114** HRW Photo Ken Karp
Chapter 5 Opener: Page 121 Focus on Sports; **135** John W. Banagan/Image Bank; **138** HRW Photo Ken Lax
Chapter 6 Opener: Page 145 Sepp Seitz; **148** Bob Gelberg/Image Bank; **154–155** HRW Photo Ken Karp; **157** Dan McCoy/Rainbow; **167** Dan McCoy/Rainbow; **170–171** HRW Photo Ken Karp
Chapter 7 Opener: Page 179 HRW Photo Richard Haynes; **189** Aram Gesar/Image Bank; **191** Audrey Gottlieb/Monkmeyer Press; **200L** Bob Daemmrich/Stock Boston; **200R** Dede Hatch/The Picture Cube; **203** David Moore/Black Star
Chapter 8 Opener: Page 209 HRW Photo Michal Heron; **214–215** HRW Photo Ken Karp; **218–219** HRW Photo Ken Karp; **221** Steve Proehl/Image Bank; **223** Bob Woodward/The Stock Market; **227** HRW Photo Ken Lax
Chapter 9 Opener: Page 233 Dave Schaefer; **235** Robert J. Capece/Monkmeyer Press; **236** HRW Photo Ken Lax; **242–243** HRW Photo Ken Karp; **245** Alvis Uptis/Image Bank; **252–253** HRW Photo Ken Karp; **254–255** HRW Photo Ken Karp; **256** HRW Photo Ken Karp; **258** HRW Photo Ken Lax; **260** HRW Photo Ken Lax
Chapter 10 Opener: Page 267 Ed Kashi; **270** HRW Photo Ken Karp; **271** HRW Photo Ken Karp; **274** Abigail Heyman; **276** HRW Photo Ken Lax; **280–281** Dan Helms/Duomo; **282–283** HRW Photo Ken Karp
Chapter 11 Opener: Page 291 James Blank; **295** HRW Photo Ken Karp; **310–311** HRW Photo Ken Karp; **312** Richard Steedman; **316** HRW Photo Richard Haynes; **323** HRW Photo Ken Karp
Chapter 12 Opener: Page 329 Michal Heron; **333** Dave Bloch; **340–341** HRW Photo Ken Karp; **349** HRW Photo Ken Lax; **354** HRW Photo Ken Karp; **357** HRW Photo Ken Lax
Chapter 13 Opener: Page 363 Frank Fournier; **366** Johnny Johnson/DRK Photo; **369** Focus West; **376** Ken Karp; **379** NASA; **382** Armando Jenik/Image Bank; **389** Charles Gupton/The Stock Market; **391** Ken Karp
Chapter 14 Opener: Page 397 HRW Photo Michal Heron; **402** Paul Chauncey; **405** Hank de Lespinasse/Image Bank; **406** Ed Kashi; **410** Gabe Palmer/The Stock Market; **415** HRW Photo Ken Lax; **416** HRW Photo Richard Haynes; **421** HRW Photo Richard Haynes